Computer Aided Engineering Design

Computer Aided Engineering Design

Anupam Saxena □ **Birendra Sahay**
Department of Mechanical Engineering
Indian Institute of Technology Kanpur, India

Springer

Anamaya

A C.I.P. catalogue record for the book is available from the Library of Congress

ISBN 1-4020-2555-6 (HB)

Copublished by Springer
233 Spring Street, New York 10013, USA
with Anamaya Publishers, New Delhi, India

Sold and distributed in North, Central and South America by Springer
233 Spring Street, New York, USA

In all the countries, except India, sold and distributed by Springer
P.O. Box 322, 3300 AH Dordrecht, The Netherlands

In India, sold and distributed by Anamaya Publishers
F-230, Lado Sarai, New Delhi-110 030, India

9 8 7 6 5 4 3 2 1

springeronline.com

To my parents, all my teachers and for my son Suved when he grows up
Anupam Saxena

To my mother, Charushila Devi, an icon of patience
To my eldest brother, Dhanendra Sahay who never let me feel the absence of my father
To my brother, Dr. Barindra Sahay for initiating me into the realm of mathematics
To my teachers at McMaster University and University of Waterloo
To my wife, Kusum, children Urvashi, Menaka, Pawan and the little fairy
Radhika (granddaughter)
Birendra Sahay

Foreword

A new discipline is said to attain maturity when the subject matter takes the shape of a textbook. Several textbooks later, the discipline tends to acquire a firm place in the curriculum for teaching and learning. Computer Aided Engineering Design (CAED), barely three decades old, is interdisciplinary in nature whose boundaries are still expanding. However, it draws its core strength from several acknowledged and diverse areas such as computer graphics, differential geometry, Boolean algebra, computational geometry, topological spaces, numerical analysis, mechanics of solids, engineering design and a few others. CAED also needs to show its strong linkages with Computer Aided Manufacturing (CAM). As is true with any growing discipline, the literature is widespread in research journals, edited books, and conference proceedings. Various textbooks have appeared with different biases, like geometric modeling, computer graphics, and CAD/CAM over the last decade.

This book goes into mathematical foundations and the core subjects of CAED without allowing itself to be overshadowed by computer graphics. It is written in a logical and thorough manner for use mainly by senior and graduate level students as well as users and developers of CAD software. The book covers

    (a) The fundamental concepts of geometric modeling so that a real understanding of designing synthetic surfaces and solid modeling can be achieved.

    (b) A wide spectrum of CAED topics such as CAD of linkages and machine elements, finite element analysis, optimization.

    (c) Application of these methods to real world problems.

In a new discipline, it is also a major contribution creating example problems and their solutions whereby these exercises can be worked out in a reasonable time by students and simultaneously encouraging them to tackle more challenging problems. Some well tried out projects are also listed which may enthuse both teachers and students to develop new projects. The writing style of the book is clear and thorough and as the student progresses through the text, a great satisfaction can be achieved by creating a software library of curve, surface, and solid modeling modules.

Dr. Anupam Saxena earned his MSME degree in 1997 at the University of Toledo, Ohio, USA. I am familiar with his work on a particularly challenging CAED problem for his thesis. He earned his Ph.D. degree from the University of Pennsylvania, USA and became a faculty member at IIT Kanpur in 2000. Dr. Sahay was Professor at IIT Kanpur where he performed research and teaching in design related fields for over the past 32 years after having earned his Ph.D. from the University of Waterloo, Canada. This textbook is a result of over ten years of teaching CAED by both authors.

The topics covered in detail in this book will, I am sure, be immensely helpful to teachers, students, practitioners and researchers.

Steven N. Kramer, PhD, PE
Professor of Mechanical and Industrial Engineering
The University of Toledo, Toledo, Ohio

Preface

The development of *computer aided engineering design* has gained momentum over the last three decades. Computer graphics, geometric modeling of curves, surfaces and solids, finite element method, optimization, computational fluid flow and heat transfer—all have now taken roots into the academic curricula as individual disciplines. Several professional softwares are now available for the design of surfaces and solids. These are very user-friendly and do not require a user to possess the intricate details of the mathematical basis that goes behind.

This book is an outcome of over a decade of teaching computer aided design to graduate and senior undergraduate students. It emphasizes the mathematical background behind geometric modeling, analysis and optimization tools incorporated within the existing software.

- Much of the material on CAD related topics is widely scattered in literature. This book is conceived with a view to arrange the source material in a logical and comprehensive sequence, to be used as a semester course text for CAD.
- The *focus* is on computer aided design. Treatment essential for geometric transformations, projective geometry, differential geometry of curves and surfaces have been dealt with in detail using examples. Only a background in elementary linear algebra, matrices and vector geometry is required to understand the material presented.
- The concepts of homogeneous transformations and affine spaces (barycentric coordinate system) have been explained with examples. This is essential to understand how a solid or surface model of an object can escape coordinate system dependence. This enables a distortion-free handling of a computer model under rigid-body transformations.
- A viewpoint that free-form solids may be regarded as composed of surface patches which instead are composed of curve segments is maintained in this book, like most other texts on CAD. Thus, geometric modeling of curve segments is discussed in detail. The basis of curve design is parametric, piecewise fitting of individual segments of low degree into a composite curve such that the desired continuity (position, slope and/or curvature) is maintained between adjacent segments. This reduces undue oscillations and provides freedom to a designer to alter the curve shape. A generic model of a curve segment is the weighted linear combination of user-specified data points where the weights are functions of a *normalized, non-negative* parameter. Further, barycentricity of weights* makes a curve segment independent of the coordinate system and provides an insight into the curve's shape. That is, the curve lies within

---

* Weights are all non-negative and for any value of the parameter, they sum to unity.

the convex hull of the data points specified. The associated variation diminishing property suggests that the curve's shape is no more complex than the polyline of the control points itself. In other words, a control polyline primitively approximates the shape of the curve. For Bézier segments, barycentricity is global in that altering any data point results in overall shape change of the segment. For B-spline curves, however, weights are locally barycentric allowing shape change only within some local region. Expressions for weights, that is, Bernstein polynomials for Bézier segments and B-spline basis functions for B-spline curves are derived and discussed in detail in this book and many examples are presented to illustrate curve design.

- With the design of free-form curve segments accomplished, surface patches can be obtained in numerous ways. With two curves, one can sweep one over the other to get a *sweep surface patch*. One of the curves can be rectilinear in shape and represent an axis about which the second curve can be revolved to get a *patch of revolution*. One can join corresponding points on the two curves using straight lines to generate a *ruled surface*. Or, if cross boundary slope information is available, one can join the corresponding points using a cubic segment to get a *lofted patch*. More involved models of surface patches are the bilinear and bicubic Coon's patches wherein four boundary curves are involved. Eventually, a direct extension of Bézier and B-spline curves is their tensor product into respective free-form Bézier and B-spline surface patches. These surface patches inherit the properties from the respective curves. That is, the surface patch lies within the control polyhedron defined by the data points, and that the polyhedron loosely represents the patch shape. The aforementioned patches are derived and discussed in detail with examples in this book. Later, methods to model composite surfaces are discussed.

- The basis for solid modeling is the extension of Jordon's curve theorem which states that a closed, simply connected** (planar) curve divides a plane into two regions; its interior and its exterior. Likewise, a closed, simply connected and orientable surface divides a three-dimensional space into regions interior and exterior to the surface. With this established, a simple, closed and connected surface constituted of various surface patches *knit* or *glued* together at their respective common boundaries encloses a finite volume within itself. The union of this interior region with the surface boundary represents a free form solid. Any solid modeler should be generic and capable of modeling unambiguous solids such that any set operation (union, intersection or difference) performed on two valid solids should yield another valid solid. With this viewpoint, the concept of geometry is relaxed to study the topological attributes of valid solids. Such properties disregard *size* (lengths and angles) and study only the *connectivity* in a solid. With these properties as basis, the three solid modeling techniques, i.e., wireframe modeling, boundary representation method and constructive solid geometry are discussed in detail with examples. Advantages and drawbacks of each method are discussed and it is emphasized that professional solid modelers utilize all three representations depending on the application. For instance, wireframe modeling is usually employed for animation as quick rendering is not possible with the boundary representation scheme.

- Determination of intersection between various curves, surfaces and solids is routinely performed by the solid modelers for curve and surface trimming and blending. Intersection determination is primarily used in computing Boolean relations between two solids in constructive solid

---

** A closed curve with no self intersection.

geometry. Computational geometry that encompasses a set of algorithms to compute various relations like proximity, intersection, decomposition and relational search (e.g., point membership classification) between geometric entities is discussed in brief in this book. The working of these algorithms is described for polygonal entities with examples for easy understanding of the subject matter.

- Reverse engineering alludes to the process of creating CAD models from existing real life components or their prototypes. Applications are prolific; some being the generation of customized fit to human surfaces, designing prostheses, and reconstruction of archaeological collections and artifacts. For an engineering component whose original data is not available, a conceptual clay or wood model is employed. A point cloud data is acquired from an existing component or its prototype using available non-contact or tactile scanning methods. Surface patches are then locally modeled over a subset of the point cloud to interpolate or best approximate the data. Reverse engineering is an important emerging application in Computer Aided Design, and various methods for surface patch fitting, depending on the scanning procedure used, are briefed in this book.

- Having discussed in detail the geometric modeling aspects in free-form design, this book provides an introductory treatment to the finite element analysis (FEM) and optimization, the other two widely employed tools in computer aided design. Using these, one can analyze and alter a design form such that the latter becomes optimal in some sense of the user specified objective. The book discusses linear elastic finite element method using some basic elements like trusses, frames, triangular and four-node elements. Discussion on optimization is restricted to some numerical methods in determining single variable extrema and classical Karush-Kuhn-Tucker necessary conditions for multi-variable unconstrained and constrained problems. Sequential Linear and Quadratic Programming, and stochastic methods like genetic algorithms and simulated annealing are given a brief mention. The intent is to introduce a student to follow-up formal courses on finite element analysis and optimization in the curricula.

This book should be used by the educators as follows:

Students from a variety of majors, e.g., mechanical engineering, computer science and engineering, aeronautical and civil engineering and mathematics are likely to credit this course. Also, students may study CAD at primarily graduate and senior undergraduate levels. Geometric modeling of curves, surfaces and solids may be relevant to all while finite element analysis and optimization may be of interest of mechanical, aeronautical and civil engineering. Discretion of the instructor may be required to cover the combination of topics for a group of students. Considering a semester course of 40 contact hours, a broad breakup of topics is suggested as follows:

- 1st hour: Introduction to computer aided design
- 3 hours: Transformations and projections
- 15 hours: Free-form curve design
- 9 hours: Surface patch modeling
- 6 hours: Solid modeling

The remaining 6 hours may be assigned as follows: for students belonging to mechanical, aeronautical and civil engineering, reverse engineering, finite element method and optimization may be introduced and for those in computer science and engineering and mathematics, computational geometry and optimization may be emphasized.

For a group of graduate students taking this course, differential geometry of curves and surfaces

(Chapters 3 and 6) may be dealt with in detail. Also, topological attributes of solids may be discussed. For only senior undergraduate students, differential geometry may be covered in brief emphasizing mainly Frenet-Serret relations, Gaussian and Mean curvatures and their importance in determining the nature of a surface. Chapters on computational geometry, reverse engineering, FEM and optimization may be omitted.

Assignments and projects form an important part of this course. Assignments may be tailored in a manner that students get a handle on manual calculations as well as code development for curve and surface design. A course project may run over a semester or can be in two parts each covering half the semester. Some example projects are mentioned in Appendix III.

Some examples presented in Chapter 1 on kinematic analysis and spring design pertain to students in mechanical engineering. For a generic class, an instructor may prefer to cover curve interpolation and fitting discussed in sections 3.1 and 3.2.

The practitioners, i.e., those developing professional software would require much deeper understanding of the design principles, mathematical foundations and computer graphics to render a robust Graphical User Interface to the software. This book would help them acquire adequate background knowledge in design principles and mathematical foundations. Those using the software may not require a deeper understanding of the mathematical principles. However, design aspects and essential properties of curve, surface and solid modeling would be needed to create the design and interpret the results.

Chapters 9 and 10 of this book on computations with geometry and modeling using point clouds has been contributed by Dr. G. Saravana Kumar, a former Ph.D. Student, Mechanical Engineering Department, IIT Kanpur. His enthusiasm as T.A. in the CAD course has also resulted in several good projects.

ANUPAM SAXENA
BIRENDRA SAHAY

Acknowledgements

The authors acknowledge the support of the Quality Improvement Program (QIP) at the Indian Institute of Technology, Kanpur for the preparation of this manuscript.

The design in Chapter 8 is attributed to three undergraduate students, Sandeep Urankar, Anurag Singh and Pranjal Jain who have accomplished their B. Tech project on *Robosloth* in 2002. G.S. Sharavanan Kumar, a Ph.D. student at IIT Kanpur has contributed Chapters 9 and 10. Other graduate students, particularly, Anupam Agarwal, Rajat Saxena and Manak Lal Jain have contributed in creating the figures and examples for some chapters. The authors also acknowledge the assignment contributions and assistance of their students in the CAD-2002(II) and 2003(I) classes, particularly, Abhishek Gupta, Anurag Tripathy, Rajkumar Singh, Gaurav Dubey, Shubham Saxena, Abhishek Luthra, Pritam Chakraborty, T.S. Sudhish Kumar and Prince Malik.

Contents

Computer Aided Engineering Design

Chapter 1

Introduction

The development of mankind has depended on the ability to modify and shape the material that nature has made available, in ways to provide them their basic needs, and security and comfort required for their survival and advancement. They have devised tools for hunting, implements for agriculture, shelter for safeguard against the vagaries of nature, and wheels for transportation, an invention mankind has always been proud of. Much of the aforementioned *design* accomplishments have resulted even before mankind may have learnt to count. The then trial-and-error and/or empirical design procedures have been systematized to a great extent using the human understanding of the laws of physics (on force, motion and/or energy transfer) with concepts from mathematics. An idea to fulfill a need and then translating the idea into an implement forms the core of activities in design. *Design and manufacture is innate to the growth of human civilization.*

1.1 Engineering Design

Design is an activity that facilitates the realization of new products and processes through which technology satisfies the needs and aspirations of the society. Engineering design of a product may be conceived and evolved in four steps:

1. *Problem definition:* Extracting a coherent appreciation of *need* or *function* of an engineering part from a fuzzy mix of facts and myths that result from an initial ill-posed problem. The data collection can be done via *observation* and/or a *detailed survey.*
2. *Creative process:* Synthesizing *form*, a design solution to satisfy the need. Multiple solutions may result (and are sought) as the creative thought process is aided by the designers' vast experience and knowledge base. *Brainstorming* is usually done in groups to arrive at various forms which are then evaluated and selected into a set of a few workable solutions.
3. *Analytical process: Sizing* the components of the designed *forms*. Requisite functionality, strength and reliability analysis, feasible manufacturing, cost determination and environmental impact may be some design goals that could be improved optimally by altering the components' dimensions and/or material. This is an iterative process requiring design changes if the analysis shows inadequacy, or scope for further improvement of a particular design. Multiple solutions may be evaluated simultaneously or separately and the *best* design satisfying most or all functional needs may be chosen.
4. *Prototype development and testing:* Providing the ultimate check through physical evaluation under, say, an actual loading condition before the design goes for production. Design changes are

needed in the step above in case the prototype fails to satisfy a set of needs in step 1. This stage forms an interface between design and manufacture. Many groups encourage prototype failure as many times as possible to quickly arrive at a successful design.

1.2 Computer as an Aid to the Design Engineer

Machines have been designed and built even before the advent of computers. During World War-II, ships, submarines, aircrafts and missiles were manufactured on a vast scale. In the significant era (19th and 20th century) of industrial revolution, steam engines, water turbines, railways, cars and power-driven textile mills were developed. The method of representing three-dimensional solid objects was soon needed and was formalized through orthographic projections by a French mathematician Gaspard Monge (1746-1818). After the military kept it a secret for nearly half a century, the approach was made available to engineers, in general, towards the end of nineteenth century.

The inception of modern computers lies in the early work by Charles Babbage (1822), punched card system developed for the US census by Herman Hollerith (1890), differential analyzer at MIT (1930), work on programmable computers by Allan Turing (1936), program storage concept and re-programmable computers by John von Neumann (1946) and micro-programmed architecture by Maurice Wilkes (1951).

The hardware went through a revolution from electronic tubes, transistors (1953), semi-conductors (1953), integrated circuits (1958) to microprocessors (1971). The first 8-bit microcomputer was introduced in 1976 with the Intel 8048 chip and subsequently 16 and 32-bit ones were introduced in 1978 and 1984. Currently, 32 bit and 64 bit PCs are used. Tremendous developments have taken place in hardware, especially in the microprocessor technology, storage devices (20 to 80 GB range), memory input/output devices, compute speed (in GHz range) and enhanced power of PCs and workstations, enabling compactness and miniaturization. The display technology has also made significant advances from its bulky Cathode Ray Tube (CRT) to Plasma Panel and LCD flat screen forms.

Interactive Computer Graphics (ICG) was developed during the 1960s. Sutherland (1962) devised the Sketchpad system with which it was possible to create simple drawings on a CRT screen and make changes interactively. By mid 1960s, General Motors (GM), Lockheed Aircraft and Bell Laboratories had developed DAC-1, CADAM and GRAPHIC-1 display systems. By late 1960s, the term Computer Aided Design (CAD) was coined in literature. During 1970s, graphics standards were introduced with the development of GKS (Graphics Kernel System), PHIGS (Programmer's Hierarchical Interface for Graphics) and IGES (Initial Graphics Exchange Specification). This facilitated the graphics file and data exchange between various computers. CAD/CAM software development occurred at a fast rate during late 1970s (GMSolid, ROMULUS, PADL-2). By 1980s and 1990s, CAD/CAM had penetrated virtually every industry including Aerospace, Automotive, Construction, Consumer products, Textiles and others. Software has been developed over the past two decades for interactive drawing and drafting, analysis, visualization and animation. A few widely used products in Computer Aided Design and drafting are Pro-Engineer™, AutoCAD™, CATIA™, IDEAS™, and in analysis are NASTRAN™, ABAQUS™, ANSYS™ and ADAMS™. Many of these softwares have/are being planned to be upgraded for potential integration of design, analysis, optimization and manufacture.

1.2.1 Computer as a Participant in a Design Team

As it stands, a computer has been rendered a major share of the design process in a man-machine team. It behooves to understand the role of a human vis-à-vis a computer in this setting:

(a) *Conceptualization*, to date, is considered still within the domain of a human designer. Product design commences with the identification of its 'need' that may be based on consumer's/market's demand. An old product may also need design revision in view of new scientific and technological developments. An expert designer or a team goes through a creative and ingenious thought process (brainstorming), mostly qualitative, to synthesize the form of a product. A computer has not been rendered the capability, as yet, to capture non-numeric, qualitative 'thought' design, though it can help a human designer by making available relevant information from its stored database.

(b) *Search, learning and intelligence* is inherent more in a human designer who can be made aware of the new technological developments useful to synthesize new products. A computer, at this time, has little learning and 'qualitative thinking' capability and is not intelligent enough to synthesize a new form on its own. However, it can passively assist a designer by making available a large set of possibilities (stored previously) from a variety of disciplines, and narrow down the search domain for the designer.

(c) *Information storage and retrieval* can be performed very efficiently by a computer that has an excellent capability to store and handle data. Human memory can fade or fail to avail appropriate information fast enough, and at the right time from diverse sources. Further, a computer can automatically create a product database in final stages of the design.

(d) *Analytical power* in a computer is remarkable in that it can perform, say, the finite element analysis of a complex mechanical part or retrieve the input/output characteristics of a designed system very efficiently, provided mathematical models are embedded. Humans usually instruct the computers, via codes or software, the requisite mathematical models employed in *geometric modeling* (modeling of curves, surfaces and solids) and *analysis* (finite element method and optimization). Geometric modeling manifests the *form* of a product that a designer has in mind (qualitatively) while analysis works towards the systematic improvement of that *form*.

(e) *Design iteration* and improvement can be performed by a computer very efficiently once the designer has offloaded his/her conception of a product via geometric modeling. Finite element analysis (or other performance evaluation routine) and optimization can be performed simultaneously with the aim to modify the dimensions/shape of a product to meet the pre-specified design goals.

(f) *Prototyping* of the optimized design can be accomplished using the tools now available for Rapid Manufacturing. The geometric information of the final product can be passed on to a manufacturing set up that would analogically *print* a three dimensional product.

Computers help in manifesting the qualitative conception of a design form a human has of a product. Further, they prove useful in iterative improvement of the design, and its eventual realization. Computers are integrated with humans in design and manufacture, and provide the scope for automation (or least human interaction) wherever needed (mainly in analysis and optimization). Computer Aided Process Planning (CAPP), scheduling (CAS), tool design (CATD), material requirement planning (MRP), tool path generation for CNC machining, flexible manufacturing system (FMS), robotic systems for assembly and manufacture, quality inspection, and many other manufacturing activities also require computers.

1.3   Computer Graphics

Computer Graphics, which is a discipline within Computer Science and Engineering, provides an important mode of interaction between a designer and computer. Sutherland developed an early form of a computer graphic system in 1963. Rogers and Adams explain computer graphics as the *use of computers to define, store, manipulate, interrogate and present pictorial output.* Computer graphics involves the creation of two and three dimensional models, shading and rendering to bring in realism to the objects, natural scene generation (sea-shores, sand dunes or hills and mountains), animation, flight simulation for training pilots, navigation using graphic images, walk through buildings, cities and highways, and creating virtual reality. War gaming, computer games, entertainment industry and advertising has immensely benefited from the developments in computer graphics. It also forms an important ingredient in Computer-Aided Manufacturing (CAM) wherein graphical data of the object is converted into machining data to operate a CNC machine for production of a component. The algorithms of computer graphics lay behind the backdrop all through the process of virtual design, analysis and manufacture of a product. Two primary constituents of computer graphics are the *hardware* and the *software*.

1.3.1 Graphics Systems and Hardware

Hardware comprises the *input*, and *display* or *output devices*. Numerous types of graphics systems are in use; those that model one-to-many interaction and others that allow one-to-one interface at a given time. *Mainframe-based systems* use a large mainframe computer on which the software, which is usually a huge code requiring large space for storage, is installed. The system is networked to many designer stations on time-sharing basis with display unit and input devices for each designer. With this setting, intricate assemblies of engineering components, say an aircraft, requiring many human designers can be handled. *Minicomputer* or *Workstation* based systems are smaller in scale than the Mainframe systems with a limited number (one or more) of display and input devices. Both systems employ one-to-many interface wherein more than one designer can interact with a computer. On the contrary, *Microcomputer* (PC) based systems allow only one-to-one interaction at a time. Between the Mainframe, Workstation and PC based systems, the Workstation based system offers advantages of distributed computing and networking potential with lower cost compared with the mainframes.

1.3.2 Input Devices

Keyboard and *mouse* are the primary input devices. In a more involved environment, digitizers, joysticks and tablets are also used. Trackballs and input dials are used to produce complex models. Data gloves, image scanners, touch screens and light pens are some other input devices. A keyboard is used for submitting alphanumeric input, three-dimensional coordinates, and other non-graphic data in 'text' form. A mouse is a small hand held pointing device used to control the position of the cursor on the screen. Below the mouse is a ball. When the mouse is moved on a surface, the amount and direction of movement of the cursor is proportional to that of the mouse. In optical mouse, an optical sensor moving on a special mouse pad having orthogonal grids detects the movements. There are push buttons on top of the mouse beneath the fingers for signaling the execution of an operation, for selecting an object created on the screen within a rectangular area, for making a selection from the pulled down menu, for dragging an object from one part of the screen to other, or for creating drawings and dimensioning. It is an important device used to expedite the drawing operations. A special *z*-mouse for CAD, animation and virtual reality includes three buttons, a thumb-wheel and a track-ball on top. It gives six degrees of freedom for spatial positioning in *x*-*y*-*z* directions. The *z*-mouse is used for rotating the object around a desired axis, moving and navigating the viewing position (observer's eye) and the object through a three-dimensional scene.

Trackballs, *space-balls* and *joysticks* are other devices used to create two and three-dimensional drawings with ease. Trackball is a 2-D positioning device whereas space-ball is used for the same in 3-D. A joystick has a vertical lever sticking out of a base box and is used to navigate the screen cursor.

Digitizers are used to create drawings by clicking input coordinates while holding the device over a given 2-D paper drawing. Maps and boundaries in a survey map, for example, can be digitized to create a computer map. *Touch panels* and *light pens* are input devices interacting directly with the computer screen. With touch panels, one can select an area on the screen and observe the details pertaining to that area. They use infrared light emitting diodes (LEDs) along vertical and horizontal edges of the screen, and go into action due to an interruption of the beam when a finger is held closer to the screen. Pencil shaped *light pens* are used to select screen position by detecting the light from the screen. They are sensitive to the short burst of light emitted from the phosphor coating as the electron beam hits the screen. *Scanners* are used to digitize and input a two-dimensional photographic data or text for computer storage or processing. The gradations of the boundaries, gray scale or color of the picture is stored as data arrays which can be used to edit, modify, crop, rotate or scale to enhance and make suitable changes in the image by software designed using geometric transformations and image processing techniques.

FaroArm®, a 3-D coordinate measuring device, is a multi-degree of freedom precision robotic arm attached to a computer. At the tip of the end-effector is attached a fine roller-tipped sensor. The tip can be contacted at several points on a curved surface to generate a point data cloud. A 3-D surface can then be fitted through the data cloud to generate the desired surface. A non-contact 3-D digitizer, Advanced Topometric Sensor (ATOS) uses optical measuring techniques. It is material independent and can scan in three-dimensions any arbitrary object such as moulds, dies, and sculptures. It is a high detailed resolution and precision machine. It uses adhesive retro targets stuck on the desired surface. Digital reflex cameras then record the positions of these retro targets from different views. The images consisting of the coordinates of targets are transferred from the digital camera to the computer. The image coordinates are then converted to the object coordinates by calculating the intersection of the rays from different camera positions. Finally, the required object surface is generated. Techniques for scanning objects in three-dimensions are very useful in reverse engineering, rapid prototyping of existing objects with complex surfaces such as sculptures and other such applications.

1.3.3 Display and Output Devices

Three types of display devices are in use: Cathode ray tube (*CRT*), Plasma Panel Display (PPD) and Liquid Crystal Display (LCD). CRT is a popular display device in use for its low cost and high-resolution color display capabilities. It is a glass tube with a front rectangular panel (screen) and a cylindrical rear tube. A cathode ray gun, when electrically heated, gives out a stream of electrons, which are then focused on the screen by means of positively charged electron-focusing lenses. The position of the focused point is controlled by orthogonal (horizontally and vertically deflecting) set of amplifiers arranged in parallel to the path of the electron beam. A popular method of CRT display is the Raster Scan. In raster scan, the entire screen is divided into a matrix of picture cells called *pixels*. The distance between pixel centers is about 0.25 mm. The total number of pixel sets is usually referred to as *resolution*. Commonly used CRTs are those with resolution of 640×480 (VGA), 1024×768 (XGA) and 1280×1024 (SXGA). With higher resolution, the picture quality is much sharper. As the focused electron beam strikes a pixel, the latter emits light, i.e. the pixel is 'on' and it becomes bright for a small duration of time. The electron beam is made to scan the entire screen line-by-line from top to bottom (525 horizontal lines in American system and 625 lines in European system) at 63.5 microseconds per scan line. The beam keeps on retracing the path. The *refresh rate* is 60Hz, implying that the screen is completely scanned in $1/60^{th}$ of a second (for European system, it is $1/50^{th}$ of a second). In a black and white display, if the pixel intensity is '0', the pixel appears black, and when '1', the pixel is bright. As the electron beam scans through the entire screen, it switches off those pixels which are supposed to be black thus creating a pattern on the screen. For the electron beam to know precisely which pixels are to be kept 'off' during scans, a *frame buffer* is used that is a hardware programmable memory. At least one memory bit ('0' or '1') is needed for each pixel, and there are as many bits allocated in the memory as the number of pixels on display. The entire memory required for displaying all the pixels is called a *bit plane* of the frame buffer.

One bit plane would create only a 'black' and 'white' image, but for a realistic picture, one would need *gray levels* or shades between black and white as well. To control the intensity (or shade) of a pixel one has to use a number of bit planes in a frame buffer. For example, if one uses 3 bit planes in single frame buffer, one can create 8 (or 2^3) combinations of intensity levels (or shades) for the same pixel- 000 (black)-001-010- 011-100-101-110-111(white). The intermediate values will control the intensity of the electron beam falling on the pixel. To have an idea about the amount of memory required for a black and white display with 256×256 (or 2^{16}) pixels, every bit plane will require a memory of $2^{16} = 65,536$ bits. If there are 3 bit planes to control the gray levels, the memory required will be 1,96,608 bits! Since memory is a digital device and the raster action is analog, one needs digital-to-analog converters (DAC). A DAC takes the signal from the frame buffer and produces an equivalent analog signal to operate the electron gun in the CRT.

For *color display*, all colors are generated by a proper combination of 3 basic colors, viz. red, green, and blue. If we assign '0' and '1' to each color in the order given, we can generate 8 colors: black (000), red (100), green (010), blue (001), yellow (110), cyan (011), magenta (101) and white (111). The frame buffer requires a minimum of 3 bit planes—one for each RGB color; this can generate 8 different colors. If more colors are desired, one needs to increase the number of bit planes for each color. For example, if each of the RGB colors has 8 bit planes (a total of 24 bit planes in the frame buffer with three 8-bit DAC), the total number of colors available for picture display would be $2^{24} = 1,67,77,216$! To further enhance the color capabilities, each 8-bit DAC is connected to a color look up memory table. Various methods are employed to decrease the access and display time and enhance the picture sharpness.

CRT displays are popular and less costly, but very bulky and suitable only for desktop PCs. Flat Panel Displays (FPD) are gaining popularity with laptop computers and other portable computers and devices. FPD belongs to one of the following two classes: (a) active FPD devices, which are primarily light emitting devices. Examples of active FPD are flat CRT, plasma gas discharge, electroluminescent and vacuum fluorescent displays. (b) Passive FPD devices are based on light modulating technologies. Liquid Crystal (LC) and Light Emitting Diodes (LED) are some examples.

Plotters and *printers* constitute the output devices. Line printers are the oldest succeeded by 9-pin and 24-pin *dot matrix plotters* and printers. *Ink jet plotters*, *laser plotters* and *thermal plotters* are used for small and medium sized plots. For large plots, *pen and ink plotters* of the flat bed, drum and pinch roller types are used.

1.4 Graphics Standards and Software

Till around 1973, software for producing graphics was mostly device dependent. Graphics software written for one type of hardware system was not portable to another type, or it became useless if the hardware was obsolete. Graphics standards were set to solve portability issues to render the application software device independent. Several standards have been developed; most popular among them are GKS (Graphics Kernel System), PHIGS (Programmer's Hierarchical Interactive Graphics System), DXF (Drawing Exchange Format), and IGES (Initial Graphics Exchange Specification).

For designing mechanical components and systems, one requires 3-D graphics capabilities for which GKS 3-D, PHIGS and DXF are suitable. For 3-D graphics and animation, PHIGS is used.

It provides high interactivity, hierarchical data structuring, real time graphic data modification, and support for geometric transformations. These standards provide the core of graphics including basic graphic primitives such as line, circle, arc, poly-lines, poly-markers, line-type and line-width, text, fill area for hatching and shading, locators for locating coordinates, valuators for real values for dimensioning, choice options and strings. Around such standard primitives, almost all standard software for CAD is written. They also include the device drivers for standard plotters and display devices.

Another comprehensive standard is IGES to enable the exchange of model databases among CAD/CAM systems. IGES contains more geometric entities such as, curves, surfaces, solid primitives, and Boolean (for Constructive Solid Geometry) operations. Wire-frame, surface modeling and solid modeling software can all be developed around IGES. It can transmit the property data associated with the drawings which helps in preparing, say, the bill of materials. Though these standards appear veiled or at the *back end*, they play a crucial role in creation of the application software.

1.5   Designer-Computer Interaction

A CAD/CAM software is designed to be primarily interactive, instructive and user-friendly wherein a designer can instruct a computer to perform a sequence of tasks ranging from designing to manufacture of an engineering component. The front end of a software is a graphical user interface or GUI while the back end comprises computation and database management routines. The front end is termed so as a user can visually observe the design operations being performed. However, computation and data storage routines are not very apparent to a designer, which is why they may be termed collectively as the back end of the software. In most CAD software, the GUI is divided into two parts (or windows) that appear on the display device or screen (Figure 1.1): (i) the visual manifestation or the

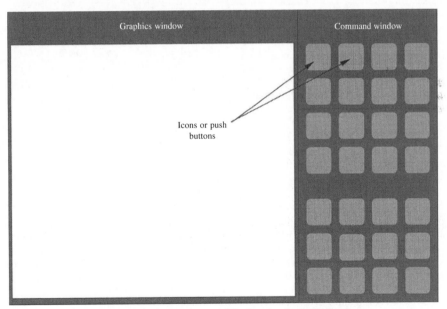

Figure 1.1   Generic appearance of the Front end of a CAD software

Graphics Window and (ii) the Command window. The Graphics window provides the visual feedback to the user detailing desired information about an object being designed. One can manipulate the position (through translation/rotation) of an object relative to another or a fixed coordinate system and visualize the changes in the Graphics window. In essence, all design operations involving transformations, curve design, design of surfaces and solids, assembly operations pertaining to relative positioning of two or more components, drafting operations that provide the engineering drawings, analysis operations that yield results pertaining to displacements and stresses, optimization operations that involve sequential alterations in design, and many others can be visualized through the Graphics window.

The design instructions are given through a user-friendly Command Window that is subdivided into several *push buttons* or *icons*. To accommodate numerous applications in CAD and to allow a guided user interface, the icons appear in groups. For instance, icons pertaining to the design of curves would be grouped in the Command window. Push buttons pertaining to curve trimming, extension, intersection and other such actions would be combined. Icons used in surface and solid design would appear in two different groups. Options under transformations, analysis, optimization and manufacture would also be clustered respectively. A user may make a design choice by clicking on an icon using the mouse. There may be many ways to design a curve, for instance. To accommodate many such possibilities, a CAD GUI employs the *pull down menus* (Figure 1.2). That is, when an icon on curve segment design is clicked on, a menu would drop down prompting the user to choose

between, say, the Ferguson, Bézier or B-spline options. Similarly, for a surface patch design, a pull down menu may have choices ranging between the analytical patches, tensor product surfaces, Coon's patches, rectangular or triangular patches, ruled or lofted patches and many others. For solid modeling, a user may have to choose between Euler operations or Boolean sequences. After a design operation is chosen using a push button and from a respective pull down menu, the user would be prompted to enter further choices through *pop up* menus. For instance, if a user chooses to sketch a line, a pop up window may appear expecting the user to feed in the start point, length and orientation of the line. Note that for a two dimensional case, a much easier option to draw a

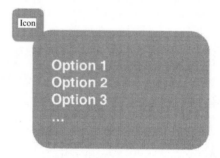

Figure 1.2 **A pull down menu that appears when clicking on an icon in the command window**

curve segment may be to select a number of points on the screen through a sequence of mouse clicks.

1.6 Motivation and Scope

Developing the front end GUI of a CAD software is an arduous and challenging task. However, it is the back end wherein the core of Computer Aided Design rests. This book discusses the design concepts based on which various modules or objects of the back end in a CAD software are written. The concepts emerge as an amalgamation of *geometry*, *mathematics* and *engineering* that renders the software the capability of *free-form* or generic design of a product, its analysis, obtaining its optimized form, if desired, and eventually its manufacture. Engineering components can be of various forms (sizes and shapes) in three-dimensions. A Solid can be thought of as composed of a simple *closed connected surface* that encloses a finite volume. The closed surface may be conceived as an interweaved arrangement of constituent surface patches, which in turn, can be individually considered as composed of a group of curves. It then behooves to discuss the generic design of curves, surfaces and solids in that order. Even before, it may be essential to understand how three-dimensional objects or geometrical entities are represented on a two-dimensional display screen, and how such entities can be positioned with respect to each other for assembly purposes or construction operations.

Engineers have converged to numerous standard ways of perceiving a three-dimensional component by way of *engineering drawings* depicted on a two-dimensional plane (conventionally blue prints, but for CAD's purpose, a display screen). The following chapter comprises a broad discussion on transformations and projections. Rotation and translation of a point (or a rigid body) with respect to the origin are discussed in two-dimensions. Both transformations are expressed in matrix notation using the homogenous coordinates. The advantage is that like rotation, translation can also be executed as a matrix multiplication operation without requiring any addition or subtraction of matrices or vectors. Performing a sequence of transformations then involves multiplying the respective transformation matrices in the same order. Rotation is next generalized about any point on the plane. The reflection transformation is discussed in two-dimensions. A property of translation, rotation and reflection matrices is that they are *orthogonal* which ensures the preservation of lengths and angles. In other words, the three transformations do not cause any deformation in a rigid body for which reason they are termed *rigid-body transformations*. Those that do affect deformations, i.e., scaling and shear, are discussed next. The aforementioned transformations are extended to use with three-dimensional solids using four-dimensional homogenous coordinates. It may be realized that these transformations help in the Computer Aided Assembly of rigid-body components. For drafting or engineering drawing applications, the geometry of perspective and parallel projections is detailed. A reader would note that the matrix forms of transformations and projections are similar. In addition to conventionally employed first (or third) angle orthographic and isometric projections to pictorially represent engineering components, perspective viewing, oblique viewing and axonometric viewing are also discussed in Chapter 2.

Chapters 3 to 5 are exclusively devoted to the design of curves. Chapter 3 commences by differentiating between curve fitting/interpolation and curve design, the latter is more generic and can be adapted to achieve the former. Among the explicit, implicit and parametric equations to describe curves, the third choice is suited best to accommodate vertical tangents, to ease the computation for intersections (for trimming purposes, for instance), and to represent curve segments by restricting the parameter range in [0, 1]. Unnecessary oscillations in curves from the design viewpoint are undesired for which reason a curve is sought to be a composite one with constituent curve segments of low degree (usually cubic) arranged end to end. The position, slope and curvature continuity at junction points of a composite curve can be addressed via the differential geometry of curves covered in this chapter. Two of the three widely used curve segment models are discussed in Chapter 4. The first is Ferguson cubic segment that requires two end points and two respective slopes to be specified by the user. For a set of data points and respective slopes, a composite Ferguson curve of degree three can be constructed. Its shape can be altered by relocating any one (or more) data point(s) and/or slopes (by changing their magnitudes and/or directions). A Ferguson curve would have the slope continuity through out, however, if one desires curvature continuity, using differential geometry, one can determine that any three consecutive slopes are related. Thus, for a given set of data points and slope information at the start and end points, intermediate slopes can be determined using the constraint equations resulting from curvature continuity. The advantage is two-fold: first, a designer need not specify all slopes which is a higher order information usually difficult for a designer to submit as input. Second, the result is a smooth, curvature continuous cubic Ferguson curve.

Higher order information, like specifying the slopes, can be avoided with Bézier curve segments that are modeled using only data points (also called control points). Bézier segments may be regarded as the geometric extension of the construction of a parabola using the three tangent theorem. The resultant algebraic equation is the weighted linear combination of data points wherein the weights are Bernstein polynomials which, in turn, are functions of the parameter. In parameter range [0, 1], Bernstein polynomials have the property of being non-negative, and that they sum to unity for any value of the parameter. These features render some interesting convex hull and variation diminishing properties to Bézier segments. The shape of the latter can be altered by relocating any data point. However, the effect is global in that the shape of the entire curve is changed. Modeling of continuous Bézier curves is also described using cubic segments. The slope and curvature continuity of composite Bézier curves at junction points restrict the placement of some data points. A designer is constrained to relocate two data points in the neighborhood of the junction point along a straight line for slope continuity. For curvature continuity, four points in the neighborhood of the junction point inclusive, need to be coplanar.

Splines, which are in a manner generalized Bézier curves, are discussed extensively in Chapter 5. The term *spline* is inspired from the draughtman's approach to pass a thin metal or wooden strip through a given set of constrained points called *ducks*. In addition to data points required to construct a spline, a set of parameter values called the *knot vector* is required. Thus, wherein primarily the number of data points determine the degree of Bézier segments, for splines, it is the number of *knots* in the knot vector. Chapter 5 discusses the modeling of polynomial splines which are then normalized to obtain *basis-splines* or *B-splines*. B-splines are basis functions similar to Bernstein polynomials in case of Bézier segments. All B-spline basis functions are non-negative, and only some among those required for curve definition, sum to unity. This renders strong convex hull property to B-Splines which provides the local shape control to a B-spline curve. Newton's divided-difference and the related Cox-de Boor recursive method to compute B-spline basis functions are described in the chapter. Generation of knot vector from given relative placement of data points, and approximation and interpolation with B-spline curves are also discussed.

Chapters 6 and 7 cover surfaces in detail. Like with curves, parametric representation of surfaces is preferred. Also, surfaces are sought as composites of patches of lower degree. There are methods to join together and to *knit* or *weave* such patches at their common boundaries to ensure tangent plane and/or curvature continuity. Chapter 6, thus details the differential geometry of surfaces. Quadric or analytical surface patches are not adequate enough to help design a free-form composite surface. Based on the principles of curve design in Chapters 4 and 5, some basic methods to design a surface patch are described in Chapter 6. These include methods to realize developable and ruled surface patches, parallel surface patches, and patches resulting from revolution and sweep. The shape of such patches can be controlled by relocating the data points and/or slopes used for the ingredient curves. Chapter 7 entails methods of surface patch design that are direct extension of the techniques described in Chapters 4 and 5. Herein, patches are treated under two groups, the tensor product patches and boundary interpolation patches. In the former, Ferguson, Bézier and B-Spline patches are covered while in the latter, bilinear and bi-cubic Coon's patches are discussed. Methods to achieve composite Ferguson, Bézier and Coons patches are also mentioned.

Discussion on curve and surface design lays the foundation for solid or volumetric modeling. Though the treatment is purely geometric when discussing curves and surfaces, it takes more than geometry alone to interpret solids. Any representation scheme for computer modeling of solids is expected to (i) be versatile and capable of modeling a generic solid, (ii) generate valid and unambiguous solids, (iii) have closure such that permitted transformations and set operations on solids always yield valid solids, and (iv) be compact and efficient in matters of information storage and retrieval. Chapter 8 commences with an understanding of solids. The Jordon's theorem establishes that a *closed connected surface* divides the Euclidean space into two subspaces, the space enclosed within the closed surface, which is the interior of a solid, and the space exterior to it. A brief discussion on topology then follows describing homeomorphism, closed-up surfaces, topological classification and invariants of surfaces. The intent is to describe solids topologically and highlight how two geometrically different solids can be topologically similar to use identical modeling methods with different geometry information. In this chapter, three solid modeling techniques, namely, *wireframe modeling*, *boundary representation method* and *Constructive Solid Geometry* are discussed. Wireframe modeling is one of the oldest ways that employs only vertex and edge information for representation of solids. The connectivity or topology is described using two tables, a vertex table that enumerates the vertices and records their coordinates, and an edge table wherein for every numbered edge, the two connecting vertices are noted. The edges can either be straight lines or curves in which case the edge table gets modified accordingly. Though the data structure is simple, wireframe models do not include the facet information and thus are ambiguous.

The boundary representation (B-rep) method is an extension of wireframe modeling in that the former includes the details of involved surface patches. A popular scheme employed is the Baumgart's *winged edge* data structure for representation of solids. Though developed for polyhedrons, the Baumgart's method is applicable to homeomorphic solids. That is, the primary B-rep data structure of a tetrahedron would be the same as that of a sphere over which a tetrahedron with curved edges is drawn. The difference would be that for a sphere, the edges and faces would be recorded as entities with finite curvature. The associated Euler-Poincaré formula is discussed next which is a topological result that ensures the validity of a wide range of polyhedral solids. Based on the Euler-Poincaré formula are the Euler operators for construction of polyhedral solids. Two groups of Euler operators are put to use, the *MAKE* and *KILL* groups for adding and deleting respectively. Euler operators are written as *Mxyz* or *Kxyz* for the Make and Kill groups respectively where *x*, *y* and *z* represent a vertex, edge, face, loop, shell or genus. Using Euler operators, every topologically valid polyhedron can be constructed from an initial polyhedron by a finite sequence of operations.

Constructive Solid Geometry (CSG) is another way for modeling solids wherein *primitives* like block, cone, cylinder, sphere, triangular prism, torus and many others can be combined using Boolean set operations like union, intersection and difference. Solids participating in CSG need not be bounded by analytical surfaces. A closed composite surface created using generic surface patches discussed in Chapters 6 and 7 can also be used to define a CSG primitive. Boolean, regularized Boolean operations and the associated construction trees are discussed in detail in Chapter 8. Other method like the Analytical Solid Modeling which is an extension of the tensor product method for surfaces to three-dimensional parametric space is also mentioned. Chapter 8 ends highlighting the importance of the *parametric modeling* for engineering components. One may require machine elements like bolts of different nominal diameters for various applications wherein parametric design helps. Also, using analysis (Chapter 11) and/or optimization (Chapter 12), one may hope to determine the optimal parameter values of an engineering component for a given application. Chapter 9 highlights some concepts from computational geometry discussing intersection problems and Boolean operations on two-dimensional polygons to consolidate the concepts in constructive solid geometry. Chapter 10 discusses different techniques to model surfaces from a set of given point cloud data, usually encountered in reverse engineering.

That analysis and optimization both play a key role in Computer Aided Design, Chapters 11 and 12 are allocated accordingly. Most engineering components are complex in shape for classical stress analysis methods to be employed. An alternative numerical approach called the Finite Element Method (FEM) is in wide use in industries and elsewhere, and is usually integrated with the CAD software. FEM is a broad field and is a result of an intensive three decade research in various areas involving stress analysis, fluid mechanics and heat transfer. The intent in Chapter 11 is to only familiarize a reader with concepts in FEM related to stress analysis. The Finite Element Method is introduced using springs and later discussed using truss, beam and frame, and triangular and four-node elements. Minimization of total potential is mainly employed when formulating the stiffness matrices for the aforementioned elements.

Chapter 12 discusses various classical and stochastic methods in optimization. Among classical methods, first, *zero-order* (function-based) and *first-order* (gradient-based) methods for objectives with single (design) variable are discussed. These include (a) the bracketing techniques wherein the search is limited to a pre-specified interval and (b) the open methods. Classical multi-variable optimization without and with constraints is discussed next. The method of Lagrange multipliers is detailed, and Karush-Kuhn Tucker necessary conditions for optimality are noted. The Simplex method and Sequential Linear Programming are briefed followed by Sequential Quadratic Programming. Among the stochastic approaches, genetic algorithm and simulated annealing are briefed.

1.7 Computer Aided Mechanism and Machine Element Design

Using existing software, solid models or engineering drawings of numerous components can be prepared. In addition, a computer can also help design machine elements like springs, bearings, shafts and fasteners. It can also help automate the design of mechanisms, for instance. A few familiar examples are presented below in this context, and many more can be similarly implemented.[1,2]

Example 1.1 A Four-Bar Mechanism
Design of mechanisms has been largely graphical or analytical. The vector loop method is a convenient tool in computer solution of planar mechanism problems such as determination of point path, velocity and acceleration. Consider a four-bar mechanism shown in Figure 1.3. OA is the crank (link-2), other links being AB (link-3) and BK (link-4). O and K are fixed to the ground forming the link-1. All joints are pin joints. Assume that the link lengths are known and that the x-axis is along OK and y-axis is perpendicular to OK. All angles are measured positive counterclockwise (CCW) with respect to the x-axis. Regard the vector $\mathbf{r}_1$ attached to the fixed link 1. Similarly, $\mathbf{r}_2$ is attached to the crank link-2 and rotates with it. Vectors $\mathbf{r}_3$ and $\mathbf{r}_4$ are similarly attached to links 3 and 4. These vectors have magnitudes equal to the link lengths to which they are attached and have directions along the instantaneous positions of the links OA, AB, and BK. Let the angle (CCW) as measure of the vector direction for $\mathbf{r}_i$ be θ_i, $i = 1, \ldots 4$. $\theta_1 = 0$ since link OK is fixed and is along the x-axis. Using vector method

$$\vec{OA} + \vec{AB} + \vec{BK} - \vec{OK} = \vec{0}$$

$$\overset{vl}{\mathbf{r}_2} + \overset{v?}{\mathbf{r}_3} + \overset{?v}{\mathbf{r}_4} - \overset{vv}{\mathbf{r}_1} = 0 \qquad (1.1)$$

where vl (magnitude and direction) on $\mathbf{r}_2$ indicates that both the magnitude and direction (input) are known, $v?$ on $\mathbf{r}_3$ shows that while the magnitude is known, the direction is yet unknown (and depends upon the present position of $\mathbf{r}_2$), vv indicates given (known) magnitude and direction, and $?v$ shows

---

[1]Nikravesh, P.E. (1988) Computer Aided Analysis of Mechanical Systems, Prentice-Hall, N.J.
[2]Hall, Jr., A.S. (1986) Notes on Mechanism Analysis, Waveland Press, Illinois.

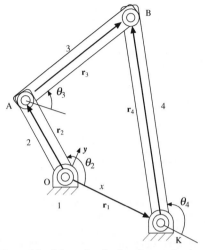

Figure 1.3   Schematic of a four-bar mechanism

unknown magnitude and known direction. The components of the vectors along x and y axes can be expressed as:

$$X : r_2 \cos \theta_2 + r_3 \cos \theta_3 + r_4 \cos \theta_4 - r_1 \cos \theta_1 = 0$$

$$Y : r_2 \sin \theta_2 + r_3 \sin \theta_3 + r_4 \sin \theta_4 - r_1 \sin \theta_1 = 0 \qquad (1.2)$$

Here, θ_2 is the known crank angle and $\omega_2 = \dot{\theta}_2$, $\alpha_2 = \ddot{\theta}_2$ are also given. Since $\theta_1 = 0$, Eq. (1.2) is reduced to

$$X : r_2 \cos \theta_2 + r_3 \cos \theta_3 + r_4 \cos \theta_4 - r_1 = 0$$

$$Y : r_2 \sin \theta_2 + r_3 \sin \theta_3 + r_4 \sin \theta_4 = 0 \qquad (1.3)$$

Evaluating Link Positions

Eq. (1.3) is nonlinear if they are to be solved for θ_3 and θ_4 for given steps of θ_2. Newton's method converts the problem into an iterative algorithm suitable for computer implementation. Let the estimated values be (θ_3', θ_4'). If the guess is not correct, Eqs. (1.3) will be different from zero, in general. Let the errors be given by:

$$X : r_2 \cos \theta_2 + r_3 \cos \theta_3' + r_4 \cos \theta_4' - r_1 = \varepsilon_1$$

$$Y : r_2 \sin \theta_2 + r_3 \sin \theta_3' + r_4 \sin \theta_4' = \varepsilon_2 \qquad (1.4)$$

For small changes $(\Delta\theta_3, \Delta\theta_4)$ in   the change in error $(\Delta\varepsilon_1, \Delta\varepsilon_2)$ is given by the Taylor's series expansion up to the first order. That is

$$\Delta\varepsilon_1 = \frac{\partial \varepsilon_1}{\partial \theta_3'} \Delta\theta_3 + \frac{\partial \varepsilon_1}{\partial \theta_4'} \Delta\theta_4 = -r_3 \sin \theta_3' \, \Delta\theta_3 - r_4 \sin \theta_4' \Delta\theta_4$$

$$\Delta \varepsilon_2 = \frac{\partial \varepsilon_2}{\partial \theta_3'} \Delta \theta_3 + \frac{\partial \varepsilon_2}{\partial \theta_4'} \Delta \theta_4 = r_3 \cos \theta_3' \, \Delta \theta_3 + r_4 \cos \theta_4' \Delta \theta_4$$

$$\begin{bmatrix} -r_3 \sin \theta_3' & -r_4 \sin \theta_4' \\ r_3 \cos \theta_3' & r_4 \cos \theta_4' \end{bmatrix} \begin{bmatrix} \Delta \theta_4 \\ \Delta \theta_4 \end{bmatrix} = \begin{bmatrix} \Delta \varepsilon_1 \\ \Delta \varepsilon_2 \end{bmatrix} \Rightarrow \begin{bmatrix} \Delta \theta_3 \\ \Delta \theta_4 \end{bmatrix} = \begin{bmatrix} -r_3 \sin \theta_3' & -r_4 \sin \theta_4' \\ r_3 \cos \theta_3' & r_4 \cos \theta_4' \end{bmatrix}^{-1} \begin{bmatrix} \Delta \varepsilon_1 \\ \Delta \varepsilon_2 \end{bmatrix}$$

(1.5)

This gives a recursive relationship

$$\begin{bmatrix} \theta_3^{\text{new}} \\ \theta_4^{\text{new}} \end{bmatrix} = \begin{bmatrix} \theta_3^{\text{old}} + \Delta \theta_3 \\ \theta_4^{\text{old}} + \Delta \theta_4 \end{bmatrix}$$

(1.6)

The iteration is started with some estimated values of (θ_3', θ_4'). From Eq. (1.4) $(\varepsilon_1, \varepsilon_2)$ is computed and in the first step $(\Delta \varepsilon_1, \Delta \varepsilon_2)$ is assigned as $(\varepsilon_1, \varepsilon_2)$. Eq. (1.5) is solved to get $(\Delta \varepsilon_3, \Delta \varepsilon_4)$ and (θ_3, θ_4) are updated using Eq. (1.6). Using the new values of (θ_3, θ_4), Eq. (1.4) is solved again to get $(\varepsilon_1, \varepsilon_2)$. This time $(\Delta \varepsilon_1, \Delta \varepsilon_2)$ is computed as the difference between the current and previous values of $(\varepsilon_1, \varepsilon_2)$. Eqs. (1.4)-(1.6) are repeatedly solved until $(\Delta \varepsilon_1, \Delta \varepsilon_2)$ and thus $(\Delta \theta_3, \Delta \theta_4)$ are desirably small. For given θ_2, therefore, positions of links 3 and 4(i.e. θ_3 and θ_4) are determined. For different values of θ_2, the procedure can be implemented to get the entire set of positions for the linkages 3 and 4.

Kinematic Coefficients, and Link Velocity and Acceleration
Consider Eq. (1.3) and note that θ_2 is the independent variable and (θ_3, θ_4) are dependent variables (link lengths are constant). On differentiating Eq. (1.3) with respect to θ_2 on both sides

$$\frac{dX}{d\theta_2} : -r_2 \sin \theta_2 - r_3 \sin \theta_3 \frac{d\theta_3}{d\theta_2} - r_4 \sin \theta_4 \frac{d\theta_4}{d\theta_2} = 0$$

$$\frac{dY}{d\theta_2} : r_2 \cos \theta_2 + r_3 \cos \theta_3 \frac{d\theta_3}{d\theta_2} + r_4 \cos \theta_4 \frac{d\theta_4}{d\theta_2} = 0$$

$$\Rightarrow \begin{bmatrix} -r_3 \sin \theta_3 & -r_4 \sin \theta_4 \\ r_3 \cos \theta_3 & r_4 \cos \theta_4 \end{bmatrix} \begin{bmatrix} \dfrac{d\theta_3}{d\theta_2} \\ \dfrac{d\theta_4}{d\theta_2} \end{bmatrix} = \begin{bmatrix} r_2 \sin \theta_2 \\ -r_2 \cos \theta_2 \end{bmatrix}$$

$$\Rightarrow \begin{bmatrix} \dfrac{d\theta_3}{d\theta_2} \\ \dfrac{d\theta_4}{d\theta_2} \end{bmatrix} = \begin{bmatrix} -r_3 \sin \theta_3 & -r_4 \sin \theta_4 \\ r_3 \cos \theta_3 & r_4 \cos \theta_4 \end{bmatrix}^{-1} \begin{bmatrix} r_2 \sin \theta_2 \\ -r_2 \cos \theta_2 \end{bmatrix} = \begin{bmatrix} h_3 \\ h_4 \end{bmatrix} \text{(say)}$$

(1.7)

$h_3 = \dfrac{d\theta_3}{d\theta_2}$ and $h_4 = \dfrac{d\theta_4}{d\theta_2}$ are called the Kinematic Coefficients (KC) of the four bar mechanism with respect to the driver crank. From Eq. (1.7), it can be observed that KC's are functions of the link lengths and instantaneous values of the angles. They are constants for a given position of the input link. At any instant of time, to get the angular velocities ω_3 and ω_4, and angular accelerations α_3 and α_4 of links 3 and 4, respectively

$$\omega_3 = \frac{d\theta_3}{dt} = \frac{d\theta_3}{d\theta_2} \frac{d\theta_2}{dt} = h_3 \omega_2 \; ; \; \omega_4 = \frac{d\theta_4}{dt} = \frac{d\theta_4}{d\theta_2} \frac{d\theta_2}{dt} = h_4 \omega_2$$

$$\alpha_3 = \frac{d}{dt}(\omega_3) = \frac{d}{dt}(h_3\,\omega_2) = h_3\dot{\omega}_2 + \frac{dh_3}{dt}\omega_2 = h_3\dot{\omega}_2 + \frac{dh_3}{d\theta_2}(\omega_2)^2 = h_3\alpha_2 + h_3'\omega_2^2$$

$$\alpha_4 = \frac{d}{dt}(\omega_4) = \frac{d}{dt}(h_4\,\omega_2) = h_4\dot{\omega}_2 + \frac{dh_4}{dt}\omega_2 = h_4\dot{\omega}_2 + \frac{dh_4}{d\theta_2}(\omega_2)^2 = h_4\alpha_2 + h_4'\omega_2^2 \quad (1.8)$$

The second order kinematic coefficients h_3', h_4' can be determined from Eq. (1.7) as follows:

$$\frac{d^2X}{d\theta_2^2}: -r_2\cos\theta_2 - r_3\cos\theta_3\left(\frac{d\theta_3}{d\theta_2}\right)^2$$

$$-r_3\sin\theta_3\left(\frac{d^2\theta_3}{d\theta_2^2}\right) - r_4\cos\theta_4\left(\frac{d\theta_4}{d\theta_2}\right)^2 - r_4\sin\theta_4\left(\frac{d^2\theta_4}{d\theta_2^2}\right) = 0$$

$$\Rightarrow -r_2\cos\theta_2 - r_3\cos\theta_3 h_3^2 - r_3\sin\theta_3 h_3' - r_4\cos\theta_4 h_4^2 - r_4\sin\theta_4 h_4' = 0$$

$$\frac{d^2Y}{d\theta_2^2}: -r_2\sin\theta_2 - r_3\sin\theta_3 h_3^2 + r_3\cos\theta_3 h_3' - r_4\sin\theta_4 h_4^2 + r_4\cos\theta_4 h_4' = 0$$

$$\begin{bmatrix} h_3' \\ h_4' \end{bmatrix} = \begin{bmatrix} -r_3\sin\theta_3 & -r_4\sin\theta_4 \\ r_3\cos\theta_3 & r_4\cos\theta_4 \end{bmatrix}^{-1} \begin{bmatrix} r_3\cos\theta_3 & r_4\cos\theta_4 \\ r_3\sin\theta_3 & r_4\sin\theta_4 \end{bmatrix} \begin{bmatrix} h_3^2 \\ h_4^2 \end{bmatrix}$$

$$+ \begin{bmatrix} -r_3\sin\theta_3 & -r_4\sin\theta_4 \\ r_3\cos\theta_3 & r_4\cos\theta_4 \end{bmatrix}^{-1} \begin{bmatrix} r_2\cos\theta_2 \\ r_2\sin\theta_2 \end{bmatrix} \quad (1.9)$$

Eqs. (1.7), (1.8) and (1.9) can be implemented into a computer code and the positions, velocities and accelerations of all the linkages can be determined at each desired instant.

Example 1.2 (Slider-Crank Mechanism)
Consider the slider-crank mechanism in Figure 1.4 with the indicated vector loop. The vector loop equation can be written as

$$\overset{v1}{\mathbf{r}_2} + \overset{v?}{\mathbf{r}_3} - \overset{vv}{\mathbf{r}_1} - \overset{?v}{\mathbf{r}_4} = 0$$

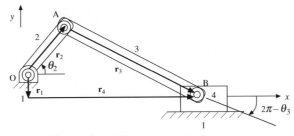

Figure 1.4 A slider-crank mechanism

The magnitude and direction (Input) for the crank $\mathbf{r}_2$ are given. The offset $\mathbf{r}_1$ has known magnitude and direction ($-90°$). The slider has known direction ($0°$) but has variable magnitude (r_4). The connecting rod $\mathbf{r}_3$ has known magnitude but variable direction (θ_3). In component form:

$$X : r_2 \cos \theta_2 + r_3 \cos \theta_3 - r_4 = 0$$

$$Y : r_2 \sin \theta_2 - r_3 \sin \theta_3 + r_1 = 0 \tag{1.10}$$

Differentiating with respect to θ_2 and using the KC's $h_3 = \dfrac{d\theta_3}{d\theta_2}, f_4 = \dfrac{dr_4}{d\theta_2}$:

$$\frac{dX}{d\theta_2} : -r_2 \sin \theta_2 - r_3 \sin \theta_3 h_3 - f_4 = 0$$

$$\frac{dY}{d\theta_2} : X : r_2 \cos \theta_2 - r_3 \cos \theta_3 h_3 = 0$$

$$\Rightarrow \begin{bmatrix} h_3 \\ f_4 \end{bmatrix} = \begin{bmatrix} -r_3 \sin \theta_3 & -1 \\ -r_3 \cos \theta_3 & 0 \end{bmatrix}^{-1} \begin{bmatrix} r_2 \sin \theta_2 \\ -r_2 \cos \theta_2 \end{bmatrix} \tag{1.11}$$

The second order KC's can be similarly determined.

$$\frac{d^2 X}{d\theta_2^2} : -r_2 \cos \theta_2 - r_3 (\sin \theta_3 h_3' + \cos \theta_3 h_3^2) - f_4' = 0$$

$$\frac{d^2 Y}{d\theta_2^2} : -r_2 \sin \theta_2 - r_3 (\cos \theta_3 h_3' - \sin \theta_3 h_3^2) = 0$$

$$\Rightarrow \begin{bmatrix} h_3' \\ f_4' \end{bmatrix} = \begin{bmatrix} -r_3 \sin \theta_3 & -1 \\ -r_3 \cos \theta_3 & 0 \end{bmatrix}^{-1} \begin{bmatrix} r_2 \cos \theta_2 + r_3 \cos \theta_3 h_3^2 \\ r_2 \sin \theta_2 - r_3 \sin \theta_3 h_3^2 \end{bmatrix} \tag{1.12}$$

Following on similar lines as in Example 1.1, the position, velocity and acceleration of other linkages with respect to the input crank (link 2) are given by

$$\sin \theta_3 = \frac{(r_1 + r_2 \sin \theta_2)}{r_3}, r_4 = (r_2 \cos \theta_2 + r_3 \cos \theta_3)$$

$$\dot{\theta}_3 = \omega_3 = h_3 \omega_2, \dot{r}_4 = \text{slider velocity} = f_4 \omega_2$$

$$\ddot{\theta}_3 = h_3' \omega_2^2 + h_3 \alpha_2, \ddot{r}_4 = \text{slider acceleration} = f_4' \omega_2^2 + f_4 \alpha_2 \tag{1.13}$$

Applying Eqs. (1.10)-(1.13), an algorithm can be developed to determine the velocity and acceleration of the connecting rod and slider for various orientations of the crank.

Example 1.3 (Design of Helical Compression Springs)
Machine component design is specially suited to computerized solutions. A computer program can help in

- looking up tables for materials and standard sizes
- iteration using a set of formulas and constraints
- logic and options
- graphical display and plots with changing parameters

Design of helical compression springs can be one such example. For given force (F) and deflection (y) characteristics, the parameters to be determined are outside diameter D_0, inside diameter D_i, wire diameter d, free length L_f, shut or solid length L_s, number of active coils N_a and spring rate k. Safety checks are to be provided for static stresses, and buckling. Designer can select from a range of spring index $C = D/d$ (ratio of mean coil diameter, $D = (D_0 + D_i)/2$, to wire diameter d), a set of standard wire diameters d, materials and their ultimate strengths S_{ut} and shear modulus G. The material strength is dependent on wire diameter d and design calculations are very sensitive to the wire size.

The relation between spring stiffness k and deflection y can be found in any machine element design book (e.g. by Norton, Shigley, and others)[3,4]

$$y = \frac{8FD^3 N_a}{Gd^4}, \quad k = \frac{F}{y} = \frac{Gd^4}{8D^3 N_a} \qquad (1.14)$$

Maximum shear stress τ is given by

$$\tau = K_w \frac{8FD}{\pi d^3}$$

where the Wahl's correction factor for stress concentration is

$$K_w = \frac{4C - 1}{4C - 4} + \frac{0.615}{C} \qquad (1.15)$$

The tensile strength S_{ut} is related to the wire diameter d as

$$S_{ut} = Ad^b \qquad (1.16)$$

Here A and b are constants depending upon the wire material and diameter. A set of typical values is given in Table 1.2[5]. All standard data has been taken from this reference.

A chronology of the design steps for static loading is described below, although there may be variations in the procedure depending on the requirements.

Problem: *Design a helical compression spring, which should apply a minimum force F_{min} and a maximum force F_{max} over a range of deflection δ. The initial compression on the spring is given to be $F_{initial}$.*

Step 1: From Table 1.1(a), a suitable material is selected. For static loading, the most commonly used, least expensive spring wire material is A227. Select a preferred wire diameter (d) from Table 1.1(b). For example, A227 is available in the diameter range from 0.70 mm to 16 mm. Select an intermediate value so that it leaves some space for iterations later, if required.
Step 2: The spring index $C=D/d$ is generally recommended to be in the range

$$12 > C > 4 \qquad (1.17)$$

---

[3]Dimarogonas, A. (1989) Computer Aided Machine Design, Prentice-Hall, N.Y.
[4]Shigley, J.E., Mischke, C.R. (2001) Mechanical Engineering Design, McGraw-Hill, Singapore.
[5]Associated Springs-Barnes Group (1987) Design Hand Book, Bristol, Conn.

Table 1.1(a) Common Spring Wire Materials

ASTM #	Material	SAE #	Description
A227	Cold-drawn wire	1066	Least expensive general-purpose spring wire. Suitable for static loading but not good for fatigue or impact. Temperature range 0°C to 120°C.
A228	Music wire	1085	Toughest, most widely used material for small coil springs. Highest tensile and fatigue strength of all spring wire. Temperature range 0°C to 120°C.
A229	Oil-tempered wire	1065	General-purpose spring steel. Less expensive and available in larger sizes than music wire. Suitable for static loading but not good for fatigue or impact. Temperature range 0°C to 180°C.
A230	Oil-tempered wire	1070	Valve-spring quality-suitable for fatigue loading.

Again, an intermediate value is preferred, rather than an extreme value. From these two assumptions, the mean coil diameter D is determined.

Step 3: From Eq. (1.15), calculate the maximum shear stress at the larger force

$$\tau_{max} = K_w \frac{8F_{max}D}{\pi d^3} \tag{1.18}$$

Step 4: Find the ultimate tensile strength (S_{ut}) of the wire material from Eq. (1.16), selecting constants A and b from Table 1.2. Usually, the torsional ultimate strength (S_{us}) and torsional yield strength (S_{ys}) are given by

$$S_{us} = 0.67S_{ut}, \; S_{ys} = 0.60S_{ut} \tag{1.19}$$

Step 5: Find the safety factor (N_s) against yielding

$$N_s = \frac{S_{ys}}{\tau_{max}} \tag{1.20}$$

For static loading, the factor of safety should be between 1 and 2.

Step 6: If the safety factor appears to be less than 1 (or less than a desired value), perform the design iteration by choosing another wire diameter from Table 1.1(b) and repeating the above steps, till an acceptable value of N_s is achieved.

Step 7: Determine the spring rate

$$k = \frac{F_{max} - F_{min}}{\delta} \tag{1.21}$$

Use Eqs. (1.14) and (1.21) to calculate the active number of coils

$$N_a = \frac{Gd^4}{8D^3 k} \tag{1.22}$$

A rounding off is done to the nearest $\frac{1}{4}$ of the coil. For example, if $N_a = 8.6$, it is rounded off to 8.5 and if it is 8.09, it is rounded off to 8.0. This will increase the stiffness slightly.

Table 1.1(b)   Preferred wire diameters

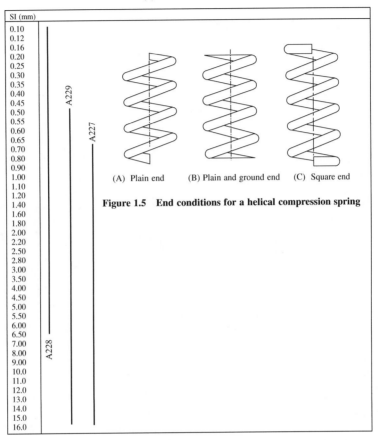

SI (mm)

| 0.10 |
| 0.12 |
| 0.16 |
| 0.20 |
| 0.25 |
| 0.30 |
| 0.35 |
| 0.40 |
| 0.45 |
| 0.50 |
| 0.55 |
| 0.60 |
| 0.65 |
| 0.70 |
| 0.80 |
| 0.90 |
| 1.00 |
| 1.10 |
| 1.20 |
| 1.40 |
| 1.60 |
| 1.80 |
| 2.00 |
| 2.20 |
| 2.50 |
| 2.80 |
| 3.00 |
| 3.50 |
| 4.00 |
| 4.50 |
| 5.00 |
| 5.50 |
| 6.00 |
| 6.50 |
| 7.00 |
| 8.00 |
| 9.00 |
| 10.0 |
| 11.0 |
| 12.0 |
| 13.0 |
| 14.0 |
| 15.0 |
| 16.0 |

A229   A227   A228

(A) Plain end   (B) Plain and ground end   (C) Square end

Figure 1.5   End conditions for a helical compression spring

Table 1.2 Coefficients and Exponents for Eq. 1.16

ASTM#	Material	Range of d (mm)	Exponent b	Coefficient A (MPa)	Correlation Factor
A227	Cold drawn	0.5-16	−0.1822	1753.3	0.998
A228	Music wire	0.3-6	−0.1625	2153.5	0.9997
A229	Oil tempered	0.5-16	−0.1833	1831.2	0.999

Step 8: The four types of end conditions for the coil, as shown in Figure 1.5, are

Plain-Ends $\Rightarrow$ $N_a = N_t$ (total number of coils)

Plain-Ground-ends $\Rightarrow$ $N_a = N_t - 1$

$$\text{Squared-ends} \qquad\qquad \Rightarrow \quad N_a = N_t - 2$$
$$\text{Squared-Ground-Ends} \qquad \Rightarrow \quad N_a = N_t - 2 \qquad\qquad (1.23)$$

The total number of coils N_t is calculated using the above.

Step 9: The shut (or solid) height L_s of the spring is calculated from

$$L_s = dN_t \qquad\qquad (1.24)$$

Step 10: The initial deflection and the clash allowance are calculated from

$$y_{\text{initial}} = \frac{F_{\text{initial}}}{k}, \quad y_{\text{clash}} = 15\% \text{ of the } \delta = 0.15\delta \qquad\qquad (1.25)$$

Step 11: The free length L_f of the spring can now be determined from

$$L_f = y_{\text{initial}} + \delta + y_{\text{clash}} + L_s \qquad\qquad (1.26)$$

Step 12: Determine the maximum force at the shut height deflection F_{shut} to check for the shear stress in the coil at this force

$$F_{\text{shut}} = k(L_f - L_s), \ \tau_{\text{shut}} = K_w \frac{8 F_{\text{shut}} D}{\pi d^3} \qquad\qquad (1.27)$$

Verify if the factor of safety $N_{\text{shut}} = \dfrac{S_{sy}}{\tau_{\text{shut}}} > 1$. If not, another iteration may be required.

Step 13: The buckling of the spring has to be checked.

$$L_{\text{critical}} \text{ (buckling)} \approx 2.63 \frac{D}{d}, L_f < L_{\text{critical}} \qquad\qquad (1.28)$$

Step 14: Now, the complete spring specifications can be written as:
Spring Material: A227 (or as selected)
Wire diameter: d Free length: L_f
Mean diameter: D Total number of coils: N_t
Outer and inner diameters: $D_0 = D + d, D_i = D - d$ Ends: As specified

Weight of the spring: $W = \left(\dfrac{\pi d}{2}\right)^2 DN_t \gamma$, where γ = material density.

EXERCISES

1. A four-bar mechanism is shown in Fig. P1.1. Fixed pivots are given to be O_2 and O_4 20 cm apart. The input crank O_2A is of 10 cm and $AB = BO_4 = 25$ cm. Trace the point path of point P for $AP = 50$ cm. All links are rigid.
2. A Chebychev's straight line linkage is shown in Fig. P1.2. Fixed pivots are given to be O_2 and O_4 20 cm apart. $O_2A = 25$ cm, $AB = 10$ cm and $BO_4 = 25$ cm. Determine the path traced by the point P for $BP = 5$ cm. All links are rigid.
3. A film advance mechanism is shown in Fig. P1.3 for a 35 mm camera. Link 2 is attached to the dc motor and rotates at a constant angular velocity. O_2O_4 are fixed pivots. Link 3 is extended and has a pin-end, which goes

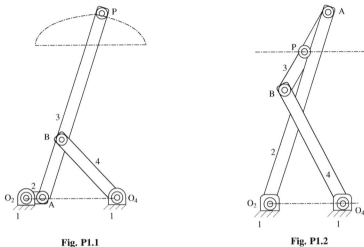

Fig. P1.1   **Fig. P1.2**

into the rectangular groove of the film, moves along a straight line by 35 mm and then lifts up to disengage from the film at the end of its motion. Design the mechanism by selecting suitable sizes of the linkages. Links 2 and 4 are parallel.

4. For the mechanism shown in Fig. P1.4, the input angle $\theta_2 = 60°$, and the constant angular velocity $\omega_2 = 5$ radians/sec (CCW). If body 4 is in rolling contact with the ground, determine the velocity and acceleration

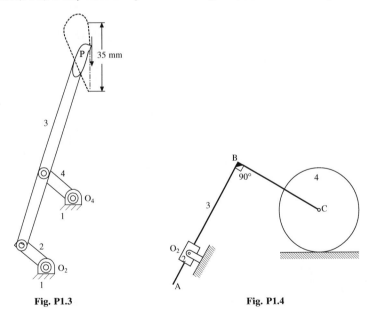

Fig. P1.3   **Fig. P1.4**

of links 3 and 4 using kinematic coefficients. Given AB = 25 cm, BC = 15 cm, and radius of the rigid roller 4 is 12.5 cm.

5. Steps for design of helical compression springs for static loading has been described earlier which can be implemented on a computer using MatLab™.

   (a) Extend the method for design of compression springs under fatigue loading. These types of springs are used in IC engines, compressors and shock absorbers in vehicles.

   (b) Add modules to your computer program (make it interactive) to include design of extension springs, torsion springs and leaf springs.

6. Write a computer program for selection of ball bearings. The program should include a look-up table (Timken or SKF) for some standard bearings. It should calculate and check the bearing life under the given loading conditions.

7. Write an interactive computer program for complete design of short journal bearings. It is easy to use Ocvirk's solution[6]. The software should take into account the bearing material, type of lubricating oils, their viscosities and thermal considerations.

8. Shafts and axles are most commonly used mechanical components designed to transmit power. They should be designed and checked for deflection and rigidity as well as static and fatigue strength for a given loading condition. Keyways, pins, splines and diameter changes introduce stress concentration. Make a computer program for the design of shafts. Look up tables for material, for instance, will be helpful to the designer.

[6]Norton, R.L. (2001) Machine Design: An Integrated Approach, Pearson Education Asia.

Chapter 2

Transformations and Projections

Geometric transformations provide *soul* or *life* to *virtual objects* created through geometric modeling discussed in later chapters. It is using transformations that one can manoeuver an object, view it from different angles, create multiple copies, create its reflected image, re-shape or scale an object, position an object with respect to the other, and much more. Projections, like orthographic and perspective on the other hand, help comprehend an object for purpose of its fabrication. Transformations have many uses, mainly pertaining *motion*, such as manipulating the relative positions of two objects in solid modeling to create a complex entity, displaying motion of mechanisms, animating an assembly to demonstrate its working or imparting motion to a virtual human for a walk through a virtual city or a building. Motion simulators for aircrafts, tanks and motor vehicles extensively employ geometric transformations.

Transformations may be employed to perform *rigid-body motion* wherein an object may be moved from one position to another without altering its shape and size. Typical rigid body transformations involve *translation*, *rotation* and *reflection*, the latter being a combination of the first two. Transformations may also cause *deformations* like *shear, scaling* and *morphing* wherein the object is altered in size and/or shape. For special effects, *free-form* deformation may be used where a geometric model is embedded inside a grid of control points, and transformations are applied to these control points to distort the object in a desired manner.

When dealing with transformations, an engineer would require a full description of the object, its position relative to a fixed point called *origin*, and a specified set of coordinate axes. An object may

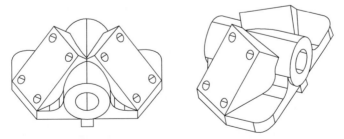

Figure 2.1   Use of transformation to view an object from different angles

be treated as an assemblage of finitely many points arranged in a non-arbitrary manner in space. The origin and coordinate axes may or may not be a part of the object. If the coordinate frame is attached to the object, it is called the *local frame* of reference. For coordinate frame not a part of the object, it is called *global frame*. Usually, since there are many objects to manoeuver at a given time, the user prefers a fixed global coordinate frame for all objects and one local coordinate system for each object. Geometric transformations may then involve: (a) moving all points of an object to a new location with respect to the global coordinate system or (b) relocating the local coordinate frame of an object to a new position without changing the object's position in the global frame. Transformations, in this chapter, are regarded as *time independent* in that the motion of an object from one position to another is *instantaneous* and does not follow a specified path in space. In other words, there can be more than one ways to manoeuver an object from its current location to a specified one.

2.1 Definition

A geometric transformation may be considered as a mapping function between a set of points both in the domain and range. The points may belong to the object or the coordinate system to be relocated. The function needs to be *one-to-one* in that any and all points in the domain (initial location) should have the corresponding images in the range (final location). Thus, if $T(P_1)$ and $T(P_2)$ represent the final locations of points P_1 and P_2 belonging to the object where T is a transformation function, then, if $P_1 \neq P_2$, $T(P_1) \neq T(P_2)$. In addition, the transformation should be *onto* in that for every final location $T(P)$, there must exist its pre-image P corresponding to the initial position of the object. In other words, any point in the newly located object must be associated with only one point belonging to the object in its original location. Thus, a one-to-one and onto map makes it possible to perform *inverse transformation*, that is, to move the object from its final to original location.

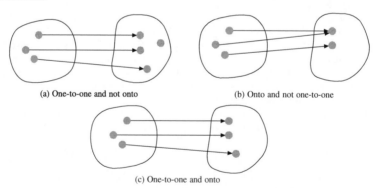

(a) One-to-one and not onto   (b) Onto and not one-to-one

(c) One-to-one and onto

Figure 2.2   Nature of geometric transformation as a function map

2.2 Rigid Body Transformations

In rigid body transformations, the geometric model stays undeformed, that is, the points constituting the model maintain the same relative positions with respect to each other. A solid model may be conceived to consist of points, curves and surfaces which should not get distorted under a rigid-body trans-formation. Rotation and translation are two transformations that can be grouped under this category. First, rotation and translation are discussed in two-dimensions. Vectors and matrices are most convenient to represent such motions. The *homogenous coordinate system*, which has some distinct advantages, is also introduced to unify the two transformations.

2.2.1 Rotation in Two-Dimensions

Consider a rigid body S packed with points P_i ($i = 1, \ldots, n$) and let a point $P_j(x_j, y_j)$ on S be rotated about the z-axis to $P_j^*(x_j^*, y_j^*)$ by an angle θ. From Figure 2.3, it can be observed that

$$x_j^* = l \cos(\theta + \alpha) = l \cos \alpha \cos \theta - l \sin \alpha \sin \theta$$

$$= x_j \cos \theta - y_j \sin \theta$$

and $\quad y_j^* = l \sin(\theta + \alpha) = l \cos \alpha \sin \theta + l \sin \alpha \cos \theta$

$$= x_j \sin \theta + y_j \cos \theta$$

Or in matrix form

$$\begin{bmatrix} x_j^* \\ y_j^* \end{bmatrix} = \begin{bmatrix} \cos \theta & -\sin \theta \\ \sin \theta & \cos \theta \end{bmatrix} \begin{bmatrix} x_j \\ y_j \end{bmatrix} \Rightarrow \mathbf{P}_j^* = \mathbf{R}\mathbf{P}_j \quad (2.1)$$

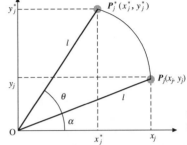

Figure 2.3   Rotation in a plane

where $\mathbf{R} = \begin{bmatrix} \cos \theta & -\sin \theta \\ \sin \theta & \cos \theta \end{bmatrix}$ is the two-dimensional
rotation matrix. For S to be rotated by an angle θ, transformation in Eq. (2.1) must be performed simultaneously for all points P_i ($i = 1, \ldots, n$) such that the entire rigid body reaches the new destination S^*.

Example 2.1 A trapezoidal lamina $ABCD$ lies in the x-y plane as shown with $A(6, 1)$, $B(8, 1)$, $C(10, 4)$ and $D(3, 4)$. The lamina is to be rotated about the z-axis by 90°. Determine the new position $A^*B^*C^*D^*$ after rotation (Figure 2.4(a)).

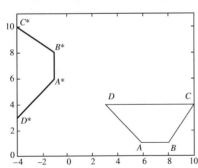

Figure 2.4 (a) Lamina rotation in Example 2.1

The transformation matrix $\mathbf{R}$ is given by Eq. (2.1) with $\theta = 90°$. Thus,

$$\begin{bmatrix} A^* \\ B^* \\ C^* \\ D^* \end{bmatrix}^T = \mathbf{R} \begin{bmatrix} A \\ B \\ C \\ D \end{bmatrix}^T = \begin{bmatrix} \cos 90° & -\sin 90° \\ \sin 90° & \cos 90° \end{bmatrix} \begin{bmatrix} 6 & 1 \\ 8 & 1 \\ 10 & 4 \\ 3 & 4 \end{bmatrix}^T$$

$$= \begin{bmatrix} 0 & -1 \\ 1 & 0 \end{bmatrix} \begin{bmatrix} 6 & 1 \\ 8 & 1 \\ 10 & 4 \\ 3 & 4 \end{bmatrix}^T = \begin{bmatrix} -1 & 6 \\ -1 & 8 \\ -4 & 10 \\ -4 & 3 \end{bmatrix}^T$$

2.2.2 Translation in Two-Dimensions: Homogeneous Coordinates

For a rigid body S to be translated along a vector $\mathbf{v}$ such that each point of S shifts by (p, q),

$$x_j^* = x_j + p, \quad y_j^* = y_j + q \Rightarrow \begin{bmatrix} x_j^* \\ y_j^* \end{bmatrix} = \begin{bmatrix} x_j \\ y_j \end{bmatrix} + \begin{bmatrix} p \\ q \end{bmatrix} \Rightarrow \mathbf{P}_j^* = \mathbf{P}_j + \mathbf{v} \quad (2.2)$$

Example 2.2 For a planar lamina *ABCD* with *A* (3, 5), *B* (2, 2), *C* (8, 2) and *D* (4, 5) in *x-y* plane and

P (4, 3) a point in the interior, the lamina is to be translated through $\mathbf{v} = \begin{bmatrix} 8 \\ 5 \end{bmatrix}$. Eq. (2.2) yields

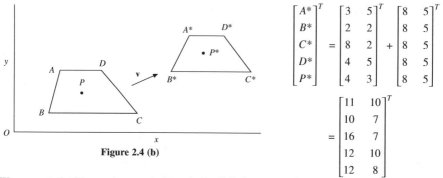

$$\begin{bmatrix} A^* \\ B^* \\ C^* \\ D^* \\ P^* \end{bmatrix}^T = \begin{bmatrix} 3 & 5 \\ 2 & 2 \\ 8 & 2 \\ 4 & 5 \\ 4 & 3 \end{bmatrix}^T + \begin{bmatrix} 8 & 5 \\ 8 & 5 \\ 8 & 5 \\ 8 & 5 \\ 8 & 5 \end{bmatrix}^T$$

$$= \begin{bmatrix} 11 & 10 \\ 10 & 7 \\ 16 & 7 \\ 12 & 10 \\ 12 & 8 \end{bmatrix}^T$$

Figure 2.4 (b)

We may note that like rotation, translation as in Eq. (2.2) does not work out to be a matrix multiplication. Instead, it is the addition of a point (position vector) and a (free) vector. One may attempt to represent translation also in the matrix multiplication form to unify the procedure for rigid body transformations. Consider Eq. (2.2) rewritten as

$$\begin{bmatrix} x_j^* \\ y_j^* \\ 1 \end{bmatrix} = \begin{bmatrix} 1 & 0 & p \\ 0 & 1 & q \\ 0 & 0 & 1 \end{bmatrix} \begin{bmatrix} x_j \\ y_j \\ 1 \end{bmatrix} = \begin{bmatrix} x_j + p \\ y_j + q \\ 1 \end{bmatrix} \tag{2.3}$$

Here, the first two rows provide the translation information while the third row gives the dummy result 1 = 1. Note also that the definition of position vector $P_j \begin{bmatrix} x_j \\ y_j \end{bmatrix}$ is altered from an ordered pair in the two-dimensional space to an ordered triplet $\begin{bmatrix} x_j \\ y_j \\ 1 \end{bmatrix}$ which are termed as the *homogenous coordinates* of P_j. We may use this new definition of position vectors to express translation in Eq. (2.3) as $\mathbf{P}_j^* = \mathbf{T}\mathbf{P}_j$ where

$$\mathbf{T} = \begin{bmatrix} 1 & 0 & p \\ 0 & 1 & q \\ 0 & 0 & 1 \end{bmatrix}$$

The rotation relation in Eq. (2.1) can be modified as well to express the result in terms of the homogeneous coordinates, that is

$$\mathbf{P}_j^* = \mathbf{R}\mathbf{P}_j \Rightarrow \begin{bmatrix} x_j^* \\ y_j^* \\ 1 \end{bmatrix} = \begin{bmatrix} \cos \theta & -\sin \theta & 0 \\ \sin \theta & \cos \theta & 0 \\ 0 & 0 & 1 \end{bmatrix} \begin{bmatrix} x_j \\ y_j \\ 1 \end{bmatrix}$$

where
$$\mathbf{R} = \begin{bmatrix} \cos\theta & -\sin\theta & 0 \\ \sin\theta & \cos\theta & 0 \\ 0 & 0 & 1 \end{bmatrix} \tag{2.4}$$

Rigid body translation and rotation thus get unified as *matrix multiplication* operations only, involving no addition or subtraction of matrices and vectors. Further, one can *concatenate* a sequence of transformations, for instance, translation of an object followed by its rotation. If one can identify the matrices for each of these transformations in the multiplication form, it becomes much easier to track the intermediate positions as well as to predict the final transformed position of the rigid body.

2.2.3 Combined Rotation and Translation

Consider a point $P\,(x, y, 1)$ in the x-y plane to be rotated by an angle θ about the z-axis to a position $P_1\,(x_1, y_1, 1)$ followed by a translation by $\mathbf{v}\,(p, q)$ to a position $P_2\,(x_2, y_2, 1)$. Using Eqs. (2.3) and (2.4), we may write

$$\mathbf{P}_1 = \mathbf{RP}, \quad \begin{bmatrix} x_1 \\ y_1 \\ 1 \end{bmatrix} = \begin{bmatrix} \cos\theta & -\sin\theta & 0 \\ \sin\theta & \cos\theta & 0 \\ 0 & 0 & 1 \end{bmatrix} \begin{bmatrix} x \\ y \\ 1 \end{bmatrix}$$

and
$$\mathbf{P}_2 = \mathbf{TP}_1, \quad \begin{bmatrix} x_2 \\ y_2 \\ 1 \end{bmatrix} = \begin{bmatrix} 1 & 0 & p \\ 0 & 1 & q \\ 0 & 0 & 1 \end{bmatrix} \begin{bmatrix} x_1 \\ y_1 \\ 1 \end{bmatrix} = \begin{bmatrix} 1 & 0 & p \\ 0 & 1 & q \\ 0 & 0 & 1 \end{bmatrix} \begin{bmatrix} \cos\theta & -\sin\theta & 0 \\ \sin\theta & \cos\theta & 0 \\ 0 & 0 & 1 \end{bmatrix} \begin{bmatrix} x \\ y \\ 1 \end{bmatrix} = \mathbf{TRP}$$

Thus,
$$\mathbf{P}_2 = \begin{bmatrix} \cos\theta & -\sin\theta & p \\ \sin\theta & \cos\theta & q \\ 0 & 0 & 1 \end{bmatrix} \begin{bmatrix} x \\ y \\ 1 \end{bmatrix} \tag{2.5}$$

On the contrary, if translation by $\mathbf{v}$ is followed by rotation about the z-axis by an angle θ to reach P_2^*, then

$$\mathbf{P}_2^* = \mathbf{RTP} = \begin{bmatrix} \cos\theta & -\sin\theta & 0 \\ \sin\theta & \cos\theta & 0 \\ 0 & 0 & 1 \end{bmatrix} \begin{bmatrix} 1 & 0 & p \\ 0 & 1 & q \\ 0 & 0 & 1 \end{bmatrix} \begin{bmatrix} x \\ y \\ 1 \end{bmatrix} = \begin{bmatrix} \cos\theta & -\sin\theta & p\cos\theta - q\sin\theta \\ \sin\theta & \cos\theta & p\sin\theta + q\sin\theta \\ 0 & 0 & 1 \end{bmatrix} \begin{bmatrix} x \\ y \\ 1 \end{bmatrix} \tag{2.6}$$

We observe from Eqs. (2.5) and (2.6) that the final positions P_2 and P_2^* are not identical. From above we can arrive at two important conclusions: (a) the homogeneous coordinate system helps to unify translation and rotation as multiplicative transformations and (b) transformations are not commutative. The sequence in which the transformations are performed is significant and must be maintained while concatenating the respective matrices. Otherwise a different orientation or position of the object is reached. If $T_1, T_2, \ldots, T_n$ are the transformations to be performed in the order, the combined transformation matrix T is given as $T = T_n\,T_{n-1}\,T_{n-2}\ldots T_2\,T_1$.

Example 2.3. Lamina *ABCD* with an inner point *P* with coordinates (4, 3), (3, 1), (8, 1), (7, 4) and (5, 2) respectively is first rotated through 60° and then translated by (5, 4). In another sequence, the trapezoid is first translated by (5, 4) and then rotated through 60°. The lamina acquires different positions and orientations given and shown below for the two sequences of transformations.

For rotation and then translation using Eq. (2.5), we have

$$
\begin{bmatrix} A' \\ B' \\ C' \\ D' \\ P' \end{bmatrix}^T = \begin{bmatrix} \cos 60° & -\sin 60° & 5 \\ \sin 60° & \cos 60° & 4 \\ 0 & 0 & 1 \end{bmatrix} \begin{bmatrix} 4 & 3 & 1 \\ 3 & 1 & 1 \\ 8 & 1 & 1 \\ 7 & 4 & 1 \\ 5 & 2 & 1 \end{bmatrix}^T = \begin{bmatrix} 4.43 & 8.96 & 1 \\ 5.63 & 7.09 & 1 \\ 8.13 & 11.42 & 1 \\ 5.03 & 12.06 & 1 \\ 5.76 & 9.33 & 1 \end{bmatrix}^T
$$

For translation and then rotation, Eq. (2.6) gives

$$
\begin{bmatrix} A* \\ B* \\ C* \\ D* \\ P* \end{bmatrix}^T = \begin{bmatrix} \cos 60° & -\sin 60° & 5\cos 60° - 4\sin 60° \\ \sin 60° & \cos 60° & 5\cos 60° + 4\sin 60° \\ 0 & 0 & 1 \end{bmatrix} \begin{bmatrix} 4 & 3 & 1 \\ 3 & 1 & 1 \\ 8 & 1 & 1 \\ 7 & 4 & 1 \\ 5 & 2 & 1 \end{bmatrix}^T
$$

$$
= \begin{bmatrix} -1.56 & 11.30 & 1 \\ -.33 & 9.43 & 1 \\ 2.17 & 13.76 & 1 \\ -.93 & 14.40 & 1 \\ -0.19 & 11.66 & 1 \end{bmatrix}^T
$$

The two different laminar positions and orientations are shown in Figure 2.5.

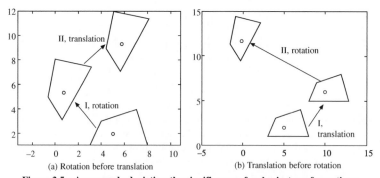

(a) Rotation before translation   (b) Translation before rotation

Figure 2.5   An example depicting the significance of order in transformations

2.2.4 Rotation of a Point Q (x_q, y_q, 1) about a Point P (p, q, 1)

Since the rotation matrix **R** about the z-axis and translation matrix **T** in the x-y plane are known from Eqs. (2.4) and (2.3) respectively, rotation of Q about P can be regarded as translating P to coincide with the origin, followed by rotation about the z-axis by an angle θ, and lastly, placing P back to its original position (Figure 2.6). These transformations can be concatenated as

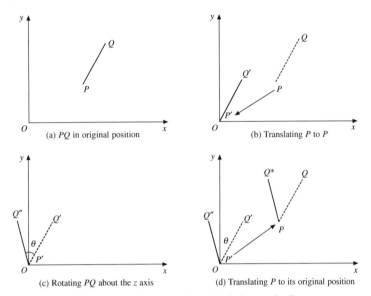

(a) PQ in original position

(b) Translating P to P

(c) Rotating PQ about the z axis

(d) Translating P to its original position

Figure 2.6   Steps to rotate point Q about point P

$$Q^* = \begin{bmatrix} x_q^* \\ y_q^* \\ 1 \end{bmatrix} = \begin{bmatrix} 1 & 0 & p \\ 0 & 1 & q \\ 0 & 0 & 1 \end{bmatrix} \begin{bmatrix} \cos\theta & -\sin\theta & 0 \\ \sin\theta & \cos\theta & 0 \\ 0 & 0 & 1 \end{bmatrix} \begin{bmatrix} 1 & 0 & -p \\ 0 & 1 & -q \\ 0 & 0 & 1 \end{bmatrix} \begin{bmatrix} x_q \\ y_q \\ 1 \end{bmatrix} \qquad (2.7)$$

2.2.5 Reflection

In 2-D, reflection of an object can be obtained by rotating it through $180°$ about the axis of reflection. For instance, if an object S in the x-y plane is to be reflected about the x-axis ($y = 0$), reflection of a point (x, y, 1) in S is given by (x^*, y^*, 1) such that

$$\begin{bmatrix} x^* \\ y^* \\ 1 \end{bmatrix} = \begin{bmatrix} x \\ -y \\ 1 \end{bmatrix} = \begin{bmatrix} 1 & 0 & 0 \\ 0 & -1 & 0 \\ 0 & 0 & 1 \end{bmatrix} \begin{bmatrix} x \\ y \\ 1 \end{bmatrix} = \mathbf{R}_{fx} \begin{bmatrix} x \\ y \\ 1 \end{bmatrix} \qquad (2.8)$$

Similarly, reflection about the y-axis is described as

$$\begin{bmatrix} x^* \\ y^* \\ 1 \end{bmatrix} = \begin{bmatrix} -x \\ y \\ 1 \end{bmatrix} = \begin{bmatrix} -1 & 0 & 0 \\ 0 & 1 & 0 \\ 0 & 0 & 1 \end{bmatrix}\begin{bmatrix} x \\ y \\ 1 \end{bmatrix} = \mathbf{R}_{fy}\begin{bmatrix} x \\ y \\ 1 \end{bmatrix} \tag{2.9}$$

Example 2.4. Consider a trapezium *ABCD* with $A = (6, 1, 1)$, $B = (8, 1, 1)$, $C = (10, 4, 1)$ and $D = (3, 4, 1)$. The entity is to be reflected through the *y*-axis. Applying $\mathbf{R}_{fy}$ in Eq. (2.9) results in

$$\begin{bmatrix} A^* \\ B^* \\ C^* \\ D^* \end{bmatrix}^T = \begin{bmatrix} -1 & 0 & 0 \\ 0 & 1 & 0 \\ 0 & 0 & 1 \end{bmatrix}\begin{bmatrix} 6 & 1 & 1 \\ 8 & 1 & 1 \\ 10 & 4 & 1 \\ 3 & 4 & 1 \end{bmatrix}^T = \begin{bmatrix} -6 & 1 & 1 \\ -8 & 1 & 1 \\ -10 & 4 & 1 \\ -3 & 4 & 1 \end{bmatrix}^T$$

The new position for the trapezium is shown as $A^*B^*C^*D^*$ in Figure 2.7. Note that identical result may be obtained by rotating the trapezium by 180° about the *y* axis. As expected there is no distortion in the shape of the trapezium. Since reflection results by combining translation and/or rotation, it is a rigid body transformation.

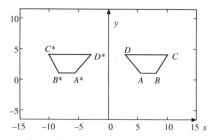

Figure 2.7   Reflection about the *y*-axis

2.2.6   Reflection About an Arbitrary Line

Let *D* be a point on line *L* and *S* be an object in two-dimensional space. It is required to reflect *S* about *L*. This reflection can be obtained as a sequence of the following transformations:

(a) Translate point *D* ($p, q, 1$) to coincide with the origin *O*, shifting the line *L* parallel to itself to a translated position L^*.
(b) Rotate L^* by an angle θ such that it coincides with the *y*-axis (new position of the line is L^{**}, say).
(c) Reflect *S* about the *y*-axis using Eq. (2.9).
(d) Rotate L^{**} through $-\theta$ to bring it back to L^*.
(e) Translate L^* to coincide with its original position *L*.

The schematic of the procedure is shown in Figure 2.8. The new image *S** is the reflection of *S* about *L* and the transformation is given by

$$\begin{bmatrix} 1 & 0 & p \\ 0 & 1 & q \\ 0 & 0 & 1 \end{bmatrix}\begin{bmatrix} \cos\theta & \sin\theta & 0 \\ -\sin\theta & \cos\theta & 0 \\ 0 & 0 & 1 \end{bmatrix}\begin{bmatrix} -1 & 0 & 0 \\ 0 & 1 & 0 \\ 0 & 0 & 1 \end{bmatrix}\begin{bmatrix} \cos\theta & -\sin\theta & 0 \\ \sin\theta & \cos\theta & 0 \\ 0 & 0 & 1 \end{bmatrix}$$

$$\times \begin{bmatrix} 1 & 0 & -p \\ 0 & 1 & -q \\ 0 & 0 & 1 \end{bmatrix} = \mathbf{T}_{OD}\mathbf{R}(-\theta)\mathbf{R}_{fy}\mathbf{R}(\theta)\mathbf{T}_{DO} \tag{2.10}$$

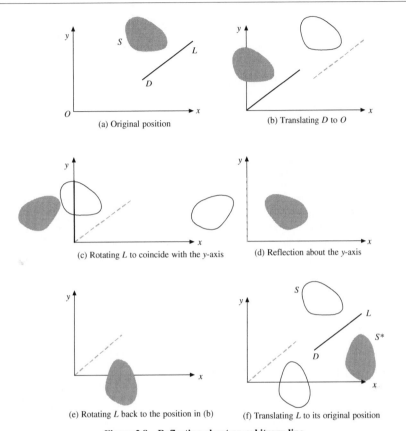

(a) Original position

(b) Translating D to O

(c) Rotating L to coincide with the y-axis

(d) Reflection about the y-axis

(e) Rotating L back to the position in (b)

(f) Translating L to its original position

Figure 2.8   Reflection about an arbitrary line

In Eq. (2.10), $\mathbf{T}_{PQ}$ represents translation from point **P** to **Q**. The above procedure is not unique in that the steps (b), (c) and (d) above can be altered so that L is made to coincide with the x-axis by rotating it through an angle α, reflection is performed about the x-axis, and the line is rotated back by $-\alpha$.

2.2.7   Reflection Through a Point
A point P $(x, y, 1)$ when reflected through the origin is written as P^* $(x^*, y^*, 1) = (-x, -y, 1)$ or

$$\begin{bmatrix} x^* \\ y^* \\ 1 \end{bmatrix} = \begin{bmatrix} -x \\ -y \\ 1 \end{bmatrix} = \begin{bmatrix} -1 & 0 & 0 \\ 0 & -1 & 0 \\ 0 & 0 & 1 \end{bmatrix} \begin{bmatrix} x \\ y \\ 1 \end{bmatrix} = \mathbf{R}_{fo} \begin{bmatrix} x \\ y \\ 1 \end{bmatrix} \tag{2.11}$$

For reflection of an object about a point P_r, we would require to shift P_r to the origin, perform the above reflection and then transform P_r back to its original position.

Example 2.5. To reflect a line with end points $P\,(2,4)$ and $Q\,(6,2)$ through the origin, from Eq. (2.11), we have

$$\begin{bmatrix} P^* \\ Q^* \end{bmatrix}^T = \begin{bmatrix} -1 & 0 & 0 \\ 0 & -1 & 0 \\ 0 & 0 & 1 \end{bmatrix} \begin{bmatrix} 2 & 4 & 1 \\ 6 & 2 & 1 \end{bmatrix}^T = \begin{bmatrix} -2 & -4 & 1 \\ -6 & -2 & 1 \end{bmatrix}^T$$

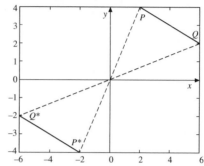

Joining P^*Q^* gives the reflection of line PQ through O as shown in Figure 2.9.

Figure 2.9   Reflection of a line through the origin

2.2.8   A Preservative for Angles!
Orthogonal Transformation Matrices

We must ensure for rigid-body transformations that if for instance a polygon is rotated, reflected or linearly shifted to a new location, the angle between the polygonal sides are preserved, that is, there is no distortion in its shape. Let $\mathbf{v}_1$ and $\mathbf{v}_2$ be vectors representing any two adjacent sides of a polygon (Figure 2.10). The angle between them is given by

$$\cos\theta = \frac{\mathbf{v}_1 \cdot \mathbf{v}_2}{|\mathbf{v}_1||\mathbf{v}_2|} \quad \text{and} \quad \sin\theta = \frac{(\mathbf{v}_1 \times \mathbf{v}_2) \cdot \mathbf{k}}{|\mathbf{v}_1||\mathbf{v}_2|} \tag{2.12}$$

where $\mathbf{v}_1 = [\mathbf{v}_{1x} \ \ \mathbf{v}_{1y} \ \ 0]$ and $\mathbf{v}_2 = [\mathbf{v}_{2x} \ \ \mathbf{v}_{2y} \ \ 0]$.

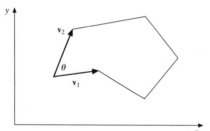

**Figure 2.10   Two adjacent sides of a polygon to be reflected,
rotated or translated to a new location**

Note the way the vectors are expressed as homogenous coordinates. For position vectors of points A and B as $[x_1, y_1, 1]^T$ and $[x_2, y_2, 1]^T$, the vector $\mathbf{AB}$ can be expressed as

$$\mathbf{AB} = \begin{bmatrix} x_2 \\ y_2 \\ 1 \end{bmatrix} - \begin{bmatrix} x_1 \\ y_1 \\ 1 \end{bmatrix} = \begin{bmatrix} x_2 - x_1 \\ y_2 - y_1 \\ 0 \end{bmatrix}$$

Thus in homogenous coordinates, free vectors have 0 as their last element. With $(\mathbf{i}, \mathbf{j}, \mathbf{k})$ as unit vectors along the coordinate axes x, y and z, respectively, applying any generic transformation $\mathbf{A}$ yields

$$\begin{bmatrix} v_1^* \\ v_2^* \end{bmatrix}^T = \mathbf{A} \begin{bmatrix} v_1 \\ v_2 \end{bmatrix}^T = \begin{bmatrix} a_{11} & a_{12} & a_{13} \\ a_{21} & a_{22} & a_{23} \\ a_{31} & a_{32} & a_{33} \end{bmatrix} \begin{bmatrix} v_{1x} & v_{1y} & 0 \\ v_{2x} & v_{2y} & 0 \end{bmatrix}^T$$

$$= \begin{bmatrix} a_{11}v_{1x} + a_{12}v_{1y} & a_{21}v_{1x} + a_{22}v_{1y} & a_{31}v_{1x} + a_{32}v_{1y} \\ a_{11}v_{2x} + a_{12}v_{2y} & a_{21}v_{2x} + a_{22}v_{2y} & a_{31}v_{2x} + a_{32}v_{2y} \end{bmatrix}^T$$

Thus,

$$\mathbf{v}_1^* \cdot \mathbf{v}_2^* = \left| \mathbf{v}_1^* \right| \left\| \mathbf{v}_2^* \right| \cos \theta^*$$

$$= (a_{11}v_{1x} + a_{12}v_{1y})(a_{11}v_{2x} + a_{12}v_{2y}) + (a_{21}v_{1x} + a_{22}v_{1y})(a_{21}v_{2x} + a_{22}v_{2y})$$

$$+ (a_{31}v_{1x} + a_{32}v_{1y})(a_{31}v_{2x} + a_{32}v_{2y})$$

$$= (a_{11}^2 + a_{21}^2 + a_{31}^2)v_{1x}v_{2x} + (a_{12}^2 + a_{22}^2 + a_{32}^2)v_{1y}v_{2y}$$

$$+ (a_{11}a_{12} + a_{21}a_{22} + a_{31}a_{32})(v_{1x}v_{2y} + v_{1y}v_{2x}) \tag{2.13}$$

Also,

$$\mathbf{v}_1^* \times \mathbf{v}_2^* = \left| \mathbf{v}_1^* \right| \left\| \mathbf{v}_2^* \right| \mathbf{k} \sin \theta^*$$

$$= (a_{11}v_{1x} + a_{12}v_{1y})(a_{21}v_{2x} + a_{22}v_{2y})\mathbf{k} - (a_{11}v_{1x} + a_{12}v_{1y})(a_{31}v_{2x} + a_{32}v_{2y})\mathbf{j}$$

$$- (a_{21}v_{1x} + a_{22}v_{1y})(a_{11}v_{2x} + a_{12}v_{2y})\mathbf{k} + (a_{21}v_{1x} + a_{22}v_{1y})(a_{31}v_{2x} + a_{32}v_{2y})\mathbf{i}$$

$$+ (a_{31}v_{1x} + a_{32}v_{1y})(a_{11}v_{2x} + a_{12}v_{2y})\mathbf{j} - (a_{31}v_{1x} + a_{32}v_{1y})(a_{21}v_{2x} + a_{22}v_{2y})\mathbf{i} \tag{2.14}$$

The angle between the original vectors $\mathbf{v}_1$ and $\mathbf{v}_2$ are given by

$$| \mathbf{v}_1 \|\mathbf{v}_2| \cos \theta = (v_{1x}\mathbf{i} + v_{1y}\mathbf{j}) \cdot (v_{2x}\mathbf{i} + v_{2y}\mathbf{j}) = (v_{1x}v_{2x} + v_{1y}v_{2y})$$

$$| \mathbf{v}_1 \|\mathbf{v}_2| \mathbf{k} \sin \theta = (v_{1x}\mathbf{i} + v_{1y}\mathbf{j}) \times (v_{2x}\mathbf{i} + v_{2y}\mathbf{j}) = (v_{1x}v_{2y} - v_{1y}v_{2x})\mathbf{k} \tag{2.15}$$

For no change in magnitude or angle, $\left| \mathbf{v}_1^* \right| \left\| \mathbf{v}_2^* \right| \cos \theta^* = | \mathbf{v}_1\|\mathbf{v}_2| \cos \theta$ and also $\mathbf{v}_1^* \times \mathbf{v}_2^* = \mathbf{v}_1 \times \mathbf{v}_2$. On comparing results, we obtain

$$(v_{1x}v_{2x} + v_{1y}v_{2y}) = (a_{11}^2 + a_{21}^2 + a_{31}^2)v_{1x}v_{2x} + (a_{12}^2 + a_{22}^2 + a_{32}^2)v_{1y}v_{2y}$$

$$+ (a_{11}a_{12} + a_{21}a_{22} + a_{31}a_{32})(v_{1x}v_{2y} + v_{1y}v_{2x}) \tag{2.16}$$

and

$$(v_{1x}v_{2y} - v_{1y}v_{2x})\mathbf{k} = (a_{11}v_{1x} + a_{12}v_{1y})(a_{21}v_{2x} + a_{22}v_{2y})\mathbf{k} - (a_{11}v_{1x} + a_{12}v_{1y})(a_{31}v_{2x} + a_{32}v_{2y})\mathbf{j}$$

$$- (a_{21}v_{1x} + a_{22}v_{1y})(a_{11}v_{2x} + a_{12}v_{2y})\mathbf{k} + (a_{21}v_{1x} + a_{22}v_{1y})(a_{31}v_{2x} + a_{32}v_{2y})\mathbf{i}$$

$$+ (a_{31}v_{1x} + a_{32}v_{1y})(a_{11}v_{2x} + a_{12}v_{2y})\mathbf{j} - (a_{31}v_{1x} + a_{32}v_{1y})(a_{21}v_{2x} + a_{22}v_{2y})\mathbf{i} \tag{2.17}$$

We can work out Eq. (2.17) and compare the coefficients of $\mathbf{i}$, $\mathbf{j}$ and $\mathbf{k}$ and further, compare the terms corresponding to $v_{1y}v_{2x}$ and $v_{1x}v_{2y}$. Finally, after comparison from Eqs. (2.16) and (2.17), we would get

$$(a_{11}^2 + a_{21}^2 + a_{31}^2) = 1 \qquad (a_{11}a_{13} + a_{21}a_{23} + a_{31}a_{33}) = 0$$

$$(a_{12}^2 + a_{22}^2 + a_{32}^2) = 1 \qquad (a_{13}a_{12} + a_{23}a_{22} + a_{33}a_{32}) = 0$$

$$(a_{11}a_{12} + a_{21}a_{22} + a_{31}a_{32}) = 0 \qquad (a_{11}a_{12} + a_{21}a_{22} + a_{31}a_{32}) = 0$$

$$(a_{11}a_{22} - a_{12}a_{21}) = 1$$

which suggests that **A** must be orthogonal having the property $\mathbf{A}^{-1} = \mathbf{A}^T$ so that $\mathbf{A}\mathbf{A}^T = \mathbf{A}^T\mathbf{A} = \mathbf{I}$, where **I** is the identity matrix of the same size as **A**. Further, $(a_{11}a_{22} - a_{12}a_{21}) = 1$ implies that the determinant of **A** should be 1. In pure rotation, the above conditions are completely met where for **R** in Eq. (2.4), $a_{13} = a_{23} = a_{31} = a_{32} = 0$, and $a_{33} = 1$. In reflection, the determinant of the transformation matrix is – 1; hence, although the matrix is orthogonal, the angle is not preserved and that it changes to $(2\pi-\theta)$ though the absolute angle between the adjacent sides of the polygon remains θ. The magnitudes of the vectors are preserved. The angle between the intersecting vectors is also preserved in case of translation, that is

$$\begin{bmatrix} v_1^* \\ v_2^* \end{bmatrix}^T = \begin{bmatrix} 1 & 0 & p \\ 0 & 1 & q \\ 0 & 0 & 1 \end{bmatrix} \begin{bmatrix} v_1 \\ v_2 \end{bmatrix}^T = \begin{bmatrix} 1 & 0 & p \\ 0 & 1 & q \\ 0 & 0 & 1 \end{bmatrix} \begin{bmatrix} v_{1x} & v_{1y} & 0 \\ v_{2x} & v_{2y} & 0 \end{bmatrix}^T = \begin{bmatrix} v_{1x} & v_{1y} & 0 \\ v_{2x} & v_{2y} & 0 \end{bmatrix}^T = \begin{bmatrix} v_1 \\ v_2 \end{bmatrix}^T$$

which implies that the translation does not alter vectors.

2.3   Deformations

Previous sections dealt with transformations wherein the object was relocated and/or reoriented without the change in its shape or size. In this section, one would deal with transformations that would alter the size and/or shape of the object. Examples involve those of *scaling* and *shear*.

2.3.1   Scaling

A point $P(x, y, 1)$ belonging to the object S can be scaled to a new position vector $P^*(x^*, y^*, 1)$ using factors μ_x and μ_y such that

$$x^* = \mu_x x \text{ and } y^* = \mu_y y$$

Or in matrix form

$$\begin{bmatrix} x^* \\ y^* \\ 1 \end{bmatrix} = \begin{bmatrix} \mu_x & 0 & 0 \\ 0 & \mu_y & 0 \\ 0 & 0 & 1 \end{bmatrix} \begin{bmatrix} x \\ y \\ 1 \end{bmatrix} = \mathbf{SP} \qquad (2.18)$$

where $\mathbf{S} = \begin{bmatrix} \mu_x & 0 & 0 \\ 0 & \mu_y & 0 \\ 0 & 0 & 1 \end{bmatrix}$ is the scaling matrix. Scale factors μ_x and μ_y are always non-zero and positive. For both μ_x and μ_y less than 1, the geometric model gets *shrunk*. In case of *uniform scaling* when $\mu_x = \mu_y = \mu$, the model gets changed uniformly in size (Figure 2.11) and there is no distortion.

Consider a curve, for instance, defined by $\mathbf{r}(u) = x(u)\mathbf{i} + y(u)\mathbf{j}$, where parameter u varies in the interval [0, 1]. The curve after scaling becomes $\mathbf{r}^*(u) = x^*(u)\mathbf{i} + y^*(u)\mathbf{j} = \mu_x x(u)\mathbf{i} + \mu_y y(u)\mathbf{j}$ and the tangent to any point on this curve is obtained by differentiating $\mathbf{r}^*(u)$ with respect to u, that is,

$$\dot{\mathbf{r}}^*(u) = \mu_x \dot{x}(u)\,\mathbf{i} + \mu_y \dot{y}(u)\,\mathbf{j}$$

Hence

$$\frac{dy}{dx} = \frac{(dy/du)}{(dx/du)} = \frac{\mu_y \dot{y}(u)}{\mu_x \dot{x}(u)} \qquad (2.19)$$

Thus, non-uniform scaling changes the tangent vector proportionally while the slope remains unaltered in uniform scaling for $\mu_x = \mu_y$.

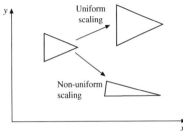

Figure 2.11   Uniform and non-uniform scaling

2.3.2 Shear

Consider a matrix $\mathbf{Sh}_x = \begin{bmatrix} 1 & sh_x & 0 \\ 0 & 1 & 0 \\ 0 & 0 & 1 \end{bmatrix}$ which when applied to a point $P\,(x, y, 1)$ results in

$$\begin{bmatrix} x^* \\ y^* \\ 1 \end{bmatrix} = \begin{bmatrix} 1 & sh_x & 0 \\ 0 & 1 & 0 \\ 0 & 0 & 1 \end{bmatrix} \begin{bmatrix} x \\ y \\ 1 \end{bmatrix} = \begin{bmatrix} x + sh_x y \\ y \\ 1 \end{bmatrix} \qquad (2.20)$$

which in effect shears the point along the x axis. Likewise, application of $\mathbf{Sh}_y = \begin{bmatrix} 1 & 0 & 0 \\ sh_y & 1 & 0 \\ 0 & 0 & 1 \end{bmatrix}$ on P yields

$$\begin{bmatrix} x^* \\ y^* \\ 1 \end{bmatrix} = \begin{bmatrix} 1 & 0 & 0 \\ sh_y & 1 & 0 \\ 0 & 0 & 1 \end{bmatrix} \begin{bmatrix} x \\ y \\ 1 \end{bmatrix} = \begin{bmatrix} x \\ sh_y x + y \\ 1 \end{bmatrix} \qquad (2.21)$$

that is, the new point gets sheared along the y direction.

Example 2.6. For a rectangle with coordinates (3, 1), (3, 4), (8, 4) and (8, 1), respectively, applying shear along the y direction (Figure 2.12) with a factor $sh_y = 1.5$ yields the new points as

$$\begin{bmatrix} P_1^* \\ P_2^* \\ P_3^* \\ P_4^* \end{bmatrix} = \begin{bmatrix} 1 & 0 & 0 \\ 1.5 & 1 & 0 \\ 0 & 0 & 1 \end{bmatrix} \begin{bmatrix} 3 & 1 & 1 \\ 3 & 4 & 1 \\ 8 & 4 & 1 \\ 8 & 1 & 1 \end{bmatrix}^T = \begin{bmatrix} 3 & 5.5 & 1 \\ 3 & 8.5 & 1 \\ 8 & 16 & 1 \\ 8 & 13 & 1 \end{bmatrix}^T$$

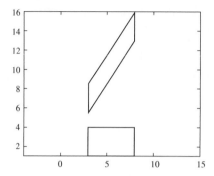

Figure 2.12   Shear along the *y* direction

2.4   Generic Transformation in Two-Dimensions

Observing the transformation matrices developed previously for translation, rotation, reflection, scaling and shear, we may realize that the matrices may be expressed generically in the partitioned form as

$$
\mathbf{A} = \begin{bmatrix} a_{11} & a_{12} & | & a_{13} \\ a_{21} & a_{22} & | & a_{23} \\ — & — & — & — \\ a_{31} & a_{32} & | & a_{33} \end{bmatrix} \tag{2.22}
$$

The top left 2×2 sub-matrix represents: (a) rotation when the elements are the sine and cosine terms of the rotation angle about the *z*-axis, (b) reflection when the diagonal elements are +1 or – 1, and the off diagonal terms are zero, (c) scaling when the diagonal elements are positive μ_x and μ_y with the off diagonal terms as zero and (d) shear when the off diagonal elements are non-zero and diagonal elements are 1. The second top-right partition of 2×1 sub-matrix represents translation. The bottom-left partition of 1×2 sub-matrix represents perspective transformation discussed later and the bottom right matrix, the diagonal element $a_{33} = 1$ represents the homogeneous coordinate scalar. Like a point in the *x*-*y* plane is represented as $(x, y, 1)$ using the homogenous system, in a three-dimensional space, the representation can be extended to $(x, y, z, 1)$. Accordingly, the matrix **A** in Eq. (2.22) gets modified to

$$
\mathbf{A} = \begin{bmatrix} a_{11} & a_{12} & a_{13} & | & a_{14} \\ a_{21} & a_{22} & a_{23} & | & a_{24} \\ a_{31} & a_{32} & a_{33} & | & a_{34} \\ — & — & — & | & — \\ a_{41} & a_{42} & a_{43} & | & a_{44} \end{bmatrix} \tag{2.23}
$$

The partitions now consist of 3×3, 3×1, 1×3, and 1×1 sub-matrices having the same role as discussed for the respective partitions above for a two-dimensional case.

2.5 Transformations in Three-Dimensions

Matrices developed for transformations in two-dimensions can be modified as per the schema in Eq. (2.23) for use in three-dimensions. For instance, the translation matrix to move a point and thus an object, e.g. in Figure 2.13, by a vector (p, q, r) may be written as

$$
\mathbf{T} = \begin{bmatrix} 1 & 0 & 0 & p \\ 0 & 1 & 0 & q \\ 0 & 0 & 1 & r \\ 0 & 0 & 0 & 1 \end{bmatrix} \qquad (2.24)
$$

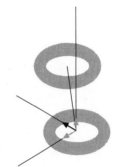

2.5.1 Rotation in Three-Dimensions

The rotation matrix in Eq. (2.4) can be modified to accommodate the three-dimensional homogenous coordinates. For rotation by angle θ about the z-axis (the z coordinate does not change), we get

Figure 2.13 Translation of a donut along an arbitrary vector

$$
\mathbf{R}_z = \begin{bmatrix} \cos\theta & -\sin\theta & 0 & 0 \\ \sin\theta & \cos\theta & 0 & 0 \\ 0 & 0 & 1 & 0 \\ 0 & 0 & 0 & 1 \end{bmatrix} \qquad (2.25)
$$

Further, using the cyclic rule for the right-handed coordinate axes, rotation matrices about the x- and y-axis for angles ψ and ϕ can be written, respectively, by inspection as

$$
\mathbf{R}_x = \begin{bmatrix} 1 & 0 & 0 & 0 \\ 0 & \cos\psi & -\sin\psi & 0 \\ 0 & \sin\psi & \cos\psi & 0 \\ 0 & 0 & 0 & 1 \end{bmatrix} \quad \text{and} \quad \mathbf{R}_y = \begin{bmatrix} \cos\phi & 0 & \sin\phi & 0 \\ 0 & 1 & 0 & 0 \\ -\sin\phi & 0 & \cos\phi & 0 \\ 0 & 0 & 0 & 1 \end{bmatrix} \qquad (2.26)
$$

Rotation of a point by angles θ, ϕ and ψ (in that order) about the z-, y- and x-axis, respectively, is a useful transformation used often for rigid body rotation. The combined rotation is given as

$$
\mathbf{R} = \mathbf{R}_x(\psi)\mathbf{R}_y(\phi)\mathbf{R}_z(\theta) = \begin{bmatrix} 1 & 0 & 0 & 0 \\ 0 & \cos\psi & -\sin\psi & 0 \\ 0 & \sin\psi & \cos\psi & 0 \\ 0 & 0 & 0 & 1 \end{bmatrix} \begin{bmatrix} \cos\phi & 0 & \sin\phi & 0 \\ 0 & 1 & 0 & 0 \\ -\sin\phi & 0 & \cos\phi & 0 \\ 0 & 0 & 0 & 1 \end{bmatrix} \begin{bmatrix} \cos\theta & -\sin\theta & 0 & 0 \\ \sin\theta & \cos\theta & 0 & 0 \\ 0 & 0 & 1 & 0 \\ 0 & 0 & 0 & 1 \end{bmatrix} \qquad (2.27)
$$

We may as well multiply the three matrices to derive the composite matrix though it is easier to express the transformation in the above form for the purpose of depicting the order of transformations. Also, it is easier to remember the individual transformation matrices than the composite matrix. We may need to rotate an object about a given line. For instance, to rotate an object in Figure 2.14 (a) by $45°$ about the line $L \equiv y = x$. One way is to rotate the object about the z-axis such that L coincides with the x-axis, perform rotation about the x-axis and then rotate L about the z-axis to its original location. The combined transformation would then be

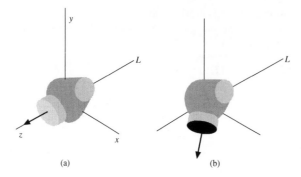

Figure 2.14   Rotation of an object: (a) about the line $y - x = 0$ and (b) rotated result

$$
\mathbf{R} = \mathbf{R}_z(45°)\mathbf{R}_x(45°)\mathbf{R}_z(-45°)
\begin{bmatrix}
\frac{1}{\sqrt{2}} & -\frac{1}{\sqrt{2}} & 0 & 0 \\[6pt]
\frac{1}{\sqrt{2}} & \frac{1}{\sqrt{2}} & 0 & 0 \\[6pt]
0 & 0 & 1 & 0 \\[6pt]
0 & 0 & 0 & 1
\end{bmatrix}
\begin{bmatrix}
1 & 0 & 0 & 0 \\[6pt]
0 & \frac{1}{\sqrt{2}} & -\frac{1}{\sqrt{2}} & 0 \\[6pt]
0 & \frac{1}{\sqrt{2}} & \frac{1}{\sqrt{2}} & 0 \\[6pt]
0 & 0 & 0 & 1
\end{bmatrix}
\begin{bmatrix}
\frac{1}{\sqrt{2}} & \frac{1}{\sqrt{2}} & 0 & 0 \\[6pt]
-\frac{1}{\sqrt{2}} & \frac{1}{\sqrt{2}} & 0 & 0 \\[6pt]
0 & 0 & 1 & 0 \\[6pt]
0 & 0 & 0 & 1
\end{bmatrix}
$$

and the result is shown in Figure 2.14 (b). An alternate way is to rotate the line about the z-axis to coincide with the y-axis, perform rotation about the y-axis and then rotate L back to its original location. Apparently, transformation procedures may not be unique though the end result would be the same if a proper transformation order is followed.

To rotate a point $\mathbf{P}$ about an axis L having direction cosines $\mathbf{n} = [n_x \ n_y \ n_z \ 0]$ that passes through a point $\mathbf{A}$ $[p \ q \ r \ 1]$, we may observe that $\mathbf{P}$ and its new location $\mathbf{P}^*$ would lie on a plane perpendicular to L and the plane would intersect L at $\mathbf{Q}$ (Figure 2.15(a)). Transformations may be composed stepwise as follows:

(i) Point $\mathbf{A}$ on L may be translated to coincide with the origin O using the transformation $\mathbf{T}_A$. The new line L' remains parallel to L.

$$
\mathbf{T}_A =
\begin{bmatrix}
1 & 0 & 0 & -p \\
0 & 1 & 0 & -q \\
0 & 0 & 1 & -r \\
0 & 0 & 0 & 1
\end{bmatrix}
$$

(ii) The unit vector $\mathbf{OU}$ (along L) projected onto the x-y and y-z planes, makes the traces OU_{xy} and OU_{yz}, respectively (Figure 2.15(b)). The magnitude of OU_{yz} is $d = \sqrt{(n_y^2 + n_z^2)} = \sqrt{(1 - n_x^2)}$. OU_{yz} makes an angle ψ with the z-axis such that

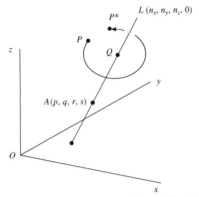

Figure 2.15(a)   Rotation of P about a line L

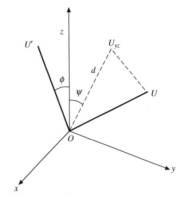

Figure 2.15(b)   Computing angles from the direction cosines

$$\cos \psi = \frac{n_z}{d}, \quad \sin \psi = \frac{n_y}{d}$$

Rotate OU about the x-axis by ψ to place it on the x-z plane (OU') in which case OU_{yz} will coincide with the z-axis. OU' makes angle ϕ with the z-axis such that $\cos \phi = d$ and $\sin \phi = n_x$. Rotate OU' about the y-axis by $-\phi$ so that in effect, OU coincides with the z-axis. The two rotation transformations are given by

$$\mathbf{R}_x = \begin{bmatrix} 1 & 0 & 0 & 0 \\ 0 & \cos \psi & -\sin \psi & 0 \\ 0 & \sin \psi & \cos \psi & 0 \\ 0 & 0 & 0 & 1 \end{bmatrix} \quad \text{and} \quad \mathbf{R}_y = \begin{bmatrix} \cos(-\phi) & 0 & \sin(-\phi) & 0 \\ 0 & 1 & 0 & 0 \\ -\sin(-\phi) & 0 & \cos(-\phi) & 0 \\ 0 & 0 & 0 & 1 \end{bmatrix}$$

(iii) The required rotation through angle α is then performed about the z-axis using

$$\mathbf{R}_z = \begin{bmatrix} \cos \alpha & -\sin \alpha & 0 & 0 \\ \sin \alpha & \cos \alpha & 0 & 0 \\ 0 & 0 & 1 & 0 \\ 0 & 0 & 0 & 1 \end{bmatrix}$$

(iv) Eventually, OU or line L is placed back to its original location by performing inverse transformations. The complete rotation transformation of point P about L can now be written as

$$\mathbf{R} = \mathbf{T}_A^{-1} \, \mathbf{R}_x^{-1}(\psi) \, \mathbf{R}_y^{-1}(-\phi) \, \mathbf{R}_z(\alpha) \, \mathbf{R}_y(-\phi) \, \mathbf{R}_x(\psi) \, \mathbf{T}_A \tag{2.28}$$

Figure 2.16 shows, as an example, the rotation of a disc about its axis placed arbitrarily in the coordinate system. Note that all matrices being orthogonal, $\mathbf{R}_y^{-1}(-\phi) = \mathbf{R}_y(\phi)$, $\mathbf{R}_x^{-1}(\psi) = \mathbf{R}_x(-\psi)$ and $\mathbf{T}_A^{-1}(-\mathbf{v}) = \mathbf{T}_A(\mathbf{v})$, where $\mathbf{v} = [p \ q \ r]^T$.

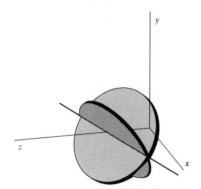

Figure 2.16   Rotation of a disc about its axis

2.5.2   Scaling in Three-Dimensions

The scaling matrix can be extended from that in a two-dimensional case (Eq. 2.18) as

$$\mathbf{S} = \begin{bmatrix} \mu_x & 0 & 0 & 0 \\ 0 & \mu_y & 0 & 0 \\ 0 & 0 & \mu_z & 0 \\ 0 & 0 & 0 & 1 \end{bmatrix} \tag{2.29}$$

where μ_x, μ_y and μ_z are the scale factors along x, y and z directions, respectively. For uniform overall scaling, $\mu_x = \mu_y = \mu_z = \mu$.

Alternatively,

$$\mathbf{S}_1 = \begin{bmatrix} 1 & 0 & 0 & 0 \\ 0 & 1 & 0 & 0 \\ 0 & 0 & 1 & 0 \\ 0 & 0 & 0 & s \end{bmatrix}$$

has the same uniform scaling effect as that of Eq. (2.29). To observe this, we may write

$$\begin{bmatrix} x^* \\ y^* \\ z^* \\ 1 \end{bmatrix} = \begin{bmatrix} 1 & 0 & 0 & 0 \\ 0 & 1 & 0 & 0 \\ 0 & 0 & 1 & 0 \\ 0 & 0 & 0 & s \end{bmatrix} \begin{bmatrix} x \\ y \\ z \\ 1 \end{bmatrix} = \begin{bmatrix} x \\ y \\ z \\ s \end{bmatrix} \equiv \begin{bmatrix} \frac{x}{s} \\ \frac{y}{s} \\ \frac{z}{s} \\ 1 \end{bmatrix} = \begin{bmatrix} \frac{1}{s} & 0 & 0 & 0 \\ 0 & \frac{1}{s} & 0 & 0 \\ 0 & 0 & \frac{1}{s} & 0 \\ 0 & 0 & 0 & 1 \end{bmatrix} \begin{bmatrix} x \\ y \\ z \\ 1 \end{bmatrix} \tag{2.30}$$

comparing which with Eq. (2.29) for $\mu_x = \mu_y = \mu_z = \mu$ yields $\mu = \dfrac{1}{s}$. Figure 2.17 shows uniform scaling of a cylindrical primitive.

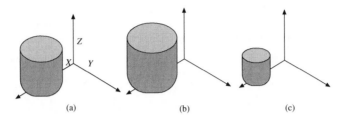

Figure 2.17   **A scaled cylinder using different factors: (a) original size, (b) twice the original size, (c) half the original size**

Eq. (2.30) uses the equivalence $[x\ y\ z\ s]^T \equiv \left[\dfrac{x}{s}\ \dfrac{y}{s}\ \dfrac{z}{s}\ 1\right]^T$ since both vectors represent the same point in the four-dimensional homogeneous coordinate system.

2.5.3   Shear in Three-Dimensions

In the 3×3 sub-matrix of the general transformation matrix (2.23), if all diagonal elements including a_{44} are 1, and the elements of 1×3 row sub-matrix and 3×1 column sub-matrix are all zero, we get the shear transformation matrix in three-dimensions, similar to the two-dimensional case. The generic form is

$$\mathbf{Sh} = \begin{bmatrix} 1 & sh_{12} & sh_{13} & 0 \\ sh_{21} & 1 & sh_{23} & 0 \\ sh_{31} & sh_{32} & 1 & 0 \\ 0 & 0 & 0 & 1 \end{bmatrix} \tag{2.31}$$

whose effect on point P is

$$\begin{bmatrix} x^* \\ y^* \\ z^* \\ 1 \end{bmatrix} = \begin{bmatrix} 1 & sh_{12} & sh_{13} & 0 \\ sh_{21} & 1 & sh_{23} & 0 \\ sh_{31} & sh_{32} & 1 & 0 \\ 0 & 0 & 0 & 1 \end{bmatrix} \begin{bmatrix} x \\ y \\ z \\ 1 \end{bmatrix} = \begin{bmatrix} x + sh_{12}y + sh_{13}z \\ sh_{21}x + y + sh_{23}z \\ sh_{31}x + sh_{32}y + z \\ 1 \end{bmatrix}$$

Thus, to shear an object only along the y direction, the entries $sh_{12} = sh_{13} = sh_{31} = sh_{32}$ would be 0 while either sh_{21} and sh_{23} or both would be non-zero.

2.5.4   Reflection in Three-Dimensions

Generic reflections about the x-y plane (z becomes $-z$), y-z plane (x becomes $-x$), and z-x plane (y becomes $-y$) can be expressed using the following respective transformations:

$$\mathbf{Rf}_{xy} = \begin{bmatrix} 1 & 0 & 0 & 0 \\ 0 & 1 & 0 & 0 \\ 0 & 0 & -1 & 0 \\ 0 & 0 & 0 & 1 \end{bmatrix}, \quad \mathbf{Rf}_{yz} = \begin{bmatrix} 1 & 0 & 0 & 0 \\ 0 & -1 & 0 & 0 \\ 0 & 0 & 1 & 0 \\ 0 & 0 & 0 & 1 \end{bmatrix} \quad \text{and} \quad \mathbf{Rf}_{zx} = \begin{bmatrix} -1 & 0 & 0 & 0 \\ 0 & 1 & 0 & 0 \\ 0 & 0 & 1 & 0 \\ 0 & 0 & 0 & 1 \end{bmatrix} \tag{2.32}$$

For reflection about a generic plane Π having the unit normal vector as $\mathbf{n} = [n_x\ n_y\ n_z\ 0]$ and for $\mathbf{A}\ [p\ q\ r\ 1]$ as any known point on it, the *modus operandi* is similar to the rotation about an arbitrary axis discussed in section 2.5.1. The steps followed are

(a) Translate Π to the new position Π' such that point $\mathbf{A}$ coincides with the origin using

$$
\mathbf{T}_A = \begin{bmatrix} 1 & 0 & 0 & -p \\ 0 & 1 & 0 & -q \\ 0 & 0 & 1 & -r \\ 0 & 0 & 0 & 1 \end{bmatrix}
$$

(b) Rotate the unit vector $\mathbf{n}$ (passing through the origin) on Π' to coincide with the z-axis. The new position of Π' will be Π'' and the reflecting plane will coincide with the x-y plane ($z = 0$). We would need the following transformations to acccomplish this step:

$$
\mathbf{R}_x = \begin{bmatrix} 1 & 0 & 0 & 0 \\ 0 & \dfrac{n_z}{\sqrt{1-n_x^2}} & -\dfrac{n_y}{\sqrt{1-n_x^2}} & 0 \\ 0 & \dfrac{n_y}{\sqrt{1-n_x^2}} & \dfrac{n_z}{\sqrt{1-n_x^2}} & 0 \\ 0 & 0 & 0 & 1 \end{bmatrix}, \quad
\mathbf{R}_y = \begin{bmatrix} \sqrt{1-n_x^2} & 0 & -n_x & 0 \\ 0 & 1 & 0 & 0 \\ n_x & 0 & \sqrt{1-n_x^2} & 0 \\ 0 & 0 & 0 & 1 \end{bmatrix}
$$

$$
\mathbf{Rf}_{xy} = \begin{bmatrix} 1 & 0 & 0 & 0 \\ 0 & 1 & 0 & 0 \\ 0 & 0 & -1 & 0 \\ 0 & 0 & 0 & 1 \end{bmatrix}
$$

(c) After reflection, the reverse order transformations need to be performed. The complete transformation would be

$$
\mathbf{R} = \mathbf{T}_A^{-1}\mathbf{R}_x^{-1}(\psi)\mathbf{R}_y^{-1}(-\phi)\ \mathbf{Rf}_{xy}\ \mathbf{R}_y(-\phi)\ \mathbf{R}_x(\psi)\ \mathbf{T}_A \tag{2.33}
$$

Example 2.7. The corners of wedge-shaped block are $A(0, 0, 2)$, $B(0, 0, 3)$, $C(0, 2, 3)$, $D(0, 2, 2)$, $E(-1, 2, 2)$ and $F(-1, 2, 3)$, and the reflection plane passes through the y-axis at $45°$ between $(-x)$ and z-axis. Determine the reflection of the wedge.

First, no translation of the reflecting plane is required as it passes through the origin. The direction cosines of the plane are $(0.707, 0, 0.707)$. We may apply Eq. (2.33) directly to get the result. Alternatively, rotate the plane about the y-axis for the reflecting plane to coincide with the y-z plane. Perform reflection about the y-z plane and thereafter, rotate the plane back to its original location.

$$
\mathbf{R}_y(-225°) = \begin{bmatrix} \cos(45°) & 0 & \sin(45°) & 0 \\ 0 & 1 & 0 & 0 \\ -\sin(45°) & 0 & \cos(45°) & 0 \\ 0 & 0 & 0 & 1 \end{bmatrix} = \begin{bmatrix} 0.707 & 0 & 0.707 & 0 \\ 0 & 1 & 0 & 0 \\ -0.707 & 0 & 0.707 & 0 \\ 0 & 0 & 0 & 1 \end{bmatrix}
$$

$$\mathbf{Rf}_{yz} = \begin{bmatrix} -1 & 0 & 0 & 0 \\ 0 & 1 & 0 & 0 \\ 0 & 0 & 1 & 0 \\ 0 & 0 & 0 & 1 \end{bmatrix}$$

The transformations are

$$\begin{bmatrix} A* \\ B* \\ C* \\ D* \\ E* \\ F* \end{bmatrix}^T = \mathbf{R}_y^{-1} \mathbf{Rf}_{yz} \, \mathbf{R}_y \begin{bmatrix} A \\ B \\ C \\ D \\ E \\ F \end{bmatrix}^T = \begin{bmatrix} 0.707 & 0 & -0.707 & 0 \\ 0 & 1 & 0 & 0 \\ 0.707 & 0 & 0.707 & 0 \\ 0 & 0 & 0 & 1 \end{bmatrix} \begin{bmatrix} -1 & 0 & 0 & 0 \\ 0 & 1 & 0 & 0 \\ 0 & 0 & 1 & 0 \\ 0 & 0 & 0 & 1 \end{bmatrix}$$

$$\times \begin{bmatrix} 0.707 & 0 & 0.707 & 0 \\ 0 & 1 & 0 & 0 \\ -0.707 & 0 & 0.707 & 0 \\ 0 & 0 & 0 & 1 \end{bmatrix} = \begin{bmatrix} -2 & 0 & 0 & 1 \\ -3 & 0 & 0 & 1 \\ -3 & 2 & 0 & 1 \\ -2 & 2 & 0 & 1 \\ -2 & 2 & 1 & 1 \\ -3 & 2 & 1 & 1 \end{bmatrix}^T$$

and the reflected object is shown in Figure 2.18.

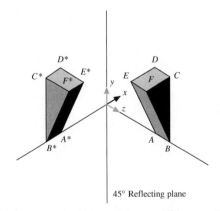

Figure 2.18   Reflection of a wedge about a plane at 45° between (– x) and z-axis.

2.6   Computer Aided Assembly of Rigid Bodies

Transformations can be used to position CAD primitives created separately and then to manipulate these using solid modeling *Boolean operations* like *join*, *cut* and *intersect*. Such operations are discussed in Chapter 8 in detail. Here, however, discussion shall be restricted to relative positioning. Consider a triangular rigid-body S_1 ($\mathbf{P_1P_2P_3}$) to be joined to another rigid body S_2 ($\mathbf{Q_1Q_2Q_3}$) such that $\mathbf{P_1}$ coincides with $\mathbf{Q_1}$ and the edge $\mathbf{P_1P_2}$ is colinear with $\mathbf{Q_1Q_2}$.

The first objective is to have both triangles in the same plane after assembly. Two local coordinate systems are constructed at the corner points $\mathbf{P_1}$ and $\mathbf{Q_1}$ with unit vectors ($\mathbf{p_1}$, $\mathbf{p_2}$, $\mathbf{p_3}$) and ($\mathbf{q_1}$, $\mathbf{q_2}$, $\mathbf{q_3}$). Here, unit vectors $\mathbf{p_1}$ and $\mathbf{p_2}$ are along the sides P_1P_2 and P_1P_3 respectively, and $\mathbf{p_3}$ is perpendicular to the plane $P_1P_2P_3$. Unit vectors $\mathbf{q_1}$ and $\mathbf{q_2}$ are along Q_1Q_2 and Q_1Q_3, respectively, and $\mathbf{q_3}$ is perpendicular to the plane $Q_1\,Q_2\,Q_3$. Thus

$$\mathbf{p}_1 = \frac{\mathbf{P}_2 - \mathbf{P}_1}{|\mathbf{P}_2 - \mathbf{P}_1|}, \ \mathbf{p}_2 = \frac{\mathbf{P}_3 - \mathbf{P}_1}{|\mathbf{P}_3 - \mathbf{P}_1|}, \ \mathbf{p}_3 = \mathbf{p}_1 \times \mathbf{p}_2$$

Vectors $\mathbf{q_1}$, $\mathbf{q_2}$, $\mathbf{q_3}$ can be determined in a similar way. Note that each of the unit vectors $\mathbf{p}$ ($\mathbf{p_1}$, $\mathbf{p_2}$, $\mathbf{p_3}$) and $\mathbf{q}$ ($\mathbf{q_1}$, $\mathbf{q_2}$, $\mathbf{q_3}$) are 4×3 matrices, the last row entries being zeros. The transformations can be constructed in the following steps:

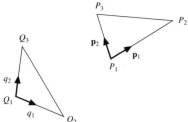

(a)   Translate P_1 to Q_1. The new set of co-ordinates for P_1, P_2 and P_3 are now P_1^*, P_2^* and P_3^* respectively, given by

$$\begin{bmatrix} P_1^* \\ P_2^* \\ P_3^* \end{bmatrix}^T = \begin{bmatrix} 1 & 0 & 0 & x_{q1} - x_{p1} \\ 0 & 1 & 0 & y_{q1} - y_{p1} \\ 0 & 0 & 1 & z_{q1} - z_{p1} \\ 0 & 0 & 0 & 1 \end{bmatrix} \begin{bmatrix} P_1 \\ P_2 \\ P_3 \end{bmatrix}^T$$

Figure 2.19(a)   Assembly of two triangular laminae

where $P_i = [x_{P_i} \ y_{P_i} \ z_{P_i} \ 1]$ and $Q_i = [x_{Q_i} \ y_{Q_i} \ z_{Q_i} \ 1]$, $i = 1, 2, 3$.

(b)   At this stage, the two planes $P_1^* P_2^* P_3^*$ and $Q_1\,Q_2\,Q_3$ are joined together at Q_1. The edge $P_1^* P_2^*$ may not be in line with Q_1Q_2. Let $\mathbf{p}_1^*$ be the unit vector along $P_1^* P_2^*$. Then

$$\mathbf{p}_1^* = \frac{P_2^* - P_1^*}{|P_2^* - P_1^*|}$$

Angle α between $\mathbf{p}_1^*$ and $\mathbf{q_1}$ can be found using $\cos \alpha = \mathbf{p}_1^* \cdot \mathbf{q}_1$. Let $\mathbf{u} = \mathbf{p}_1^* \times \mathbf{q}_1 = [u_x \ u_y \ u_z \ 0]$ be a unit vector passing through P_1^* (which is coincident with Q_1) and perpendicular to the plane containing $\mathbf{p}_1^*$ and $\mathbf{q}_1$. Rotating $P_1^* P_2^*$ to coincide with Q_1Q_2 involves rotating P_2^* about $\mathbf{u}$ through an angle α for P_2^* to finally lie on Q_1Q_2. Let the new position of $P_1P_2P_3$ be $P_1' P_2' P_3'$.

(c)   At this time, the two edges $P_1' P_2'$ and Q_1Q_2 are coincident. However, angle between the triangular planes may not be the desired angle. To rotate $P_1' P_2' P_3'$ about Q_1Q_2 would require knowing the angle between the planes $P_1' P_2' P_3'$ and $Q_1\,Q_2\,Q_3$. This is given by the angle between the normal vectors to the two planes. The unit normal to $Q_1\,Q_2\,Q_3$ is $\mathbf{q}_3$. For $\mathbf{p}_3'$, the unit normal to $P_1' P_2' P_3'$, we compute the unit vectors along $P_1' P_2'$ and $P_1' P_3'$. With $\mathbf{p}_1'$ known (as q_1), $\mathbf{p}_2'$ as $\dfrac{\mathbf{P}_3' - \mathbf{P}_1'}{|\mathbf{P}_3' - \mathbf{P}_1'|}$, $\mathbf{p}_3'$

can be determined as $\mathbf{p}_1' \times \mathbf{p}_2'$. The angle β between the planes $P_1' P_2' P_3'$ and $Q_1 Q_2 Q_3$ is now given by $\cos \beta = \mathbf{p}_3' \cdot \mathbf{q}_3$. To orient the plane $P_1' P_2' P_3'$ with respect to $Q_1 Q_2 Q_3$ at any desired angle θ, we can rotate point P_3' about $P_1' P_2'$ (or $\mathbf{p}_1'$) through an angle $\theta - \beta$ as discussed in section 2.5.1.

Example 2.8. Given two triangular objects, S_1 {P_1 (0, 0, 1), P_2 (1, 0, 0), P_3 (0, 0, 0)} and S_2 {Q_1 (0, 0, 2), Q_2(0, 2, 0), Q_3(2, 0, 0)}, it is required that after assembly, point P_1 coincides with Q_1 and edge $P_1 P_2$ lie on $Q_1 Q_3$. Determine the transformations if (i) S_1 is required to be in the same plane as S_2 and (ii) S_1 is perpendicular to S_2.

Translation of $P_1 P_2 P_3$ to a new position $P_1^* P_2^* P_3^*$ with P_1 to coincide with Q_1 is obtained by

$$\begin{bmatrix} P_1^* \\ P_2^* \\ P_3^* \end{bmatrix}^T = \mathbf{T}\begin{bmatrix} P_1 \\ P_2 \\ P_3 \end{bmatrix}^T = \begin{bmatrix} 1 & 0 & 0 & 0 \\ 0 & 1 & 0 & 0 \\ 0 & 0 & 1 & (2-1) \\ 0 & 0 & 0 & 1 \end{bmatrix} \begin{bmatrix} 0 & 0 & 1 & 1 \\ 1 & 0 & 0 & 1 \\ 0 & 0 & 0 & 1 \end{bmatrix}^T = \begin{bmatrix} 0 & 0 & 2 & 1 \\ 1 & 0 & 1 & 1 \\ 0 & 0 & 1 & 1 \end{bmatrix}^T$$

It can be verified that P_2^* lies on line $Q_1 Q_3$ and thus one does not need to perform step (b) above. It is now required to determine the angle between the lamina $P_1^* P_2^* P_3^*$ and $Q_1 Q_2 Q_3$ which can be obtained using step (c).

$$\mathbf{p}_1^* = \frac{\mathbf{P}_2^* - \mathbf{P}_1^*}{|\mathbf{P}_2^* - \mathbf{P}_1^*|} = \frac{[(1-0), (0), (1-2), (1-1)]}{\sqrt{1^2 + (-1)^2}} = \left(\frac{1}{\sqrt{2}} \quad 0 \quad -\frac{1}{\sqrt{2}} \quad 0 \right)$$

Similarly, $\quad \mathbf{p}_2^* = \dfrac{\mathbf{P}_3^* - \mathbf{P}_1^*}{|\mathbf{P}_3^* - \mathbf{P}_1^*|} = (0 \quad 0 \quad -1 \quad 0), \quad \mathbf{p}_3^* = \dfrac{(\mathbf{P}_3^* - \mathbf{P}_1^*) \times (\mathbf{P}_2^* - \mathbf{P}_1^*)}{|(\mathbf{P}_3^* - \mathbf{P}_1^*) \times (\mathbf{P}_2^* - \mathbf{P}_1^*)|} = (0 \quad -1 \quad 0 \quad 0)$

$$\mathbf{q}_2 = \frac{\mathbf{Q}_3 - \mathbf{Q}_1}{|\mathbf{Q}_3 - \mathbf{Q}_1|} = \frac{[(2-0), (0), (0-2), (1-1)]}{\sqrt{2^2 + (-2)^2}} = \left(\frac{1}{\sqrt{2}} \quad 0 \quad -\frac{1}{\sqrt{2}} \quad 0 \right)$$

$$\mathbf{q}_1 = \left(0 \quad \frac{1}{\sqrt{2}} \quad -\frac{1}{\sqrt{2}} \quad 0 \right), \mathbf{q}_3 = \frac{(\mathbf{Q}_3 - \mathbf{Q}_1) \times (\mathbf{Q}_2 - \mathbf{Q}_1)}{|(\mathbf{Q}_3 - \mathbf{Q}_1) \times (\mathbf{Q}_2 - \mathbf{Q}_1)|} = \left(\frac{1}{\sqrt{3}} \quad \frac{1}{\sqrt{3}} \quad \frac{1}{\sqrt{3}} \quad 0 \right)$$

Therefore, $\cos \beta = \mathbf{q}_3 \cdot \mathbf{p}_3^* = \left(\dfrac{1}{\sqrt{3}} \quad \dfrac{1}{\sqrt{3}} \quad \dfrac{1}{\sqrt{3}} \quad 0 \right) \cdot (0 \quad -1 \quad 0 \quad 0) = -\dfrac{1}{\sqrt{3}} \Rightarrow \beta = 125.26°$

Angle β (or, $180° - \beta$) is the angle between the planes S_1^* and S_2, and $Q_1 Q_3$ is the line about which P_3^* is to be rotated to bring S_1^* to be either: (i) in plane with S_2, or (ii) perpendicular to S_2.

The direction cosines of $Q_1 Q_3$ are given by

$$\mathbf{q}_2 = \left(\frac{1}{\sqrt{2}} \quad 0 \quad -\frac{1}{\sqrt{2}} \quad 0 \right) = (n_x \ n_y \ n_z \ 0)$$

Following section 2.5.1, where rotation of a point about an arbitrary line is discussed, we shift Q_1 to the origin, rotate line $Q_1 Q_3$ about the x-axis and then about y-axis

$$d = \sqrt{1 - n_x^2} = \frac{1}{\sqrt{2}}, \quad \frac{n_z}{d} = -1, \quad \frac{n_y}{d} = 0$$

$$\mathbf{R}_x = \begin{bmatrix} 1 & 0 & 0 & 0 \\ 0 & -1 & 0 & 0 \\ 0 & 0 & -1 & 0 \\ 0 & 0 & 0 & 1 \end{bmatrix}, \quad \mathbf{R}_y = \begin{bmatrix} 0.707 & 0 & -0.707 & 0 \\ 0 & 1 & 0 & 0 \\ 0.707 & 0 & 0.707 & 0 \\ 0 & 0 & 0 & 1 \end{bmatrix}, \quad \mathbf{T}_{Q_1 \to 0} = \begin{bmatrix} 1 & 0 & 0 & 0 \\ 0 & 1 & 0 & 0 \\ 0 & 0 & 1 & -2 \\ 0 & 0 & 0 & 1 \end{bmatrix}$$

$$\mathbf{R}_\alpha(\alpha) = \begin{bmatrix} \cos\alpha & -\sin\alpha & 0 & 0 \\ \sin\alpha & \cos\alpha & 0 & 0 \\ 0 & 0 & 1 & 0 \\ 0 & 0 & 0 & 1 \end{bmatrix}$$

Finally,

$$\begin{bmatrix} P_1' \\ P_2' \\ P_3' \end{bmatrix}^T = \begin{bmatrix} 1 & 0 & 0 & 0 \\ 0 & 1 & 0 & 0 \\ 0 & 0 & 1 & -2 \\ 0 & 0 & 0 & 1 \end{bmatrix}^{-1} \begin{bmatrix} 1 & 0 & 0 & 0 \\ 0 & -1 & 0 & 0 \\ 0 & 0 & -1 & 0 \\ 0 & 0 & 0 & 1 \end{bmatrix}^{-1} \begin{bmatrix} 0.707 & 0 & -0.707 & 0 \\ 0 & 1 & 0 & 0 \\ 0.707 & 0 & 0.707 & 0 \\ 0 & 0 & 0 & 1 \end{bmatrix}^{-1} \begin{bmatrix} \cos\alpha & -\sin\alpha & 0 & 0 \\ \sin\alpha & \cos\alpha & 0 & 0 \\ 0 & 0 & 1 & 0 \\ 0 & 0 & 0 & 1 \end{bmatrix}$$

$$\begin{bmatrix} 0.707 & 0 & -0.707 & 0 \\ 0 & 1 & 0 & 0 \\ 0.707 & 0 & 0.707 & 0 \\ 0 & 0 & 0 & 1 \end{bmatrix} \begin{bmatrix} 1 & 0 & 0 & 0 \\ 0 & -1 & 0 & 0 \\ 0 & 0 & -1 & 0 \\ 0 & 0 & 0 & 1 \end{bmatrix} \begin{bmatrix} 1 & 0 & 0 & 0 \\ 0 & 1 & 0 & 0 \\ 0 & 0 & 1 & -2 \\ 0 & 0 & 0 & 1 \end{bmatrix} \begin{bmatrix} 0 & 0 & 2 & 1 \\ 1 & 0 & 1 & 1 \\ 0 & 0 & 1 & 1 \end{bmatrix}^T$$

(i) When $\alpha = \beta = 54.73°$, rotation of P_3^* around Q_1Q_3 by α will bring the plane S_1^* on the top of S_2 as shown in Figure 2.19 (b). Hence

$$\begin{bmatrix} P_1' \\ P_2' \\ P_3' \end{bmatrix} = \begin{bmatrix} 0 & 0 & 2 & 1 \\ 1 & 0 & 1 & 1 \\ 0 & 0 & 1 & 1 \end{bmatrix} \begin{bmatrix} .7887 & -.5773 & -.2113 & 0 \\ .5773 & .5773 & .5773 & 0 \\ -.2113 & -.5773 & .7887 & 0 \\ .4226 & 1.1547 & .4226 & 1 \end{bmatrix} = \begin{bmatrix} 0 & 0 & 2 & 1 \\ 1 & 0 & 1 & 1 \\ .2113 & .5773 & 1.2113 & 1 \end{bmatrix}$$

(ii) For $\alpha = -(180° - \beta)$, the two planes S_1^* and S_2 are hinged about Q_1Q_3 such that they are in the same plane as shown in Figure 2.19 (c)

$$\begin{bmatrix} P_1' \\ P_2' \\ P_3' \end{bmatrix} = \begin{bmatrix} 0 & 0 & 2 & 1 \\ 1 & 0 & 1 & 1 \\ 0 & 0 & 1 & 1 \end{bmatrix} \begin{bmatrix} .2113 & .5773 & -.7887 & 0 \\ -.5773 & -.5773 & -.5773 & 0 \\ -.7887 & .5773 & .2113 & 0 \\ 1.5773 & -1.1547 & 1.5773 & 1 \end{bmatrix} = \begin{bmatrix} 0 & 0 & 2 & 1 \\ 1 & 0 & 1 & 1 \\ .7887 & -.5773 & 1.7887 & 1 \end{bmatrix}$$

(iii) For $\alpha = -(90° - \beta)$, S_1^* and S_2 are hinged about Q_1Q_3 and are perpendicular to each other (Figure 2.19d)

$$\begin{bmatrix} P_1' \\ P_2' \\ P_3' \end{bmatrix} = \begin{bmatrix} 0 & 0 & 2 & 1 \\ 1 & 0 & 1 & 1 \\ 0 & 0 & 1 & 1 \end{bmatrix} \begin{bmatrix} .9082 & .4083 & -.0918 & 0 \\ -.4083 & .8165 & -.4083 & 0 \\ -.0918 & .4083 & .9082 & 0 \\ .1835 & -.8165 & .1835 & 1 \end{bmatrix} = \begin{bmatrix} 0 & 0 & 2 & 1 \\ 1 & 0 & 1 & 1 \\ .0918 & -.4083 & 1.0918 & 1 \end{bmatrix}$$

(iv) For $\alpha = (90° + \beta)$, S_1^* and S_2 are hinged about Q_1Q_3 and are perpendicular to each other (Figure 2.19e).

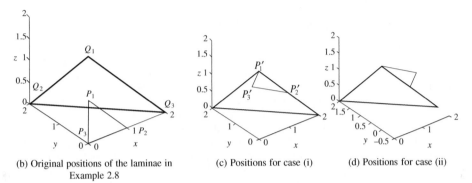

(b) Original positions of the laminae in Example 2.8

(c) Positions for case (i)

(d) Positions for case (ii)

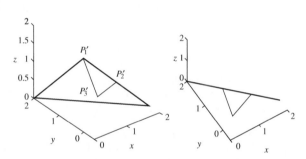

(e) Positions for case (iii), figure in right is edge view for $Q_1Q_2Q_3$

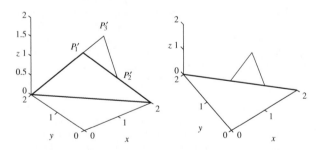

(f) Positions for case (iv), figure in right is edge view for $Q_1Q_2Q_3$

Figure 2.19   Example for positioning two triangles relative to each other using transformations

$$\begin{bmatrix} P_1' \\ P_2' \\ P_3' \end{bmatrix} = \begin{bmatrix} 0 & 0 & 2 & 1 \\ 1 & 0 & 1 & 1 \\ 0 & 0 & 1 & 1 \end{bmatrix} \begin{bmatrix} .0918 & -.4083 & -.9082 & 0 \\ .4083 & -.8165 & .4083 & 0 \\ -.9082 & -.4083 & .0912 & 0 \\ 1.8165 & .8165 & 1.8165 & 1 \end{bmatrix} = \begin{bmatrix} 0 & 0 & 2 & 1 \\ 1 & 0 & 1 & 1 \\ 0.9082 & 0.4083 & 1.9082 & 1 \end{bmatrix}$$

Previous sections discussed five kinds of transformations, namely, translation, rotation, reflection, scaling and shear, in both two and three dimensions. Homogeneous coordinates were introduced to unify all transformations into matrix multiplication operations. Of the five, the first three are rigid-body transformations while the other two cause change in the shape of the object and/or size. It is apparent from the examples that the mathematics of transformations at the back end of the CAD software is quite involved. At the front end, however, a user barely feels the rigor as the operations are hidden behind the graphical user interface that is designed to be very user-friendly. Transformations are not only applied in Computer Aided Assembly of many engineering components, but are also, significantly used in design operations. Interactively repositioning a data point in free-form curve/surface design requires translation. In constructive solid geometry (Chapter 8), many primitives (cylinders, blocks and others) require scaling and repositioning before they can be combined using Boolean operations (cut, join and intersect) to obtain a desired solid model. Transformations form an integral part of a CAD software.

2.7 Projections

Over a long period of time, designers and engineers have developed visualization techniques for three-dimensional objects that have helped in their representation, comprehension, communication and viewing. Pyramids, chariots, temples, canals, planned cities (Harappa-Mohen-jo-daro, for example), cave paintings, all suggest that architects, city planners and designers may have used projections to explain their ideas to the supervisors or artisans to execute the plan appropriately. A floor plan of a building, belonging to 2150 B.C. has been found in Mesopotamia as a part of the statue of King Gudea of the city of Lagash. Some temples and structures in South-East Asia have been carved out of a single piece of rock, suggesting a remarkable sense of three-dimensional geometry and precision in chipping off the stone pieces. Likewise, developments have also been observed in Roman and Greek architectures. Some 15th century artists, namely, Brunellesci, Leone Alberti, and Leonardo da Vinci, who were mathematicians as well, introduced *perspective* in their two-dimensional renderings. In 17th century, Pascal, DesCarte and Kepler developed analytical tools for *projective geometry*. The method of *orthographic projections*, as every engineer knows today, was developed by a French engineer, Gaspard Monge (founding member of Ecole Polytechnique, 1746-1818). *Engineering drawing* was further developed during the industrial revolution in 19th and 20th century, and since then, this mode of representation for engineering components has been in wide use. The conventional paper and pencil approach to represent engineering drawings is gradually paving way to *computer graphics* that has been in use since the 1970s.

Visual communication has two aspects: (a) the information that a two-dimensional picture of a three-dimensional component is trying to communicate and (b) how it communicates. Till recently, two-dimensional drawings were the only means to reveal engineering ideas but now, with better comprehension capabilities in three-dimensions, relatively cheap prototypes of machine parts designed with intricate shapes can be manufactured with great precision. Numerically controlled manufacturing machine tools can be programmed for a given geometry. Rapid prototyping machines can print physical models after acquiring the instructions directly from the geometric model created using the computer. Though recent developments in computer graphics facilitate better appreciation of an object in three-dimensions, projective geometry or *engineering graphics* still plays a vital role in visual communication. Engineering graphics is developed using the theory of projections that allow representing three-dimensional objects on two-dimensional planes.

Projections can be primarily classified as *perspective* and *parallel*. Projective geometry operates using: (a) location of the eye in three-dimensional space with respect to the object, also called the *view point* and (b) location of the *plane of projection* or the *image* or *picture plane*, in relation to the object. A *line of sight* is an imaginary ray of light between the view point and the object. In perspective projection, all lines of sight commence from a single point. The view point is at a finite distance from the object, and the lines of sight connecting the view point to the boundaries of the object are not parallel. On the contrary, in parallel projection, the lines of sight are parallel, or the view point is stationed at infinity in relation to the object. The plane of projection is imaginary upon which the rays along the lines of sight impinge and create points corresponding to the boundaries or the interior features of the object. Joining such points on the plane systematically creates a trace or image of the object. This plane may either be the computer screen (in modern day practice) or a piece of paper (in a conventional set-up).

Perspective projections are closest to what a human eye visualizes. However, they are difficult to construct, and it is also difficult to obtain realistic dimensions of the object for its creation or manufacture. Parallel projections are less realistic, but are easier to draw. It is easier to communicate the actual dimensions and manufacturing details through parallel projections. *Orthographic projection* is a parallel projection technique in which the plane of projection is positioned perpendicular to the lines of sight. Orthographic projections can either be *axonometric* or *multi-view*. Axonometric projections provide a three-dimensional view of the object and can be classified into *isometric, dimetric* or *trimetric*. Multi-view orthographic projections provide two-dimensional views of the object, and many such views are required to obtain its comprehensive three-dimensional appreciation. This method is more popular in engineering as multi-view projections give true dimensions without much further calculations. They provide an accurate description for manufacturing and construction. A technician can easily be trained to read multi-view orthographic drawings without requiring of him to have an artistic acumen. The rest of the chapter discusses the theoretical aspects of generic perspective and parallel projections, with emphasis on orthographic projections. Aerial perspective is beyond the scope of this chapter. Classification of projections is provided in Figure 2.20.

2.7.1 Geometry of Perspective Viewing

In perspective viewing, the image plane is placed between the view point and the object. Although, this is not a restriction, for the object between the view point and image plane, a reversed image is formed. The eye should neither lie on the object nor on the image plane. The image plane need not, in general, be perpendicular to the object plane. For the image plane as planar, we obtain a *linear perspective projection* though the image plane may as well be spherical, cylindrical or a part of any generic curved surface.

Consider a point P (x, y, z) on the object (Figure 2.21) and E as the observer's eye located at $(0, 0, -w)$ on the z-axis. Let the image plane be the x-y plane and the line segment EP intersect the image plane at P^* $(x^*, y^*, 0)$. Let $P^*B = x^*$ and $P^*D = y^*$. For P' as the foot of the perpendicular from P to the x-y plane, $P'C = y$ and $P'A = x$. For similar triangles P^*OE and $P^*P'P$

$$\frac{|\mathbf{OE}|}{|\mathbf{PP'}|} = \frac{|\mathbf{OP^*}|}{|\mathbf{P^*P'}|} = \frac{|\mathbf{EP^*}|}{|\mathbf{P^*P}|}$$

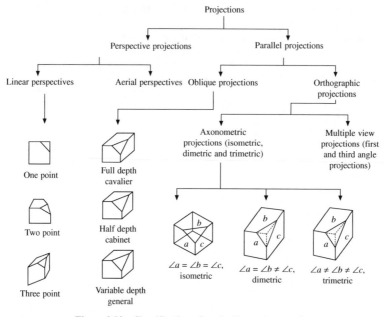

Figure 2.20   Classification of projections using a cube

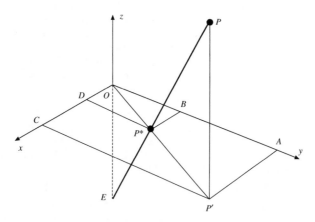

Figure 2.21   Perspective projection of *P* on the *x*-*y* plane

Thus,
$$|\mathbf{EP^*}| = \left(\frac{w}{z}\right)|\mathbf{P^*P}|, \quad \text{or} \quad \mathbf{EP^*} = \left(\frac{w}{z}\right)\mathbf{P^*P}$$

since both vectors are colinear. From vector geometry, we get

$$\mathbf{OP^*} = \mathbf{OE} + \mathbf{EP^*} = x^*\mathbf{i} + y^*\mathbf{j} + z^*\mathbf{k}$$

$$= -w\mathbf{k} + \left(\frac{w}{z}\right)\mathbf{P^*P} = -w\mathbf{k} + \left(\frac{w}{z}\right)(x - x^*)\mathbf{i} + \left(\frac{w}{z}\right)(y - y^*)\mathbf{j} + \left(\frac{w}{z}\right)(z - z^*)\mathbf{k}$$

Thus, $x^* = \mathbf{OP^*} \cdot \mathbf{i} = \frac{w}{z}(x - x^*)$, $y^* = \mathbf{OP^*} \cdot \mathbf{j} = \frac{w}{z}(y - y^*)$ and $\mathbf{OP^*} \cdot \mathbf{k} = \frac{w}{z}(z - z^*)$ yielding

$$x^* = \frac{wx}{z + w}, \quad y^* = \frac{wy}{z + w}, \quad \text{and } z^* = 0$$

This suggests that the image of P as seen from E on the plane of projection ($z = 0$) is given by $P^* = \left[\frac{wx}{z+w}, \frac{wy}{z+w}, 0, 1\right]$ which can be expressed using the 4×4 matrix as

$$\mathbf{P^*} = \begin{bmatrix} x^* \\ y^* \\ 0 \\ 1 \end{bmatrix} = \begin{bmatrix} \frac{wx}{z+w} \\ \frac{wy}{z+w} \\ 0 \\ 1 \end{bmatrix} \equiv \begin{bmatrix} x \\ y \\ 0 \\ \frac{z}{w}+1 \end{bmatrix} = \mathbf{P}_{\text{ers}}\mathbf{P} = \begin{bmatrix} 1 & 0 & 0 & 0 \\ 0 & 1 & 0 & 0 \\ 0 & 0 & 0 & 0 \\ 0 & 0 & \frac{1}{w} & 1 \end{bmatrix}\begin{bmatrix} x \\ y \\ z \\ 1 \end{bmatrix} \tag{2.34}$$

We can develop similar perspective projection matrices for the human eye to be on the x- and y-axis, respectively, using cyclic symmetry. For the view point E_x at $x = -w$ on the x-axis, a line joining E_x and P will intersect the y-z image plane at $P_{yz}^*\left[0 \quad \frac{wy}{x+w} \quad \frac{wz}{x+w} \quad 1\right]$. Similarly, if the view point is shifted to E_y at $y = -w$ on y-axis, the line joining E_y and P will intersect z-x image plane at $P_{zx}^*\left[\frac{wx}{y+w} \quad 0 \quad \frac{wz}{y+w} \quad 1\right]$.

Example 2.9. A line P_1P_2 has coordinates $\mathbf{P}_1(4, 4, 10)$ and $\mathbf{P}_2(8, 2, 4)$ and the observer's eye E_z is located at $(0, 0, -4)$. Find the perspective projection of the line on the x-y plane.

Any point $\mathbf{P}$ on a given line can be written in the parametric form $\mathbf{P} = (1 - u)\mathbf{P}_1 + u\mathbf{P}_2$, where $u \in [0\ 1]$. When $u = 0$, $\mathbf{P} = \mathbf{P}_1$ and when $u = 1$, $\mathbf{P} = \mathbf{P}_2$. The perspective image of P on the x-y image plane as seen from E_z can be obtained as follows:

$$\mathbf{P} = (1 - u)[4\ 4\ 10] + u[8\ 2\ 4] = [4(1 + u)\ 2(2 - u)\ 2(5 - 3u)]$$

Using the transformation in Eq. (2.34), the perspective image of $\mathbf{P}$ on x-y plane is

$$\mathbf{P^*} = \begin{bmatrix} 1 & 0 & 0 & 0 \\ 0 & 1 & 0 & 0 \\ 0 & 0 & 0 & 0 \\ 0 & 0 & \frac{1}{4} & 1 \end{bmatrix}\begin{bmatrix} 4(1+u) \\ 2(2-u) \\ 2(5-3u) \\ 1 \end{bmatrix} = \begin{bmatrix} 4(1+u) \\ 2(2-u) \\ 0 \\ \frac{7-3u}{2} \end{bmatrix} \equiv \begin{bmatrix} \frac{8(1+u)}{(7-3u)} \\ \frac{4(2-u)}{(7-3u)} \\ 0 \\ 1 \end{bmatrix}$$

The perspective image $P_1^* P_2^*$ of $P_1 P_2$ as seen from E_z is obtained by substituting $u = 0$ and $u = 1$. The resulting coordinates are $P_1^* = (8/7 \ 8/7 \ 0)$ and $P_2^* = (4 \ 1 \ 0)$ as shown in Figure 2.22.

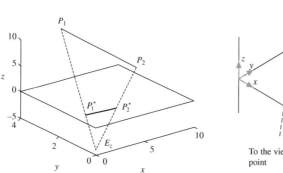

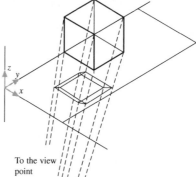

Figure 2.22   **Perspective image of a line on the x-y plane in example 2.9**

Figure 2.23   **Perspective image of a cube in example 2.10 on the x-y plane**

Example 2.10. A unit cube is placed in the first octant, as shown in Figure 2.23, such that its edges are parallel to the axes and one of the vertices is shifted from $(0, 0, 0)$ to $(1, 1, 1)$. Determine the perspective projection of the cube on the x-y plane as seen by the observer at $z = -10$.

The coordinates of the corners of the unit cube, with one corner at $(0, 0, 0)$, are easily obtainable. However, its perspective image on the x-y plane will be a unit square with one of the vertices at the origin. After shifting the $(0, 0, 0)$ vertex of the cube to $(1, 1, 1)$ with the translation matrix $\mathbf{T}$, we get the new coordinates as

$$
\begin{bmatrix} A' \\ B' \\ C' \\ D' \\ E' \\ F' \\ G' \\ H' \end{bmatrix}^T
=
\begin{bmatrix} 1 & 0 & 0 & 1 \\ 0 & 1 & 0 & 1 \\ 0 & 0 & 1 & 1 \\ 0 & 0 & 0 & 1 \end{bmatrix}
\begin{bmatrix} 1 & 0 & 1 & 1 \\ 1 & 1 & 1 & 1 \\ 0 & 1 & 1 & 1 \\ 0 & 0 & 1 & 1 \\ 1 & 0 & 0 & 1 \\ 1 & 1 & 0 & 1 \\ 0 & 1 & 0 & 1 \\ 0 & 0 & 0 & 1 \end{bmatrix}^T
=
\begin{bmatrix} 2 & 1 & 2 & 1 \\ 2 & 2 & 2 & 1 \\ 1 & 2 & 2 & 1 \\ 1 & 1 & 2 & 1 \\ 2 & 1 & 1 & 1 \\ 2 & 2 & 1 & 1 \\ 1 & 2 & 1 & 1 \\ 1 & 1 & 1 & 1 \end{bmatrix}^T
$$

Using the perspective transformation matrix with $w = 10$, or $1/w = 0.1$, the perspective projections of the vertices $A'B'C'D'E'F'G'H'$ of the cube can be computed as

$$
\begin{bmatrix} A^* \\ B^* \\ C^* \\ D^* \\ E^* \\ F^* \\ G^* \\ H^* \end{bmatrix}^T
=
\begin{bmatrix} 1 & 0 & 0 & 0 \\ 0 & 1 & 0 & 0 \\ 0 & 0 & 0 & 0 \\ 0 & 0 & \dfrac{1}{w} & 1 \end{bmatrix}
\begin{bmatrix} A' \\ B' \\ C' \\ D' \\ E' \\ F' \\ G' \\ H' \end{bmatrix}^T
=
\begin{bmatrix} 1 & 0 & 0 & 0 \\ 0 & 1 & 0 & 0 \\ 0 & 0 & 1 & 0 \\ 0 & 0 & 0.1 & 1 \end{bmatrix}
\begin{bmatrix} 2 & 1 & 2 & 1 \\ 2 & 2 & 2 & 1 \\ 1 & 2 & 2 & 1 \\ 1 & 1 & 2 & 1 \\ 2 & 1 & 1 & 1 \\ 2 & 2 & 1 & 1 \\ 1 & 2 & 1 & 1 \\ 1 & 1 & 1 & 1 \end{bmatrix}^T
$$

$$= \begin{bmatrix} 2 & 1 & 0 & 1.2 \\ 2 & 2 & 0 & 1.2 \\ 1 & 2 & 0 & 1.2 \\ 1 & 1 & 0 & 1.2 \\ 2 & 1 & 0 & 1.1 \\ 2 & 2 & 0 & 1.1 \\ 1 & 2 & 0 & 1.1 \\ 1 & 1 & 0 & 1.1 \end{bmatrix}^T \equiv \begin{bmatrix} 1.67 & .833 & 0 & 1 \\ 1.67 & 1.67 & 0 & 1 \\ .833 & 1.67 & 0 & 1 \\ .833 & .833 & 0 & 1 \\ 1.82 & .91 & 0 & 1 \\ 1.82 & 1.82 & 0 & 1 \\ .91 & 1.82 & 0 & 1 \\ .91 & .91 & 0 & 1 \end{bmatrix}^T$$

We observe all twelve edges of the cube in its perspective projection in Figure 2.23.

2.7.2 Two Point Perspective Projection

Example 2.10 suggests that translating the object may show up its multiple faces giving a three-dimensional effect on the plane of projection. Rotating an object about an axis also reveals two or more faces. A rotation about z-axis by an angle θ followed by a single point perspective projection on $y = 0$ plane with center of projection at $y = y_p$ gives the following transformation matrix:

$$\mathbf{M}_1 = \begin{bmatrix} 1 & 0 & 0 & 0 \\ 0 & 0 & 0 & 0 \\ 0 & 0 & 1 & 0 \\ 0 & -\dfrac{1}{y_p} & 0 & 1 \end{bmatrix} \begin{bmatrix} \cos\theta & -\sin\theta & 0 & 0 \\ \sin\theta & \cos\theta & 0 & 0 \\ 0 & 0 & 1 & 0 \\ 0 & 0 & 0 & 1 \end{bmatrix} = \begin{bmatrix} \cos\theta & -\sin\theta & 0 & 0 \\ 0 & 0 & 0 & 0 \\ 0 & 0 & 1 & 0 \\ -\dfrac{\sin\theta}{y_p} & -\dfrac{\cos\theta}{y_p} & 0 & 1 \end{bmatrix} \quad (2.35a)$$

A rotation about the x-axis by an angle ψ followed by a single point perspective projection on $y = 0$ plane with center of projection at $y = y_p$ gives another transformation matrix $\mathbf{M}_2$, where

$$\mathbf{M}_2 = \begin{bmatrix} 1 & 0 & 0 & 0 \\ 0 & 0 & 0 & 0 \\ 0 & 0 & 1 & 0 \\ 0 & -\dfrac{1}{y_p} & 0 & 1 \end{bmatrix} \begin{bmatrix} 1 & 0 & 0 & 0 \\ 0 & \cos\psi & -\sin\psi & 0 \\ 0 & \sin\psi & \cos\psi & 0 \\ 0 & 0 & 0 & 1 \end{bmatrix} = \begin{bmatrix} 1 & 0 & 0 & 0 \\ 0 & 0 & 0 & 0 \\ 0 & \sin\psi & \cos\psi & 0 \\ 0 & -\dfrac{\cos\psi}{y_p} & \dfrac{\sin\psi}{y_p} & 1 \end{bmatrix} \quad (2.35b)$$

Example 2.11. Given a square planar sheet $ABCD$ in the x-y plane with A (1, 0, 0), B (1, 1, 0), C (0, 1, 0) and D (0, 0, 0), find the perspective image of the sheet on $y = 0$ plane, with the view point at $y_p = 2$. The sheet is rotated $60°$ about the z-axis and translated -2 units along z-axis.

The transformation matrix $\mathbf{M}$ is given by $\mathbf{P}_{ers}\,(y_p = 2)\ \mathbf{T}(z = -2)\ \mathbf{R}_z(60°)$, that is

$$\begin{bmatrix} 1 & 0 & 0 & 0 \\ 0 & 0 & 0 & 0 \\ 0 & 0 & 1 & 0 \\ 0 & -.5 & 0 & 1 \end{bmatrix} \begin{bmatrix} 1 & 0 & 0 & 0 \\ 0 & 1 & 0 & 0 \\ 0 & 0 & 1 & -2 \\ 0 & 0 & 0 & 1 \end{bmatrix} \begin{bmatrix} \cos 60° & -\sin 60° & 0 & 0 \\ \sin 60° & \cos 60° & 0 & 0 \\ 0 & 0 & 1 & 0 \\ 0 & 0 & 0 & 1 \end{bmatrix} = \begin{bmatrix} 0.5 & -0.866 & 0 & 0 \\ 0 & 0 & 0 & 0 \\ 0 & 0 & 1 & -2 \\ -0.433 & -0.25 & 0 & 1 \end{bmatrix}$$

The vertices of the square $ABCD$ are now transformed to get the perspective image $A^*B^*C^*D^*$ in Figure 2.24.

$$
\begin{bmatrix} A^* \\ B^* \\ C^* \\ D^* \end{bmatrix}^T = \mathbf{M} \begin{bmatrix} A \\ B \\ C \\ D \end{bmatrix}^T = \begin{bmatrix} 0.5 & -0.866 & 0 & 0 \\ 0 & 0 & 0 & 0 \\ 0 & 0 & 1 & -2 \\ -0.433 & -0.25 & 0 & 1 \end{bmatrix} \begin{bmatrix} 1 & 0 & 0 & 1 \\ 1 & 1 & 0 & 1 \\ 0 & 1 & 0 & 1 \\ 0 & 0 & 0 & 1 \end{bmatrix}^T
$$

$$
= \begin{bmatrix} 0.5 & 0 & -2 & 0.567 \\ -0.366 & 0 & -2 & 0.317 \\ -0.866 & 0 & -2 & 0.750 \\ 0 & 0 & -2 & 1 \end{bmatrix}^T \equiv \begin{bmatrix} 0.882 & 0 & -3.527 & 1 \\ -1.115 & 0 & -6.309 & 1 \\ -1.155 & 0 & -2.667 & 1 \\ 0 & 0 & -2 & 1 \end{bmatrix}
$$

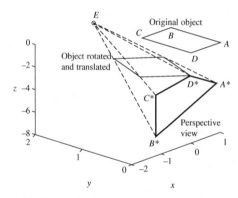

Figure 2.24   Perspective image on the x-z plane for Example 2.11

2.8   Orthographic Projections

Orthographic projections have been universally adopted for engineering drawings, especially for machine parts. They are simplest among parallel projections and are popular in all manufacturing industries because they accurately depict the true size and shape of a planar-faced object. In an orthographic projection, the projectors are perpendicular to the view plane. *Multi-view projections* are a set of orthographic images, usually on the coordinate planes, generated with direction of projections perpendicular to different faces of the object. The following transformation matrices obtain parallel projections on the x-y, y-z and z-x planes.

$$
\mathbf{Pr}_{xy} = \begin{bmatrix} 1 & 0 & 0 & 0 \\ 0 & 1 & 0 & 0 \\ 0 & 0 & 0 & 0 \\ 0 & 0 & 0 & 1 \end{bmatrix}, \quad \mathbf{Pr}_{yz} = \begin{bmatrix} 0 & 0 & 0 & 0 \\ 0 & 1 & 0 & 0 \\ 0 & 0 & 1 & 0 \\ 0 & 0 & 0 & 1 \end{bmatrix}, \quad \mathbf{Pr}_{zx} = \begin{bmatrix} 1 & 0 & 0 & 0 \\ 0 & 0 & 0 & 0 \\ 0 & 0 & 1 & 0 \\ 0 & 0 & 0 & 1 \end{bmatrix} \quad (2.36)
$$

We may observe that for projection on the *x-y* plane, the entire third row of $\mathbf{Pr}_{xy}$ is 0. Similarly, for projections on the *y-z* and *z-x* planes, the entire first and second rows of $\mathbf{Pr}_{yz}$ and $\mathbf{Pr}_{zx}$, respectively, are zero. We can obtain six views for six sides of the object enclosed inside an imaginary cube using these transformations. Usually, two or three views are adequate to show all features of the object. However, for some objects with curved surfaces, projections on *auxiliary* planes may be required. An auxiliary plane is not parallel to any coordinate plane and a unit normal for it is first obtained. The object is then manoeuvered till the normal to the auxiliary plane is coincident with one of the coordinate axes. The respective projection transformation in Eq. (2.36) is applied, and then concatenating the inverse transformations from the left places the object back to its original location.

Two schemas in wide use in orthographic projections are: (a) the first angle and (b) the third angle projections. The object is enclosed in an imaginary cube and parallel projections are taken from the object to planes of the cube. In the first angle projection, the projections pass through the object to intersect the plane behind while in the third angle schema, projections reflect back onto the plane in front. Planes with projections are then unfolded to show the required views in two dimensions. The first angle projections of an object are shown in Figure 2.25 as an example.

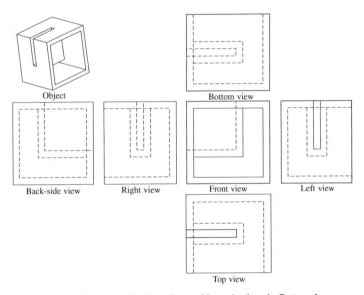

Figure 2.25   An object's orthographic projections in first angle

2.8.1   Axonometric Projections

In an axonometric projection, the object is rotated about any axis and translated, if desired, till the view reveals more than two faces of the object. Projections are then made with the eye at infinity (parallel rays of projection) and one of the coordinate planes as the plane of projection. Any other plane of projection may also be chosen so long as the rays are perpendicular to it. Axonometric projection contains more geometric information about the object than an orthographic projection in a single view. However, true dimensions are not shown as there is a foreshortening of the dimensions depending upon the placement of the object. Three types of axonometric projections of interest are: (a) trimetric, (b) dimetric and (c) isometric; the latter being more popular in use.

2.8.1.1 Trimetric Projection

Consider a cube placed with one corner at the origin and three of its orthogonal edges coincident with the coordinate axes. The cube is rotated by an angle ϕ about the y-axis and ψ about the x-axis, and its projection is taken on the x-y plane with the eye placed at infinity along the z-axis (parallel projection-rays). Matrix $\mathbf{M} = \mathbf{Pr}_{xy}\mathbf{R}_x(\psi)\mathbf{R}_y(\phi)$ provides the final transformation with

$$\mathbf{M} = \begin{bmatrix} 1 & 0 & 0 & 0 \\ 0 & 1 & 0 & 0 \\ 0 & 0 & 0 & 0 \\ 0 & 0 & 0 & 1 \end{bmatrix}\begin{bmatrix} 1 & 0 & 0 & 0 \\ 0 & \cos\psi & -\sin\psi & 0 \\ 0 & \sin\psi & \cos\psi & 0 \\ 0 & 0 & 0 & 1 \end{bmatrix}\begin{bmatrix} \cos\phi & 0 & \sin\phi & 0 \\ 0 & 1 & 0 & 0 \\ -\sin\phi & 0 & \cos\phi & 0 \\ 0 & 0 & 0 & 1 \end{bmatrix} = \begin{bmatrix} \cos\phi & 0 & \sin\phi & 0 \\ \sin\phi\sin\psi & \cos\psi & -\cos\phi\sin\psi & 0 \\ 0 & 0 & 0 & 0 \\ 0 & 0 & 0 & 1 \end{bmatrix}$$

$$(2.37)$$

Since the cube rests at a corner on the x-y plane, the projections of the sides are no longer of original length as they are foreshortened. The foreshortening ratios sh_x, sh_y and sh_z can be determined as the magnitudes of resultant vectors after transformation in Eq. (2.37) are applied to the three edges of the cube, that is

$$\begin{bmatrix} \cos\phi & 0 & \sin\phi & 0 \\ \sin\phi\sin\psi & \cos\psi & -\cos\phi\sin\psi & 0 \\ 0 & 0 & 0 & 0 \\ 0 & 0 & 0 & 1 \end{bmatrix}\begin{bmatrix} 1 & 0 & 0 & 1 \\ 0 & 1 & 0 & 1 \\ 0 & 0 & 1 & 1 \end{bmatrix}^T = \begin{bmatrix} \cos\phi & \sin\phi\sin\psi & 0 & 1 \\ 0 & \cos\psi & 0 & 1 \\ \sin\phi & -\cos\phi\sin\psi & 0 & 1 \end{bmatrix}^T$$

which gives the respective foreshortened ratios as

$$sh_x = \sqrt{\cos^2\phi + (\sin\phi\sin\psi)^2}, \quad sh_y = |\cos\psi|, \quad sh_z = \sqrt{\sin^2\phi + (-\cos\phi\sin\psi)^2}$$

For a trimetric projection, all three foreshortening factors are unequal.

2.8.1.2 Dimetric Projection

In a dimetric projection, any two foreshortening factors are equal. Thus, for $sh_y = sh_z$

$$\sin^2\phi + (-\cos\phi\sin\psi)^2 = \cos^2\psi, \quad \text{also,} \quad sh_x^2 = \cos^2\phi + (\sin\phi\sin\psi)^2$$

Adding together, we get

$$1 + \sin^2\psi = \cos^2\psi + sh_x^2 \Rightarrow \sin^2\psi = \frac{sh_x^2}{2} \Rightarrow \sin\psi = \pm\frac{sh_x}{\sqrt{2}} \Rightarrow \cos\phi = \pm\frac{sh_x}{\sqrt{2 - sh_x^2}}$$

The result suggests that for a value of a given foreshortening factor, there are four possible combinations of ϕ and ψ and thus four possible diametric projections.

2.8.1.3 Isometric Projection

In engineering drawings, especially in mechanical engineering, isometric projections are used extensively.

If all three foreshortening factors are equal, we get an isometric projection. For $sh_y = sh_z = sh_x$ and using the above equations

$$\sin^2\phi = \frac{\sin^2\psi}{1 - \sin^2\psi}, \quad \text{also} \quad \sin^2\phi = \frac{1 - 2\sin^2\psi}{1 - \sin^2\psi} \Rightarrow \sin\psi = \pm\frac{1}{\sqrt{3}} \Rightarrow \psi = \pm35.26° \Rightarrow \phi = \pm45°$$

Thus, the foreshortening factor for an isometric projection is given by $sh_y = sh_z = sh_x = sh = \sqrt{1 - \sin^2\psi} = \frac{\sqrt{2}}{\sqrt{3}} = 0.8165$. For an isometric projection of a machine part, we can measure the dimensions on the figure and divide it by 0.8165 to obtain the actual dimensions of the object. A rotation of $\pm45°$ about the y-axis and $\pm35.26°$ about the x-axis gives a tilted object with respect to the x-y plane. The object is placed such that its principal edges or axes make equal angles with the x-y plane. The edges are thus foreshortened in equal proportions to 81.65%. Thus, for a cube, the edges will appear to be at 120° (or 60°) with respect to each other in the projection. Projecting the unit vector $\mathbf{i} = [1\ 0\ 0\ 0]$ along x^*-axis attached to the tilted cube on to the plane of projection gives

$$\begin{bmatrix} \cos\phi & 0 & \sin\phi & 0 \\ \sin\phi\sin\psi & \cos\psi & -\cos\phi\sin\psi & 0 \\ 0 & 0 & 0 & 0 \\ 0 & 0 & 0 & 1 \end{bmatrix} [1\ 0\ 0\ 0]^T = [\cos\phi \quad \sin\phi\sin\psi\ 0\ 0]^T$$

This is a vector on the plane (x-y) of projection and passing through the origin O^*. The angle α between O^*x^* and the projected line on the plane of projection is given by

$$\tan\alpha = \frac{\sin\phi\sin\psi}{\cos\phi} = \frac{\sin 45°\sin\psi}{\cos 45°} = \pm\sin\psi \Rightarrow \alpha = \tan^{-1}(\pm\sin 35.26°) = \pm30°$$

In drawing an isometric scale, first a base line L is made and then a line l at 45° to the base line. The true scale is drawn on l. Another line m is drawn at 30° to L and the true scale is projected from l to m. This is called the *isometric scale*. Isometric projections have the following general characteristics: (a) parallel edges on the object remain parallel in the isometric projection, (b) vertical edges of the object remain vertical in the projection and (c) all horizontal lines appear at 30° with the horizontal.

Example 2.12. A prismatic machine block is composed of 10 planar surfaces with vertices having the following homogenous coordinates:

P1 = [0 0 0 1; 6 0 0 1; 6 3 0 1; 0 3 0 1; 0 0 0 1]
P2 = [0 0 0 1; 0 0 3 1; 2 0 3 1; 2 0 2 1; 6 0 2 1; 6 0 0 1; 0 0 0 1]
P3 = [0 0 3 1; 2 0 3 1; 2 1 5 1; 0 1 5 1; 0 0 3 1]
P4 = [0 1 5 1; 2 1 5 1; 2 2 5 1; 0 2 5 1; 0 1 5 1]
P5 = [2 2 5 1; 2 3 3 1; 0 3 3 1; 0 2 5 1; 2 2 5 1
P6 = [2 3 3 1; 0 3 3 1; 0 3 0 1; 6 3 0 1; 6 3 2 1; 2 3 2 1; 2 3 3 1]
P7 = [2 0 2 1; 2 3 2 1; 6 3 2 1; 6 0 2 1; 2 0 2 1]
P8 = [0 0 0 1; 0 0 3 1; 0 1 5 1; 0 2 5 1; 0 3 3 1; 0 3 01; 0 0 0 1]
P9 = [2 0 2 1; 2 0 3 1; 2 1 5 1; 2 2 5 1; 2 3 3 1; 2 3 2 1; 2 0 2 1]
P10 = [6 0 0 1; 6 0 2 1; 6 3 2 1; 6 3 0 1; 6 0 0 1]

Draw

(a) An isometric projection of the block.
(b) A dimetric projection of the block for $sh_z = 0.5$ ($\psi = \pm 20.7$, $\phi = \pm 22.21$ for $sh_x = sh_y$).
(c) A trimetric projection with $\phi = 30°$, $\psi = 45°$ (rotations about y- and x- axes, respectively) and projection on the $z = 0$ plane.
(d) Orthographic projections on the three coordinate planes.

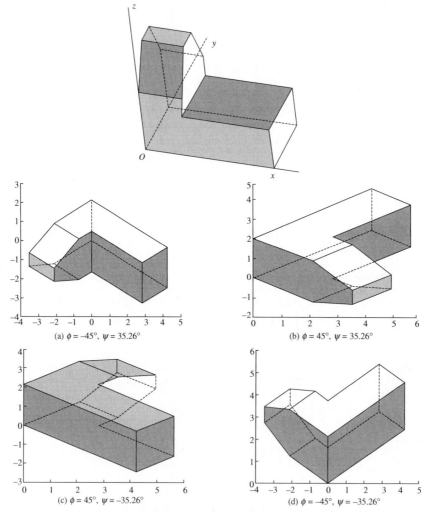

Figure 2.26 Isometric projections for various views (rotation angles) of the block

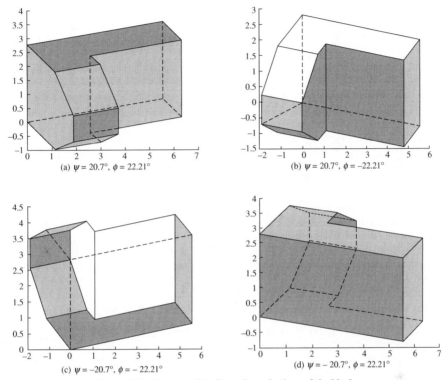

(a) $\psi = 20.7°$, $\phi = 22.21°$ (b) $\psi = 20.7°$, $\phi = -22.21°$

(c) $\psi = -20.7°$, $\phi = -22.21°$ (d) $\psi = -20.7°$, $\phi = 22.21°$

Figure 2.27   Four possible dimetric projections of the block

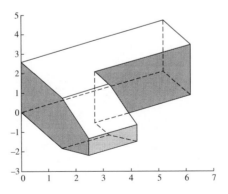

Figure 2.28   A trimetric projection of the block

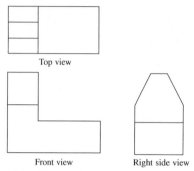

Top view

Front view   Right side view

Figure 2.29   Three orthographic views in third angle

2.9   Oblique Projections

In axonometric projections, the parallel rays or projectors are perpendicular to the plane of projection. If these projectors are inclined at an angle to the plane of projection, the image obtained is the oblique projection. It is sometimes found useful to show the three-dimensional details of the object through oblique projections. Faces parallel to the plane of projection are not foreshortened and the angles between the edges of the parallel faces are also preserved. However, faces not parallel to the plane of projection, get distorted.

Two types of oblique projections popular in engineering drawing are: (a) cavalier and (b) cabinet. The geometry of oblique projections can be explained by assuming three orthogonal axes and unit vectors $\mathbf{i}\,(1, 0, 0), \mathbf{j}\,(0, 1, 0), \mathbf{k}\,(0, 0, 1)$. Let the tip of $\mathbf{k}$ be designated by the point A and the projection plane be the x-y plane. Consider a set of parallel rays at an angle θ to the x-y plane (from behind the z-axis as shown in Figure 2.30). A parallel ray through A intersects the x-y plane at $B\,(b_x, b_y, 0)$. Let angle BOX be ψ. Consider any point $C\,(0, 0, z)$ on the z-axis such that a parallel ray through C intersects the x-y plane at $D\,(d_x, d_y, 0)$. Here $OB = f$ is the shrink factor. From triangles AOB and COD, we can get the following relationships:

$$b_x = f \cos \psi, \quad b_y = f \sin \psi, \quad f = \cot \theta, \quad d_x = fz \cos \psi, \quad d_y = fz \sin \psi \qquad (2.38)$$

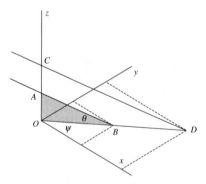

Figure 2.30   Geometry for oblique projections

For a parallel ray passing through any general point $P(x, y, z)$, we can obtain the image $Q(q_x, q_y, 0)$ on the x-y plane as $(x + fz \cos \psi, y + fz \sin \psi, 0)$. Or

$$\begin{bmatrix} q_x \\ q_y \\ 0 \\ 1 \end{bmatrix} = \begin{bmatrix} 1 & 0 & f \cos \psi & 0 \\ 0 & 1 & f \sin \psi & 0 \\ 0 & 0 & 0 & 0 \\ 0 & 0 & 0 & 1 \end{bmatrix} \begin{bmatrix} x \\ y \\ z \\ 1 \end{bmatrix} \tag{2.39}$$

This is incorporated by shifting $C(0, 0, z)$ by $(x, y, 0)$ to $P(x, y, z)$ on a plane parallel to the x-y plane and at a distance z above it. The consequence is the corresponding shift in the image from $D(d_x, d_y, 0)$ to $Q(q_x, q_y, 0)$ is computed above. The oblique projections are cavalier for $f = 1$ and cabinet for $f = \frac{1}{2}$.

Example 2.13. For the block shown in Example 2.12,

(a) Draw its cabinet projections on x-y plane (for $f = \frac{1}{2}$) for $\psi = 0°$, $\psi = 15°$, $\psi = 30°$ and $\psi = 45°$
(b) Draw its cavalier projections on x-y plane (for $f = 1$) for $\psi = 0°$, $\psi = 15°$, $\psi = 30°$ and $\psi = 45°$
(c) Draw projections for $\psi = 45°$ when $f = 1$, $f = 3/4$, and $f = 1/2$.

Part (a) is shown in Figure 2.31.

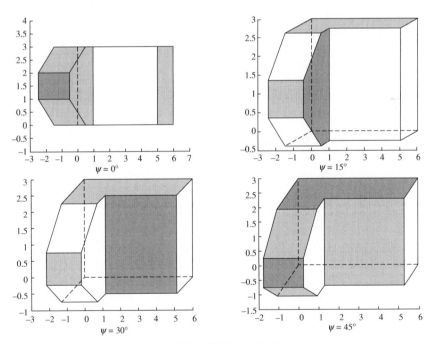

Figure 2.31 Cabinet projections

Parts (b) and (c) are shown in Figures 2.32 and 2.33, respectively.

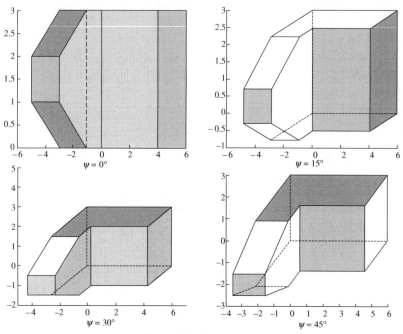

Figure 2.32   Cavalier projections

EXERCISES

1. For the points $p_1(1, 1)$, $p_2(3, 1)$, $p_3(4, 2)$, $p_4(2, 3)$, that defines a 2-D polygon, develop a single transformation matrix that

   (a) reflects about the line $x = 0$,
   (b) translates by -1 in both x and y directions, and
   (c) rotates about the z-axis by $180°$

   Using the transformations, determine the new position vectors.
2. Develop an algorithm to find a set of vertices making a regular 2-D polygon. You may use only transformations on points. Input parameters are the starting point $\mathbf{p}_0$ $(0, 0)$, number of edges n, and length of edge l.
3. Prove that the transformation matrix

$$\mathbf{R} = \begin{bmatrix} \dfrac{1 - t^2}{1 + t^2} & \dfrac{2t}{1 + t^2} & 0 \\ \dfrac{-2t}{1 + t^2} & \dfrac{1 - t^2}{1 + t^2} & 0 \\ 0 & 0 & 1 \end{bmatrix}$$

   produces pure rotation. Find the equivalent rotation angle.

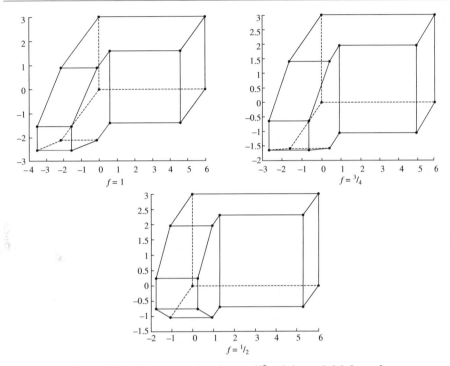

Figure 2.33   Oblique projections for $\psi = 45°$ and shown shrink factor f

4. Show that the reflection about an arbitrary line $ax + by + c = 0$ is given by

$$
\begin{bmatrix}
b^2 - a^2 & -2ab & 0 \\
-2ab & a^2 - b^2 & 0 \\
-2ac & -2bc & \dfrac{1}{a^2 + b^2}
\end{bmatrix}
$$

5. Consider two lines L1: $y = c$ and L2: $y = mx + c$ which intersect at point C on y-axis. The angle θ between these lines can be found easily. A point $P\,(x_1, y_1)$ is first reflected through L1 and subsequently through L2. Show that this is equivalent to rotating the point P about the intersection point C by 2θ.

6. A point $P\,(x, y)$ has been transformed to $P^*(x^*, y^*)$ by a transformation $\mathbf{M}$. Find the matrix $\mathbf{M}$.

7. Matrix $\mathbf{M} = \begin{bmatrix} 1 & b & 0 \\ c & 1 & 0 \\ 0 & 0 & 1 \end{bmatrix}$ shears an object by factors c and b along the Ox and Oy axes respectively.

Determine the matrix that shears the object by the same factors, but along Ox_1 and Oy_1 axes inclined at an angle θ to the original axes.

8. Scaling of a point $P(x, y)$ relative to a point $P_0(x_0, y_0)$ is defined as

$$x^* = x_0 + (x - x_0)s_x = xs_x + x_0(1 - s_x)$$

$$y^* = y_0 + (y - y_0)s_y = ys_y + y_0(1 - s_y)$$

$$[x^* \quad y^* \quad 1]^T = \begin{bmatrix} s_x & 0 & 0 \\ 0 & s_y & 0 \\ x_0(1 - s_x) & y_0(1 - s_y) & 1 \end{bmatrix} [x \quad y \quad 1]^T$$

Find the resulting matrix for two consecutive scaling transformations about points $P_1(x_1, y_1)$ and $P_2(x_2, y_2)$ by scaling factors k_1 and k_2, respectively. Show that the product of two scalings is a third scaling; but about what point?

9. Reflection through the origin $(0, 0)$ in 2-D is given by

$$\mathbf{Rf}_0 = \begin{bmatrix} -1 & 0 & 0 \\ 0 & -1 & 0 \\ 0 & 0 & 1 \end{bmatrix}$$

Reflect a line PQ given by $P(x_1, y_1)$ and $Q(x_2, y_2)$ through a point $A(a, b)$. Check the result for $P(2, 4)$, $Q(6, 2)$ and $A(1, 3)$.

10. The corners of a wedge shaped block are $(0\ 0\ 2;\ 0\ 0\ 3;\ 0\ 2\ 3;\ 0\ 2\ 2;\ -1\ 2\ 2;\ -1\ 2\ 3)$. A plane passes through $(0\ 0\ 1)$ and its equation is given by $3x + 4y + z - 1 = 0$. Find the reflection of the wedge through this plane.

11. Develop a computer program for reflecting a polygonal object through a given plane in 3-D. Test your program for Problem 10.

12. A prismatic solid S has a square base lying in the $y = 0$ plane as shown in Figure P 2.1. The vertices are $B(a, 0, -a)$, $C(-a, 0, -a)$, $D(-a, 0, a)$, $E(a, 0, a)$. The apex of the solid is at $A(b, b, b)$. The solid S is now linearly translated to S^* such that vertex C coincides with a point $P(p, q, r)$, where p, q, and r are all greater than a.

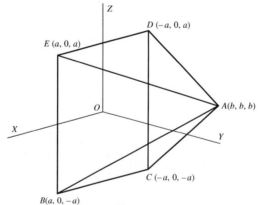

Figure P2.1

(a) If the observer's eye is situated at $z = -z_c$, find perspective projection of the solid on $z = 0$ plane. Solve the problem for $y = -y_c$ and $x = -x_c$ with the image plane as $y = 0$ and $x = 0$, respectively. Assume your own values for the required parameters. Show stepwise numerical results with matrices at all the intermediate steps along with projected images.

(b) The solid S is chopped off by a plane $y = d$ $(d < b)$ and part containing with the vertex A is removed.

You can calculate the coordinates of the rectangular section *FGHI* thus created. This frustum is now translated to *S*** as before with *C* coinciding with *P*. If *J* is the center of the rectangle *FGHI*, find the direction cosines of vector *O***J*, *O*** is the center of the square *BCDE***. Rotate the frustum by an angle a about a line *L* through *O***, where *L* is parallel to *x*-axis in the plane of *BCDE***. Show calculations and graphical results for $\alpha = 30°$ and $45°$.

13. A machine block is shown in Figure P2.2. Using transformations, show the following graphical results:
  (a) Orthographic projections.
  (b) The object is rotated about the *y*-axis by an angle φ and then about the *x*-axis through ψ. This is followed by a parallel projection on $z = 0$ plane to get a trimetric projection. For $\varphi = 30°$ and $45°$, draw figures for trimetric projections when ψ takes on the values $30°$, $45°$, $60°$ and $90°$. Calculate the foreshortening factors for each of the positions.

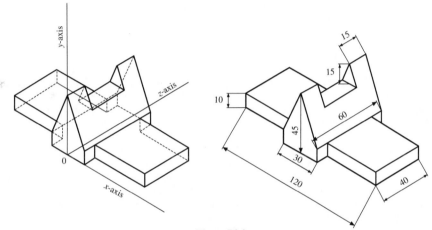

Figure P2.2

14. For the component shown in Figure P2.2, show the cavalier and cabinet projections for $\alpha = 30°$, $0°$ and $-45°$.
15. Develop a procedure to handle transformations and projections in general of polyhederal solids.

Chapter 3

Differential Geometry of Curves

The form of a real world object is often represented using points, curves, surfaces and solids. Although, a form can be sketched manually, it may be useful to construct a mathematical or computer model for a detailed description. In engineering design, we need to represent an object with precise drawings, perform requisite analysis to possibly optimize the form and finally, manufacture it. Geometric modeling of curves provides an invaluable tool for representation or visualization, analysis and manufacture of any machine part by providing the basis for the representation of surfaces and solids, and thus the real world objects (Figure 3.1). That curve design is fundamentally significant in computer-aided design, it behoves to understand and layout the underlining intent: *to seek a generic mathematical representation of a curve in a manner that renders absolute shape control to the user.* In other words, the curve definition must be general to encompass any possible shape and also, the user should be able to manipulate a curve's shape locally without altering the curve overall.

There are, therefore, two issues to address in curve design: (a) representation and (b) shape control. Analytic curves like conics (pair of straight lines, circles, ellipses, parabolae and hyperbolae in two dimensions) and helix, helical spiral and many more in three dimensions are all well-defined and well-studied. However, shapes and forms obtained using analytic curves are limited in engineering applications and pose restrictions to curve design in real situations. Moreover, local shape control with these curves is usually not possible. For instance, the coefficients a, b and c in the equation $ax + by + c = 0$ of a straight line L determine the slope and intercept of the line and changing their

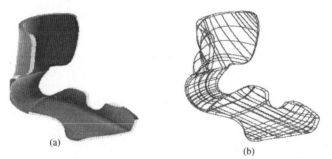

(a)

(b)

Figure 3.1   A solid (a) represented as a network of curves (b)

values implies only reorienting the line. Another example is of a second-degree polynomial $S \equiv ax^2$ $+ \ 2hxy + by^2 + 2gx + 2fy + c = 0$ which is representative of all conic sections in the x-y plane. Coefficients a, h, b, g, f and c in some combination represent a class of conic sections. Arbitrary values of these coefficients would yield only a few shapes mentioned above. More precisely, the shape of S depends on two invariants, D and I_2, where

$$D = \begin{vmatrix} a & h & g \\ h & b & f \\ g & f & c \end{vmatrix} \qquad (3.1)$$

and $I_2 = h^2 - ab$. One reason D and I_2 are called invariants is that their values remain unaltered if a translation $x = x' + p$, $y = y' + q$ and/or rotation ($x = x' \cos \theta - y' \sin \theta$, $y = x' \sin \theta + y' \cos \theta$) is applied to S. (a) For $D = 0$, if $I_2 = 0$, S represents a set of parallel lines. For positive I_2, S is a pair of intersecting lines while for negative I_2, S is a point. (b) For $D \neq 0$, if $I_2 = 0$, S represents a parabola, if $I_2 > 0$, S is a hyperbola and if $I_2 < 0$, S is an ellipse or a circle. Apart from the limited shapes analytic curves have to offer, direct or active control on their shape is not available to a user. However, segments of analytic curves like an arc, an elliptic or parabolic segment, if so desired, are often used in the wireframe modeling of solids (see Chapter 8).

Shapes of the reaction turbine blades, car windshields, aircraft fuselage, potteries, temple minarets, kitchenware, cathode ray tubes, air-conditioning ducts, seats for cars, scooters or bicycles, instrument panels for aircrafts provide many examples of some household and industrial products where a free-form surface is desired. This surface may be composed of a network of curves, and a designer requires an active control to arrive at a desired shape of a curve. It is interesting to observe how a potter creating a clay pot on a rotating wheel merely adjusts and manipulates his finger pressure at a few points to obtain a desired shape.

The way active control on curve's shape can be sought is by choosing a set of data points and requiring to *interpolate* or *best fit* a curve through it. *Curve interpolation* and *curve fitting* methods have been two of the oldest methods available in curve design. For given n data points (x_i, y_i), $i = 0, ..., n - 1$, interpolation requires to pass the curve through all the points by choosing a polynomial $g(x)$ of degree $n - 1$ and determining the unknown coefficients. Alternatively, in curve fitting, one may choose a polynomial of a smaller degree m ($< n - 1$) such that the curve depicts the best possible trend or distribution of data points. In both methods, a user gets a distinct advantage in that the shape of the curve is governed by the placement of data points, that is, the user may actively control the position of data points to affect the change in shape of the interpolated or best fit curve. Curve interpolation and fitting lay the groundwork for curve design and thus are discussed in detail below.

3.1 Curve Interpolation

Given a set of n ordered data points (x_i, y_i), $i = 0, ..., n - 1$, let $y = p(x)$ be a polynomial of degree $n - 1$ in x with unknowns $a_0, a_1, ..., a_{n-1}$. That $p(x)$ traverses through data points above implies

$$y_0 = p(x_0) = a_0 + a_1 x_0 + a_2 x_0^2 + a_3 x_0^3 + ... + a_{n-1} x_0^{n-1}$$

$$y_1 = p(x_1) = a_0 + a_1 x_1 + a_2 x_1^2 + a_3 x_1^3 + ... + a_{n-1} x_1^{n-1}$$

$$y_2 = p(x_2) = a_0 + a_1 x_2 + a_2 x_2^2 + a_3 x_2^3 + ... + a_{n-1} x_2^{n-1}$$

$$...$$

$$y_{n-1} = p(x_{n-1}) = a_0 + a_1 x_{n-1} + a_2 x_{n-1}^2 + ... + a_{n-1} x_{n-1}^{n-1} \qquad (3.2)$$

which is a system of n linear equations in a_i, $i = 0, ..., n - 1$, and can be solved by inverting an $n \times n$ matrix. This inversion may prove cumbersome if the number of data points is large. It is possible to reduce some effort in computation by posing the interpolating polynomial in a slightly different manner. For example, in the Newton's *divided differences* approach, the polynomial is posed as

$$y \equiv p(x) = \alpha_0 + \alpha_1 (x - x_0) + \alpha_2 (x - x_0) (x - x_1) + \ldots + \alpha_{n-1} (x - x_0) (x - x_1)\ldots (x - x_{n-2}) \quad (3.3)$$

so that at the data points, the equations in unknowns α_i take the form

$$y_0 = \alpha_0$$
$$y_1 = \alpha_0 + \alpha_1(x_1 - x_0)$$
$$y_2 = \alpha_0 + \alpha_1(x_2 - x_0)(x_2 - x_1)$$
$$\ldots$$
$$y_{n-1} = \alpha_0 + \alpha_1(x_{n-1} - x_0) + \ldots + \alpha_{n-1}(x_{n-1} - x_0) (x_{n-1} - x_1) \ldots (x_{n-1} - x_{n-2}) \quad (3.4)$$

The unknowns α_i can be determined by a series of forward substitutions. Note that α_0 depends only on y_0, α_1 depends on y_0 and y_1, α_2 depends on y_0, y_1 and y_2, and so on. If, additionally, a new data point, (x_n, y_n) is introduced, an equation is further added with only one unknown α_n to be determined, that is, the addition of a new data point does not alter the previously calculated coefficients. This is in contrast to the system of equations in (3.2) wherein the addition of a data point requires all the $n + 1$ coefficients to be recomputed by inverting an $(n + 1) \times (n + 1)$ matrix.

The third possibility in curve interpolation due to Lagrange does not require computing the coefficients and can be elucidated using the following example. For three data points (x_0, y_0), (x_1, y_1) and (x_2, y_2), consider the expression

$$L_0^2 (x) = \frac{(x - x_1)(x - x_2)}{(x_0 - x_1)(x_0 - x_2)} \quad (3.5)$$

On setting $x = x_0$, $L_0^2(x)$ becomes unity, however, for $x = x_1$ or x_2, $L_0^2(x) = 0$. Similarly, $L_1^2(x) = \frac{(x - x_0)(x - x_2)}{(x_1 - x_0)(x_1 - x_2)}$ is 1 for $x = x_1$ and 0 for $x = x_0$ and $x = x_2$ and that $L_2^2(x) = \frac{(x - x_0)(x - x_1)}{(x_2 - x_0)(x_2 - x_1)}$ is 1 for $x = x_2$ and 0 for $x = x_0$ and $x = x_1$. Using the functions $L_0^2(x), L_1^2(x)$ and $L_2^2(x)$, which are all quadratic in x (the superscript denotes the degree in x), and the y values, we can construct a quadratic function

$$P_L^2 (x) = L_0^2 (x)y_0 + L_1^2(x)y_1 + L_2^2 (x)y_2 \quad (3.6)$$

which passes through the three data points. In general, $L_i^{n-1}(x)$ are termed as Lagrangian interpolation coefficients where the subscript i signifies that $L_i^{n-1}(x)$ is the weight of y_i in Eq. (3.8). The superscript $n-1$ denotes the degree of interpolating polynomial. By inspection from Eq. (3.5) and the related expressions, $L_i^{n-1}(x)$ may be written as

$$L_i^{n-1}(x) = \frac{(x - x_0) \ldots (x - x_{i-1})(x - x_{i+1}) \ldots (x - x_{n-1})}{(x_i - x_0) \ldots (x_i - x_{i-1})(x_i - x_{i+1}) \ldots (x_i - x_{n-1})} = \prod_{\substack{j=0 \\ j \neq i}}^{n-1} \frac{(x - x_j)}{(x_i - x_j)} \quad (3.7)$$

The interpolating polynomial then becomes

$$P_L{}^{n-1}(x) = \sum_{i=0}^{n-1} L_i^{n-1}(x) y_i \tag{3.8}$$

Example 3.1. Construct a polynomial to interpolate through the data points (0, 0), (1, 2), (3, 2) and (6, −1) using the Newton's divided difference and Lagrangian approaches. Perturb point (3, 2) to (1.5, 4) and observe the change in the curve shape.

Using Newton's divided difference approach, since there are four data points, the interpolating polynomial is a cubic, that is

$$y = \alpha_0 + \alpha_1 (x - x_0) + \alpha_2 (x - x_0)(x - x_1) + \alpha_3 (x - x_0)(x - x_1)(x - x_2)$$

Now

$$\alpha_0 = y_0 = 0$$

$$\alpha_1 = \frac{y_1 - \alpha_0}{x_1 - x_0} = \frac{2 - 0}{1 - 0} = 2$$

$$\alpha_2 = \frac{(y_2 - \alpha_0) - \alpha_1 (x_2 - x_0)}{(x_2 - x_0)(x_2 - x_1)} = \frac{(2 - 0) - 2(3 - 0)}{(3 - 0)(3 - 1)} = -\frac{2}{3}$$

$$\alpha_3 = \frac{(y_3 - \alpha_0) - \alpha_1 (x_3 - x_0) - \alpha_2 (x_3 - x_0)(x_3 - x_1)}{(x_3 - x_0)(x_3 - x_1)(x_3 - x_2)} = \frac{(-1) - 2(6) + \frac{2}{3}(6)(5)}{(6)(5)(3)} = \frac{7}{90}$$

The polynomial becomes

$$y = 2x - \frac{2}{3}x(x - 1) + \frac{7}{90}x(x - 1)(x - 3)$$

Using the Lagrangian approach, the polynomial is

$$y = L_0^3(x) y_0 + L_1^3(x) y_1 + L_2^3(x) y_2 + L_3^3(x) y_3$$

or

$$y = \frac{(x - x_1)(x - x_2)(x - x_3)}{(x_0 - x_1)(x_0 - x_2)(x_0 - x_3)} y_0 + \frac{(x - x_0)(x - x_2)(x - x_3)}{(x_1 - x_0)(x_1 - x_2)(x_1 - x_3)} y_1$$

$$+ \frac{(x - x_0)(x - x_1)(x - x_3)}{(x_2 - x_0)(x_2 - x_1)(x_2 - x_3)} y_2 + \frac{(x - x_0)(x - x_1)(x - x_2)}{(x_3 - x_0)(x_3 - x_1)(x_3 - x_2)} y_3$$

$$= \frac{(x - 1)(x - 3)(x - 6)}{(-1)(-3)(-6)} 0 + \frac{(x - 0)(x - 3)(x - 6)}{(1)(-2)(-5)} 2$$

$$+ \frac{(x - 0)(x - 1)(x - 6)}{(3)(2)(-3)} 2 + \frac{(x - 0)(x - 1)(x - 3)}{(6)(5)(3)} (-1)$$

$$= \frac{(x)(x - 3)(x - 6)}{5} - \frac{(x)(x - 1)(x - 6)}{9} - \frac{(x)(x - 1)(x - 3)}{90}$$

On simplification, the above yields the same result as that from the Newton's divided differences method. Moving data point (3, 2) to (1.5, 4) requires re-computing the curve. With the divided difference approach, only the last two coefficients need to be computed, that is

$$\alpha_0 = y_0 = 0$$

$$\alpha_1 = \frac{y_1 - \alpha_0}{x_1 - x_0} = \frac{2 - 0}{1 - 0} = 2$$

$$\alpha_2 = \frac{(y_2 - \alpha_0) - \alpha_1(x_2 - x_0)}{(x_2 - x_0)(x_2 - x_1)} = \frac{(4 - 0) - 2(1.5 - 0)}{(1.5 - 0)(1.5 - 1)} = \frac{4}{3}$$

$$\alpha_3 = \frac{(y_3 - \alpha_0) - \alpha_1(x_3 - x_0) - \alpha_2(x_3 - x_0)(x_3 - x_1)}{(x_3 - x_0)(x_3 - x_1)(x_3 - x_2)} = \frac{(-1) - 2(6) - \frac{4}{3}(6)(5)}{(6)(5)(4.5)} = -\frac{53}{135}$$

Thus, the new polynomial becomes

$$y_n = 2x + \frac{4}{3}x(x - 1) - \frac{53}{135}x(x - 1)\left(x - \frac{3}{2}\right)$$

Comparative plots of y and y_n are provided in Figure 3.2 which shows the change in curve shape. Moving a data point results in the shape change of the entire interpolated curve.

Curve interpolation provides a simple tool for curve design with data points governing the curve shape. The degree of the interpolating polynomial is dependent on the number of data points specified. Note that an $n-1$ degree polynomial has at most $n-1$ roots and thus crosses the x axis at most $n-1$ times. In cases where the number of data points is large, the number of real roots of a high degree polynomial would be large. Such polynomials would then exhibit many oscillations or fluctuations undesirable from the view point of curve design. It is required, therefore, to choose a polynomial of a lower order (a polynomial of order m is of degree $m-1$) known *a priori* and determine its unknown coefficients for which the polynomial best fits the given design points.

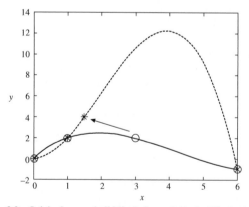

Figure 3.2 Original curve (solid line) changed (dashed line) when a data
point is moved. Change is global with curve interpolation

3.2 Curve Fitting

Consider a set of n data points (x_i, y_i), $i = 0, \ldots, n - 1$ which are to be best fitted, say, by a quadratic polynomial

$$p(x) = a_0 + a_1 x + a_2 x^2 \tag{3.9}$$

with coefficients a_0, a_1 and a_2 unknown. At $x = x_i$, the ordinate value from Eq. (3.9) is $p(x_i)$ and therefore the error in the ordinate values is $\delta_i = y_i - (a_0 + a_1 x_i + a_2 x_i^2)$. Squaring and adding the error values for all the data points, we get

$$\Delta = \sum_{i=0}^{n-1} \delta_i^2 = \sum_{i=0}^{n-1} [y_i - (a_0 + a_1 x_i + a_2 x_i^2)]^2 \tag{3.10}$$

The unknowns can be determined by minimizing Δ by differentiating Eq. (3.10) with respect to a_r, $r = 0$, 1 and 2, that is

$$\frac{\partial \Delta}{\partial a_r} = -2 \sum_{i=0}^{n-1} x_i^r [y_i - (a_0 + a_1 x_i + a_2 x_i^2)] = 0 \tag{3.11}$$

which after slight rearrangement leads to a symmetric 3×3 system of linear equations.

$$a_0 \sum_{i=0}^{n-1} 1 + a_1 \sum_{i=0}^{n-1} x_i + a_2 \sum_{i=0}^{n-1} x_i^2 = \sum_{i=0}^{n-1} y_i$$

$$a_0 \sum_{i=0}^{n-1} x_i + a_1 \sum_{i=0}^{n-1} x_i^2 + a_2 \sum_{i=0}^{n-1} x_i^3 = \sum_{i=0}^{n-1} x_i y_i$$

$$a_0 \sum_{i=0}^{n-1} x_i^2 + a_1 \sum_{i=0}^{n-1} x_i^3 + a_2 \sum_{i=0}^{n-1} x_i^4 = \sum_{i=0}^{n-1} x_i^2 y_i \tag{3.12}$$

Example 3.2. With the four data points in Example 3.1, i.e. (0, 0), (1, 2), (3, 2) and (6, −1), obtain the best quadratic fit through them. Comment on the change in curve shape when point (3, 2) is moved to (1.5, 4).

Using Eq. (3.12), we have

$$\sum_{i=0}^{3} 1 = 4, \quad \sum_{i=0}^{3} x_i = 10, \quad \sum_{i=0}^{3} x_i^2 = 46, \quad \sum_{i=0}^{3} y_i = 3$$

$$\sum_{i=0}^{3} x_i^3 = 244, \quad \sum_{i=0}^{3} x_i y_i = 2, \quad \sum_{i=0}^{3} x_i^4 = 1378, \quad \sum_{i=0}^{3} x_i^2 y_i = -16$$

The 3×3 system becomes

$$\begin{bmatrix} 4 & 10 & 46 \\ 10 & 46 & 244 \\ 46 & 244 & 1378 \end{bmatrix} \begin{bmatrix} a_0 \\ a_1 \\ a_2 \end{bmatrix} = \begin{bmatrix} 3 \\ 2 \\ -16 \end{bmatrix} \quad \text{or} \quad \begin{bmatrix} a_0 \\ a_1 \\ a_2 \end{bmatrix} = \begin{bmatrix} 0.27 \\ 1.55 \\ -0.30 \end{bmatrix}$$

and the quadratic is $y = 0.27 + 1.55x - 0.30x^2$, the plot of which is shown in Figure 3.3 along with data points (solid lines). Although the curve does not pass through the data points, note its proximity with the latter. For a new set of data points, namely, (0, 0), (1, 2), (1.5, 4) and (6, −1), we have

$$\sum_{i=0}^{3} 1 = 4, \quad \sum_{i=0}^{3} x_i = 8.5, \quad \sum_{i=0}^{3} x_i^2 = 39.25, \quad \sum_{i=0}^{3} y_i = 5$$

$$\sum_{i=0}^{3} x_i^3 = 220.37, \quad \sum_{i=0}^{3} x_i y_i = 2, \quad \sum_{i=0}^{3} x_i^4 = 1302.06, \quad \sum_{i=0}^{3} x_i^2 y_i = -25$$

and the linear system is

$$\begin{bmatrix} 4 & 8.5 & 39.25 \\ 8.5 & 39.25 & 220.37 \\ 39.25 & 220.37 & 1302.06 \end{bmatrix} \begin{bmatrix} a_0 \\ a_1 \\ a_2 \end{bmatrix} = \begin{bmatrix} 5 \\ 2 \\ -25 \end{bmatrix}, \text{ or } \begin{bmatrix} a_0 \\ a_1 \\ a_2 \end{bmatrix} = \begin{bmatrix} -0.16 \\ 3.35 \\ -0.58 \end{bmatrix}$$

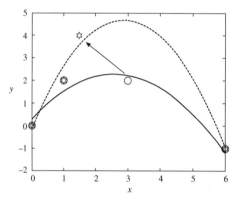

Figure 3.3   Results of curve fitting using a quadratic polynomial. Curve shape is changed globally when a data point is relocated

The resultant quadratic $y = -0.16 + 3.35x - 0.58x^2$ is plotted (dashed lines) and compared with the previous curve to note that the shape change is global.

The method described above is known as *least square* fitting. It is not mandatory always to employ a polynomial for the purpose. Instead, any non-polynomial (trigonometric or exponential) function may be chosen to best fit the given data. With polynomials of a chosen degree, although curve fitting may resolve unwarranted fluctuations as is the case with curve interpolation, it would still not impart local shape control to the designer (e.g. Example (3.2)) and changing a data point would require re-computing the entire curve.

Both curve interpolation and fitting methods discussed above have some limitations with regard to curve design. Before alternative methods are explored, it is first imperative to understand and choose the best possible mathematical representation for curves to particularly suit their design in three-dimensions. Based on the forgoing discussion, we may surmise that the use of low degree polynomials (usually cubic) is preferable over the high degree ones to avoid unwarranted fluctuations. Further, to allow local control on curve shape and close proximity to data points, *piecewise fitting* of the consecutive subsets of data points could be considered. For instance, a cubic polynomial can interpolate four data points, we may choose to interpolate four consecutive points at a time from a given set. The entire set of data points may then be piecewise interpolated and the resultant would be a composite curve with cubic segments juxtaposed sequentially (Figure 3.4). There, however, would be continuity related issues at a data point common to the two adjacent cubic segments. Though both segments would pass through the data point (position continuity), the slope and/or curvature would be discontinuous and the composite curve may not be smooth overall. It is here that an insight into the differential properties of curves would be of help.

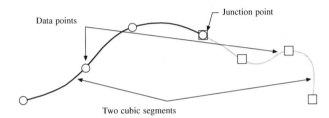

Figure 3.4   A schematic showing a composite curve with two cubic segments interpolating the data points

3.3   Representing Curves

Curves may be expressed mathematically using one of the three forms, viz. explicit, implicit or parametric. In two dimensions, explicit equations are of the form

$$y = f(x)$$

wherein the slope at a point (x, y) is given as $\dfrac{\partial y}{\partial x}$ or $\dfrac{\partial f}{\partial x}$. Consider, for instance, the equation of a straight line in the two-point form

$$y - y_1 = \frac{(y_2 - y_1)}{(x_2 - x_1)}(x - x_1)$$

So long as x_1 and x_2 are not equal, the representation works well in that there is a unique value of y for every x. However, as x_2 approaches x_1, the slope approaches infinity. Thus, for $x_1 = x_2$, even though the line is vertical, that the y value can be non-unique is not apparent, that is, explicit representations by themselves cannot accommodate vertical lines or tangents.

Implicit equations are of the type

$$g(x, y) = 0$$

for instance the equation of a straight line $ax + by + c = 0$ or the circle $x^2 + y^2 - r^2 = 0$. To determine the intersection of the line and circle above, the implicit forms need to be first converted into the respective explicit versions. Two possibilities would exist for the roots; either they both are complex or both are real. In case the roots are real and equal, the line would be tangent to the circle. For unequal and real roots, the line will intersect the circle at two points. In general, additional processing is required to determine the intersection points for any two curves. Moreover, a concern with both explicit and implicit form of representations is that they cannot, by themselves, represent a curve segment which is what the designers are usually interested in. For instance, it would be very difficult to represent a circle in the first quadrant. Herein, the parametric form of representation becomes useful. For two-dimensional curves, parametric equations can be written as

$$x = f(u)$$

$$y = g(u) \tag{3.13}$$

where u is the parameter. Note that the issue of vertical tangents easily gets resolved by using $f(u) = x_0 = $ constant and $g(u) = u$, u ranging from $-\infty$ to ∞. For two curves, $[f_1(u_1), g_1(u_1)]$ and $[f_2(u_2), g_2(u_2)]$ to intersect, the equations

$$f_1(u_1) = f_2(u_2) \quad \text{and} \quad g_1(u_1) = g_2(u_2) \tag{3.14}$$

can be solved for u_1 and u_2. Finally, curve segments can be represented by imposing the bounds on the parameters. Thus, for a straight line segment between (x_1, y_1) and (x_2, y_2)

$$x = (1 - u)x_1 + ux_2$$

$$y = (1 - u)y_1 + uy_2 \qquad 0 \le u \le 1 \tag{3.15a}$$

and for a circular arc of radius r between arguments θ_1 and θ_2

$$x = r \cos \theta$$

$$y = r \sin \theta \qquad \theta_1 \le \theta \le \theta_2 \tag{3.15b}$$

With u as the parameter, the equation of a curve in three-dimensions can be written in compact vector form as

$$\boldsymbol{r}(u) = x(u)\mathbf{i} + y(u)\mathbf{j} + z(u)\mathbf{k} \tag{3.16}$$

where $x(u)$, $y(u)$ and $z(u)$ are scalar functions of u. Many analytic curves may be represented in the above parametric form. For instance, the equation of a circle of radius a in terms of parameter $u = \omega t$ is given by

$$\boldsymbol{r}(t) = a \cos(\omega t)\mathbf{i} + a \sin(\omega t)\mathbf{j} \tag{3.17a}$$

where a particle may be considered traversing on the circumference with an angular velocity ω at time t. Similarly, parametric equations for an ellipse, parabola, hyperbola and cylindrical helix can be expressed, respectively, by

$$\boldsymbol{r}(u) = a \cos(u)\mathbf{i} + b \sin(u)\mathbf{j}$$

$$\boldsymbol{r}(u) = u^2\mathbf{i} + 2a^{1/2}u \, \mathbf{j}$$

$$\boldsymbol{r}(u) = a \sec(u) \, \mathbf{i} + b \tan(u)\mathbf{j}$$

$$\boldsymbol{r}(u) = a \cos(u)\mathbf{i} + a \sin(u)\mathbf{j} + bu \, \mathbf{k} \tag{3.17b}$$

Curves of intersection between solids like cylinders, cones and spheres are often encountered in engineering design. One such example is the intersection curve between a cylinder $[(x - a)^2 + y^2 = a^2]$ and a sphere $[x^2 + y^2 + z^2 = 4a^2]$ known as *Viviani*'s curve (Figure 3.5) whose parametric equation may be written as

$$\boldsymbol{r}(u) = a(1 + \cos u)\mathbf{i} + a \sin u \, \mathbf{j} + 2a \sin \frac{1}{2} u \, \mathbf{k} \tag{3.17c}$$

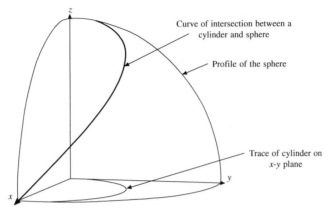

Figure 3.5   Viviani's curve shown in one octant

For a generic curve seqment, the scalar functions $x(u)$, $y(u)$ and $z(u)$ are preferred to be polynomials of a lower degree. With regard to position, slope and/or curvature continuity of a composite curve overall, differential properties of curves in parametric form are discussed below.

3.4   Differential Geometry of Curves

Consider two closely adjacent points $P(\mathbf{r}(u))$ and $Q(\mathbf{r}(u+\Delta u))$ on a parametric curve $\mathbf{r} = \mathbf{r}(u)$ in Figure 3.6, Δu, the change in parameter being small. The length of the segment Δs between P and Q may be approximated by the chord length $|\Delta \mathbf{r}| = |\mathbf{r}(u+\Delta u) - \mathbf{r}(u)|$. Taylor series expansion gives

$$\mathbf{r}(u + \Delta u) = \mathbf{r}(u) + \frac{d\mathbf{r}}{du}\Delta u + \frac{1}{2!}\frac{d^2\mathbf{r}}{du^2}(\Delta u)^2 + \ldots \text{ higher order terms} \ldots \qquad (3.18)$$

For very small Δu, only the first order term may be retained. Thus

$$\Delta s \approx |\Delta \mathbf{r}| = |\mathbf{r}(u + \Delta u) - \mathbf{r}(u)| \approx \left|\frac{d\mathbf{r}}{du}\right|\Delta u \qquad (3.19)$$

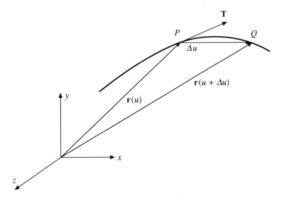

Figure 3.6   Parametric curve represented in vector form

As Q approaches P, i.e. in the limit $\Delta u \rightarrow 0$, the length Δs becomes the differential arc length ds of the curve, that is

$$ds = \left| \frac{d\mathbf{r}}{du} \right| du = |\dot{\mathbf{r}}| \, du = \sqrt{\dot{\mathbf{r}} \cdot \dot{\mathbf{r}}} \, du \qquad (3.20)$$

For a reference value u_0, the arc length $s(u)$ at parameter value u may be computed from Eq. (3.20) as

$$s(u) = \int_{u_0}^{u} \sqrt{\dot{\mathbf{r}} \cdot \dot{\mathbf{r}}} \, du = \int_{u_0}^{u} \sqrt{\dot{x}^2 + \dot{y}^2 + \dot{z}^2} \, du \qquad (3.21)$$

The parametric velocity $\boldsymbol{v}$ may be defined as

$$\boldsymbol{v} = \frac{d\boldsymbol{r}}{du} = \dot{\mathbf{r}}(u) \qquad (3.22)$$

A unit tangent $\mathbf{T}$ at point P is along the direction of the parametric velocity, that is, $\mathbf{T} = \boldsymbol{v}/|\boldsymbol{v}|$, where $|\boldsymbol{v}| = |d\mathbf{r}/du| = ds/du$ from Eq. (3.20). Thus

$$\mathbf{T} = \frac{\dot{\mathbf{r}}(u)}{|\dot{\mathbf{r}}(u)|} = \frac{d\mathbf{r}(s)}{ds} = \mathbf{r}'(s) \qquad (3.23)$$

Therefore,
$$\dot{\mathbf{r}} = \frac{d\mathbf{r}}{du} = \frac{d\mathbf{r}}{ds}\frac{ds}{du} = \mathbf{r}'(s)\boldsymbol{v} = \mathbf{T}\boldsymbol{v}$$

where $\boldsymbol{v}$ is the parametric speed equal to $|\boldsymbol{v}|$. The unit tangent $\mathbf{T}$ is expressed above as a function of the arc length. On a parametric curve $\mathbf{r} = \mathbf{r}(u)$, P is said to be a *regular point* if $|\dot{\mathbf{r}}| \neq 0$. If P is not regular, it is termed *singular*. The curve can be represented either in the form $\mathbf{r} \equiv \mathbf{r}(u)$, or $\mathbf{r} \equiv \mathbf{r}(s)$; the first is dependent on the parameter u and thus on the co-ordinate axes chosen while the second is independent of the co-ordinate axes and is a function of the *natural parameter* or the arc length s.

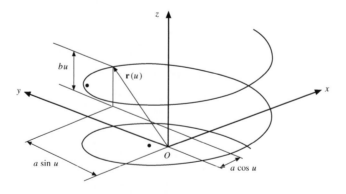

Figure 3.7 Cylindrical helix

Example 3.3. Find the length of a portion of the helix $x = a \cos u$, $y = a \sin u$, $z = bu$.
To use Eq. (3.21)

$$\dot{x} = -a \sin u, \quad \dot{y} = a \cos u, \quad \dot{z} = b, \quad \mathbf{r}(u) = a \cos u \mathbf{i} + a \sin u \mathbf{j} + bu \mathbf{k}$$

Therefore,

$$\dot{x}^2 + \dot{y}^2 + \dot{z}^2 = a^2 + b^2$$

Hence

$$s = \int_0^u \sqrt{a^2 + b^2} \, du = (a^2 + b^2) u$$

Since the length s is independent of the co-ordinate axis chosen, another representation of the helical curve in terms of natural parameter would be

$$\mathbf{r}(s) = a \cos \left[\frac{s}{\sqrt{a^2 + b^2}} \right] \mathbf{i} + a \sin \left[\frac{s}{\sqrt{a^2 + b^2}} \right] \mathbf{j} + b \left[\frac{s}{\sqrt{a^2 + b^2}} \right] \mathbf{k}$$

The normal at a point on a two-dimentional curve is unique. However, in three-dimensions, there exists a plane of vectors perpendicular to the slope $\mathbf{T}$. This plane is often referred to as the *normal plane* (Figure 3.8 (a)). To span the vectors that are orthogonal to $\mathbf{T}$, two unique vectors are identified in the normal plane. The first is the *principal* normal $\mathbf{N}$, while the second is the *binormal* $\mathbf{B}$. To determine $\mathbf{N}$, consider

$$\mathbf{T}(s) = \frac{d\mathbf{r}(s)}{ds} = \mathbf{r}'(s) \quad \text{and} \quad \mathbf{T}(s + \Delta s) = \mathbf{r}'(s + \Delta s)$$

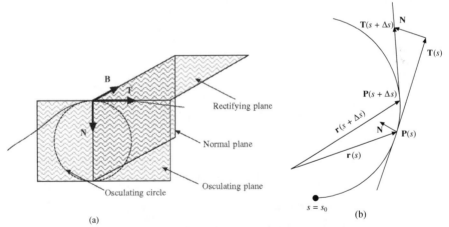

(a)

(b)

Figure 3.8   (a) The normal plane and (b) definition of a unit normal N

The net change in the direction of unit tangent in moving from P to a neighboring point Q (Figures 3.6 and 3.8b) is given by

$$\Delta \mathbf{T}(s) = \mathbf{T}(s + \Delta s) - \mathbf{T}(s) = \left\{ \mathbf{T}(s) + \frac{d\mathbf{T}}{ds} \Delta s + \frac{1}{2!} \frac{d^2 \mathbf{T}}{ds^2} (\Delta s)^2 + \ldots \right\} - \mathbf{T}(s) \approx \frac{d\mathbf{T}(s)}{ds} \Delta s \quad (3.24)$$

As $\Delta s \to 0$,
$$\frac{\Delta \mathbf{T}}{\Delta s} \to \frac{d\mathbf{T}}{ds} = \mathbf{r}''(s) \tag{3.25}$$

To determine the direction of $\mathbf{T}'(s)$ or $\mathbf{r}''(s)$, consider $\mathbf{r}'(s) \cdot \mathbf{r}'(s) = 1$ differentiating which with respect to s yields

$$\mathbf{r}' \cdot \mathbf{r}'' + \mathbf{r}'' \cdot \mathbf{r}' = 0 \Rightarrow 2\mathbf{r}' \cdot \mathbf{r}'' = 0 \Rightarrow \mathbf{r}' \cdot \mathbf{r}'' = 0$$

Thus, $\mathbf{r}'$ and $\mathbf{r}''$ are orthogonal to each other implying that $\mathbf{r}''$ is perpendicular to $\mathbf{T}$. We may, therefore, define $\mathbf{N}$ (a unit normal vector) such that

$$\kappa \mathbf{N} = \frac{d\mathbf{T}}{ds} \tag{3.26}$$

where $\kappa = \left| \dfrac{d\mathbf{T}}{ds} \right|$ is the scaling factor to ensure that $\mathbf{N}$ is a unit vector. Also note that

$$\mathbf{N} = \mathbf{r}''/|\mathbf{r}''| = \frac{(d\mathbf{T}/ds)}{\left| \dfrac{d\mathbf{T}}{ds} \right|} = \frac{d\mathbf{T}/du(du/ds)}{\left| \dfrac{d\mathbf{T}}{du} \dfrac{du}{ds} \right|} = \frac{(d\mathbf{T}/du)}{\left| \dfrac{d\mathbf{T}}{du} \right|}$$

The binormal $\mathbf{B}$ can then be defined as

$$\mathbf{B} = \mathbf{T} \times \mathbf{N} \tag{3.27}$$

The plane containing $\mathbf{T}$ and $\mathbf{B}$ is termed the *rectifying* plane while that containing $\mathbf{T}$ and $\mathbf{N}$ is referred to as the *osculating* plane. To interpret the scalar κ physically, let P, Q and W be three points on the curve in close vicinity with values $\mathbf{r}(u)$, $\mathbf{r}(u + \Delta u)$ and $\mathbf{r}(u - \Delta u)$, respectively, as shown in Figure 3.9.

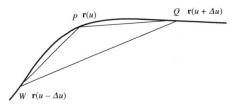

Figure 3.9   Determination of κ

The vector $\mathbf{QP} \times \mathbf{QW}$ can be computed as

$$\mathbf{QP} \times \mathbf{QW} = [\mathbf{r}(u + \Delta u) - \mathbf{r}(u)] \times [\mathbf{r}(u + \Delta u) - \mathbf{r}(u - \Delta u)] \tag{3.28}$$

Using the first order Taylor series expansion and ignoring higher order terms in Δu, Eq. (3.28) can be rewritten as

$$\mathbf{QP} \times \mathbf{QW} = \left[\frac{d\mathbf{r}}{du} \Delta u + \frac{1}{2} \frac{d^2\mathbf{r}}{du^2} \Delta u^2 \right] \times \left[2\frac{d\mathbf{r}}{du} \Delta u + 2\frac{d^2\mathbf{r}}{du^2} \Delta u^2 \right] = \left[\frac{d\mathbf{r}}{du} \times \frac{d^2\mathbf{r}}{du^2} \right] \Delta u^3 \tag{3.29}$$

From Eq. (3.23), $\mathbf{T} = \dfrac{d\mathbf{r}(s)}{ds}$ so that $\dfrac{d\mathbf{r}}{du} = \mathbf{T}\dfrac{ds}{du}$ using chain rule. Differentiating further gives

$$\frac{d^2\mathbf{r}}{du^2} = \frac{d\mathbf{T}}{du}\frac{ds}{du} + \mathbf{T}\frac{d^2s}{du^2}$$

Implementing the above result in Eq. (3.29) yields

$$\mathbf{QP} \times \mathbf{QW} = \left[\mathbf{T}\frac{ds}{du} \times \left(\frac{d\mathbf{T}}{du}\frac{ds}{du} + \mathbf{T}\frac{d^2s}{du^2} \right) \right]\Delta u^3$$

$$= \left[\mathbf{T} \times \frac{d\mathbf{T}}{du}\left(\frac{ds}{du}\right)^2 \right]\Delta u^3 = \left[\mathbf{T} \times \frac{d\mathbf{T}}{ds}\left(\frac{ds}{du}\right)^3 \right]\Delta u^3$$

Using Eqs. (3.26) and (3.27), we get

$$\mathbf{QP} \times \mathbf{QW} = \kappa\mathbf{T} \times \mathbf{N}\left[\left(\frac{ds}{du}\right)^3 \right]\Delta u^3 = \kappa\mathbf{B}\left[\left(\frac{ds}{du}\right)^3 \right]\Delta u^3 \tag{3.30a}$$

From Eqs. (3.29) and (3.30a)

$$\left[\frac{d\mathbf{r}}{du} \times \frac{d^2\mathbf{r}}{du^2} \right] = \kappa\mathbf{B}\left[\left(\frac{ds}{du}\right)^3 \right] \tag{3.30b}$$

From Eq. (3.30a), $\mathbf{QP} \times \mathbf{QW}$ is parallel to $\mathbf{B}$ implying that if a circle is drawn through P, Q and W, the normal to the plane containing the circle would be parallel to $\mathbf{B}$, that is, the circle would be contained in the osculating plane for which reason it is termed as the *osculating circle* (Figure 3.8). From vector algebra, the radius of curvature ρ at P is given as

$$\rho = \frac{|\,\mathbf{WP}\,\|\,\mathbf{WQ}\,\|\,\mathbf{WP} - \mathbf{WQ}\,|}{2\,|\,\mathbf{WP} \times \mathbf{WQ}\,|} = \frac{|\,\mathbf{WP}\,\|\,\mathbf{WQ}\,\|\,\mathbf{WP} - \mathbf{WQ}\,|}{2\,|\,(\mathbf{WQ} - \mathbf{PQ}) \times \mathbf{WQ}\,|} = \frac{|\,\mathbf{WP}\,\|\,\mathbf{WQ}\,\|\,\mathbf{WP} - \mathbf{WQ}\,|}{2\,|\,\mathbf{PQ} \times \mathbf{WQ}\,|}$$

or $\qquad \rho = \dfrac{\left| \Delta u\dfrac{d\mathbf{r}}{du} + \dfrac{1}{2}\dfrac{d^2\mathbf{r}}{du^2}\Delta u^2 + \ldots \right|\left| 2\Delta u\dfrac{d\mathbf{r}}{du} + \ldots \right|\left| \Delta u\dfrac{d\mathbf{r}}{du} - \dfrac{1}{2}\dfrac{d^2\mathbf{r}}{du^2}\Delta u^2 + \ldots \right|}{2\left| \dfrac{d\mathbf{r}}{du} \times \dfrac{d^2\mathbf{r}}{du^2} \right|\Delta u^3}$

$$= \frac{\Delta u^3 \left| \dfrac{d\mathbf{r}}{du} \right|^3}{\left| \dfrac{d\mathbf{r}}{du} \times \dfrac{d^2\mathbf{r}}{du^2} \right|\Delta u^3}$$

or using Eq. (3.30b)

$$\rho = \frac{\left| \dfrac{d\mathbf{r}}{du} \right|^3}{\left| \dfrac{d\mathbf{r}}{du} \times \dfrac{d^2\mathbf{r}}{du^2} \right|} = \frac{1}{\kappa} \frac{\left| \dfrac{d\mathbf{r}}{du} \right|^3}{\left(\dfrac{ds}{du} \right)^3} = \frac{1}{\kappa} \tag{3.30c}$$

In other words, the scalar κ is the inverse of the radius of curvature, ρ for which reason κ is referred to as curvature. Note that if the curve lies on a plane, so does the osculating circle and hence **B** is invariant, that is, $d\mathbf{B}/ds = 0$. Otherwise, $d\mathbf{B}/ds$ can be computed in the following manner. Noting that $\mathbf{B} \cdot \mathbf{T} = 0$,

$$\mathbf{T} \cdot (d\mathbf{B}/ds) + \mathbf{B} \cdot (d\mathbf{T}/ds) = 0$$

From Eq. (3.26), $(d\mathbf{T}/ds) = \kappa \mathbf{N}$ and since **B** is orthogonal to **N**, $\mathbf{T} \cdot (d\mathbf{B}/ds) = 0$, implying that $d\mathbf{B}/ds$ is perpendicular to **T**. Moreover, since **B** is a unit vector, $\mathbf{B} \cdot (d\mathbf{B}/ds) = 0$ and thus $d\mathbf{B}/ds$ is parallel to $\mathbf{B} \times \mathbf{T}$ or **N**. Define $d\mathbf{B}/ds$ as

$$d\mathbf{B}/ds = -\tau \mathbf{N} \tag{3.31}$$

where τ is termed as the *torsion* of the curve. Now, since **T**, **N** and **B** are mutually orthogonal, using $\mathbf{N} = \mathbf{B} \times \mathbf{T}$ and differentiating, we get

$$\begin{aligned}
d\mathbf{N}/ds &= (d\mathbf{B}/ds) \times \mathbf{T} + \mathbf{B} \times (d\mathbf{T}/ds) \\
&= (d\mathbf{B}/ds) \times \mathbf{T} + \kappa \mathbf{B} \times \mathbf{N} \\
&= -\tau \mathbf{N} \times \mathbf{T} + \kappa \mathbf{B} \times \mathbf{N} \\
&= \tau \mathbf{B} - \kappa \mathbf{T}
\end{aligned} \tag{3.32}$$

Eqs. (3.23), (3.26), (3.31) and (3.32) are collectively termed as the Frenet-Serret formulae summarized as follows:

$$\begin{aligned}
d\mathbf{r}/ds &= \mathbf{T} \\
d\mathbf{T}/ds &= \kappa \mathbf{N} \\
d\mathbf{B}/ds &= -\tau \mathbf{N} \\
d\mathbf{N}/ds &= \tau \mathbf{B} - \kappa \mathbf{T}
\end{aligned} \tag{3.33}$$

Most often, it is easier to work with parameter u as opposed to the natural or arc length parameter s for which the Frenet-Serret formulae can be modified accordingly using Eq. (3.20). Such conditions provide useful information on the slope and curvature of the segments which is very helpful when implementing the continuity requirements at the common data points (or junction points) of piecewise composite curves.

Example 3.4. Consider the helix $\mathbf{r}(t) = a \cos t\, \mathbf{i} + a \sin t\, \mathbf{j} + bt\, \mathbf{k}$ in parameter t. Determine the tangent, normal, bi-normal, radius of curvature, curvature and torsion at a point on the helix.

The unit tangent vector **T** is given by

$$\dot{\mathbf{r}}(t) = -a \sin t\, \mathbf{i} + a \cos t\, \mathbf{j} + b\, \mathbf{k}$$

$$\mathbf{T} = \frac{\dot{\mathbf{r}}(t)}{|\dot{\mathbf{r}}(t)|} = \frac{-a \sin t\, \mathbf{i} + a \cos t\, \mathbf{j} + b\, \mathbf{k}}{\sqrt{(-a \sin t)^2 + (a \cos t)^2 + b^2}} = \frac{-a \sin t\, \mathbf{i} + a \cos t\, \mathbf{j} + b\, \mathbf{k}}{\sqrt{a^2 + b^2}}$$

The unit bi-normal vector **B** may be obtained from Eq. (3.30b) as

$$\mathbf{B} = \frac{\dot{\mathbf{r}} \times \ddot{\mathbf{r}}}{|\dot{\mathbf{r}} \times \ddot{\mathbf{r}}|}$$

$$\ddot{\mathbf{r}} = -a \cos t\,\mathbf{i} - a \sin t\,\mathbf{j} + 0\mathbf{k}$$

$$\dot{\mathbf{r}} \times \ddot{\mathbf{r}} = \begin{bmatrix} \mathbf{i} & \mathbf{j} & \mathbf{k} \\ -a \sin t & a \cos t & b \\ -a \cos t & -a \sin t & 0 \end{bmatrix} = ab \sin t\,\mathbf{i} - ab \cos t\,\mathbf{j} + a^2 \mathbf{k}$$

$$|\dot{\mathbf{r}} \times \ddot{\mathbf{r}}| = \sqrt{(ab \sin t)^2 + (-ab \cos t)^2 + (a^2)^2} = a\sqrt{a^2 + b^2}$$

Therefore,
$$\mathbf{B} = \frac{b \sin t\,\mathbf{i} - b\cos t\,\mathbf{j} + a\mathbf{k}}{\sqrt{a^2 + b^2}}$$

The normal vector **N** is given by

$$\mathbf{N} = \mathbf{B} \times \mathbf{T}$$

Therefore
$$\mathbf{N} = \frac{1}{a^2 + b^2} \begin{vmatrix} \mathbf{i} & \mathbf{j} & \mathbf{k} \\ b \sin t & -b \cos t & a \\ -a \sin t & a \cos t & b \end{vmatrix} = -(\cos t\,\mathbf{i} + \sin t\,\mathbf{j} + 0\mathbf{k})$$

The curvature is given by Eq. 3.30(c).

$$\kappa = \frac{|\dot{\mathbf{r}} \times \ddot{\mathbf{r}}|}{|\dot{\mathbf{r}}|^3} = \frac{a\sqrt{a^2 + b^2}}{(a^2 + b^2)^{3/2}} = \frac{a}{(a^2 + b^2)}.$$

Therefore, the radius of curvature $\rho = \dfrac{(a^2 + b^2)}{a}$.

From Eq. (3.33), the torsion τ is given as

$$\tau = \left| \frac{d\mathbf{B}}{ds} \right|$$

$$\frac{d\mathbf{B}}{ds} = \frac{d\mathbf{B}}{dt} \Big/ \left(\frac{ds}{dt} \right) = \frac{d\mathbf{B}}{dt} \Big/ \left| \frac{d\mathbf{r}}{dt} \right|$$

$$\frac{d\mathbf{B}}{dt} = \frac{b\cos t\,\mathbf{i} + b \sin t\,\mathbf{j}}{\sqrt{a^2 + b^2}}$$

$$\left| \frac{d\mathbf{r}}{dt} \right| = \sqrt{a^2 + b^2}$$

$\Rightarrow$
$$\tau = \left| \frac{b \cos t\,\mathbf{i} + b \sin t\,\mathbf{j}}{(a^2 + b^2)} \right| = \frac{b}{(a^2 + b^2)}$$

EXERCISES

1. Find the parametric equation of an Archimedean spiral in a polar form. The largest and the smallest radii of the spiral are 100 mm and 20 mm, respectively. The spiral has two convolutions to reduce the radius from the largest to the smallest value.

2. Derive the equation in parametric form of a cycloid. A cycloid is obtained as the locus of a point on the circumference of a circle when the circle rolls without slipping on a straight line for one complete revolution. Assume the diameter of the circle to be 50 mm. Also, derive the parametric equation for the tangent and the normal at any generic point on the curve. Furthermore, find the coordinates of the center of curvature.

3. Find the curvature and torsion of the following curves.
   (a) $x = u$, $y = u^2$, $z = u^3$
   (b) $x = u$, $y = (1 + u)/u$, $z = (1 - u^2)/u$
   (c) $x = a(u - \sin u)$, $y = a(1 - \cos u)$, $z = bu$

4. Derive the parametric equation of parabolic arch whose span is 150 mm and rise is 65 mm.

5. Derive the parametric equation of an equilateral hyperbola passing through a point P (15, 65).

6. Find the parametric equation of a circle passing through three points $\mathbf{p}_0$, $\mathbf{p}_1$ and $\mathbf{p}_2$ lying on the XY plane. Discuss under what conditions the equation will fail to define a circle.

7. Find the equation for the skew distance (shortest distance) as well as the skew angle between a pair of skew lines AB and CD.

8. For a line AB, specified in space, find the angle of this line from the XOY plane. Also, find the angle that the projection of this line in the XOY plane makes with the x-axis.

9. Find the osculating, tangent (rectifying) and normal planes for the following curves:
   (a) $x(u) = 3u$, $y(u) = 3u^2$, $z(u) = 2u^3$
   (b) $x(u) = a \cos u$, $y(u) = a \sin u$, $z(u) = b\,u$
   Show these planes using plots for $-1 \le u \le 1$.

10. If $\mathbf{r}(s)$ is an arc length parametrized curve such that the torsion $\tau = 0$, and curvature κ is a constant, show that $\mathbf{r}(s)$ is a circle.

11. Calculate the moving trihedron values (tangent, normal and bi-normal) as functions of u and plot the curvature and torsion for $\mathbf{r}(u) = (3u - u^3, 3u^2, 3u + u^3)$ shown in Figure P3.1.

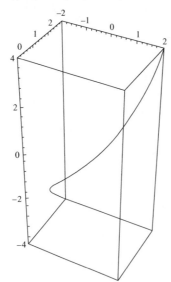

Figure P3.1

12. Plot the curvature and torsion of the Viviani's curve (intersection of a cylinder and a sphere).

13. Write a procedure to create Frenet-Frame (defined by the tangent, normal and bi-normal) at any given point of a 3-D curve. Also write a procedure to calculate curvature and torsion of the curve at a given point. Any symbolic math package like Maple, Mathematica may be used for the purpose. It would be helpful to incorporate plotting in the procedure.

14. Given three points A, B and C on a curve with position veotors **a**, **b** and **c**, respectively (Figure P3.2), determine the radius of curvature ρ using vector algebra.

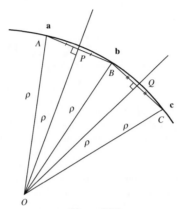

Figure P3.2

Chapter 4

Design of Curves

Shapes are created by curves in that a surface, such as the rooftop of a car, the fuselage of an aircraft, or a washbasin can be created by motion of curves in space in a specified manner. This may involve sweep, revolution, deformation, contraction or expansion, and forming joints with other curves.

The *analytical properties* of curves were derived in Chapter 3, where it was assumed that the equation of the curve is *known*. However, in *design*, an engineer first creates a *shape* using his imagination, without knowing the equation of the curve at this stage. The computer should help the designer in *synthesizing* the *curve shape* so that one can (a) *replicate the imagined shape* without worrying about the equations of the curve (b) change or fine tune the shape to conform to technical, manufacturing, aesthetic and other requirements.

The principles of curve design envisages the following:

1. The *shape* of the curve should be controlled by placing only a few number of data points. The curve thus created should behave like an elastic string that a designer can manipulate to give a desired shape.
2. The curve should be *synthetically* composed of polytnomial segments of lower degree to avoid undue oscillations and minimize computation time and complexity.
3. The curve model should have "*affine*" properties ensuring *shape independence from the co-ordinate frame of reference*. This makes it possible to treat the curve model as a *real* object in space that does not get distorted because of different frame of reference.
4. Since, in real design, complex shapes have to be created, it is more suitable to join together several segments of curves, fulfilling position, slope and/or curvature continuities at the joints. If we look at the profile of a car or an aircraft at any cross section, we can appreciate the smoothness with which various curve segments are joined together. Thus, curve models are developed which form simple *building blocks* for piecing them together to create a desired shape.
5. *Parametric* description is preferred over the implicit or explicit forms as it provides an articulate representation of curve segments in three dimensions. In additio, and trimming like operations can be handled with relative ease.

Synthetic curves are suitable for designing generic forms that may not be represented by analytic curves. Mathematically, though both synthetic and analytic curves are polynomial representations, the former provides more control in that they may be derived from a given set of data points and/or slopes via interpolation or curve fitting. Since the aim is to use low order parametric segments, two models of cubic segments are discussed in detail in this chapter, namely (a) Ferguson's or Hermite cubic segments and (b) Bézier segments. Cubic segments are usually a good compromise for form representation in most engineering applications. Differential properties like the tangents, normals and curvatures are easy to compute. A cubic form of the segment ensures continuity of a composite curve up to second order. It is computationally more efficient than the higher degree polynomials. Linear or quadratic forms of curve segments, on the other hand, are incapable of modeling inflexions in the curve.

We may commence with the parametric representation of a three-dimensional curve $\mathbf{r}(u)$ in Eq. (6), that is

$$\mathbf{r}(u) = x(u)\mathbf{i} + y(u)\mathbf{j} + z(u)\mathbf{k} \equiv [x(u), y(u), z(u)]$$

For $\mathbf{r}(u)$ to be cubic in u, the scalar polynomials $x(u)$, $y(u)$ and $z(u)$ can be correspondingly expanded as

$$x(u) = a_{0x} + a_{1x}u + a_{2x}u^2 + a_{3x}u^3$$
$$y(u) = a_{0y} + a_{1y}u + a_{2y}u^2 + a_{3y}u^3 \tag{4.1}$$
$$z(u) = a_{0z} + a_{1z}u + a_{2z}u^2 + a_{3z}u^3$$

where a_{0x}, a_{1x}, a_{2x} and a_{3x} are all unknowns in $x(u)$ and likewise for $y(u)$ and $z(u)$. Alternatively, $\mathbf{r}(u)$ may be written in the matrix form $[x(u), y(u), z(u)]$ as

$$r(u) \equiv [x(u), y(u), z(u)] = [u^3 \quad u^2 \quad u \quad 1] \begin{bmatrix} a_{3x} & a_{3y} & a_{3z} \\ a_{2x} & a_{2y} & a_{2z} \\ a_{1x} & a_{1y} & a_{1z} \\ a_{0x} & a_{0y} & a_{0z} \end{bmatrix} = \mathbf{UA} \tag{4.2}$$

where $\mathbf{U}$ is the *power basis vector* and $\mathbf{A}$ the *algebraic coefficient matrix* determined by the conditions imposed on the curve segment to acquire a desired shape.

Example 4.1. Find a parametric cubic curve that starts at P_0 (−1, 2), ends at P_3 (8, 4) and passes through two prescribed points $P_1(2, 4)$ and P_2 (6, 6).

We use Eq. (4.2) for this 2-D curve, ignoring the third column in $\mathbf{A}$. Let $u_0 = 0$ at P_0 and $u_3 = 1$ at P_3. Different ways may be employed to determine parameter values u_1 and u_2 corresponding to the points P_1 and P_2, respectively. For instance, we may place them uniformly in [0, 1] by setting $u_1 = 0.33$ and $u_2 = 0.66$. The other alternative is to determine them in a manner suggestive of the relative position of data points. We can find the chord lengths P_0P_1, P_1P_2 and P_2P_3 as

$$d_1 = P_0P_1 = \sqrt{\{(2 - (-1))^2 + (4 - 2)^2\}} = 3.61$$
$$d_2 = P_1P_2 = \sqrt{\{(6 - 2)^2 + (6 - 4)^2\}} = 4.47$$
$$d_3 = P_2P_3 = \sqrt{\{(8 - 6)^2 + (4 - 6)^2\}} = 2.83$$

and determine parameter values as

$$u_0 = 0, \quad u_1 = \frac{d_1}{\sum\limits_{j=1}^{3} d_j} = \frac{3.61}{3.61 + 4.47 + 2.83} = 0.33$$

$$u_2 = \frac{d_1 + d_2}{\sum\limits_{j=1}^{3} d_j} = \frac{3.61 + 4.47}{10.91} = 0.74, \quad u_3 = 1$$

Eq. (4.2) for the four data points becomes

$$\begin{bmatrix} u_0^3 & u_0^2 & u_0^1 & 1 \\ u_1^3 & u_1^2 & u_1^1 & 1 \\ u_2^3 & u_2^2 & u_2^1 & 1 \\ u_3^3 & u_3^2 & u_3^1 & 1 \end{bmatrix} \begin{bmatrix} a_{3x} & a_{3y} \\ a_{2x} & a_{2y} \\ a_{1x} & a_{1y} \\ a_{0x} & a_{0y} \end{bmatrix} = \begin{bmatrix} x_0 & y_0 \\ x_1 & y_1 \\ x_2 & y_2 \\ x_3 & y_3 \end{bmatrix}$$

$$\Rightarrow \begin{bmatrix} 0 & 0 & 0 & 1 \\ 0.33^3 & 0.33^2 & 0.33 & 1 \\ 0.74^3 & 0.74^2 & 0.74 & 1 \\ 1 & 1 & 1 & 1 \end{bmatrix} \begin{bmatrix} a_{3x} & a_{3y} \\ a_{2x} & a_{2y} \\ a_{1x} & a_{1y} \\ a_{0x} & a_{0y} \end{bmatrix} = \begin{bmatrix} -1 & 2 \\ 2 & 4 \\ 6 & 6 \\ 8 & 4 \end{bmatrix}$$

which gives

$$\begin{bmatrix} a_{3x} & a_{3y} \\ a_{2x} & a_{2y} \\ a_{1x} & a_{1y} \\ a_{0x} & a_{0y} \end{bmatrix} = \begin{bmatrix} -3.9 & -17.15 \\ 5.15 & 16.74 \\ 7.83 & 2.4 \\ -1 & 2 \end{bmatrix}$$

The equation of the required curve is

$$\mathbf{r}(u) = x(u)\mathbf{i} + y(u)\mathbf{j} = [-3.9u^3 + 5.15u^2 + 7.83u - 1]\mathbf{i} + [-17.15u^3 + 16.74u^2 + 2.4u + 2]\mathbf{j}$$

the plots of which ae shown in Figure 4.1.

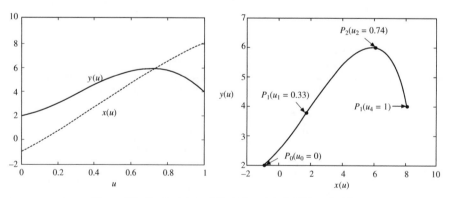

Figure 4.1   Parametric and Cartesian plots for Example 4.1

4.1 Ferguson's or Hermite Cubic Segments

A cubic Ferguson's segment is designed like an arch with two end points and two respective slopes known (Figure 4.2). Let the end points be $\mathbf{P}_i(x_i, y_i, z_i)$ at $u = 0$ and $\mathbf{P}_{i+1}$ $(x_{i+1}, y_{i+1}, z_{i+1})$ at $u = 1$. Also, let the respective slopes be $\mathbf{T}_i(p_i, q_i, r_i)$ and $\mathbf{T}_{i+1}(p_{i+1}, q_{i+1}, r_{i+1})$. $\mathbf{T}_i$ and $\mathbf{T}_{i+1}$ need not be of unit magnitude and may be written in terms of respective unit vectors $\mathbf{t}_i$ and $\mathbf{t}_{i+1}$ as $\mathbf{T}_i = c_i\mathbf{t}_i$ and $\mathbf{T}_{i+1} = c_{i+1}$ $\mathbf{t}_{i+1}$ for some scalars c_i and c_{i+1}. Consider the parametric variation only along the x coordinate, that is

$$x(u) = a_{0x} + a_{1x}u + a_{2x}u^2 + a_{3x}u^3 \tag{4.3}$$

with $x(0) = x_i$, $x(1) = x_{i+1}$, and also $dx(0)/dt = p_i$ and $dx(1)/dt = p_{i+1}$. We get

$$x_i = a_{0x}$$

$$x_{i+1} = a_{0x} + a_{1x} + a_{2x} + a_{3x}$$

$$p_i = a_{1x}$$

$$p_{i+1} = a_{1x} + 2a_{2x} + 3a_{3x}$$

solving which gives

Figure 4.2   A cubic Ferguson segment

$$a_{0x} = x_i$$

$$a_{1x} = p_i$$

$$a_{2x} = 3\Delta x_i - 3p_i - \Delta p_i$$

$$a_{3x} = \Delta p_i - 2\Delta x_i + 2p_i \tag{4.4}$$

where $\Delta x_i = x_{i+1} - x_i$ and $\Delta p_i = p_{i+1} - p_i$. The polynomial in Eq. (4.3) becomes

$$x(u) = x_i + p_iu + (3\Delta x_i - 3p_i - \Delta p_i)u^2 + (\Delta p_i - 2\Delta x_i + 2p_i)\, u^3$$

$$= (1 - 3u^2 + 2u^3)x_i + (3u^2 - 2u^3)\, x_{i+1} + (u - 2u^2 + u^3)\, p_i + (-u^2 + u^3)\, p_{i+1}$$

$$= H_0^3(u)x_i + H_1^3(u)x_{i+1} + H_2^3(u)p_i + H_3^3(u)p_{i+1} \tag{4.5}$$

$H_i^3(u)$, $i = 0, \ldots, 3$ are functions of parameter u and are termed as Hermite polynomials. They serve as *blending functions* or *basis functions* or *weights* to combine the end point and slope information to generate the shape. At $u = (0$ and $1)$, $H_0^3(u) = (1$ and $0)$ while $H_1^3(u) = (0$ and $1)$, and $H_2^3(u)$ and $H_3^3(u)$ are both $(0$ and $0)$. This implies that at the end points, the slope information is not used. We can further compute the first derivatives of Hermite basis functions to find that both $\dfrac{dH_0^3(u)}{du} = \dfrac{dH_3^3(u)}{du} = 0$ at $u = 0$ and 1. This decouples the data points with slopes and allows their selective modification to change the shape of the cubic segment.

In matrix form, Eq. (4.5) can be expressed as

$$x(u) = [u^3 \quad u^2 \quad u \quad 1] \begin{bmatrix} 2 & -2 & 1 & 1 \\ -3 & 3 & -2 & -1 \\ 0 & 0 & 1 & 0 \\ 1 & 0 & 0 & 0 \end{bmatrix} \begin{bmatrix} x_i \\ x_{i+1} \\ p_i \\ P_{i+1} \end{bmatrix} \tag{4.6}$$

Similar treatment can be employed for $y(u)$ and $z(u)$ starting with the cubic form in Eq. (4.3) to obtain results analogous to Eq. (4.6). The combined result for $\mathbf{r}(u) \equiv [x(u), y(u), z(u)]$ may be expressed as

$$\mathbf{r}(u) = [x(u) \ y(u) \ z(u)] = [u^3 \ u^2 \ u \ 1] \begin{bmatrix} 2 & -2 & 1 & 1 \\ -3 & 3 & -2 & -1 \\ 0 & 0 & 1 & 0 \\ 1 & 0 & 0 & 0 \end{bmatrix} \begin{bmatrix} x_i & y_i & z_i \\ x_{i+1} & y_{i+1} & z_{i+1} \\ p_i & q_i & r_i \\ p_{i+1} & q_{i+1} & r_{i+1} \end{bmatrix}$$

or

$$\mathbf{r}(u) = [u^3 \ u^2 \ u \ 1] \begin{bmatrix} 2 & -2 & 1 & 1 \\ -3 & 3 & -2 & -1 \\ 0 & 0 & 1 & 0 \\ 1 & 0 & 0 & 0 \end{bmatrix} \begin{bmatrix} \mathbf{P}_i \\ \mathbf{P}_{i+1} \\ \mathbf{T}_i \\ \mathbf{T}_{i+1} \end{bmatrix} = \mathbf{UMG} \tag{4.7}$$

where $\mathbf{U} = [u^3 \ u^2 \ u \ 1]$ is the row matrix, $\mathbf{M}$ is a 4×4 square Hermite matrix and $\mathbf{G}$ is a 4×3 geometric matrix containing the end point and slope information Matrices $\mathbf{U}$ and $\mathbf{M}$ are identical for all Hermite cubic segments, but the geometric matrix $\mathbf{G}$ is *user defined,* that is, to alter the curve's shape, we need to alter the entries in the geometric matrix. We may relocate $\mathbf{P}_i$ or $\mathbf{P}_{i+1}$, or alter the end tangents $\mathbf{T}_i$ and $\mathbf{T}_{i+1}$, both in magnitude and direction, to effect shape change. Keeping the end points and directions of the end tangents the same, we may as well observe the change in shape when altering the magnitudes of the end tangents. For a Ferguson segment given by Eq. (4.7), we can compute the first and second derivatives by differentiating $\mathbf{r}(u)$ with respect to u (Eq. 4.8). This would help in computing differential properties like tangents and end curvatures when imposing continuity conditions at the junction points.

$$\mathbf{r}^u(u) = \frac{d\mathbf{r}(u)}{du} = (6u^2 - 6u)\mathbf{P}_i + (-6u^2 + 6u)\mathbf{P}_{i+1} + (3u^2 - 4u + 1)\mathbf{T}_i + (3u^2 - 2u)\mathbf{T}_{i+1}$$

$$\mathbf{r}^{uu}(u) = \frac{d^2\mathbf{r}(u)}{du^2} = (12u - 6)\mathbf{P}_i + (-12u + 6)\mathbf{P}_{i+1} + (6u - 4)\mathbf{T}_i + (6u - 2)\mathbf{T}_{i+1} \tag{4.8}$$

The above equations can be written in the matrix form as

$$\mathbf{r}^u(u) = [u^3 \ u^2 \ u \ 1] \begin{bmatrix} 0 & 0 & 0 & 0 \\ 6 & -6 & 3 & 3 \\ -6 & 6 & -4 & -2 \\ 0 & 0 & 1 & 0 \end{bmatrix} \begin{bmatrix} \mathbf{P}_i \\ \mathbf{P}_{i+1} \\ \mathbf{T}_i \\ \mathbf{T}_{i+1} \end{bmatrix} = \mathbf{UM}_1\mathbf{G}$$

$$\mathbf{r}^{uu}(u) = [u^3 \ u^2 \ u \ 1] \begin{bmatrix} 0 & 0 & 0 & 0 \\ 0 & 0 & 0 & 0 \\ 12 & -12 & 6 & 6 \\ -6 & 6 & -4 & -2 \end{bmatrix} \begin{bmatrix} \mathbf{P}_i \\ \mathbf{P}_{i+1} \\ \mathbf{T}_i \\ \mathbf{T}_{i+1} \end{bmatrix} = \mathbf{UM}_2\mathbf{G} \tag{4.9}$$

Example 4.2. The starting and end points of a planar curve segment are $\mathbf{P}_i = (-1, 2)$ and $\mathbf{P}_{i+1} = (8, 5)$. The unit tangent vectors at the ends are $\mathbf{t}_i = (0.94, 0.35)$ and $\mathbf{t}_{i+1} = (0.39, -0.92)$ with tangent magnitudes as $c_i = 8.5$ and $c_{i+1} = 15.2$. Determine the tangent, radius of curvature, normal, and binormal for a point on the curve at $u = 0.5$.

From Eq. (4.9), $\mathbf{T}(u) = \mathbf{r}^u(u) = \mathbf{U}\mathbf{M}_1\mathbf{G}$ with $\mathbf{T}_i = c_i\mathbf{t}_i$ and $\mathbf{T}_{i+1} = c_{i+1}\mathbf{t}_{i+1}$. Therefore

$$\mathbf{T}(u) = [u^3 \quad u^2 \quad u \quad 1]\begin{bmatrix} 0 & 0 & 0 & 0 \\ 6 & -6 & 3 & 3 \\ -6 & 6 & -4 & -2 \\ 0 & 0 & 1 & 0 \end{bmatrix}\begin{bmatrix} -1 & 2 \\ 8 & 5 \\ 8 & 3 \\ 6 & -14 \end{bmatrix}$$

$$= (-12u^2 + 10u + 8)\mathbf{i} + (-51u^2 + 34u + 3)\mathbf{j} = \dot{\mathbf{r}}$$

$$\frac{d\mathbf{T}}{du} = \mathbf{r}^{uu}(u) = (-24u + 10)\mathbf{i} + (-102u + 34)\mathbf{j} = \ddot{\mathbf{r}}$$

Thus,

$$\mathbf{T}(u = 0.5) = \dot{\mathbf{r}}(0.5) = (-3 + 5 + 8)\mathbf{i} + (-12.75 + 17 + 3)\mathbf{j} = 10\mathbf{i} + 7.25\mathbf{j}$$

$$|\mathbf{T}| = \sqrt{(10^2 + 7.25^2)} = 12.35$$

$$\ddot{\mathbf{r}}(u = 0.5) = -2\mathbf{i} - 17\mathbf{j}$$

The curvature is given by

$$\kappa = |\dot{\mathbf{r}} \times \ddot{\mathbf{r}}| / |\dot{\mathbf{r}}|^3 = \frac{155.5}{(12.35)^3}. \text{ So } \rho = 1/\kappa = 1/0.083 = 12.11$$

The bi-normal $\mathbf{B}$ is

$$\dot{\mathbf{r}} \times \ddot{\mathbf{r}} / |\dot{\mathbf{r}} \times \ddot{\mathbf{r}}| = -155.5\mathbf{k}/155.5 = -\mathbf{k}$$

The principal normal

$$\mathbf{N} = \mathbf{B} \times \frac{\mathbf{T}}{|\mathbf{T}|} = -\mathbf{k} \times (0.81\mathbf{i} + 0.59\mathbf{j}) = -0.81\mathbf{j} + 0.59\mathbf{i}$$

It may be left as an exercise to show that the torsion $\tau = \dfrac{(\dot{\mathbf{r}} \times \ddot{\mathbf{r}}) \cdot \dddot{\mathbf{r}}}{|\dot{\mathbf{r}} \times \ddot{\mathbf{r}}|^2}$ at $u = 0.5$ is zero.

Effect of the tangent magnitudes c_i and c_{i+1}

Keeping the end points fixed, Figure 4.3 shows the change in curve shape when the end tangent magnitudes are altered. For increase in c_i, the curve leans towards $\mathbf{P}_{i+1}$ and eventually forms a cusp. At higher values, a loop is formed. For a two-dimensional cubic segment, there are 8 unknowns (a_{0x}, $a_{1x}, a_{2x}, a_{3x}, a_{0y}, a_{1y}, a_{2y}$ and a_{3y}) needing eight conditions for evaluation. Four conditions are available from the given end locations ($x(0), y(0); x(1), y(1)$). The direction cosines of unit tangents $\mathbf{t}_i = (t_{ix}, t_{iy})$ and $\mathbf{t}_{i+1} = (t_{i+1x}, t_{i+1y})$ provide only two of the other four conditions. This is because $t_{ix}^2 + t_{iy}^2 = 1$ and $t_{i+1x}^2 + t_{i+1y}^2 = 1$ implying that only two among the parameters ($t_{ix}, t_{iy}, t_{i+1x}, t_{i+1y}$) may be supplied while normalization constraints the other two. The remaining two conditions are supplied as the magnitudes with vectors $\mathbf{t}_i$ and $\mathbf{t}_{i+1}$, that is, c_i and c_{i+1}. The tangent magnitudes, therefore, play a vital role in shaping a Hermite-Ferguson curve while preserving the location of the end points and the direction of tangent vectors.

4.1.1   Composite Ferguson Curves

Following on the notion that the overall curve is a piecewise fit of individual cubic Ferguson segments, consider any two neighboring segments of a composite curve, $\mathbf{r}^{(1)}(u_1)$ and $\mathbf{r}^{(2)}(u_2)$ with $0 \leq u_1, u_2 \leq 1$. Here, the superscripts (1) and (2) refer to the curve segments and u_1 and u_2 are the

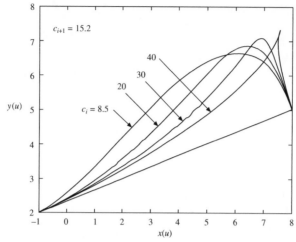

Figure 4.3   Change of shape in a Ferguson segment for different end tangent magnitudes

respective independent parameters, that is, a point on the composite curve belongs to $\mathbf{r}^{(1)}(u_1)$ for $0 \le u_1 \le 1$.

The composite curve is said to be *position* continuous or C^0 continuous if the end point of the first segment $\mathbf{r}^{(1)}$ and the start point of the second segment $\mathbf{r}^{(2)}$ are coincident, that is

$$\mathbf{r}^{(1)}(u = 1) = \mathbf{r}^{(2)}(u_2 = 0) \tag{4.10}$$

For slope or C^1 continuity at the junction point, the respective tangents of the two segments should have the same direction (not necessarily the same magnitude) for which

$$\alpha_1 \frac{d}{du}\mathbf{r}^{(1)}(1) = \alpha_2 \frac{d}{du}\mathbf{r}^{(2)}(0) = \mathbf{t} \tag{4.11}$$

where α_1 and α_2 are the normalizing scalars and $\mathbf{t}$ is the unit tangent vector at the junction point. In Figure 4.4, $\mathbf{r}^{(1)}(u_1)$ is a Ferguson segment with end points (x_i, y_i, z_i), $(x_{i+1}, y_{i+1}, z_{i+1})$ and end slopes

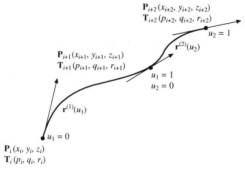

Figure 4.4 A composite Ferguson curve

(p_i, q_i, r_i) and $(p_{i+1}, q_{i+1}, r_{i+1})$. Also, $\mathbf{r}^{(2)}(u_2)$ is modeled with end points $(x_{i+1}, y_{i+1}, z_{i+1})$, $(x_{i+2}, y_{i+2}, z_{i+2})$ and end slopes $(p_{i+1}, q_{i+1}, r_{i+1})$ and $(p_{i+2}, q_{i+2}, r_{i+2})$. From Eq. (4.7) for continuity of the two curves at the common joint:

$$C^0 \text{ continuity: } \mathbf{r}^{(1)}(1) = \mathbf{r}^{(2)}(0) = (x_{i+1}, y_{i+1}, z_{i+1})$$

$$C^1 \text{ continuity: } \frac{d}{du}\mathbf{r}^{(1)} = \frac{d}{du}\mathbf{r}^{(2)}(0) \; (p_{i+1}, q_{i+1}, r_{r+1})$$

Thus, from the way the Ferguson segments result, the composite Ferguson curve is always C^1 continuous at the junction points. C^1 continuity is guaranteed at other intermediate points on the curve as well since the segments are differentiable indefinitely, the segments being cubic in degree.

It is often difficult to conjecture the geometric interpretation of the slope $d\mathbf{r}(u)/du$ or (p_i, q_i, r_i) at the junction points in terms of their use as design parameters. A designer would desire to specify only data points in curve design and evade the specifications pertaining to the slope, curvature or higher order information. To avoid the slope specification from the user at intermediate data points, we can additionally impose curvature or C^2 continuity at the junction points for which it would require that

$$\kappa^{(1)}(1) = \kappa^{(2)}(0)$$

or

$$\frac{\frac{d}{du}\mathbf{r}^{(1)}(1) \times \frac{d^2}{du^2}\mathbf{r}^{(1)}(1)}{\left|\frac{d}{du}\mathbf{r}^{(1)}(1)\right|^3} = \frac{\frac{d}{du}\mathbf{r}^{(2)}(0) \times \frac{d^2}{du^2}\mathbf{r}^{(2)}(0)}{\left|\frac{d}{du}\mathbf{r}^{(2)}(0)\right|^3} \tag{4.12}$$

Eq. (4.12) on substituting the conditions for position and slope continuity from Eqs. (4.10) and (4.11) becomes

$$\mathbf{t} \times \frac{d^2}{du^2}\mathbf{r}^{(1)}(1) = \left(\frac{\alpha_2}{\alpha_1}\right)^2 \mathbf{t} \times \frac{d^2}{du^2}\mathbf{r}^{(2)}(0) \tag{4.13}$$

An equation that satisfies the condition above is

$$\frac{d^2}{du^2}\mathbf{r}^{(1)}(1) = (\alpha_2/\alpha_1)^2 \frac{d^2}{du^2}\mathbf{r}^{(2)}(0) + \mu \frac{d}{du}\mathbf{r}^{(2)}(0) \tag{4.14}$$

where μ is some arbitrary scalar. Note that $\frac{d^2}{du^2}\mathbf{r}^{(1)}(0)$ and $\frac{d}{du}\mathbf{r}^{(2)}(0)$ have the same direction for which theis class product is zero. For Ferguson's composite curve, $\frac{d}{du}\mathbf{r}^{(1)}(1) = \frac{d}{du}\mathbf{r}^{(2)}(0)$ for which $\alpha_1 = \alpha_2$ from Eq. (4.11). Assuming $\mu = 0$, Eq. (4.14) becomes

$$\frac{d^2}{du^2}\mathbf{r}^{(1)}(1) = \frac{d^2}{du^2}\mathbf{r}^{(2)}(0) \tag{4.15}$$

The second derivative of $\mathbf{r}(u)$ from Eq. (4.9) is given by

$$\frac{d^2}{du^2}\mathbf{r}(u) = \mathbf{P}_i(-6 + 12u) + \mathbf{P}_{i+1}(6 - 12u) + \mathbf{T}_i(-4 + 6u) + \mathbf{T}_{i+1}(-2 + 6u) \tag{4.16}$$

From Eqs. (4.15) and (4.16), we have

$$6\mathbf{P}_i - 6\mathbf{P}_{i+1} + 2\mathbf{T}_i + 4\mathbf{T}_{i+1} = -6\mathbf{P}_{i+1} + 6\mathbf{P}_{i+2} - 4\mathbf{T}_{i+1} - 2\mathbf{T}_{i+2}$$

or

$$\mathbf{T}_i + 4\mathbf{T}_{i+1} + \mathbf{T}_{i+2} = 3\mathbf{P}_{i+2} - 3\mathbf{P}_i, \; i = 0, 1, \ldots, n - 2 \tag{4.17}$$

where $n + 1$ is the number of data points. Eq. (4.17) suggests that for a cubic composite curve to be curvature continuous throughout, the intermediate slopes $\mathbf{T}_i$ are related and thus need not be specified. For given $n + 1$ data points $\mathbf{P}_i$, $i = 0, 1, \ldots, n$, Eq. (4.17) provides $n - 1$ relations, one for each intermediate junction point, in $n + 1$ unknown slopes $\mathbf{T}_i$, $i = 0, 1, \ldots, n$. Thus, two additional conditions are required can be the slopes $\mathbf{T}_0$ and $\mathbf{T}_n$ specified at the two ends of the composite Ferguson's curve. Once all the slopes are determined, a C^2 continuous Ferguson's composite curve is obtained.

Example 4.3. For data points $A(0, 0)$, $B(1, 2)$, $C(3, 2)$ and $D(6, -1)$, determine C^1 and C^2 continuous Ferguson curves. For the first case, use slopes as $45°$, $30°$, $0°$ and $-45°$ at the data points. For a C^2 continuous curve, use end slopes as $45°$ and $-45°$, respectively. Comment on the variation in the shapes of the composite curves if (a) data point $C(3, 2)$ is relocated to $(1.5, 4)$ and (b) slope at point $(0, 0)$ is modified to $90°$.

Using chain rule, $\dfrac{dy}{dx} = \dfrac{(dy/du)}{(dx/du)} = \tan \theta$, where θ is the slope at a data point. For $\dfrac{dx}{du} = 1$, at all data points, the following table summarizes the end tangent computation.

i	Data point	θ	$\dfrac{dy}{du} = \dfrac{dx}{du} \tan \theta$	$\mathbf{T}_i$
0	$A(0, 0)$	$45°$	1	$(1, 1)$
1	$B(1, 2)$	$30°$	$\dfrac{1}{\sqrt{3}}$	$\left(1, \dfrac{1}{\sqrt{3}}\right)$
2	$C(3, 2)$	$0°$	0	$(1, 0)$
3	$D(6, -1)$	$-45°$	-1	$(1, -1)$

Recursive use of Eq. (4.7) would result in the Ferguson segments between two successive data points. For instance, the segment between A and B is

$$\mathbf{r}_1(u_1) = [u_1^3 \; u_1^2 \; u_1 \; 1] \begin{bmatrix} 2 & -2 & 1 & 1 \\ -3 & 3 & -2 & -1 \\ 0 & 0 & 1 & 0 \\ 1 & 0 & 0 & 0 \end{bmatrix} \begin{bmatrix} 0 & 0 \\ 1 & 2 \\ 1 & 1 \\ 1 & 1/\sqrt{3} \end{bmatrix}, \quad 0 \le u_1 \le 1$$

or $\mathbf{r}_1(u_1) = [u_1, -2.42u_1^3 + 3.42u_1^2 + u_1]$

Similarly, the curve segment BC is

$$\mathbf{r}_2(u_2) = [u_2^3 \; u_2^2 \; u_2 \; 1] \begin{bmatrix} 2 & -2 & 1 & 1 \\ -3 & 3 & -2 & -1 \\ 0 & 0 & 1 & 0 \\ 1 & 0 & 0 & 0 \end{bmatrix} \begin{bmatrix} 1 & 2 \\ 3 & 2 \\ 1 & 1/\sqrt{3} \\ 1 & 0 \end{bmatrix}, \quad 0 \le u_2 \le 1$$

or $\mathbf{r}_2(u_2) = [-2u_2^3 + 3u_2^2 + u_2 + 1, 0.58(u_2^3 - 2u_2^2 + u_2) + 2]$

Finally, segment CD is

$$\mathbf{r}_3(u_3) = \begin{bmatrix} u_3^3 & u_3^2 & u_3 & 1 \end{bmatrix} \begin{bmatrix} 2 & -2 & 1 & 1 \\ -3 & 3 & -2 & -1 \\ 0 & 0 & 1 & 0 \\ 1 & 0 & 0 & 0 \end{bmatrix} \begin{bmatrix} 3 & 2 \\ 6 & -1 \\ 1 & 0 \\ 1 & -1 \end{bmatrix}, \quad 0 \le u_3 \le 1$$

or
$$\mathbf{r}_3(u_3) = [-4u_3^3 + 6u_3^2 + u_3 + 3,\ 5u_3^3 - 8u_3^2 + 2]$$

Point $C(3, 2)$ appears as a junction point between segments $\mathbf{r}_2(u)$ and $\mathbf{r}_3(u)$ and both segments would get modified after C is relocated to (1.5, 4). The new segments would be

$$\mathbf{r}_2^{\text{new}}(u_2) = \begin{bmatrix} u_2^3 & u_2^2 & u_2 & 1 \end{bmatrix} \begin{bmatrix} 2 & -2 & 1 & 1 \\ -3 & 3 & -2 & -1 \\ 0 & 0 & 1 & 0 \\ 1 & 0 & 0 & 0 \end{bmatrix} \begin{bmatrix} 1 & 2 \\ 1.5 & 4 \\ 1 & 1/\sqrt{3} \\ 1 & 0 \end{bmatrix}, 0 \le u_2 \le 1$$

or
$$\mathbf{r}_2^{\text{new}}(u_2) = [u_2^3 - 1.5u_2^2 + u_2 + 1,\ -3.42u_2^3 + 4.85u_2^2 + 0.58u_2 + 2]$$

and
$$\mathbf{r}_3^{\text{new}}(u_3) = \begin{bmatrix} u_3^3 & u_3^2 & u_3 & 1 \end{bmatrix} \begin{bmatrix} 2 & -2 & 1 & 1 \\ -3 & 3 & -2 & -1 \\ 0 & 0 & 1 & 0 \\ 1 & 0 & 0 & 0 \end{bmatrix} \begin{bmatrix} 1.5 & 4 \\ 6 & -1 \\ 1 & 0 \\ 1 & -1 \end{bmatrix}, 0 \le u_3 \le 1$$

or
$$\mathbf{r}_3^{\text{new}}(u_3) = [-7u_3^3 + 10.5u_3^2 + u_3 + 1.5,\ 9u_3^3 - 14u_3^2 + 4]$$

The original and modified composite C^1 continuous Ferguson curves are shown in Figure 4.5(a). Note that the composite curve gets modified only partially.

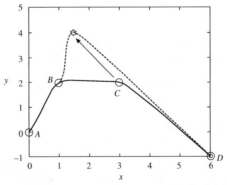

Figure 4.5 **(a) Original (solid) and modified (dashed) C^1 continuous Ferguson curves in Example 4.3 showing local control in shape change**

To obtain C^2 continuous curve, Eq. (4.17) may be employed to get the intermediate slopes. Using the end slopes as $\mathbf{T}_0 = (1, 1)$ and $\mathbf{T}_3 = (1, -1)$, the system in Eq. (4.17) results in

$$(1, 1) + 4\mathbf{T}_1 + \mathbf{T}_2 = 3(3, 2) - 3(0, 0)$$

$$\mathbf{T}_1 + 4\mathbf{T}_2 + (1, -1) = 3(6, -1) - 3(1, 2)$$

$$4\mathbf{T}_1 + \mathbf{T}_2 = 3(3, 2) - 3(0, 0) - (1, 1) = (8, 5)$$

or
$$\mathbf{T}_1 + 4\mathbf{T}_2 = 3(6, -1) - 3(1, 2) - (1, -1) = (14, -8)$$

We can solve the above system individually for x and y components of the slopes to get $\mathbf{T}_1 = (1.2, 1.87)$ and $\mathbf{T}_2 = (3.2, -2.47)$. The process of obtaining the individual segments is then identical to that described for C^1 continuous Ferguson curves. The polynomials for the curve segments are

$$\mathbf{r}_1(u_1) = [0.2u_1^3 - 0.4u_1^2 + u_1, -1.13u_1^3 + 2.13u_1^2 + u_1]$$

$$\mathbf{r}_2(u_2) = [0.4u_2^3 + 0.4u_2^2 + 1.2u_2 + 1, -0.6u_2^3 - 1.27u_2^2 + 1.87u_2 + 2]$$

$$\mathbf{r}_3(u_3) = [-1.8u_3^3 + 1.6u_3^2 + 3.2u_3 + 3, 2.53u_3^3 - 3.06u_3^2 - 2.47u_3 + 2]$$

For data point $C(3, 2)$ to be relocated to $(1.5, 4)$, the intermediate tangents must be re-computed. Thus,

$$4\mathbf{T}_1 + \mathbf{T}_2 = 3(1.5, 4) - 3(0, 0) - (1, 1) = (3.5, 11)$$

$$\mathbf{T}_1 + 4\mathbf{T}_2 = 3(6, -1) - 3(1, 2) - (1, -1) = (14, -8)$$

the solution for which is $\mathbf{T}_1 = (0.0, 3.47)$ and $\mathbf{T}_2 = (3.5, -2.87)$. Using these slopes, Eq. (4.17) can be employed to generate the new Ferguson segments as

$$\mathbf{r}_1^{new}(u_1) = [-3.5u_1^3 + 3.5u_1^2 + u_1, 0.47u_1^3 + 0.53u_1^2 + u_1]$$

$$\mathbf{r}_2^{new}(u_2) = [2.5u_2^3 - 2.0u_2^2 + 1, -3.4u_2^3 + 1.93u_2^2 + 3.47u_2 + 2]$$

$$\mathbf{r}_3^{new}(u_3) = [-4.5u_3^3 + 5.5u_3^2 + 3.5u_3 + 1.5, 6.13u_3^3 - 8.26u_3^2 - 2.87u_3 + 4]$$

Figure 4.5 (b) shows C^2 continuous composite curves before (solid lines) and after (dashed lines) point $C(3, 2)$ is moved to a new location. The recorded change in curve shape is global.

When a data point (or a slope vector) in a C^1 continuous composite Ferguson curve is altered, a maximum of two segments having that data point (or slope) at the junction get reshaped. In other words, a C^1 continuous composite Ferguson curve possesses local shape control properties. For a C^2 continuous curve, however, altering a data point requires re-computing the slopes using Eq. (4.17). There is an overall change in the composite curve.

4.1.2 Curve Trimming and Re-parameterization

In many design situations, a user may require to trim a curve that has been sketched. Commonly, trimming is performed at the intersection of two curves. Figure 4.6 shows a segment to be trimmed with the geometric matrix $\mathbf{G}$ as $[\mathbf{P}_i \ \mathbf{P}_{i+1} \ \mathbf{T}_i \ \mathbf{T}_{i+1}]^T$. Let trimming be performed at $u = u_i$ $(0 < u_i < 1)$ and $u = u_j$ $(0 < u_j < 1)$. The segments $0 \le u < u_i$ and $u_j < u \le 1$ are to be removed. The resultant curve BC lies in the parametric interval $u_i \le u \le u_j$. Sometimes, it is useful to express this trimmed curve using a new parameter interval $0 \le v \le 1$.

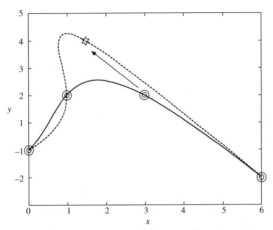

Figure 4.5 **(b) Original (solid) and modified (dashed) C^2 continuous Ferguson curves for Example 4.3. The shape change is global for a shift in data point**

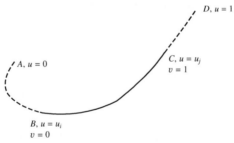

Figure 4.6 Trimming of a curve

To maintain the same parametric (cubic) form, a linear relation between u and v is sought. An advantage is that the directions of the tangent vectors at the ends are preserved. Thus,

$$v = \alpha u + \beta, \quad v_i = \alpha u_i + \beta \quad \text{and} \quad v_j = \alpha u_j + \beta \tag{4.18}$$

Here, v_i and v_j are values for v at the two ends of the trimmed segment. For re-parameterization, $v_i = 0$ and $v_j = 1$ and hence from Eq. (4.18)

$$\alpha = \frac{1}{(u_j - u_i)} \quad \text{and} \quad \beta = \frac{-u_i}{u_j - u_i} \tag{4.19}$$

Let $\mathbf{rv}(v)$ represent the retained segment in Figure 4.6 noting that $\mathbf{rv}(v)$ in $0 \le v \le 1$ is identical to $\mathbf{r}(u)$ in $u_i \le u \le u_j$. Then

$$\mathbf{rv}(0) = \mathbf{r}(u_i) \quad \text{and} \quad \mathbf{rv}(1) = \mathbf{r}(u_j)$$

Also $\quad \dfrac{d\mathbf{r}\mathbf{v}(v)}{dv} = \dfrac{d\mathbf{r}\mathbf{v}(v)}{du}\dfrac{du}{dv} \Rightarrow \mathbf{r}\mathbf{v}^v(v) = (u_j - u_i)\dfrac{d\mathbf{r}(u)}{du} = (u_j - u_i)\mathbf{r}^u(u)$

The tangents of $\mathbf{r}\mathbf{v}(v)$ at $u = u_i$ and $u = u_j$ are $\dfrac{d\mathbf{r}\mathbf{v}(0)}{dv} = (u_j - u_i)\dfrac{d\mathbf{r}(u_i)}{du}$ and $\dfrac{d\mathbf{r}\mathbf{v}(1)}{dv} = (u_j - u_i)\dfrac{d\mathbf{r}(u_j)}{du}$. The geometric matrix $\mathbf{G}\mathbf{v}$ for the trimmed segment becomes

$$\mathbf{G}\mathbf{v} = [\mathbf{r}\mathbf{v}(0) \quad \mathbf{r}\mathbf{v}(1) \quad \mathbf{r}\mathbf{v}^v(0) \quad \mathbf{r}\mathbf{v}^v(1)]^T$$

$$= [\mathbf{r}(u_i) \quad \mathbf{r}(u_j) \quad (u_j - u_i)\mathbf{r}^u(u_i) \quad (u_j - u_i)\mathbf{r}^u(u_j)]^T \tag{4.20}$$

and the equation for the same is

$$\mathbf{r}\mathbf{v}(v) = \mathbf{V}\mathbf{M}\mathbf{G}\mathbf{v} \tag{4.21}$$

where $\mathbf{V} = [v^3 \quad v^2 \quad v \quad 1]$. A similar approach can be used to re-parameterize two Ferguson segments joined together with C^1 continuity.

4.1.3   Blending of Curve Segments

Curve blending is quite common in design and can be easily accomplished between two Ferguson segments. Consider two curve segments AB and CD ($\mathbf{r}^{(1)}(u_1)$ and $\mathbf{r}^{(3)}(u_3)$) shown in Figure 4.7. The gap between B and C is filled by a blending curve $\mathbf{r}^{(2)}(u_2)$ which can be determined as follows:

Let $\mathbf{G}_1$ and $\mathbf{G}_3$ be the geometric matrices of $\mathbf{r}^{(1)}(u_1)$ and $\mathbf{r}^{(3)}(u_3)$, respectively. From Figure 4.7

$$\mathbf{G}_1 = [\mathbf{P}_i \quad \mathbf{P}_{i+1} \quad \mathbf{T}_i \quad \mathbf{T}_{i+1}]^T$$

$$\mathbf{G}_3 = [\mathbf{P}_{i+2} \quad \mathbf{P}_{i+3} \quad \mathbf{T}_{i+2} \quad \mathbf{T}_{i+3}]^T$$

The geometric matrix $\mathbf{G}_2$ for the blending curve between B and C can be written as

$$\mathbf{G}_2 = [\mathbf{P}_{i+1} \quad \mathbf{P}_{i+2} \quad \alpha\mathbf{T}_{i+1} \quad \beta\mathbf{T}_{i+2}]^T \tag{4.22}$$

where parameters α and β can be suitably varied to blend a large variety of curves maintaining C^1 continuity at B and C.

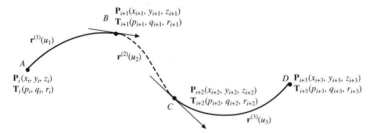

Figure 4.7   Blending of two Ferguson curve segments

Example 4.4. A planar Hermite-Ferguson curve (1) starts at A $(0, 0)$ and ends at B $(4, 2)$. The tangent vectors are given as $\mathbf{T}_i = (7, 7)$ and $\mathbf{T}_{i+1} = (5, -8)$. Another curve (3) starts at C $(8, 4)$ and ends at D

(12, 6) and the end tangents are given by $\mathbf{T}_{i+2} = (-8, 5)$ and $\mathbf{T}_{i+3} = (7, -7)$. Blend a curve between B and C to ensure C^1 continuity.

From the given data, the geometric matrices $\mathbf{G}_1$ and $\mathbf{G}_3$ for curves (1) and (3) can be formed and $\mathbf{G}_2$ for the blending curve can be determined. Thus

$$\mathbf{G}_1 = \begin{bmatrix} 0 & 0 \\ 4 & 2 \\ 7 & 7 \\ 5 & -8 \end{bmatrix}, \quad \mathbf{G}_3 = \begin{bmatrix} 8 & 4 \\ 12 & 6 \\ -8 & 5 \\ 7 & -7 \end{bmatrix}$$

Therefore,
$$\mathbf{G}_2 = \begin{bmatrix} 4 & 2 \\ 8 & 4 \\ \alpha(5) & \alpha(-8) \\ \beta(-8) & \beta(5) \end{bmatrix}$$

Coefficients α and β can be used to attenuate the magnitudes of the tangents while maintaining their directions. Figure 4.8 shows two candidate blending curves for ($\alpha = 1$, $\beta = 1$) and ($\alpha = 1$, $\beta = 4$).

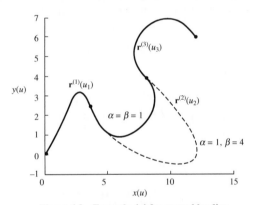

Figure 4.8   Example 4.4 for curve blending

4.1.4   Lines and Conics with Ferguson Segments

The end points and tangents can be chosen such that one can generate curves of degree less than 3 with Hermite cubic curves. Recall from Eq. (4.7) that

$$\mathbf{r}(u) = [u^3 \quad u^2 \quad u \quad 1] \begin{bmatrix} 2 & -2 & 1 & 1 \\ -3 & 3 & -2 & -1 \\ 0 & 0 & 1 & 0 \\ 1 & 0 & 0 & 0 \end{bmatrix} \begin{bmatrix} \mathbf{P}_i \\ \mathbf{P}_{i+1} \\ \mathbf{T}_i \\ \mathbf{T}_{i+1} \end{bmatrix} = \mathbf{UMG}$$

For $\mathbf{T}_i = \mathbf{T}_{i+1} = \mathbf{P}_{i+1} - \mathbf{P}_i$, we get

$$\mathbf{r}(u) = (1 - 3u^2 + 2u^3)\ \mathbf{P}_i + (3u^2 - 2u^3)\ \mathbf{P}_{i+1} + (u - 2u^2 + u^3)\mathbf{T}_i + (-u^2 + u^3)\mathbf{T}_{i+1}$$

$$= (1 - 3u^2 + 2u^3)\mathbf{P}_i + (3u^2 - 2u^3)\ \mathbf{P}_{i+1} + (u - 3u^2 + 2u^3)\ (\mathbf{P}_{i+1} - \mathbf{P}_i)$$

$$= (1 - u)\mathbf{P}_i + (u)\mathbf{P}_{i+1}$$

which is linear in u and thus is a line segment $\mathbf{P}_i\mathbf{P}_{i+1}$ for u in [0, 1]. More generally, if $\mathbf{T}_i = c_i(\mathbf{P}_{i+1} - \mathbf{P}_i)$ and $\mathbf{T}_{i+1} = c_{i+1}(\mathbf{P}_{i+1} - \mathbf{P}_i)$, then

$$\mathbf{r}(u) = \mathbf{P}_i + (\mathbf{P}_{i+1} - \mathbf{P}_i)\{3u^2 - 2u^3 + c_i(u - 2u^2 + u^3) + c_{i+1}(-u^2 + u^3)\}$$

Observe that for $c_i = c_{i+1} = 1$, we get $\mathbf{r}(u) = (1 - u)\mathbf{P}_i + (u)\mathbf{P}_{i+1}$. For $c_i = c_{i+1} = c$, we have

$$\mathbf{r}(u) = \mathbf{P}_i + (\mathbf{P}_{i+1} - \mathbf{P}_i)\ \{3u^2 - 2u^3 + c(u - 3u^2 + 2u^3)\}$$

$$= \mathbf{P}_i + (\mathbf{P}_{i+1} - \mathbf{P}_i)\ \{(1 - c)\ (3u^2 - 2u^3) + cu\}$$

If $c = 0$, then $\mathbf{r}(u) = \mathbf{P}_i + (\mathbf{P}_{i+1} - \mathbf{P}_i)\ (3u^2 - 2u^3)$ which is also a line with $v = 3u^2 - 2u^3$. However, while the point on $\mathbf{r}(u) = (1 - u)\ \mathbf{P}_i + (u)\mathbf{P}_{i+1}$ moves with a constant speed with respect to u, the point on $\mathbf{r}(u) = \mathbf{P}_i + (\mathbf{P}_{i+1} - \mathbf{P}_i)\ (3u^2 - 2u^3)$ moves with a variable speed. Note that for the latter

$$\mathbf{r}^u(u) = (-6u^2 + 6u)\ (\mathbf{P}_{i+1} - \mathbf{P}_i) \quad \text{and} \quad \mathbf{r}^{uu}(u) = (-12u + 6)\ (\mathbf{P}_{i+1} - \mathbf{P}_i)$$

Thus, the tangents at the start and end points of this line are zero, and the point on the line accelerates till $u = \frac{1}{2}$, when $\mathbf{r}(u = \frac{1}{2}) = \frac{1}{2}\ (\mathbf{P}_{i+1} + \mathbf{P}_i)$, and then decelerates. For $c = -1$, it can be determined that

$$\mathbf{r}(u) = \mathbf{P}_i + (\mathbf{P}_{i+1} - \mathbf{P}_i)(6u^2 - 4u^3 - u)$$

Special Cases of Ferguson Curves

1. $\mathbf{r}^u(u) = 0$ for all u, it can be shown that the curve reduces to a point and $\mathbf{r}(u) = \mathbf{P}_i$.

2. $\mathbf{r}^u \times \mathbf{r}^{uu} \neq 0$ and $\begin{vmatrix} r_x^u & r_x^{uu} & r_x^{uuu} \\ r_y^u & r_y^{uu} & r_y^{uuu} \\ r_z^u & r_z^{uu} & r_z^{uuu} \end{vmatrix} = 0$, then $\mathbf{r}(u)$ is a planar curve.

3. If $\mathbf{r}^u \neq 0$, and $\mathbf{r}^u \times \mathbf{r}^{uu} = 0$, then $\mathbf{r}(u)$ is a straight line.

4. If $\mathbf{r}^u \times \mathbf{r}^{uu} \neq 0$, and $\begin{vmatrix} r_x^u & r_x^{uu} & r_x^{uuu} \\ r_y^u & r_y^{uu} & r_y^{uuu} \\ r_z^u & r_z^{uu} & r_z^{uuu} \end{vmatrix} \neq 0$, , $\mathbf{r}(u)$ is a three-dimensional curve.

To generate conics using Ferguson segments, consider

$$\mathbf{r}(u) = (1 - 3u^2 + 2u^3)\ \mathbf{P}_i + (3u^2 - 2u^3)\ \mathbf{P}_{i+1} + (u - 2u^2 + u^3)\mathbf{T}_i + (-u^2 + u^3)\mathbf{T}_{i+1}$$

which, for $u = \frac{1}{2}$ gives

$$\mathbf{r}(\tfrac{1}{2}) = \frac{1}{2}(\mathbf{P}_i + \mathbf{P}_{i+1}) + \frac{1}{8}(\mathbf{T}_i - \mathbf{T}_{i+1})$$

To determine the geometric matrix $\mathbf{G}$ for a conic section, consider the following construction in Figure 4.9. Let the end tangents at points A ($\mathbf{P}_i$) and B ($\mathbf{P}_{i+1}$) on a conic section meet at point $C(\mathbf{r}_2)$ and let D be the mid point of AB, that is, $\frac{1}{2}$ ($\mathbf{P}_i + \mathbf{P}_{i+1}$). Let CD intersect the curve at P. Draw DQ parallel to AC and PQ parallel to BC.

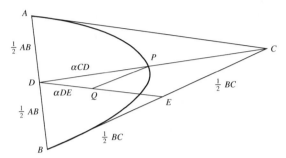

Figure 4.9   Construction for a conic section

In similar triangles EBD and CBA, since D is the mid point of AB, the ratio $BE/BD = BC/BA = \frac{1}{2}$ which implies $BE = \frac{1}{2} BC = EC$. In similar triangles PDQ and CDE,

$$\frac{DP}{DC} = \frac{DQ}{DE} = \frac{PQ}{CE} = \alpha, \text{ thus } \mathbf{DQ} = \alpha \mathbf{DE} = \frac{1}{2}\alpha \mathbf{AC}, \quad \mathbf{QP} = \alpha \mathbf{EC} = -\frac{1}{2}\alpha \mathbf{CB}$$

Now, $\mathbf{DP} = \mathbf{DQ} + \mathbf{QP} = \frac{1}{2}\alpha\,(\mathbf{AC} - \mathbf{CB})$

and $\mathbf{r}_p = \mathbf{r}_D + \mathbf{DP} = \frac{1}{2}(\mathbf{P}_i + \mathbf{P}_{i+1}) + \frac{1}{2}\alpha[(\mathbf{r}_2 - \mathbf{P}_i) - (\mathbf{P}_{i+1} - \mathbf{r}_2)]$

Since $\mathbf{r}_p = \mathbf{r}(u = 1/2) = \frac{1}{2}(\mathbf{P}_i + \mathbf{P}_{i+1}) + \frac{1}{8}(\mathbf{T}_i - \mathbf{T}_{i+1})$

On comparison

$$\mathbf{T}_i = 4\alpha\,(\mathbf{r}_2 - \mathbf{P}_i), \quad \mathbf{T}_{i+1} = 4\alpha\,(\mathbf{P}_{i+1} - \mathbf{r}_2)$$

Therefore, the geometric matrix **G** for the Ferguson segment of a conic section (except the circle) is given by

$$\mathbf{G} = [\mathbf{P}_i \quad \mathbf{P}_{i+1} \quad 4\alpha\,(\mathbf{r}_2 - \mathbf{P}_i) \quad 4\alpha\,(\mathbf{P}_{i+1} - \mathbf{r}_2)]^T \tag{4.23}$$

With the above matrix: (a) if $\alpha < 0.5$, the curve is an elliptical segment, (b) if $\alpha = 0.5$, the curve is a parabolic segment while for (c) $0.5 < \alpha < 1$, the curve represents a hyperbolic segment.

Example 4.5. Design a conic with end points $\mathbf{P}_i = (4, -8)$ and $\mathbf{P}_{i+1} = (4, 8)$ when the end tangents meet at $\mathbf{r}_2 = (-4, 0)$.

For known α, we can compute the end tangents using Eq. (4.23) as $\mathbf{T}_i = 4\alpha\,(-8, 8)$ and $\mathbf{T}_{i+1} = 4\alpha\,(8, 8)$. The Ferguson's segment is

$$\mathbf{r}(u) = [u^3 \quad u^2 \quad u \quad 1]\begin{bmatrix} 2 & -2 & 1 & 1 \\ -3 & 3 & -2 & -1 \\ 0 & 0 & 1 & 0 \\ 1 & 0 & 0 & 0 \end{bmatrix}\begin{bmatrix} 4 & -8 \\ 4 & 8 \\ -32\alpha & 32\alpha \\ 32\alpha & 32\alpha \end{bmatrix}$$

or $\mathbf{r}(u) = (32\,\alpha u^2 - 32\alpha u + 4)\mathbf{i} + [(-32 + 64\alpha)u^3 + (48 - 96\alpha)\,u^2 + 32\alpha u - 8]\mathbf{j} = x(u)\mathbf{i} + y(u)\mathbf{j}$

Note that for $\alpha = \frac{1}{2}$, the coefficient of u^3 is zero and $\mathbf{r}(u)$ is of degree 2. In fact, we can show that

$$\mathbf{r}(u) = (16u^2 - 16u + 4)\mathbf{i} + (16u - 8)\mathbf{j} = x(u)\mathbf{i} + y(u)\mathbf{j}$$

For $y(u) = 16u - 8$ or $u = \dfrac{y + 8}{16}$, substituting into $x(u)$, the above results in

$$x = 16\left(\frac{y + 8}{16}\right)^2 - 16\left(\frac{y + 8}{16}\right) + 4$$

or $16x = y^2$ which is a parabola. Without proof, for $\alpha = 0.3$ and 0.8, respectively, the ellipse and hyperbola are also shown in Figure 4.10.

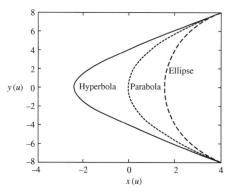

Figure 4.10   Conics with Ferguson's segments

4.1.5   Need for Other Geometric Models for the Curve

Relocating the end points or altering the magnitudes and/or direction of the end tangents results in shape change of a Ferguson segment. However, (a) it is not as intuitive to specify the tangent information, and the designer is more comfortable in specifying the data points. (b) For C^1 continuous composite Ferguson curves, modifying a data point or its slope would result in local shape change of the curve as discussed in Section 4.1.1. For C^2 continuous composite Ferguson curves, however, modifying any data point would result in re-computation of slopes (Eq. 4.17) resulting in an overall shape change of the composite curve. A user would therefore seek a design method that allows specifying only data points while maintaining local shape control properties for the entire curve.

Pierre Etienne Bézier , who worked with Renault, a French car manufacturer in 1970s, developed a method to mathematically describe the curves and surfaces of an automobile body using data points (henceforth referred to as *control points*). By shifting these points, the shape of the curve contained within some *local region* could be changed predictably. Bézier created the UNISURF CAD system for designing car bodies which utilized his curve theories. Paul de Faget de Casteljau's work with car manufacturer Citroen had similar results earlier than Bézier (1960s). Both works on Bézier curves are based on *Bernstein polynomials* developed much earlier by the Russian mathematician Sergei Natanovich Bernstein in 1912 as his research on approximation theory. The geometric construction of a parabola using the three tangent theorem is first discussed. The construction is generalized to generate a curve of any degree, attributed to de Casteljau. The resultant Bernstein polynomials have certain properties useful in predicting shape change in Bézier curves.

4.2 Three-Tangent Theorem

Stated without proof, consider three points *a*, *r* and *c* on a parabola. Let the tangents at *a* and *c* intersect at *d*. Also, let the tangent at *r* intersect the previous two tangents at *e* and *f*, respectively. Then

$$\frac{|\,ae\,|}{|\,ed\,|} = \frac{|\,er\,|}{|\,rf\,|} = \frac{|\,df\,|}{|\,fc\,|} \tag{4.24}$$

Based on this theorem, a parabola may be constructed to verify the aforementioned conditions. Let point *e* be chosen on *ad* such that

$$e = (1 - u)\,a + ud \text{ for some } u \in [0, 1] \tag{4.25}$$

This implies $\frac{|\,ae\,|}{|\,ed\,|} = \frac{u}{1-u}$, so that Eq. (4.24) is satisfied. Choose *f* and *r* on *dc* and *ef*, respectively, such that

$$f = (1 - u)d + uc$$

$$r = (1 - u)e + uf \tag{4.26}$$

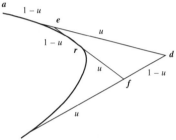

Substituting for *e* and *f* in Eq. (4.26) in terms of *a*, *c* and *d*, we have

$$r = (1 - u)\,\{(1 - u)\,a + ud\} + u\{(1 - u)\,d + uc\}$$

$$= (1 - u)^2\,a + 2u\,(1 - u)d + u^2 c \tag{4.27}$$

This is the equation of a parabola on which *r* lies for some parameter value of *u*. From Eq. (4.27),

Figure 4.11 Geometric construction of a parabola

at $u = 0$, $r = a$ and at $u = 1$, $r = c$ implying that *a* and *c* lie on the parabola as well. To find the tangents at *a* and *c*, Eq. (4.27) may be differentiated with respect to *u* as

$$dr/du = -2(1 - u)a + (2 - 4u)d + 2uc \tag{4.28}$$

using which dr/du at $u = 0$ is $2(d - a)$ and that at $u = 1$ is $2(c - d)$. Thus, *ad* and *dc* are tangents to the parabola at *a* and *c*, respectively. Rearranging Eq. (4.28) yields

$$dr/du = -2\{(1 - u)a + ud\} + 2\{(1 - u)d + uc\}$$

$$= -2e + 2f \tag{4.29}$$

This implies that *ef* is a tangent to the parabola at *r* for some value of *u*. Thus, the three tangent theorem for a parabola is verified and in the process, a procedure for constructing a parabola with three given points (*a*, *d* and *c*) is evolved. The construction which is known as the *de Casteljau's algorithm* involves two levels of repeated linear interpolation given by Eqs. (4.25) and (4.26).

4.2.1 Generalized de Casteljau's Algorithm

The above algorithm can be generalized for use with $n + 1$ data points to generate a curve of degree *n*. Given data points $b_0, b_1, \ldots, b_n$, compute b_i^j such that

$$\mathbf{b}_i^j = (1 - u)\mathbf{b}_i^{j-1} + u\mathbf{b}_{i+1}^{j-1}, \quad j = 1, \ldots, n; \quad i = 0, \ldots, n - j; \quad u \in [0, 1] \tag{4.30}$$

with $b_i^0 = b_i$. Here b_i^j, called *de Casteljau points*, represent intermediate points like e, f and r in Figure 4.11, for the nth degree curve. For instance, b_3^2 represents the fourth point in the second level of linear interpolation. After linear interpolation is exhausted, the final point b_0^n lies on the nth degree curve for some parameter u limited in the range [0, 1]. The line segments b_0b_1, b_1b_2, . . . , $b_{n-1}b_n$, called *legs,* when joined in this order form a *polyline* mostly referred to as the *control polyline*. The working of the algorithm is illustrated for $n = 2$ and 3, respectively, and the schematic of geometric construction is given in Figure 4.12 (a) and (b). For $n = 2$

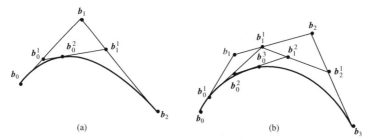

Figure 4.12 **Generalized de Casteljau's algorithm for: (a)** $n = 2$ **and (b)** $n = 3$.
All line segments are divided in the ratio u**: (1 − u)**

$j = 1$, $i = 0$; $b_0^1 = (1 - u)b_0^0 + ub_1^0 = (1 - u)b_0 + ub_1$

$j = 1$, $i = 1$: $b_1^1 = (1 - u)b_1^0 + ub_2^0 = (1 - u)b_1 + ub_2$

$j = 2$, $i = 0$: $b_0^2 = (1 - u)b_0^1 + ub_1^1$

$\qquad\qquad\qquad = (1 - u)\{(1 - u)b_0 + ub_1\} + u\{(1 - u)b_1 + ub_2\}$

$\qquad\qquad\qquad = (1 - u)^2 b_0 + 2u(1 - u)b_1 + u^2 b_2$

$\qquad\qquad\qquad = {}^2C_0(1 - u)^2\, u^0 b_0 + {}^2C_1(1 - u)^1\, u^1 b_1 + {}^2C_2(1 - u)^0 u^2 b_2$ (4.31)

For $n = 3$

$j = 1$, $i = 0$: $b_0^1 = (1 - u)b_0^0 + ub_1^0 = (1 - u)b_0 + ub_1$

$j = 1$, $i = 1$: $b_1^1 = (1 - u)b_1^0 + ub_2^0 = (1 - u)b_1 + ub_2$

$j = 1$, $i = 2$: $b_2^1 = (1 - u)b_2^0 + ub_3^0 = (1 - u)b_2 + ub_3$

$j = 2$, $i = 0$: $b_0^2 = (1 - u)b_0^1 + ub_1^1 = (1 - u)^2\, b_0 + 2u(1 - u)b_1 + u^2 b_2$

$j = 2$, $i = 1$: $b_1^2 = (1 - u)b_1^1 + ub_2^1 = (1 - u)^2\, b_1 + 2u(1 - u)b_2 + u^2 b_3$

$j = 3$, $i = 0$: $b_0^3 = (1 - u)b_0^2 + ub_1^2 = (1 - u)^3\, b_0 + 3u(1 - u)^2\, b_1 + 3u^2\,(1 - u)b_2 + u^3 b_3$

$\qquad\qquad\qquad = {}^3C_0\,(1 - u)^3\, u^0 b_0 + {}^3C_1\,(1 - u)^2\, u^1 b_1 + {}^3C_2\,(1 - u)^1 u^2 b_2 + {}^3C_3\,(1 - u)^0\, u^3 b_3$

(4.32)

The computation of intermediate de Casteljau's points for degree n Bézier curve can be illustrated by the triangular scheme in Figure 4.13. de Casteljau arrived at this result in 1959 at Citroen (a French car company), where he was working on the shape design of curves. He, however, never published his procedure until the internal reports were discovered in 1975. From Eqs. (4.31) and (4.32), by inspection b_0^n can be written as

$$b_0^n = \sum_{i=0}^{n} {}^{n}C_i (1-u)^{n-i} u^i b_i = \sum_{i=0}^{n} B_i^n(u) b_i \qquad (4.33)$$

where $B_i^n(u) = {}^{n}C_i (1-u)^{n-i} u^i$ are termed as *Bernstein polynomials*. Eq. (4.33) defines Bézier curves (discussed in Section 4.4) that Bézier proposed independently using Bernstein polynomials. In de Casteljau's work, no assumption about the Bernstein type blending functions was made, yet repeated linear subdivisions of the control polylines resulted in the same conclusion as Bézier's work.

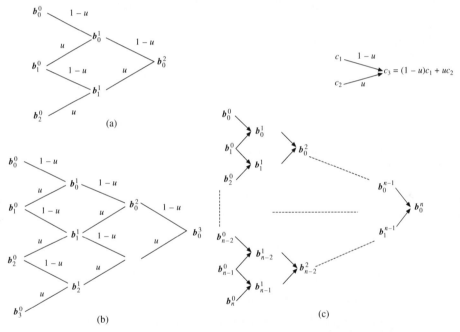

Figure 4.13 Triangular scheme for (a) $n = 2$, (b) $n = 3$ and (c) to compute intermediate de Casteljau points. Schema on top right is representative of the weighted linear combination of de Casteljau's points

4.2.2 Properties of Bernstein Polynomials
Bernstein polynomials play a significant role in predicting the segment's shape and by understanding their behavior, a great deal can be learnt about the Bézier curves.

(a) Non-negativity: For $0 \le u \le 1$, $B_i^n(u)$ are all non-negative. $\qquad (4.34)$

Since $0 \leq 1 - u \leq 1$ as well,

$$B_i^n(u) = {}^nC_i u^i (1-u)^{n-i} = \frac{n!}{i!(n-i)!} u^i (1-u)^{n-i} \geq 0$$

Non-negativity can be appreciated by the plots of $B_i^3(u)$, $i = 0, \ldots, 3$ and $B_i^4(u)$, $i = 0, \ldots, 4$ in Figure 4.14.

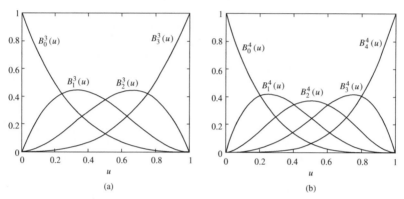

Figure 4.14   **Plot of Bernstein polynomials** $B_i^n(u)$ **for: (a)** $n = 3$ **and (b)** $n = 4$

(b) Partition of Unity and Barycentric Coordinates: Irrespective of the values of u, the Bernstein polynomials sum to unity, that is

$$\sum_{i=0}^{n} B_i^n(u) = 1 \tag{4.35}$$

From Binomial expansion

$$[(1-u) + u]^n = (1-u)^n + n(1-u)^{n-1}u + \ldots + \frac{n!}{r!(n-r)!}(1-u)^{n-r}u^r + \ldots + u^n$$

$$1 = {}^nC_0(1-u)^n u^0 + {}^nC_1(1-u)^{n-1}u + {}^nC_2(1-u)^{n-2}u^2 + {}^nC_n(1-u)^0 u^n$$

or $\qquad 1 = B_0^n(u) + B_1^n(u) + \ldots + B_n^n(u)$

Eqs. (4.33)-(4.35) suggest that the point b_0^n on the nth degree Bézier curve is the weighted linear combination of the $n+1$ data points $b_0, b_1, \ldots, b_n$ with respective weights as $B_i^n(u)$, where $B_i^n(u)$ are all non-negative and sum to unity. These weights are analogous to the point masses placed, respectively, at $b_0, b_1, \ldots, b_n$ whose center of mass is located at b_0^n. For this reason, such weights are known as barycentric coordinates, the term *barycenter* implying the center of gravity. Note that as the center of mass always lies within the convex hull of the locations of individual point masses, so does b_0^n lie in the convex hull of data points for values of u between 0 and 1. The convex hull of a set of points is the smallest convex set that contains all given points. Any line segment joining two arbitrary points in a convex set also lies in that set.

(c) Symmetry:

$$B_i^n(u) = B_{n-i}^n(1-u) \tag{4.36}$$

Though suggested in Figure 4.14, the property is shown as follows:

$$B_i^n(u) = {}^nC_i(1-u)^{n-i}u^i = \frac{n!}{i!(n-i)!}(1-u)^{n-i}u^i$$

$$= \frac{n!}{p!(n-p)!}(t)^p(1-t)^{n-p} \quad \text{for } t = (1-u), \; n-i = p$$

$$= {}^nC_p(1-t)^{n-p}t^p = B_p^n(t) = B_{n-i}^n(1-u)$$

(d) Recursion: The polynomials can be computed by the recursive relationship

$$B_i^n(u) = (1-u)B_i^{n-1}(u) + uB_{i-1}^{n-1}(u) \tag{4.37}$$

This is expected inherently from the de Casteljau's algorithm. We can show Eq. (4.37) to be true using the definition of Bernstein polynomials. Considering the right hand side

$$(1-u)B_i^{n-1}(u) + uB_{i-1}^{n-1}(u) = \frac{(n-1)!}{(i)!(n-1-i)!}(1-u)^{n-i}u^i + \frac{(n-1)!}{(i-1)!(n-i)!}(1-u)^{n-i}u^i$$

$$= \frac{(n-1)!}{(i-1)!(n-1-i)!}(1-u)^{n-i}u^i\left(\frac{1}{i} + \frac{1}{n-i}\right)$$

$$= \frac{(n-1)!}{(i-1)!(n-1-i)!}(1-u)^{n-i}u^i\left(\frac{n}{i(n-i)}\right)$$

$$= \frac{(n)!}{(i)!(n-i)!}(1-u)^{n-i}u^i = B_i^n(u)$$

(e) Derivative: The derivative with respect to u has a recursive form

$$\frac{dB_i^n(u)}{du} = n\left[B_{i-1}^{n-1}(u) - B_i^{n-1}(u)\right]$$

where

$$B_{-1}^{n-1}(u) = B_n^{n-1}(u) = 0 \tag{4.38}$$

By definition

$$B_i^n(u) = \frac{n!}{i!(n-i)!}(1-u)^{n-i}u^i$$

$$\frac{dB_i^n(u)}{du} = \frac{n!}{i!(n-i)!}[-(n-i)(1-u)^{n-1-i}u^i + i(1-u)^{n-i}u^{i-1}]$$

$$= -\frac{n!}{i!(n-1-i)!}(1-u)^{n-1-i}u^i + \frac{n!}{(i-1)!(n-i)!}(1-u)^{n-i}u^{i-1}$$

$$= n \left[\frac{(n-1)!}{(i-1)!(n-i)!} (1-u)^{n-i} u^{i-1} - \frac{(n-1)!}{i!(n-1-i)!} (1-u)^{n-1-i} u^i \right]$$

$$= n \left[{}^{n-1}C_{i-1}(1-u)^{n-i}u^{i-1} - {}^{n-1}C_i(1-u)^{n-1-i} u^i \right]$$

$$= n[B_{i-1}^{n-1}(u) - B_i^{n-1}(u)]$$

4.3 Barycentric Coordinates and Affine Transformation

In addition to constraining a Bézier curve to lie within the convex hull of the control polyline, Bernstein polynomials also allow to describe the curve in space independent of the coordinate frame. The shape of a given curve, surface, or solid should not depend on the choice of the coordinate system. In other words, the relative positions of points describing a curve, surface, or solid should remain unaltered during rotation or translation of the chosen axes. Consider for instance, two points $A(x_1, y_1)$ and $B(x_2, y_2)$ in a two-dimensional space defined by an origin O and a set of axes Ox-Oy with unit vectors $(\mathbf{i}, \mathbf{j})$. Let point C be defined as a linear combination of position vectors $\mathbf{OA}$ and $\mathbf{OB}$, that is, $\mathbf{OC} = \lambda \mathbf{OA} + \mu \mathbf{OB}$, where λ and μ are scalars. In terms of the ordered pair, C is then $(\lambda x_1 + \mu x_2, \lambda y_1, + \mu y_2)$.

The axes Ox-Oy are rotated through an angle θ about the z-axis to form a new set of axes Ox'-Oy' with unit vectors $(\mathbf{i}', \mathbf{j}')$. Let A and B be described by (x_1', y_1') and (x_2', y_2') under the new coordinate system for which the new definition of C is $C'(\lambda x_1' + \mu x_2', \lambda y_1' + \mu y_2')$. From Chapter 2, the rotation matrix transforming $(\mathbf{i}, \mathbf{j})$ to $(\mathbf{i}', \mathbf{j}')$ is given by

$$R_z = \begin{bmatrix} \cos\theta & -\sin\theta \\ \sin\theta & \cos\theta \end{bmatrix}$$

A and B ae placed at the same location in space. However, their new coordinates are now

$$A' \equiv \begin{bmatrix} x_1' \\ y_1' \end{bmatrix} = \begin{bmatrix} \cos\theta & -\sin\theta \\ \sin\theta & \cos\theta \end{bmatrix} \begin{bmatrix} x_1 \\ y_1 \end{bmatrix} \text{ and } B' \equiv \begin{bmatrix} x_2' \\ y_2' \end{bmatrix} = \begin{bmatrix} \cos\theta & -\sin\theta \\ \sin\theta & \cos\theta \end{bmatrix} \begin{bmatrix} x_2 \\ y_2 \end{bmatrix}$$

Now, let us define

$$C'^* \equiv (\lambda x_1' + \mu x_2', \lambda y_1' + \mu y') = \lambda \begin{bmatrix} x_1' \\ y_1' \end{bmatrix} + \mu \begin{bmatrix} x_2' \\ y_2' \end{bmatrix}$$

$$= \lambda \begin{bmatrix} \cos\theta & -\sin\theta \\ \sin\theta & \cos\theta \end{bmatrix} \begin{bmatrix} x_1 \\ y_1 \end{bmatrix} + \mu \begin{bmatrix} \cos\theta & -\sin\theta \\ \sin\theta & \cos\theta \end{bmatrix} \begin{bmatrix} x_2 \\ y_2 \end{bmatrix}$$

$$= \begin{bmatrix} \cos\theta & -\sin\theta \\ \sin\theta & \cos\theta \end{bmatrix} \begin{bmatrix} \lambda x_1 + \mu x_2 \\ \lambda y_1 + \mu y_2 \end{bmatrix} = \begin{bmatrix} \cos\theta & -\sin\theta \\ \sin\theta & \cos\theta \end{bmatrix} C = C'$$

This implies that the relative positions of A, B and C remain unaltered after rotation and thus rotation transformation is *affine*.

Next, consider a new set of axes $O'x'$-$O'y'$ formed by shifting the origin O to O' by a vector (p, q) as in Fig. 4.15(b). The set $O'x'$-$O'y'$ is parallel to Ox-Oy and thus the unit vectors stay the same, i.e., $(\mathbf{i}', \mathbf{j}') = (\mathbf{i}, \mathbf{j})$. The coordinates of points A and B in the transformed system is given by

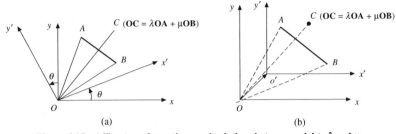

Figure 4.15 Affine transformations and relation between weights λ and μ

$$A' \equiv \begin{bmatrix} x'_1 \\ y'_1 \end{bmatrix} = \begin{bmatrix} x_1 - p \\ y_1 - q \end{bmatrix} \quad \text{and} \quad B' \equiv \begin{bmatrix} x'_2 \\ y'_2 \end{bmatrix} = \begin{bmatrix} x_2 - p \\ y_2 - q \end{bmatrix}$$

Let

$$C'^* = \lambda \begin{bmatrix} x'_1 \\ y'_1 \end{bmatrix} + \mu \begin{bmatrix} x'_2 \\ y'_2 \end{bmatrix}$$

$$= \lambda \begin{bmatrix} x_1 - p \\ y_1 - q \end{bmatrix} + \mu \begin{bmatrix} x_2 - p \\ y_2 - q \end{bmatrix} = \begin{bmatrix} \lambda x_1 + \mu x_2 - (\lambda + \mu)p \\ \lambda y_1 + \mu y_2 - (\lambda + \mu)q \end{bmatrix}$$

$$= \begin{bmatrix} \{\lambda x_1 + \mu x_2 - p\} - (\lambda + \mu - 1)p \\ \{\lambda y_1 + \mu y_2 - q\} - (\lambda + \mu - 1)q \end{bmatrix}$$

Like points A and B, if C was expressed in the new coordinate system without any change in its position due to the changed axes, then

$$C' = \begin{bmatrix} \lambda x_1 + \mu x_2 - p \\ \lambda y_1 + \mu y_2 - q \end{bmatrix}$$

Note that C' and C'^* are not identical, and there is a change in the position of C due to the shift in the origin O to O' by (p, q). To make this transformation also affine, the arbitrary scalars λ and μ need to be constrained as

$$\lambda + \mu - 1 = 0 \quad \text{or} \quad \lambda + \mu = 1 \tag{4.39}$$

Affine combination is therefore a *type* of linear combination wherein the respective weights sum to unity. It is needed to preserve the relative positions of points (describing a geometric entity) during transformation (change of coordinae axes), which is ensured by Bernstein polynomials in case of Bézier curves.

4.4 Bézier Segments

For $n + 1$ data points P_i, $i = 0, \dots, n$, a Bézier segment is defined as their weighted linear combination using Bernstein polynomials (Eq. (4.33))

$$\mathbf{r}(u) = \sum_{i=0}^{n} {}^nC_i (1 - u)^{n-i} u^i P_i = \sum_{i=0}^{n} B_i^n(u) P_i, \quad 0 \le u \le 1 \tag{4.40a}$$

For a composite curve, individual segments need to be of lower order, preferably cubic. Thus, a cubic Bézier segment in algebraic and matrix forms for data points $\mathbf{P}_0$, $\mathbf{P}_1$, $\mathbf{P}_2$ and $\mathbf{P}_3$ is given by

$$\mathbf{r}(u) = (1 - u)^3\mathbf{P}_0 + 3u(1 - u)^2\,\mathbf{P}_1 + 3u^2\,(1 - u)\mathbf{P}_2 + u^3\mathbf{P}_3$$

$$= (1 - 3u + 3u^2 - u^3)\,\mathbf{P}_0 + (3u - 6u^2 + 3u^3)\,\mathbf{P}_1 + (3u^2 - 3u^3)\mathbf{P}_2 + u^3\mathbf{P}_3$$

$$= u^3(-\mathbf{P}_0 + 3\mathbf{P}_1 - 3\mathbf{P}_2 + \mathbf{P}_3) + u^2(3\mathbf{P}_0 - 6\mathbf{P}_1 + 3\mathbf{P}_2) + u(-3\mathbf{P}_0 + 3\mathbf{P}_1) + \mathbf{P}_0$$

$$= [u^3 \ u^2 \ u \ 1] \begin{bmatrix} -1 & 3 & -3 & 1 \\ 3 & -6 & 3 & 0 \\ -3 & 3 & 0 & 0 \\ 1 & 0 & 0 & 0 \end{bmatrix} \begin{bmatrix} \mathbf{P}_0 \\ \mathbf{P}_1 \\ \mathbf{P}_2 \\ \mathbf{P}_3 \end{bmatrix} = \mathbf{U}\mathbf{M}_B\mathbf{G} \qquad (4.40b)$$

Eq. (4.40b) is similar to the Hermite cubic segment (Eq. (4.7)) with the parametric vector $\mathbf{U}$, the 4×4 Bézier matrix $\mathbf{M}_B$ and the geometric matrix $\mathbf{G}$ of size 4×3 which is an array of data points.

The geometric matrix $\mathbf{G}$ is to be defined by the user ($\mathbf{U}$ and $\mathbf{M}_g$ remaining the same for all cubic Bézier curves). Note that the curve does not pass through the points $\mathbf{P}_1$ and $\mathbf{P}_2$. To change the curve's shape, the user may relocate any of the control point $\mathbf{P}_0, \mathbf{P}_1, \mathbf{P}_2$ or $\mathbf{P}_3$. Recall that for Fergulon's segments, the user had to specify the end slopes for a particular shape which is difficult to speculate a peiori. A Bézier curve more or lees mimics the shape of the control polyline, which is easier to specify.

Example 4.6. A set of control points is given by $\mathbf{P}_0 = (4, 4)$, $\mathbf{P}_1 = (6, 8)$, $\mathbf{P}_2 = (8, 9)$ and $\mathbf{P}_3 = (10, 3)$. Compute the Bézier curve. Let the coordinate axes be moved to $(2, 2)$ and then rotated by $30°$ counter-clockwise. What is the effect on the shape of the curve? Observe the shape change when: (a) $\mathbf{P}_2$ is moved to $(12, 12)$ and (b) when $\mathbf{P}_1$ is located at $(3, 10)$.

From Eq. (4.40b),

$$[x(u) \ y(u)] = [u^3 \ u^2 \ u \ 1] \begin{bmatrix} -1 & 3 & -3 & 1 \\ 3 & -6 & 3 & 0 \\ -3 & 3 & 0 & 0 \\ 1 & 0 & 0 & 0 \end{bmatrix} \begin{bmatrix} 4 & 4 \\ 6 & 8 \\ 8 & 9 \\ 10 & 3 \end{bmatrix}$$

$$= [6u + 4, \ -4u^3 - 9u^2 + 12u + 4]$$

plot of which is given in Figure 4.16(a). The transformation given is equivalent to moving the curve towards the origin by $(-2, -2)$ and then rotating the curve by $-30°$ clockwise. The net transformation is (see Chapter 3).

$$\mathbf{T} = \begin{bmatrix} \cos(-30°) & -\sin(-30°) & 0 \\ \sin(-30°) & \cos(-30°) & 0 \\ 0 & 0 & 1 \end{bmatrix} \begin{bmatrix} 1 & 0 & -2 \\ 0 & 1 & -2 \\ 0 & 0 & 1 \end{bmatrix} = \begin{bmatrix} 0.866 & 0.500 & -2.732 \\ -0.500 & 0.866 & -0.732 \\ 0 & 0 & 1 \end{bmatrix}$$

which when applied to the original segment expressed in homogenous coordinates gives

$$\begin{bmatrix} x(u)' \\ y(u)' \\ 1 \end{bmatrix} = \begin{bmatrix} 0.866 & 0.500 & -2.732 \\ -0.500 & 0.866 & -0.732 \\ 0 & 0 & 1 \end{bmatrix} \begin{bmatrix} 6u + 4 \\ -4u^3 - 9u^2 + 12u + 4 \\ 1 \end{bmatrix}$$

The transformed Bézier segment is plotted in Figure 4.16 (b). Observe that shape of the segment does not change. This is due to the affine properties of Bernstein polynomials as weighting functions. Modified curves for $\mathbf{P}_2$ relocated to (12, 12) and $\mathbf{P}_1$ repositioned at (3, 10) are shown in Figure 4.16 (c) comparing with the original curve. Also note that moving a single data point affects an overall change in the curve segment.

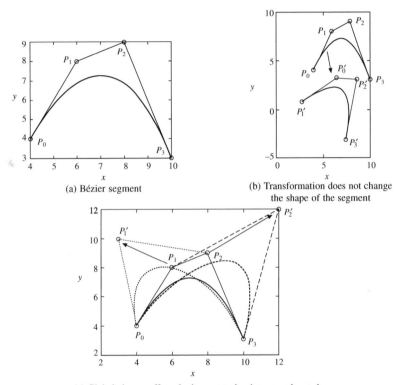

(a) Bézier segment

(b) Transformation does not change the shape of the segment

(c) Global change affected when control points are relocated

Figure 4.16   Bézier segments for Example 4.6

4.4.1 Properties of Bézier Segments

Based on the properties of Bernstein polynomials, much can be known about the Bézier segments. These properties are discussed for a general Bézier segment of degree n.

(a) End Points: Note that at $u = 0$, $B_0^n(0) = 1$ while all the other polynomials $B_i^n(0)$ are zero from the non-negativity and partition of unity properties of Bernstein polynomials. Thus, $\boldsymbol{P}_0$ is an end point on the Bézier segment. Also, at $u = 1$, $B_n^n(1) = 1$ while all other Bernstein coefficients are zero, implying that $\boldsymbol{P}_n$ is another end point on the segment.

(b) End Tangents: The end tangents have the directions of $\boldsymbol{P}_1 - \boldsymbol{P}_0$ and $\boldsymbol{P}_n - \boldsymbol{P}_{n-1}$, respectively. From Eq. (4.38),

$$\dot{\mathbf{r}}(u) = \sum_{j=0}^{n} \mathbf{P}_j \, \frac{d}{du} \, B_j^n (u) = \sum_{j=0}^{n} \mathbf{P}_j n[B_{j-1}^{n-1} (u) - B_j^{n-1} (u)]$$

$$= n \left[\sum_{j=0}^{n-1} \mathbf{P}_j B_{j-1}^{n-1} (u) - \sum_{j=0}^{n-1} \mathbf{P}_j B_j^{n-1} (u) \right] = n \left[\sum_{j=0}^{n-1} \mathbf{P}_{j+1} B_j^{n-1} (u) - \sum_{j=0}^{n-1} \mathbf{P}_j B_j^{n-1} (u) \right]$$

noting that $B_{-1}^{n-1} (u) = 0$.

$$\dot{\mathbf{r}}(u) = n \sum_{j=0}^{n-1} (\mathbf{P}_{j+1} - \mathbf{P}_j) B_j^{n-1} (u)$$

Thus, $\dot{\mathbf{r}}(0) = n(\mathbf{P}_1 - \mathbf{P}_0)$ and $\dot{\mathbf{r}}(1) = n(\mathbf{P}_n - \mathbf{P}_{n-1})$

noting that only $B_0^{n-1} (0)$ and $B_{n-1}^{n-1} (1)$ are 1.

(c) Geometry Invariance: Due to the partition of unity property of the Bernstein polynomials, the shape of the curve is invariant under rotation and translation of the coordinate frame. This is illustrated in Section 4.3 and shown as an example in Figure 4.16 (b).

(d) Convex Hull Property: The barycentric nature of Bernstein polynomials ensures that the Bézier segment lies within the convex hull of the control points. The property is useful in intersection problems, detection of interference, and provides estimates of the position of the curve by computing the bounds of the polygon. Figure 4.17 shows an example.

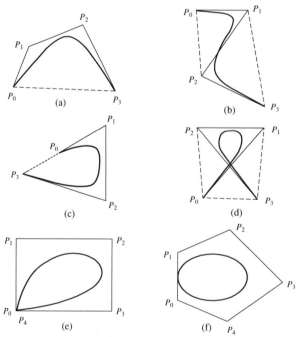

Figure 4.17   Bézier curves and the associated convex hulls (closed convex polygons)

(e) Variation Diminishing: For a planar Bézier segment, it can be verified geometrically that no straight line on that plane intersects with the segment more times than it does with the corresponding control polyline. Similarly, for a spatial Bézier curve, the property holds for a plane intersecting the curve and its control polyline. Note that special cases may occur when one or more legs of the control polyline may coincide with the intersecting line or a plane for which the property holds true. This is because the number of intersection points of the curve with the intersecting line/plane would be finite. However, with the control polyline, they would be infinitely many. The property suggests that the shape of the curve is no more complex compared with its control polyline. In some sense, the convex hull and variation diminishing properties, together, suggest that the shape of a Bézier segment is predictable and is roughly depicted by the control polyline. In a singular case where the control polyline is a straight line (control line), so is the Bézier segment from the convex hull property. Here, cae, the variation diminishing property may not be used as it is inconclusive especially when the intersecting line/plane happens to coincide with/contain the control line.

(f) Symmetry: Due to the symmetry in of Bernstein polynomials (Eq. (4.36)), if the sequence of the control points is reversed, i.e. $\mathbf{P}_{n-r}^{*} = \mathbf{P}_r$, the symmetry of the curve is preserved, that is

$$\sum_{r=0}^{n} \mathbf{P}_r B_r^n(u) = \sum_{r=0}^{n} \mathbf{P}_{n-r} B_r^n(1-u)$$

(g) Parameter Transformation: At times, we may have to express a Bézier segment as a non-normalized parameter u' between a and b. In such a case, set

$$u = \frac{u' - a}{b - a} \qquad (4.41)$$

to use Eq. (4.39).

(h) No Local Control: The shape of a Bézier segment changes globally if any data point is moved to a new location. To see this, let a control point $\boldsymbol{P}_k$ be moved along a specified vector $\boldsymbol{v}$. The original Bézier segment changes to

$$\mathbf{r}^{\text{new}}(u) = \sum_{\substack{i=0 \\ i \neq k}}^{n} B_i^n(u)\mathbf{P}_i + B_k^n(u)(\mathbf{P}_k + \boldsymbol{v}) = \sum_{i=0}^{n} B_i^n(u)\mathbf{P}_i + B_k^n(u)\boldsymbol{v} = \mathbf{r}(u) + B_k^n(u)\boldsymbol{v} \qquad (4.42)$$

For u between 0 and 1, every point on the old Bézier segment $\mathbf{r}(u)$ gets translated by $B_k^n(u)\boldsymbol{v}$ implying that the shape of the entire curve is changed.

(i) Derivative of a Bézier Curve
Using Eq. (4.38)

$$d\mathbf{r}(u)/du = \sum_{i=0}^{n} n \left[B_{i-1}^{n-1}(u) - B_i^{n-1}(u) \right]\mathbf{P}_i = n \sum_{i=0}^{n} B_{i-1}^{n-1}(u)\,\mathbf{P}_i - n \sum_{i=0}^{n} B_i^{n-1}(u)\mathbf{P}_i$$

$$= n B_{-1}^{n-1}(u)\mathbf{P}_0 + n \sum_{i=1}^{n} B_{i-1}^{n-1}(u)\mathbf{P}_{i-1} - n \sum_{i=0}^{n-1} B_i^{n-1}(u)\mathbf{P}_i - n B_n^{n-1}(u)\mathbf{P}_n$$

$$= n \sum_{i=0}^{n-1} B_i^{n-1}(u)\mathbf{P}_{i+1} - n \sum_{i=0}^{n-1} B_i^{n-1}(u)\mathbf{P}_i$$

$$= n \sum_{i=0}^{n-1} B_i^{n-1}(u)(\mathbf{P}_{i+1} - \mathbf{P}_i) \qquad (4.43)$$

Thus, the derivative of a Bézier segment is a degree $n-1$ Bézier segment with control points $n(\mathbf{P}_1 - \mathbf{P}_0)$, $n(\mathbf{P}_2 - \mathbf{P}_1)$, ..., $n(\mathbf{P}_n - \mathbf{P}_{n-1})$. Alternatively, it is the difference of two Bézier segments of degree $n-1$, times n. This derivative is usually referred to as the *hodograph* of the original Bézier segment. Note that $n(\mathbf{P}_i - \mathbf{P}_{i-1})$ are no longer position vectors but are free vectors instead. It is when the tails of the vectors are made to coincide with the origin that they may be termed as the control points for the hodograph. If $\boldsymbol{a}$ is any vector along which the original control points, $\mathbf{P}_0$, $\mathbf{P}_1$, . . . , $\mathbf{P}_n$ are displaced, the original Bézier segment gets displaced by $\boldsymbol{a}$ (Eq. 4.42). However, its hodograph remains unchanged.

(j) Higher Order Derivatives
From Eq. (4.43)

$$d^2\mathbf{r}(u)/du^2 = n \sum_{i=0}^{n-1} \left[dB_i^{n-1}(u)/du \right](\mathbf{P}_{i+1} - \mathbf{P}_i)$$

$$= n \sum_{i=0}^{n-1} (n-1)[B_{i-1}^{n-2}(u) - B_i^{n-2}(u)] \, (\mathbf{P}_{i+1} - \mathbf{P}_i)$$

$$= n(n-1) \sum_{i=0}^{n-1} B_{i-1}^{n-2}(u)(\mathbf{P}_{i+1} - \mathbf{P}_i) - n(n-1) \sum_{i=0}^{n-1} B_i^{n-2}(u)(\mathbf{P}_{i+1} - \mathbf{P}_i)$$

$$= n(n-1) \sum_{i=0}^{n-2} B_i^{n-2}(u)(\mathbf{P}_{i+2} - \mathbf{P}_{i+1}) - n(n-1) \sum_{i=0}^{n-2} B_i^{n-2}(u)(\mathbf{P}_{i+1} - \mathbf{P}_i)$$

$(B_{-1}^{n-2}(u) = B_{n-1}^{n-2}(u) = 0$; index i shifted accordingly)

$$= n(n-1) \sum_{i=0}^{n-2} B_i^{n-2}(u) \, [(\mathbf{P}_{i+2} - \mathbf{P}_{i+1}) - (\mathbf{P}_{i+1} - \mathbf{P}_i)] \tag{4.44a}$$

$$= n(n-1) \sum_{i=0}^{n-2} B_i^{n-2}(u) \, [\mathbf{P}_{i+2} - 2\mathbf{P}_{i+1} + \mathbf{P}_i] \tag{4.44b}$$

We may find by inspection, from Eqs. (4.44a and b) that

$$d^3\mathbf{r}(u)/du^3 = n(n-1)(n-2) \sum_{i=0}^{n-3} B_i^{n-3}(u)[(\mathbf{P}_{i+3} - 2\mathbf{P}_{i+2} + \mathbf{P}_{i+1}) - (P_{i+2} - 2\mathbf{P}_{i+1} + \mathbf{P}_i)\}$$

$$= n(n-1)(n-2) \sum_{i=0}^{n-3} B_i^{n-3}(u)[\mathbf{P}_{i+3} - 3\mathbf{P}_{i+2} + 3\mathbf{P}_{i+1} - \mathbf{P}_i] \tag{4.44c}$$

To express higher order derivatives more concisely, a finite difference scheme may be employed.

$$\mathbf{D}_i^j = \mathbf{D}_{i+1}^{j-1} - \mathbf{D}_i^{j-1}, j = 1, \ldots, n; i = 0, \ldots, n-j$$

$$\mathbf{D}_i^0 = \mathbf{P}_i \tag{4.44d}$$

Thus

$$\mathbf{D}_i^1 = \mathbf{D}_{i+1}^0 - \mathbf{D}_i^0 = \mathbf{P}_{i+1} - \mathbf{P}_i$$

$$\mathbf{D}_i^2 = \mathbf{D}_{i+1}^1 - \mathbf{D}_i^1 = (\mathbf{P}_{i+2} - \mathbf{P}_{i+1}) - (\mathbf{P}_{i+1} - \mathbf{P}_i) = \mathbf{P}_{i+2} - 2\mathbf{P}_{i+1} + \mathbf{P}_i$$

$$\mathbf{D}_i^3 = \mathbf{D}_{i+1}^2 - \mathbf{D}_i^2 = (\mathbf{P}_{i+3} - 2\mathbf{P}_{i+2} + \mathbf{P}_{i+1}) - (\mathbf{P}_{i+2} - 2\mathbf{P}_{i+1} + \mathbf{P}_i)$$

$$= \mathbf{P}_{i+3} - 3\mathbf{P}_{i+2} + 3\mathbf{P}_{i+1} - \mathbf{P}_i$$

using which the kth derivative of a Bézier segment can be written as

$$d^k \mathbf{r}(u)/du^k = n(n-1)(n-2)\ldots(n-k+1)\sum_{i=0}^{n-k} B_i^{n-k}(u)\mathbf{D}_i^k \qquad (4.45)$$

4.4.2 Subdivision of a Bézier Segment

Subdivision may have many applications in curve design. We may desire to trim a curve at the subdivision point retaining only a part, or, may subdivide a curve and design a segment separately without changing the shape of the other segment thus gaining additional flexibility in design. Subdivision may be performed as many times as desired. It involves partitioning a Bézier segment $\mathbf{r}(u)$ at some point $u = c$ into two segments each of which by itself is a Bézier segment. The resulting segments have their own control polylines and each are of the same degree as the parent curve. With $n + 1$ control points $\boldsymbol{b}_0, \boldsymbol{b}_1, \ldots \boldsymbol{b}_n$, and a parameter value $u = c$, $0 < c < 1$, two new sets of control points $\boldsymbol{p}_0, \boldsymbol{p}_1, \ldots, \boldsymbol{p}_n$ and $\boldsymbol{q}_0, \boldsymbol{q}_1, \ldots, \boldsymbol{q}_n$ are required so that the two Bézier segments span the original segment in the parameter range $0 \le u \le c$ and $c \le u \le 1$, respectively. To find the control points for the first segment, the parent Bézier segment for $0 \le u \le c$ may be re-parameterized with $u' = u/c$ so that when $u = 0$, $u' = 0$ and when $u = c$, $u' = 1$. The segment $\mathbf{r}_1(u)$ for $0 \le u \le c$ ($0 \le u' \le 1$) becomes

$$\mathbf{r}_1(u') = \sum_{i=0}^{n} B_i^n(cu')\boldsymbol{b}_i = \sum_{i=0}^{n} B_i^n(u')\boldsymbol{p}_i \qquad (4.46)$$

Since the two curves are identical, so are their derivatives at $u = 0$. Thus:

$$\frac{d^k}{du'^k}\left[\sum_{i=0}^{n} B_i^n(cu')\boldsymbol{b}_i\right] = \left\{(c)(n-1)\ldots(n-k+1)\sum_{i=0}^{n-k} B_i^{n-k}(cu')\mathbf{D}_i^k\right\}\left\{\frac{d(cu')}{du'}\right\}^k$$

$$= c^k n(n-1)\ldots(n-k+1)\sum_{i=0}^{n-k} B_i^{n-k}(cu')\mathbf{D}_i^k \qquad (4.47)$$

while

$$\frac{d^k}{du'^k}\left[\sum_{i=0}^{n} B_i^n(u')\boldsymbol{p}_i\right] = n(n-1)\ldots(n-k+1)\sum_{i=0}^{n-k} B_i^{n-k}(cu')\mathbf{P}_i^k \qquad (4.48)$$

where $\mathbf{P}_i^k$ are the differences in control points $\mathbf{P}_i$ related in a manner similar to Eq. (4.44d) as

$$\mathbf{P}_i^j = \mathbf{P}_{i+1}^{j-1} - \mathbf{P}_i^{j-1}, \; j = 1, \ldots, n; i = 0, \ldots, n-j, \text{ with } \mathbf{P}_i^0 = \boldsymbol{p}_i \qquad (4.49)$$

Note that $\mathbf{D}_i^0 = \boldsymbol{b}_i$ using which $\mathbf{D}_i^k$ can be computed accordingly from Eq. (4.44d).

Comparing Eqs. (4.47a) and (4.47b) for $u' = 0$ gives.

$$c^k \mathbf{D}_0^k = \mathbf{P}_0^k, \, k = 0, \ldots, n \qquad (4.50)$$

As an example, the control points for a cubic Bézier segment in $0 \le u \le c$ are determined. From Eq. (4.50)

$k = 0$: $\mathbf{D}_0^0 = \mathbf{P}_0^0$ $\Rightarrow p_0 = b_0$

$k = 1$: $c\mathbf{D}_0^1 = \mathbf{P}_0^1$ $\Rightarrow c(b_1 - b_0) = (p_1 - p_0)$

$\qquad\qquad\qquad\qquad \Rightarrow p_1 = (1 - c)\,b_0 + cb_1$

$k = 2$: $c^2\mathbf{D}_0^2 = \mathbf{P}_0^2$ $\Rightarrow c^2(b_2 - 2b_1 + b_0) = (p_2 - 2p_1 + p_0)$

$\qquad\qquad\qquad\qquad \Rightarrow p_2 = (1 - c)^2\,b_0 + 2c(1 - c)\,b_1 + c^2 b_2$

$k = 3$: $c^2\mathbf{D}_0^3 = \mathbf{P}_0^3$ $\Rightarrow c^3(b_3 - 3b_2 + 3b_1 - b_0) = (p_3 - 3p_2 + 3p_1 - p_0)$

$\qquad\qquad\qquad\qquad \Rightarrow p_3 = (1 - c)^3\,b_0 + 3c(1 - c)^2\,b_1 + 3c^2(1 - c)\,b_2 + c^3 b_3$ (4.51)

Comparing Eq. (4.51) with (4.32), it can be observed that $p_0 = b_0^0, p_1 = b_0^1, p_2 = b_0^2$ and $p_3 = b_0^3$ for $u = c$. In general, $p_k = b_0^k, k = 0, \ldots, n$. Geometrically, therefore, the new control points $p_0, p_1, \ldots, p_n$ for the first Bézier segment are the intermediate de Casteljau points, $b_0^0, b_0^1, \ldots, b_0^n$ for $u = c$ which in Figure 4.13, in the triangular schema to compute the de Casteljau points, correspond to the top edge of the triangle.

For control points $q_0, q_1, \ldots, q_n$ of the second segment, the Bézier curve for $c \le u \le 1$ may be re-parameterized with u' such that $u = 1 - (1 - c)(1 - u)$. Thus, when $u' = c$, $u = 0$ and for $u' = 1$, $u = 1$. The Bézier segment $\mathbf{r}_2(u)$ for $c \le u \le 1$ is

$$\mathbf{r}_2(u) = \sum_{i=0}^{n} B_i^n(1 - (1 - c)(1 - u))\,b_i = \sum_{i=0}^{n} B_i^n(u)q_i \qquad (4.52)$$

Identical to the treatment for the first segment, here, the derivatives of the two curves can be matched at $u' = 1$. Consider the kth derivative as per Eq. (4.45) and after implementing the chain rule

$$d^k\mathbf{r}_2/du'^k = \left[n(n - 1) \ldots (n - k + 1) \sum_{i=0}^{n-k} B_i^{n-k}(u)\mathbf{D}_i^k \right](1 - c)^k$$

$$= d^k\mathbf{r}_2/du^k = n(n - 1) \ldots (n - k + 1) \sum_{i=0}^{n-k} B_i^{n-k}(u')\mathbf{Q}_i^k \qquad (4.53a)$$

where $\mathbf{Q}_i^j$ are the differences in q_i related in a manner similar in Eqs. (4.44d) and (4.49) as

$$\mathbf{Q}_i^j = \mathbf{Q}_{i+1}^{j-1} - \mathbf{Q}_i^{j-1}, j = 1, \ldots, n; i = 0, \ldots, n - j, \text{ with } \mathbf{Q}_i^0 = \mathbf{q}_i \qquad (4.53b)$$

At $u' = 1$, Eq. (4.53a) becomes

$$(1 - c)^k\mathbf{D}_{n-k}^k = \mathbf{Q}_{n-k}^k, k = 0, \ldots, n \qquad (4.54)$$

To illustrate the computations, the control points for a cubic Bézier segment in $c \le u \le 1$ can be determined as

$k = 0: \ \mathbf{D}_n^0 = \mathbf{Q}_n^0 \qquad\qquad \Rightarrow q_n = b_n$

$k = 1: \ (1 - c) \, \mathbf{D}_{n-1}^1 = \mathbf{Q}_{n-1}^1 \ \Rightarrow (1 - c) \, (b_n - b_{n-1}) = (q_n - q_{n-1})$

$\qquad\qquad\qquad\qquad\qquad \Rightarrow q_{n-1} = (1 - c)b_{n-1} + cb_n$

$k = 2: \ (1 - c)^2 \mathbf{D}_{n-2}^2 = \mathbf{Q}_{n-2}^2 \Rightarrow (1 - c)^2(b_n - 2b_{n-1} + b_{n-2}) = (q_n - 2q_{n-1} + q_{n-2})$

$\qquad\qquad\qquad\qquad \Rightarrow q_{n-2} = (1 - c)^2 \, b_{n-2} + 2c \, (1 - c)b_{n-1} + c^2 b_n$

$k = 3: \ (1 - c)^3 \mathbf{D}_{n-3}^3 = \mathbf{Q}_{n-3}^3 \Rightarrow (1 - c)^3 \, (b_n - 3b_{n-1} + 3b_{n-2} - b_{n-3}) = (q_n - 3q_{n-1} + 3q_{n-2} - q_{n-3})$

$\qquad\qquad\qquad\qquad \Rightarrow q_{n-3} = (1 - c)^3 b_{n-3} + 3c(1 - c)^2 \, b_{n-2} + 3c^2(1 - c)b_{n-1} + c^3 b_n$

$$(4.55)$$

Eqs. (4.55) and (4.32) show that $q_n = b_n^0$, $q_{n-1} = b_{n-1}^1$, $q_{n-2} = b_{n-2}^2$ and $q_{n-3} = b_{n-3}^3$ for $u = c$. In general, $q_{n-k} = b_{n-k}^k$, $k = 0, \ldots n$, that is, the control points for the second Bézier segment are the intermediate de Casteljau points, $b_n^0, b_{n-1}^1, \ldots, b_0^n$ for $u = c$ which correspond to the bottom edge in the triangular scheme in Figure 4.13. Figure 4.18 depicts the control polylines for the two subdivided cubic segments. Note that b_0^3 is a common end point for both segments, and $b_0^2 b_0^3$ and $b_0^3 b_1^2$ are tangents to the respective segments.

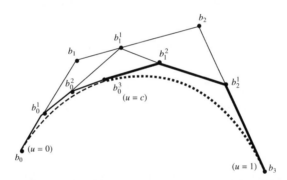

Figure 4.18 **Control polylines for the two subdivided curves (thick and thicker solid lines) and subdivided curves (dashed lines)**

A procedure reverse to subdivision may also be employed to extend a curve. For given c between 0 and 1 and for given control points $p_0, p_1, \ldots, p_n$, the control polyline $b_0, b_1, \ldots, b_n$ for the extended curve (for $0 \le u \le 1$) can be computed using Eqs. (4.51) by a series of forward substitutions.

Example 4.7. The equation for a Bézier curve with the control points $\mathbf{P}_0, \mathbf{P}_1, \mathbf{P}_2, \mathbf{P}_3$ is given by

$$\mathbf{r}(u) = \sum_{i=0}^{3} \mathbf{P}_i B_i^3(u)$$

The curve is required to be subdivided at $u = 1/2$. Develop a formulation for subdivision into two Bézier segments (a) in the interval $u \in [0, 1/2]$ and (b) $u \in [1/2, 1]$.

Let the two *segments* be represented as:

$$\mathbf{r}(u_1) = \sum_{i=1}^{3} \mathbf{Q}_i B_i^3(u_1), \quad \mathbf{r}_2(u_2) = \sum_{i=1}^{3} \mathbf{R}_i B_i^3(u_2)$$

Defining the parameters $u_1 = 2u$, $u_2 = 2u - 1$, they satisfy the requirement that $u_1 \in [0, 1]$, when $u \in [0, 1/2]$, and $u_2 \in [0, 1]$, when $u \in [1/2, 1]$. The new control points of the two curve segments are $\mathbf{Q}_0$, $\mathbf{Q}_1$, $\mathbf{Q}_2$, $\mathbf{Q}_3$ and $\mathbf{R}_0$, $\mathbf{R}_1$, $\mathbf{R}_2$, $\mathbf{R}_3$. These are yet to be determined. The original curve and its segments can be written as:

(a) $\mathbf{r}(u) = (1 - u)^3\mathbf{P}_0 + 3u(1 - u)^2\mathbf{P}_1 + 3u^2(1 - u)\mathbf{P}_2 + u^3\mathbf{P}_3$

(b) $\mathbf{r}_1(u_1) = (1 - u_1)^3\mathbf{Q}_0 + 3u_1(1 - u_1)^2\mathbf{Q}_1 + 3u_1^2(1 - u_1)\mathbf{Q}_2 + u_1^3\mathbf{Q}_3$

(c) $\mathbf{r}_2(u_2) = (1 - u_1)^3\mathbf{R}_0 + 3u_2(1 - u_2)^2\mathbf{R}_1 + 3u_2^2(1 - u_2)\mathbf{R}_2 + u_2^3\mathbf{R}_3$

(d) $\dot{\mathbf{r}}(u) = -3(1 - u)^2\mathbf{P}_0 + 3[(1 - u)^2 - 2u(1 - u)]\mathbf{P}_1 + 3[2u(1 - u) - u^2]\mathbf{P}_2 + 3u^2\mathbf{P}_3$

(e) $\dot{\mathbf{r}}_1(u) = \dfrac{d\mathbf{r}_1}{du_1}\dfrac{du_1}{du}$

$$= 2[-3(1-u_1)^2\mathbf{Q}_0 + 3[(1 - u_1)^2 - 2u_1(1 - u_1)]\mathbf{Q}_1 + 3[2u_1(1 - u_1) - u_1^2]\mathbf{Q}_2 + 3u_1^2\mathbf{Q}_3]$$

(f) $\ddot{\mathbf{r}}(u) = 6(1 - u)\mathbf{P}_0 + 3[-4 + 6u]\mathbf{P}_1 + 3[2 - 6u]\mathbf{P}_2 + 6u\mathbf{P}_3$

(g) $\dddot{\mathbf{r}}(u) = -6\mathbf{P}_0 + 18\mathbf{P}_1 - 18\mathbf{P}_2 + 6\mathbf{P}_3$

(h) $\ddot{\mathbf{r}}_1(u_1) = 4[6(1 - u_1)\mathbf{Q}_0 + 3[-4 + 6u_1]\mathbf{Q}_1 + 3[2 - 6u_1]\mathbf{Q}_2 + 6u_1\mathbf{Q}_3]$

(i) $\dddot{\mathbf{r}}_1(u_1) = 8[-6\mathbf{Q}_0 + 18\mathbf{Q}_1 - 18\mathbf{Q}_2 + 6\mathbf{Q}_3]$

Evaluate at $u = 0$, where the parameter $u_1 = 0$. From (a) and (b) we can find that (i) $\mathbf{Q}_0 = \mathbf{P}_0$.
    From (d) and (e), since the slopes are equal

$$\dot{\mathbf{r}}(0) = \dot{\mathbf{r}}_1(0) \Rightarrow 3(\mathbf{P}_1 - \mathbf{P}_0) = 2 * 3(\mathbf{Q}_1 - \mathbf{Q}_0) \Rightarrow \text{(ii)} \ (\mathbf{P}_1 - \mathbf{P}_0) = 2(\mathbf{Q}_1 - \mathbf{Q}_0)$$

From (f) and (h), since the second derivatives are the same

$$\ddot{\mathbf{r}}(0) = \ddot{\mathbf{r}}_1(0) \Rightarrow 6(\mathbf{P}_0 - 2\mathbf{P}_1 + \mathbf{P}_2) = 4 * 6(\mathbf{Q}_0 - 2\mathbf{Q}_1 + \mathbf{Q}_2) \Rightarrow \text{(iii)} \ (\mathbf{P}_0 - 2\mathbf{P}_1 + \mathbf{P}_2) = 4(\mathbf{Q}_0 - 2\mathbf{Q}_1 + \mathbf{Q}_2)$$

From (g) and (i) one obtains

$$\dddot{\mathbf{r}}(0) = \dddot{\mathbf{r}}_1(0) \Rightarrow \text{(iv)} \ (\mathbf{P}_3 - 3\mathbf{P}_2 + 3\mathbf{P}_1 - \mathbf{P}_0) = 8(\mathbf{Q}_3 - 3\mathbf{Q}_2 + 3\mathbf{Q}_1 - \mathbf{Q}_0)$$

From the above equations (i) to (iv)

$$\mathbf{Q}_1 = \mathbf{P}_0$$

$$\mathbf{Q}_1 = (\mathbf{P}_0 + \mathbf{P}_1)/2$$

$$\mathbf{Q}_2 = (\mathbf{P}_2 + 2\mathbf{P}_1 + \mathbf{P}_0)/4$$

$$\mathbf{Q}_3 = (\mathbf{P}_0 + 3\mathbf{P}_1 + 3\mathbf{P}_2 + \mathbf{P}_3)/8$$

Determine $\mathbf{R}_0$, $\mathbf{R}_1$, $\mathbf{R}_2$, $\mathbf{R}_3$ in a similar manner in terms of $\mathbf{P}_0$, $\mathbf{P}_1$, $\mathbf{P}_2$, $\mathbf{P}_3$. this is left as an exercise.

4.4.3 Degree-Elevation of a Bézier Segment

The flexibility in designing a Bézier segment may also be improved by increasing its degree which results in an addition of a data point. The shape of the segment, however, remains unchanged. Thus, for Bézier curve of degree n defined by control points $b_0, b_1, \ldots, b_n$, to raise its degree by one requires finding a new set of $n + 2$ control points $q_0, q_1, \ldots, q_{n+1}$. Since the two segments are identical

$$\sum_{i=0}^{n+1} {}^{n+1}C_i (1 - u)^{n+1-i} u^i q_i = \sum_{i=0}^{n} {}^{n}C_i (1 - u)^{n-i} u^i b_i = (1 - u + u) \sum_{i=0}^{n} {}^{n}C_i (1 - u)^{n-i} u^i b_i$$

$$= \sum_{i=0}^{n} {}^{n}C_i (1 - u)^{n-i+1} u^i b_i + \sum_{i=0}^{n} {}^{n}C_i (1 - u)^{n-i} u^{i+1} b_i$$

Comparing the coefficients of $(1 - u)^{n+1-i} u^i$ yields

$$^{n+1}C_i q_i = {}^{n}C_i b_i + {}^{n}C_{i-1} b_{i-1}$$

or
$$q_i = \left(1 - \frac{i}{n + 1} \right) b_i + \left(\frac{i}{n + 1} \right) b_{i-1}, i = 0, \ldots, n + 1 \qquad (4.56)$$

Note that for $i = 0$, $q_0 = b_0$ and when $i = n + 1$, $q_{n+1} = b_n$. Even though expressions involving b_{-1} and b_{n+1} may appear, they may not be required as the respective coefficients are 0 at $i = 0$ and $i = n + 1$. Eq. (4.56) suggests that q_i is the weighted linear combination of b_{i-1} and b_i with non-negative weights that add to 1. Thus, q_i lies in the convex hull of b_{i-1} and b_i. More precisely, the new control polyline lies within the convex hull of the old polyline and the Bézier segment lies within the convex hulls of both polylines. The process of degree elevation may be repeated as many times as desired. Each time the degree elevation is performed, the resultant control polyline moves closer to the Bézier segment, adding one control point at a time. When the number of times the degree elevation is performed approaches infinity, the control polyline approaches the Bézier segment.

Example 4.8 Given data points $A(0, 0)$, $B(1, 2)$, $C(3, 2)$ and $D(6, -1)$, elevate the degree of this cubic Bézier segment to four and five and show the new control polylines.

Using Eq. (4.56), data points for the degree 4 segment can be computed as

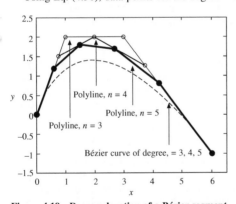

Figure 4.19 **Degree elevation of a Bézier segment**

$$q_0 = b_0 = (0, 0)$$

$$q_1 = \left(1 - \frac{1}{4} \right) b_1 + \left(\frac{1}{4} \right) b_0 = (0.75, 1.50)$$

$$q_2 = \left(1 - \frac{2}{4} \right) b_2 + \left(\frac{2}{4} \right) b_1 = (2.00, 2.00)$$

$$q_3 = \left(1 - \frac{3}{4} \right) b_3 + \left(\frac{3}{4} \right) b_2 = (3.75, 1.25)$$

$$q_4 = b_3 = (6, -1)$$

Similarly, using q_i, $i = 0, \ldots, 4$ above, data points can be computed for degree 5 segment. Figure 4.19 shows the control polylines for degree 3, 4 and 5 Bézier segments.

4.4.4   Relationship between Bézier and Ferguson Segments

That Bézier and Ferguson cubic segments have similar matrix forms (Eqs. (4.7) and (4.40)), we may realize that the two geometric matrices may be related. In other words, a Ferguson's segment may be converted to the cubic Bézier form and vice-versa. Given control points $\mathbf{P}_i$, $i = 0, \ldots, 3$, and realizing that a Ferguson's segment would pass through $\mathbf{P}_0$ and $\mathbf{P}_3$, equating the two forms results in

$$\mathbf{r}(u) = [u^3 \quad u^2 \quad u \quad 1] \begin{bmatrix} 2 & -2 & 1 & 1 \\ -3 & 3 & -2 & -1 \\ 0 & 0 & 1 & 0 \\ 1 & 0 & 0 & 0 \end{bmatrix} \begin{bmatrix} \mathbf{P}_0 \\ \mathbf{P}_3 \\ \mathbf{T}_0 \\ \mathbf{T}_3 \end{bmatrix}$$

$$= [u^3 \quad u^2 \quad u \quad 1] \begin{bmatrix} -1 & 3 & -3 & 1 \\ 3 & -6 & 3 & 0 \\ -3 & 3 & 0 & 0 \\ 1 & 0 & 0 & 0 \end{bmatrix} \begin{bmatrix} \mathbf{P}_0 \\ \mathbf{P}_1 \\ \mathbf{P}_2 \\ \mathbf{P}_3 \end{bmatrix} \qquad (4.57a)$$

or $[u^3 \quad u^2 \quad u \quad 1] \begin{bmatrix} 2\mathbf{P}_0 - 2\mathbf{P}_3 + \mathbf{T}_0 + \mathbf{T}_3 \\ -3\mathbf{P}_0 + 3\mathbf{P}_3 - 2\mathbf{T}_0 - \mathbf{T}_3 \\ \mathbf{T}_0 \\ \mathbf{T}_3 \end{bmatrix} = [u^3 \quad u^2 \quad u \quad 1] \begin{bmatrix} -\mathbf{P}_0 + 3\mathbf{P}_1 - 3\mathbf{P}_2 + \mathbf{P}_3 \\ 3\mathbf{P}_0 - 6\mathbf{P}_1 + 3\mathbf{P}_2 \\ -3\mathbf{P}_0 + 3\mathbf{P}_1 \\ \mathbf{P}_0 \end{bmatrix}$

Comparing the coefficients of u gives $\mathbf{T}_0 = 3(\mathbf{P}_1 - \mathbf{P}_0)$ while comparing those of u^2 results in

$$-3\mathbf{P}_0 + 3\mathbf{P}_3 - 2\mathbf{T}_0 - \mathbf{T}_3 = 3\mathbf{P}_0 - 6\mathbf{P}_1 + 3\mathbf{P}_2$$

or $\qquad \mathbf{T}_3 = -6\mathbf{P}_0 + 6\mathbf{P}_1 - 3\mathbf{P}_2 + 3\mathbf{P}_3 - 2\mathbf{T}_0$

$$= 6(\mathbf{P}_1 - \mathbf{P}_0) + 3(\mathbf{P}_3 - \mathbf{P}_2) - 6(\mathbf{P}_1 - \mathbf{P}_0) = 3(\mathbf{P}_3 - \mathbf{P}_2)$$

Equating the coefficients of u^3 thereafter becomes redundant. Thus, given control points $\mathbf{P}_i$, $i = 0, \ldots, 3$, the geometric matrix for the Ferguson's segment can be written as

$$\mathbf{G} = [\mathbf{P}_0 \ \mathbf{P}_3 \ 3(\mathbf{P}_1 - \mathbf{P}_0) \ 3(\mathbf{P}_3 - \mathbf{P}_2)]^T.$$

Likewise, for given two end points $\mathbf{P}_i$ and $\mathbf{P}_{i+1}$, and end tangents, $\mathbf{T}_i$ and $\mathbf{T}_{i+1}$ for Ferguson's model, the geometric matrix for the Bézier segment can be constructed as

$$\mathbf{G} = \left[\mathbf{P}_i \quad \left(\frac{\mathbf{T}_i}{3} + \mathbf{P}_i \right) \left(\mathbf{P}_{i+1} - \frac{\mathbf{T}_{i+1}}{3} \right) \ \mathbf{P}_{i+1} \right]^T \qquad (4.57c)$$

4.5 Composite Bézier Curves

In foregoing sections, Bézier segment of a generic degree was considered and its properties were discussed in detail. Consider, in a composite curve, any two contiguous Bézier segments, $\mathbf{r}_1(u_1)$ of degree m with data points $p_0, p_1, \ldots, p_m$, and $\mathbf{r}_2(u_2)$ of degree n with data points $q_0, q_1, \ldots, q_n$. For position (C^0) continuity, since the segments pass through the end points, the last point in $\mathbf{r}_1(u_1)$ should coincide with the first point in $\mathbf{r}_2(u_2)$, that is (Figure 4.20)

$$\mathbf{r}_1(u_1 = 1) = \mathbf{r}_2(u_2 = 0)$$

or $\qquad\qquad\qquad\qquad p_m = q_0 \qquad\qquad\qquad\qquad (4.58)$

Eq. (4.58) implies that with the first segment given, the position continuity constraints the first control point $\boldsymbol{q}_0$ of the neighboring segment. For slope (C^1) continuity at the junction point $\boldsymbol{p}_m = \boldsymbol{q}_0$ (Figure 4.20b),

$$\alpha_1 \frac{d\mathbf{r}_1(1)}{du_1} = \alpha_2 \frac{d\mathbf{r}_2(0)}{du_2} = \mathbf{t}$$

or

$$\alpha_1 m(\boldsymbol{p}_m - \boldsymbol{p}_{m-1}) = \alpha_2 n(\boldsymbol{q}_1 - \boldsymbol{q}_0)$$

or

$$\boldsymbol{q}_1 = \lambda(\boldsymbol{p}_m - \boldsymbol{p}_{m-1}) + \boldsymbol{p}_m = (\lambda + 1)\boldsymbol{p}_m - \lambda\boldsymbol{p}_{m-1} \tag{4.59}$$

where α_1 and α_2 are normalizing scalars for the slope along the unit tangent vector $\mathbf{t}$ and $\lambda = \dfrac{\alpha_1 m}{\alpha_2 n}$.

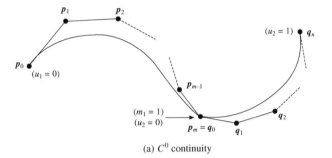

(a) C^0 continuity

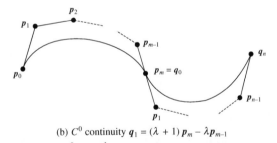

(b) C^0 continuity $\boldsymbol{q}_1 = (\lambda + 1)\boldsymbol{p}_m - \lambda\boldsymbol{p}_{m-1}$

Figure 4.20 C^0 and C^1 continuous Composite Bézice curves

Thus, for two Bézier segments with position continuity at the junction point, slope continuity further constraints the second control point $\boldsymbol{q}_1$ of $\mathbf{r}_2(u_2)$ to be collinear with the last leg $\boldsymbol{p}_{m-1}\boldsymbol{p}_m$ of the first polyline. For the segments to have the curvature (C^2) continuity at the junction point $\kappa_1(1) = \kappa_2(0)$, where $\kappa_1(u_1)$ and $\kappa_2(u_2)$ are curvature expressions for the two segments. Or

$$\frac{\dfrac{d}{du_1}\mathbf{r}_1(1) \times \dfrac{d^2}{du_1^2}\mathbf{r}_1(1)}{\left| \dfrac{d}{du_1}\mathbf{r}_1(1) \right|^3} = \frac{\dfrac{d}{du_2}\mathbf{r}_2(0) \times \dfrac{d^2}{du_2^2}\mathbf{r}_2(0)}{\left| \dfrac{d}{du_2}\mathbf{r}_2(0) \right|^3} \tag{4.60}$$

Using position and slope continuity conditions in Eq. (4.60) yields

$$\mathbf{t} \times \frac{d^2}{du_1^2} \, \mathbf{r}_1(1) = \left(\frac{\alpha_2}{\alpha_1} \right)^2 \mathbf{t} \times \frac{d^2}{du_2^2} \, \mathbf{r}_2(0) \qquad (4.61)$$

Equation that satisfies the condition above is

$$\frac{d^2}{du_1^2} \, \mathbf{r}_1(1) = \left(\frac{\alpha_2}{\alpha_1} \right)^2 \frac{d^2}{du_2^2} \, \mathbf{r}_2(0) + \mu \frac{d}{du_2} \, \mathbf{r}_2(0)$$

$$= \frac{m^2}{(n\lambda)^2} \frac{d^2}{du_2^2} \, \mathbf{r}_2(0) + \mu \frac{d}{du_2} \, \mathbf{r}_2(0) \qquad (4.62)$$

for some scalar μ. Using Eq. (4.44b),

$$\frac{d^2}{du_1^2} \, \mathbf{r}_1(1) = m(m-1) \sum_{i=0}^{m-2} B_i^{m-2}(1)[\boldsymbol{p}_{i+2} - 2\boldsymbol{p}_{i+1} + \boldsymbol{p}_i] = m(m-1)[\boldsymbol{p}_m - 2\boldsymbol{p}_{m-1} + \boldsymbol{p}_{m-2}] \quad (4.63a)$$

while $\quad \dfrac{d^2}{du_2^2} \, \mathbf{r}_1(0) = n(n-1) \sum\limits_{i=0}^{n-2} B_i^{n-2}(0)[\boldsymbol{q}_{i+2} - 2\boldsymbol{q}_{i+1} + \boldsymbol{q}_i] = n(n-1)[\boldsymbol{q}_2 - 2\boldsymbol{q}_1 + \boldsymbol{q}_0] \quad (4.63b)$

Substituting Eq. (4.63) into (4.62) and also using $\dfrac{d}{du_2} \, \mathbf{r}_2(0) = n(\boldsymbol{q}_1 - \boldsymbol{q}_0)$ from Eq. (4.43), we get

$$m(m-1)[\boldsymbol{p}_m - 2\boldsymbol{p}_{m-1} + \boldsymbol{p}_{m-2}] = \frac{m^2}{(n\lambda)^2} \, n(n-1) \, [\boldsymbol{q}_2 - 2\boldsymbol{q}_1 + \boldsymbol{q}_0] + \mu n(\boldsymbol{q}_1 - \boldsymbol{q}_0)$$

or $\quad m(m-1)(\boldsymbol{p}_m - \boldsymbol{p}_{m-1}) - m(m-1)(\boldsymbol{p}_{m-1} - \boldsymbol{p}_{m-2}) + \left[\dfrac{m^2(n-1)}{n\lambda^2} - \mu n \right](\boldsymbol{q}_1 - \boldsymbol{q}_0)$

$$= \left[\frac{m^2(n-1)}{n\lambda^2} \right](\boldsymbol{q}_2 - \boldsymbol{q}_1)$$

Using C^1 continuity (Eq. 4.59), we gets

$$\left[\frac{m^2(n-1)}{n\lambda^2} - \mu n\lambda + m(m-1) \right](\boldsymbol{p}_m - \boldsymbol{p}_{m-1}) - m(m-1)(\boldsymbol{p}_{m-1} - \boldsymbol{p}_{m-2})$$

$$= \left[\frac{m^2(n-1)}{n\lambda^2} \right](\boldsymbol{q}_2 - \boldsymbol{q}_1) \qquad (4.64)$$

This implies that $(\boldsymbol{q}_2 - \boldsymbol{q}_1)$ expressed as a linear combination of vectors $(\boldsymbol{p}_m - \boldsymbol{p}_{m-1})$ and $(\boldsymbol{p}_{m-1} - \boldsymbol{p}_{m-2})$ lies in the plane containing the latter two. In other words, $\boldsymbol{p}_{m-2}, \boldsymbol{p}_{m-1}, \boldsymbol{p}_m = \boldsymbol{q}_0, \boldsymbol{q}_1$ and $\boldsymbol{q}_2$ are *coplanar*. Note that for a composite, C^1 continuous planar Bézier curve, this condition is inherently satisfied. However, for a spatial, C^2 continuous composite curve, $\boldsymbol{q}_2$ is constrained to lie in the same plane as $\boldsymbol{p}_{m-2}, \boldsymbol{p}_{m-1}, \boldsymbol{p}_m = \boldsymbol{q}_0$ and $\boldsymbol{q}_1$. The foregoing generalized analysis was for two Bézier segments of degrees m and n. To design a C^2 continuous composite Bézier curve with cubic segments, the first four data points can be chosen freely. For the second segment, three of the four points, namely q_0, q_1 and q_2 are constrained by the three continuity conditions; q_0 becomes the fourth point p_3 of the first segment, q_1 is constrained to be placed along the vector $p_2 \, p_3$, and q_2 must be placed on a plane defined by the four data points previous to it. It is only the fourth point q_3 of the subsequent segment that can be chosen *freely*. Note that different values of scalars λ and μ may be specified to choose q_1 and q_2 to satisfy slope and curvature continuities. Nevertheless, this freedom is indirect. This restricts the flexibility in curve design for which reason, designers tend to prefer degree 5 or 7 Bézier segments. When working with degree 3 segments, if a user seeks more flexibility in design, subdivision (Section 4.4.2) or degree elevation (section 4.4.3) can be incorporated to generate more data points.

Example 4.9. For a two segment C^2 continuous composite Bézier curve, data points for the first cubic segment are given as $p_0 \, (0, 0, 0)$, $p_1(1, 2, 0)$, $p_2(3, 2, 0)$ and $p_3(6, -1, 0)$. Generate the second cubic segment with some chosen values of scalars λ and μ as they appear in Eqs. (4.59) and (4.62).

Let the data points for the second cubic segment be q_0, q_1, q_2 and q_3. For position continuity (Eq. 4.58), $q_0 \equiv p_3 = (6, -1, 0)$. For slope continuity from Eq. (4.59), we have

$$q_1 = (\lambda + 1)p_3 - \lambda p_2 = (\lambda + 1) \, (6, -1, 0) - \lambda(3, 2, 0) = (6 + 3\lambda, -1 -3\lambda, 0)$$

while for curvature continuity, using Eq. (4.64) yields

$$q_2 = \left(1 - \frac{\mu\lambda^2}{2} + 2\lambda\right)(p_3 - p_2) - (\lambda)(p_2 - p_1)$$

$$+ \left(1 - \frac{\mu\lambda^2}{2} + 2\lambda\right)[3, -3, 0] - (\lambda)\,[2, 0, 0] + [6, -1, 0]$$

$$= \left(9 + 4\lambda - \frac{3}{2}\mu\lambda^2, -4 - 6\lambda + \frac{3}{2}\mu\lambda^2, 0\right)$$

q_3 being a free choice, it is assumed as $(5, 2, 2)$ in this example. Figure 4.21 shows the composite Bézier curve for different values of the pair (λ, μ). Note that five control points around the junction point and including it lie on the *x-y* plane.

4.6 Rational Bézier Curves

Bézier segments, by themselves, do not have any local control in that change in the position of a data point causes the shape of the entire segment to change. Achieving local shape control is the prime motivation to discuss B-spline curves in Chapter 5. However, in this section, we discuss Rational Bézier curves that provide more freedom to a designer in defining the shape of a segment/curve.

In Chapter 2, homogenous coordinates were introduced that helped in unifying rotation and translation as matrix multiplication opertions. In essence, $\mathbf{P}_i \equiv [x_i, y_i, z_i, 1]$ and $\mathbf{P}_i^H \equiv [X_i = w_i y_i, Y_i = w_i z_i, Z_i = w_i z_i, W_i = w_i]$ represent the same poolnt in the Euclidean space $\mathbf{E}^3$. $\mathbf{P}_i$ is, in a way, the projection of $\mathbf{P}_i^H$ on the $w_i = 1$ hyperplane. Since a curve (surface or solid) may need to be gtransformed at some intermediatge stage in a design operation, it behooves to work with the generalized homogenous coordinates of data points. This provides more freedom to a designer in that the user now needs to specify weight w_i as an additional parameter with the Euclidean coordinates $[x_i, y_i, z_i]$ of a data point. With $n + 1$ data points $\mathbf{P}_i^H$, $i = 0, ..., n$, the nth degree Bézier segment can be defined as

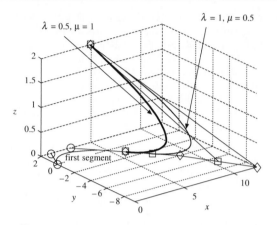

Figure 4.21   A C^2 continuous composite Bézier curve

$$\mathbf{P}^H(t) \equiv \begin{bmatrix} X(t) \\ Y(t) \\ Z(t) \\ W(t) \end{bmatrix} = \sum_{i=0}^{n} B_i^n(t) \mathbf{P}_i^H = \sum_{i=0}^{n} B_i^n(t) \begin{bmatrix} w_i x_i \\ w_i y_i \\ w_i z_i \\ w_i \end{bmatrix} \tag{4.65}$$

The corresponding Euclidean coordinates can then be computed as

$$x(t) = \frac{X(t)}{W(t)} = \frac{\sum_{i=0}^{n} B_i^n(t) w_i x_i}{\sum_{i=0}^{n} B_i^n(t) w_i} = \sum_{i=0}^{n} Br_i^n(t) x_i$$

$$x(t) = \frac{Y(t)}{W(t)} = \frac{\sum_{i=0}^{n} B_i^n(t) w_i y_i}{\sum_{i=0}^{n} B_i^n(t) w_i} = \sum_{i=0}^{n} Br_i^n(t) y_i$$

$$x(t) = \frac{Z(t)}{W(t)} = \frac{\sum_{i=0}^{n} B_i^n(t) w_i z_i}{\sum_{i=0}^{n} B_i^n(t) w_i} = \sum_{i=0}^{n} Br_i^n(t) z_i \tag{4.66}$$

where $Br_i^n(t) = \dfrac{w_i B_i^n(t)}{\sum_{i=0}^{n} w_i B_i^n(t)}$ are the rational Bernstein polynomials in t for which reason $\mathbf{P}(t) =$

$[x(t), y(t), z(t)]$ is termed as the rational Bézier segment. Note that $Br_i^n(t)$ are barycentric, that is, for $w_i \geq 0$, the rational functions are all nonzero and they sum to 1. For a special case when $w_i = 1$, $i = 0, \ldots, n$, Eq. (4.66) yields a Bézier segment. An advantage when using rational Bézier segments is the design freedom a user achieves by specifying weights w_i to data points $[x_i y_i z_i]$ at will. For $w_i = 0$, $\mathbf{P}_i(x_i, y_i, z_i)$ has no effect on the shape of the curve since its corresponding coefficient $Br_i^n(t)$ is zero. As w_i approaches infinity, all other $Br_i^n(t)$ approach zero for which the curve converges to $\mathbf{P}_i$. A rational Bézier segment has all the properties of a Bézier segment. That is, a rational Bézier segment passes through the end points, it lies within the convex hull defined by the control points and it has the variation diminishing property. Further, by modifying weights appropriately, a rational Bézier segment can be made more proximal to a chosen control point.

Example 4.10. For a set of control point $\mathbf{P}_0 = (4, 4)$, $\mathbf{P}_1 = (6, 8)$, $\mathbf{P}_2 = (8, 9)$ and $\mathbf{P}_3 = (10, 3)$ of Example 4.6, compute the rational Bézier segment initially for all weights $w_0 = w_1 = w_2 = w_3 = 1$. Alter the values of w_2 to realize the change in the curve shape.

The x and y coordinates of points on the rational Bézier segment can be computed as

$$x(t) = \frac{w_0(1-t)^3 x_0 + 3w_1 t(1-t)^2 x_1 + 3w_2 t^2(1-t) x_2 + t^3 x_3}{w_0(1-t)^3 + 3w_1 t(1-t)^2 + 3w_2 t^2(1-t) + t^3}$$

$$y(t) = \frac{w_0(1-t)^3 y_0 + 3w_1 t(1-t)^2 y_1 + 3w_2 t^2(1-t) y_2 + t^3 y_3}{w_0(1-t)^3 + 3w_1 t(1-t)^2 + 3w_2 t^2(1-t) + t^3}$$

and the rational Bézier segments are shown in Figure 4.22 for $w_0 = w_1 = w_3 = 1$ and for different values w_2. As w_2 is increased, the segment shapes towards $\mathbf{P}_2 = (8, 9)$.

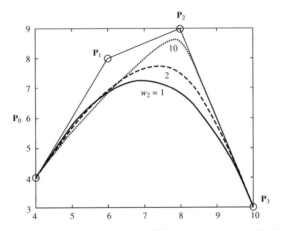

Figure 4.22   Change in curve shape of a rational Bézier segment due to the change in weight

Another advantage when using rational Bézier segments is the ability to design conics precisely, especially a circular arc which cannot be designed accurately using a polynomial Bézier segment of even higher degrees like cubic, quadric or quintic. We consider designing conics using a rational Bézier segment of degree 2 which is given as

$$\mathbf{P}(t) = \frac{(1 - t)^2 \, \mathbf{P}_0 + 2wt(1 - t)\mathbf{P}_1 + t^2 \mathbf{P}_2}{(1 - t)^2 + 2wt(1 - t) + t^2} \tag{4.67}$$

where $w_0 = w_1 = 1$ and $w_2 = w$ to retain symmetry in the rational polynomials like in Bernstein polynomials. Referring to Figure 4.9, the midpoint $\mathbf{P}$ of the curve is given by

$$\mathbf{P} = \frac{\mathbf{P}_0 + 2w\mathbf{P}_1 + \mathbf{P}_2}{2(1 + w)}$$

Now,

$$\mathbf{DP} = \mathbf{P} - \mathbf{D} = \frac{\mathbf{P}_0 + 2w\mathbf{P}_1 + \mathbf{P}_2}{2(1 + w)} - \frac{1}{2}(\mathbf{P}_0 + \mathbf{P}_2) = \frac{w(2\mathbf{P}_1 - \mathbf{P}_0 - \mathbf{P}_2)}{2(1 + w)}$$

while

$$\mathbf{DP}_1 = \mathbf{P}_1 - \mathbf{D} = \mathbf{P}_1 - \frac{1}{2}(\mathbf{P}_0 + \mathbf{P}_2) = \frac{(2\mathbf{P}_1 - \mathbf{P}_0 - \mathbf{P}_2)}{2}$$

and so

$$\frac{|\mathbf{DP}|}{|\mathbf{DP}_1|} = \alpha = \frac{w}{1 + w} \tag{4.68}$$

We can achieve different conic sections as follows. For $\alpha < 0.5$ (or $w < 1$), one gets an elliptic segment. For $\alpha = 0.5$ (or $w = 1$), the segment is parabolic while for $\alpha > 0.5$ (or $w > 1$), the segment is hyperbolic.

To draw a circular arc with an included angle 2θ (Figure 4.23) using the rational quadratic Bézier segment, we have

$$|\mathbf{OP}_1| = \frac{|\mathbf{OP}_0|}{\sin \theta}$$

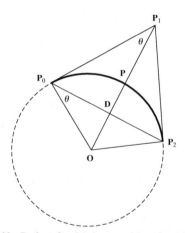

Figure 4.23   Design of a circular arc with rational Bézier curve

$$| \text{ OD } | = | \text{ OP}_0 | \sin \theta$$

and so

$$| \text{ DP}_1 | = | \text{ OP}_1 | - | \text{ OD } | = | \text{ OP}_0 | \left(\frac{1}{\sin \theta} - \sin \theta \right)$$

Also,

$$| \text{ DP } | = | \text{ OP } | - | \text{ OP } | = | \text{ OP}_0 | - | \text{ OD } | = | \text{OP}_0 | (1 - \sin \theta)$$

Thus

$$\frac{| \text{ DP } |}{| \text{ DP}_1 |} = \frac{(1 - \sin \theta) \sin \theta}{(1 - \sin^2 \theta)} = \frac{\sin \theta}{(1 + \sin \theta)} \tag{4.69}$$

Comparing Eqs. (4.68) and (4.68), we have $w = \sin \theta$.

Example 4.11. For given data points $\mathbf{P}_0 = (1, 0)$, $\mathbf{P}_1 = (a, a)$ and $\mathbf{P}_2 = (0, 1)$, determine the circular arcs using rational quadratic Bézier curves for different values of a. Also, draw the corresponding circles of which the arcs are a part.

The included angle is given by

$$2\theta = \cos^{-1} \left\{ \frac{(\mathbf{P}_2 - \mathbf{P}_1) \cdot (\mathbf{P}_0 - \mathbf{P}_1)}{| \mathbf{P}_2 - \mathbf{P}_1 | | \mathbf{P}_2 - \mathbf{P}_1 |} \right\}$$

$$= \cos^{-1} \left\{ \frac{-2a(1 - a)}{a^2 + (1 - a^2)} \right\}$$

using which θ can be computed and the weight $w = \sin \theta$ can be assigned to $\mathbf{P}_1$. The center $\mathbf{O}$ of the circle lies on the lines perpendicular to $\mathbf{P}_0 \mathbf{P}_1$ and $\mathbf{P}_2 \mathbf{P}_1$ with $\mathbf{P}_0$ and $\mathbf{P}_2$ as two points on the circle. The equations of the lines containing the center are

$$y - 1 = \frac{a}{1 - a} x$$

$$y = \frac{1 - a}{a} (x - 1)$$

solving which gives the coordinates of the center as $\left(\frac{1 - a}{1 - 2a}, \frac{1 - a}{1 - 2a} \right)$. The radius r of the circle is

$$| \text{ OP}_0 | = | \text{ OP}_2 | = \sqrt{\frac{(1 - a)^2 + a^2}{(1 - 2a)^2}}$$

Figure 4.24 depicts the circular arcs (thick lines) and the corresponding circles (dashed lines) for different positions of $\mathbf{P}_1$ on the line $y = x$. Note Figure 4.24 (e) for $a = \frac{1}{2}$ when the three points are collinear and the circular arc degenerates to a straight line.

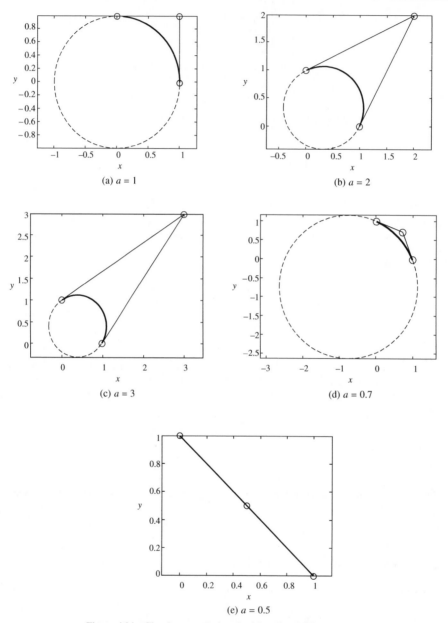

Figure 4.24   Circular arcs designed with rational Bézier segments for different positions of P_1 on $y = x$

EXERCISES

1. Consider a parametric cubic curve $\mathbf{r}(u)$ where

$$\mathbf{r}(u) = F_1\mathbf{P}_0 + F_2\mathbf{P}_1 + F_3\mathbf{P}_0' + F_4\mathbf{P}_1' \quad 0 \le u \le 1$$

   where $F_1 = 1 - 3u^2 + 2u^3$, $F_2 = 3u^2 - 2u^3$, $F_3 = u - 2u^2 + u^3$, $F_4 = -u^2 + u^3$.
   In some situations, data about $\mathbf{P}_0'$ and $\mathbf{P}_1'$ is not available. Instead, vectors $\mathbf{P}_0''$ and $\mathbf{P}_1''$ are known. In such cases, derive the expressions for all elements of $\mathbf{K}$ for the parametric equation to be written in the form

$$\mathbf{r}(u) = \mathbf{U}\,\mathbf{K}\,\mathbf{C}$$

   where $\mathbf{U} = [u^3 \; u^2 \; u \; 1]$, $\mathbf{C}^T = [\mathbf{P}_0 \; \mathbf{P}_1 \; \mathbf{P}_0'' \; \mathbf{P}_1'']$ and $\mathbf{K}$ is the 4×4 matrix.

2. Given a parametric cubic curve whose geometric coefficients are $[\mathbf{P}_0 \; \mathbf{P}_1 \; \mathbf{P}_0' \; \mathbf{P}_1']^T$ snip or trim the curve at $u = 0.7$ and reparametrize this segment so that $0 \le u \le 1$. Find the relationship between the geometric coefficients of the snipped and original curves.

3. Derive the cubic Bézier curve in the matrix form, illustrating the control points, the curve shape, and the blending functions through sketches. Derive also the expression for the tangent at any given point on the curve. Write a computer code to display a 3D cubic Bézier curve. The input shall be the control point coordinates. Shift any one of the given control points to a new location and show the change in shape using a plot. Output also the tangent at any given u value.

4. Consider a Bézier cubic curve obtained by a set of points $\mathbf{P}_0$, $\mathbf{P}_1$, $\mathbf{P}_2$ and $\mathbf{P}_3$. Assume that it is not possible to specify $\mathbf{P}_1$ and $\mathbf{P}_2$ but one can specify $\mathbf{P}_*$, the point of intersection of $\mathbf{P}_0\mathbf{P}_1$ and $\mathbf{P}_2\mathbf{P}_3$. The Bézier curve for $\mathbf{P}_0$, $\mathbf{P}^*$, $\mathbf{P}_2$ will be quadratic one. What will be the relation between $\mathbf{P}^*$, $\mathbf{P}_0$, $\mathbf{P}_1$, $\mathbf{P}_2$ and $\mathbf{P}_3$ so that the cubic as well the quadratic Bézier curves are identical.

5. A parametric cubic curve is to be fitted to pass through (interpolate) four points $\mathbf{P}_0$, $\mathbf{P}_1$, $\mathbf{P}_2$, $\mathbf{P}_3$. The first and last points $\mathbf{P}_0$, $\mathbf{P}_3$ are to be at $u = 0$ and $u = 1$, respectively. Points $\mathbf{P}_1$ and $\mathbf{P}_2$ are at $u = 1/3$ and $u = 2/3$, respectively. The equation of the curve is to be written in the form

$$\mathbf{r}(u) = \mathbf{U}\mathbf{M}_p\mathbf{P} = [u^3 \;\; u^2 \;\; u \;\; 1] \begin{bmatrix} m_{11} & m_{12} & m_{13} & m_{14} \\ m_{21} & m_{22} & m_{23} & m_{24} \\ m_{31} & m_{32} & m_{33} & m_{34} \\ m_{41} & m_{42} & m_{43} & m_{44} \end{bmatrix} \begin{bmatrix} \mathbf{P}_0 \\ \mathbf{P}_1 \\ \mathbf{P}_2 \\ \mathbf{P}_3 \end{bmatrix}$$

   Show that $\mathbf{M}_p$ is given by

$$\mathbf{M}_p = \begin{bmatrix} -4.5 & 13.5 & -13.5 & 4.5 \\ 9.0 & -22.5 & 18 & -4.5 \\ -5.5 & 9.0 & -4.5 & 1.0 \\ 1.0 & 0 & 0 & 0 \end{bmatrix}$$

   (a) Plot the curve passing through $(0, 0)$, $(1, 0)$, $(1, 1)$, $(0, 1)$.
   (b) A circular arc of radius 2 lies in the first quadrant. Write the coordinates of the 4 points that are equally spaced on this arc. Determine the point on the arc at $u = \frac{1}{2}$ using $\mathbf{r}(u)$ above. How far does it deviate from the midpoint of the true quarter circle?

6. In Exercise 5, let $\mathbf{P}_2$ and $\mathbf{P}_3$ be at $u = \alpha$ and $u = \beta(\alpha < \beta < 1)$. Re-derive the expression for the basis matrix $\mathbf{M}_p$.

7. A 3-D parametric cubic curve has the start and end points at $\mathbf{P}_0$ $(0, 0, 0)$ and $\mathbf{P}_1(1, 1, 1)$, and the end tangents are $(1, 0, 0)$ and $(0, 1, 0)$.
   (a) Find and draw the parametric equation of the curve segment.

(b) If the end tangents have the magnitudes as α and β, show some results of the variation in curve shape due to the changes in α, β.

8. The geometric matrix $\mathbf{G}$ of a parametric cubic curve defines a straight line segment if

$$\mathbf{G} = [\mathbf{P}_0 \quad \mathbf{P}_1 \quad \alpha(\mathbf{P}_1 - \mathbf{P}_0) \quad \beta(\mathbf{P}_1 - \mathbf{P}_0)]^{\mathrm{T}}$$

Express the equation of the straight line as a cubic function in u. Tabulate and draw the points on the straight lines at intervals of $\Delta u = 0.01$ from $u = 0$ to $u = 1$ in the following cases:

(a) $\alpha = \beta = 1$
(c) $\alpha = 2, \beta = 4$

(b) $\alpha = \beta = -1$
(d) $\alpha = -2, \beta = -4$

At what values of u the trace of the line changes directions in each of the cases?

9. Write a procedure to truncate a parametric Ferguson segment curve at two specified values of u and subsequently reparametrize it. Test your program for a parametric cubic curve with a given set of end points $\mathbf{P}_0(1, 1, 1)$ and $\mathbf{P}_1$ (4, 2, 4) and the end tangents $\mathbf{r}^u(0) = (1, 1, 0)$ and $\mathbf{r}^u(1) = (1, 1, 1)$ truncated at: (a) $u = 0.25$ and $u = 0.75$, (b) $u = 0.333$ and $u = 0.667$.

10. Write a procedure for blending a Ferguson segment between two given such segments. Create a 2-D numerical example to test your algorithm. Show the effect of changing the magnitudes of the tangent vectors at curve joints.

11. Find the expressions for the curvature at a point on a Ferguson and Bézier segment. Calculate the curvatures at the end points of a Bézier segment having the control points (1, 1), (2, 3), (4, 6), (7, 1). Plot the Bézier curve along with its convex polygon.

12. A composite Bézier curve is to be obtained by joining two Bézier curves with control points at $\mathbf{P}_0$, $\mathbf{P}_1$, $\mathbf{P}_2$, $\mathbf{P}_3$ and $\mathbf{Q}_0$, $\mathbf{Q}_1$, $\mathbf{Q}_2$, $\mathbf{Q}_3$. Develop a procedure and check your results by taking a 2-D example. Modify your results by taking $\mathbf{Q}_0$, $\mathbf{Q}_1$, $\mathbf{Q}_2$, $\mathbf{Q}_3$, $\mathbf{Q}_4$ as control polyline for the second curve.

13. Enumerate conditions to obtain a closed, C^1 continuous Bézier curve.

14. Write a computer program implementing de Casteljau's algorithm for cubic curves, over some interval u_1 and u_2. Test your program with points $\mathbf{P}_0 = (6, -5)$, $\mathbf{P}_1 = (-6, 12)$, $\mathbf{P}_2 = (-6, -14)$, and $\mathbf{P}_3 = (6, 5)$. Use de Casteljau's algorithm to find the coordinates of points on the curve at $u = 0.25, 1/3, 0.5, 2/3, 0.75$ and 1. Plot the cubic curve.

15. Show, through an example, that a Bézier curve is affine under both translation and rotation. You can choose the control points in Exercise 15 and rotate the axes by 45 degrees or translate the origin to $(-2, -2)$ for demonstration.

16. Given a set of control points $\mathbf{P}_0$, $\mathbf{P}_1$, $\mathbf{P}_2$, $\mathbf{P}_3$ explain what happens to a Bézier segment when two of the control points are coincident. Give an example. Does the degree of the curve drop? Does the curve have a cusp at some control point? Does the curve have an inflexion at some control point?

17. Show that the curvature of a planar curve is independent of the parametrization, that is, if $\mathbf{r}(u) = [x(u)\ y(u)]$ is the curve, then a change of variables $u = \varphi(v)$, where $\dot{\varphi}(v) \neq 0$ does not affect the curvature.

18. Let $\mathbf{P}_0$, $\mathbf{P}_1$, $\mathbf{P}_2$, $\mathbf{P}_3$ be given control points. Construct two quadratic segments $\mathbf{Q}_1\mathbf{Q}_2$ $(\mathbf{r}_1(u))$ and $\mathbf{Q}_2\mathbf{Q}_3$ $(\mathbf{r}_2(u))$ such that, for $u \in [0, 1]$ as shown in Figure P4.1.

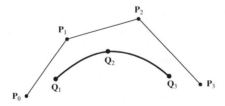

Figure P4.1

$$\mathbf{r}_1(u) = [u^2 \quad u \quad 1] \begin{bmatrix} a_2 & b_2 & c_2 \\ a_1 & b_1 & c_1 \\ a_0 & b_0 & c_0 \end{bmatrix} \begin{bmatrix} \mathbf{P}_0 \\ \mathbf{P}_1 \\ \mathbf{P}_2 \end{bmatrix}$$

$$\mathbf{r}_2(u) = [u^2 \quad u \quad 1] \begin{bmatrix} a_2 & b_2 & c_2 \\ a_1 & b_1 & c_1 \\ a_0 & b_0 & c_0 \end{bmatrix} \begin{bmatrix} \mathbf{P}_1 \\ \mathbf{P}_2 \\ \mathbf{P}_3 \end{bmatrix}$$

$$[u^2 \quad u \quad 1] \begin{bmatrix} a_2 & b_2 & c_2 \\ a_1 & b_1 & c_1 \\ a_0 & b_0 & c_0 \end{bmatrix} = (a_2 u^2 + a_1 u + a_0, b_2 u^2 + b_1 u + b_0, c_2 u^2 + c_1 u + c_0) = \{a(u), b(u), c(u)\}$$

The nine elements of the 3×3 matrix are unknowns and are to be calculated from the following conditions:

(a) The two tangents are to meet at the common point $\mathbf{Q}_2$ with C^1 continuity, that is

$$\mathbf{r}_1(u = 1) = \mathbf{r}_2(u = 0) \quad \text{and} \quad \dot{\mathbf{r}}_1(u = 1) = \dot{\mathbf{r}}_2(u = 0)$$

(b) The entire curve should be independent of the coordinate system used which means that the weights should sum to unity, that is, $a(u) + b(u) + c(u) = 1$.

Show that the matrix is given by

$$\frac{1}{2} \begin{bmatrix} 1 & -2 & 1 \\ -2 & 2 & 0 \\ 1 & 1 & 0 \end{bmatrix}$$

Determine the start and end points $\mathbf{Q}_1, \mathbf{Q}_2, \mathbf{Q}_3$. Draw the curve with the control points given as (1, 2), (3, 6), (7, 10), (12, 3).

Chapter 5

Splines

A natural way of designing a curve (or a surface) is to first sketch a general contour of the curve (or surface) and then make *local changes* in the curve to achieve the required shape. This chapter addresses the following issues:

(1) Local modification over any segment of the curve: One should be able to change the position of a control point in an intuitive way without changing the overall (global) shape of the entire curve.

(2) Delink the number of control points and the degree of the polynomial: One should be able to use lower degree polynomial segments and still maintain a large number of control points to help in shape refinement.

(3) Finer shape control by "knot" insertions: This provides additional tool for designing and local editing of the curve shape.

In Chapter 4 we studied curve design with parametric piecewise curves using Ferguson and Bézier segments. Composite Ferguson curves are naturally C^1 continuous at junction points. However, their design requires specifying the first order (slope) information along with data points which most often is non-intuitive from the designers' perspective. For C^2 or curvature continuous composite Ferguson curves, the slope information is reduced to specification only at the two end points. This curve has no local control for if one changes the location of a data point, the entire curve is altered and needs to be re-computed along with the intermediate slopes. With Bézier segments, only data points are specified. However, individual segments have no local control. Composite Bézier curves further tend to constrain the position of data points of the subsequent segments. For instance, slope continuity at the junction point requires the junction point and its two immediate neighbors to be collinear. Further, C^2 continuity requires four data points around the junction to lie on a plane that contains the junction point itself. Choosing data points freely for composite Bézier curves with relatively lower degree segments therefore is difficult. For this reason, Bézier segments with orders 6 or 8 (degree 5 or 7, respectively) are employed by most CAD softwares. In addition to being parametric piecewise fits, it is also desired for a curve to be inherently C^2 continuous everywhere with local control properties. These design requirements are met by a class of curves called *splines* which are discussed in detail in this chapter.

5.1 Definition

The term *spline* is derived from the analogy to a draughtsman's approach to pass a thin metal or wooden strip through a given set of constrained points called *ducks* (Figure 5.1). We can imagine any

Figure 5.1   **Schematic of the draughtsman's approach and the simply supported beam model**

segment between two consecutive ducks to be a thin simply supported beam across which the bending moment varies linearly. Applying the linearized Euler-Bernoulli beam equation for small deformation

$$EI\chi = EI \frac{d^2 y}{dx^2} = Ax + B \tag{5.1}$$

where EI is the flexural rigidity of the beam, χ the curvature, y the vertical deflection and A and B are known constants. Solving for deflection yields

$$y = \frac{Ax^3}{6EI} + \frac{Bx^2}{2EI} + C_1 x + C_2 \tag{5.2}$$

where C_1 and C_2 are unknown constants. Let l be the length of the beam segment. As the segment is simply supported, at $x = 0$ and l, $y = 0$. Thus

$$C_2 = 0$$

$$C_1 = -\frac{Al^2}{6EI} - \frac{Bl}{2EI}$$

Hence
$$y = \frac{A(l^3 x - x^2)}{6EI} + \frac{B(lx^2 - x)}{2EI} \tag{5.3}$$

This is a cubic equation in $0 \le x \le l$. For continuity of the second derivative at junction points $x = 0$ and l, it is required from Eq. (5.1) that the bending moment, $Ax + B$ compares with the neighboring segments at those points. This is ensured by the equilibrium condition and thus the resulting deflection curve inclusive of all segments is inherently a C^2 continuous curve. A *cubic spline*, therefore, is a *curve for which the second derivative is continuous throughout in the interval of definition*. Note that Eq. (5.1) represents the strong form of the equilibrium condition. Alternatively, the weak form in terms of the strain energy stored in the beam may be written as

$$\text{Minimize: Strain Energy} = \frac{1}{2} \int EI\chi^2 dx; \quad y = 0 \text{ at } x = 0 \text{ and } l \tag{5.4}$$

Eq. (5.4) provides an alternative description of a spline, that is, the *resulting physical spline is a smooth curve for which the strain energy* or the *mean squared curvature is a minimum*. The general mathematical definition of a spline, however, can be extended as:

An nth order ($n - 1$ degree) spline is a curve which is C^{n-2} continuous in the domain of definition, that is, the $(n - 2)$th derivative of the curve exists everywhere in the above domain.

5.2   Why Splines?

The motivation is to develop Bernstein polynomials like basis functions $\Psi_i(t)$ inheriting advantages of barycentric (non-negativity and partition of unity) properties, with a difference that such properties be local, that is, for parameter values of $t \notin [t_0, t_1]$, it is desired that $\Psi_i(t) = 0$ while for $t \in [t_0, t_1]$, $\Psi_i(t) > 0$. By intuition, we may expect $\Psi_i(t)$ to be like a *bell-shaped* function shown in Figure 5.2. Further, if $\Psi_i(t)$ is an *nth* order spline, a linear combination of such weights will inherently be C^{n-2} continuous. Below are discussed various ways of computing the splines in an attempt to mould them into basis functions with local control properties. The treatment and notation of B-spline basis functions, to a large extent, follows from [27].

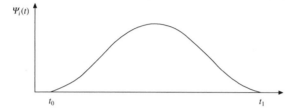

Figure 5.2   Schematic of the basis function as a spline curve $\Psi_i(t)$

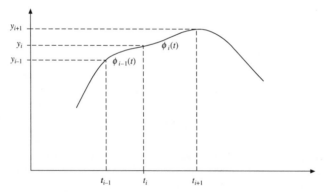

Figure 5.3   Schematic of a polynomial spline

5.3   Polynomial Splines

Let $\Phi(t)$ be a polynomial spline that has values y_i at parameter values t_i, $i = 0, 1, \ldots , n$, with $t_{i-1} < t_i < t_{i+1}$. Further, let $\Phi(t)$ be a cubic spline in each subinterval $[t_{i-1}, t_i]$, with $\Phi(t)$ and its derivatives, $\Phi'(t)$ and $\Phi''(t)$ all continuous at the junction points (t_i, y_i). The t_i, $i = 0, 1, \ldots , n$ are termed as *knots* and $[t_{i-1}, t_i]$, $i = 1, \ldots , n$ as *knot spans*. If the knots are equally spaced (i.e., $t_{i+1} - t_i$ is a constant for $i = 0, 1, \ldots , n{-}1$), the knot vector or the knot sequence is said to be *uniform*; otherwise, it is *non-uniform*.

One way to construct a polynomial spline is as follows. Let $\Phi_i(t)$ represent the spline in the i th span, $t_i \leq t \leq t_{i+1}$. For the first span $t_0 \leq t \leq t_1$, $\Phi_0(t_0)$ and $\Phi_0(t_1)$ are known as y_0 and y_1, respectively. To get a cubic spline, however, two more conditions are required for which let $\Phi_0'(t_0)$ and $\Phi_0''(t_0)$

be known. For the second span, $t_1 \le t \le t_2$, $\Phi_1(t_1) = y_1$ and $\Phi_1(t_2) = y_2$ are known. The remaining two conditions may be obtained by incorporating C^1 and C^2 continuity at $t = t_1$. That is, $\Phi'_1(t_1) = \Phi'_0(t_1)$ and $\Phi''_1(t_1) = \Phi''_0(t_1)$.

Proceeding likewise, cubic segments $\Phi_i(t)$, $i = 0, \ldots, n-1$ over all the knot spans can be determined. In practice however, polynomial splines are not computed in this manner. First, it is not recommended to specify second or higher order derivatives as input since they are usually not very accurate. Second, there may be a possibility for accumulation of errors especially when the number of knot spans is large.

Alternatively, a polynomial spline may be computed as follows. Since $\Phi_i(t)$ is cubic, let

$$\Phi_i(t) = a_0 + a_1 t + a_2 t^2 + a_3 t^3 \tag{5.5}$$

in $t_i \le t \le t_{i+1}$. Also, let s_i and s_{i+1} be the unknown slopes at $t = t_i$ and $t = t_{i+1}$ respectively. The unknowns a_1, $i = 0, \ldots, 3$ can be determined using the following conditions.

$$
\begin{bmatrix}
1 & t_i & t_i^2 & t_i^3 \\
1 & t_{i+1} & t_{i+1}^2 & t_{i+1}^3 \\
0 & 1 & 2t_i & 3t_i^2 \\
0 & 1 & 2t_{i+1} & 3t_{i+1}^2
\end{bmatrix}
\begin{bmatrix} a_0 \\ a_1 \\ a_2 \\ a_3 \end{bmatrix}
=
\begin{bmatrix} y_i \\ y_{i+1} \\ s_i \\ s_{i+1} \end{bmatrix}
\quad \text{or} \quad
\begin{bmatrix} a_0 \\ a_1 \\ a_2 \\ a_3 \end{bmatrix}
=
\begin{bmatrix}
1 & t_i & t_i^2 & t_i^3 \\
1 & t_{i+1} & t_{i+1}^2 & t_{i+1}^3 \\
0 & 1 & 2t_i & 3t_i^2 \\
0 & 1 & 2t_{i+1} & 3t_{i+1}^2
\end{bmatrix}^{-1}
\begin{bmatrix} y_i \\ y_{i+1} \\ s_i \\ s_{i+1} \end{bmatrix}
\tag{5.6}
$$

for which Eq. (5.5) becomes

$$
\Phi_i(t) = \begin{bmatrix} 1 & t & t^2 & t^3 \end{bmatrix}
\begin{bmatrix}
1 & t_i & t_i^2 & t_i^3 \\
1 & t_{i+1} & t_{i+1}^2 & t_{i+1}^3 \\
0 & 1 & 2t_i & 3t_i^2 \\
0 & 1 & 2t_{i+1} & 3t_{i+1}^2
\end{bmatrix}^{-1}
\begin{bmatrix} y_i \\ y_{i+1} \\ s_i \\ s_{i+1} \end{bmatrix}
$$

$$
= \begin{bmatrix} 1 & t & t^2 & t^3 \end{bmatrix}
\begin{bmatrix}
\dfrac{(t_{i+1} - 3t_i)t_{i+1}^2}{h_i^3} & \dfrac{(3t_{i+1} - t_i)t_i^2}{h_i^3} & \dfrac{-t_i t_{i+1}^2}{h_i^2} & \dfrac{-t_i^2 t_{i+1}}{h_i^2} \\
\dfrac{6t_i t_{i+1}}{h_i^3} & \dfrac{-6t_i t_{i+1}}{h_i^3} & \dfrac{(2t_i + t_{i+1})t_{i+1}}{h_i^2} & \dfrac{(t_i + 2t_{i+1})t_i}{h_i^2} \\
\dfrac{-3(t_i + t_{i+1})}{h_i^3} & \dfrac{3(t_i + t_{i+1})}{h_i^3} & \dfrac{-(t_i + 2t_{i+1})}{h_i^2} & \dfrac{-(2t_i + t_{i+1})}{h_i^2} \\
\dfrac{2}{h_i^3} & \dfrac{-2}{h_i^3} & \dfrac{1}{h_i^2} & \dfrac{1}{h_i^2}
\end{bmatrix}
\begin{bmatrix} y_i \\ y_{i+1} \\ s_i \\ s_{i+1} \end{bmatrix}
$$

$$\tag{5.7}$$

Eq. (5.7) ensures that the curve $\Phi(t) = \{\Phi_i(t), i = 0, \ldots, n-1\}$ is position and slope continuous for $t_0 \le t \le t_n$. For continuity of the second derivative, one must impose $\Phi''_{i-1}(t_i) = \Phi''_i(t_i)$. Differentiating Eq. (5.7) twice gives

$$\Phi_i''(t) = [0 \quad 0 \quad 2 \quad 6t] \begin{bmatrix} \dfrac{(t_{i+1} - 3t_i)t_{i+1}^2}{h_i^3} & \dfrac{(3t_{i+1} - t_i)t_i^2}{h_i^3} & \dfrac{-t_i t_{i+1}^2}{h_i^2} & \dfrac{-t_i^2 t_{i+1}}{h_i^2} \\[2ex] \dfrac{6t_i t_{i+1}}{h_i^3} & \dfrac{-6t_i t_{i+1}}{h_i^3} & \dfrac{(2t_i + t_{i+1})t_{i+1}}{h_i^2} & \dfrac{(t_i + 2t_{i+1})t_i}{h_i^2} \\[2ex] \dfrac{-3(t_i + t_{i+1})}{h_i^3} & \dfrac{3(t_i + t_{i+1})}{h_i^3} & \dfrac{-(t_i + 2t_{i+1})}{h_i^2} & \dfrac{-(2t_i + t_{i+1})}{h_i^2} \\[2ex] \dfrac{2}{h_i^3} & \dfrac{-2}{h_i^3} & \dfrac{1}{h_i^2} & \dfrac{1}{h_i^2} \end{bmatrix} \begin{bmatrix} y_i \\ y_{i+1} \\ s_i \\ s_{i+1} \end{bmatrix}$$

$$= \begin{bmatrix} \dfrac{6(2t - t_i - t_{i+1})}{h_i^3} & \dfrac{-6(2t - t_i - t_{i+1})}{h_i^3} & \dfrac{2(3t - t_i - 2t_{i+1})}{h_i^2} & \dfrac{2(3t - 2t_i - t_{i+1})}{h_i^2} \end{bmatrix} \begin{bmatrix} y_i \\ y_{i+1} \\ s_i \\ s_{i+1} \end{bmatrix} \tag{5.8}$$

Likewise

$$\Phi_{i-1}''(t) = \begin{bmatrix} \dfrac{6(2t - t_{i-1} - t_i)}{h_{i-1}^3} & \dfrac{-6(2t - t_{i-1} - t_i)}{h_{i-1}^3} & \dfrac{2(3t - t_{i-1} - 2t_i)}{h_{i-1}^2} & \dfrac{2(3t - 2t_{i-1} - t_i)}{h_{i-1}^2} \end{bmatrix} \begin{bmatrix} y_{i-1} \\ y_i \\ s_{i-1} \\ s_i \end{bmatrix} \tag{5.9}$$

Now $\Phi_{i-1}''(t_i) = \Phi_i''(t_i)$ results in

$$\begin{bmatrix} \dfrac{6}{h_{i-1}^2} & \dfrac{-6}{h_{i-1}^2} & \dfrac{2}{h_{i-1}} & \dfrac{4}{h_{i-1}} \end{bmatrix} \begin{bmatrix} y_{i-1} \\ y_i \\ s_{i-1} \\ s_i \end{bmatrix} = \begin{bmatrix} \dfrac{-6}{h_i^2} & \dfrac{6}{h_i^2} & \dfrac{-4}{h_i} & \dfrac{-2}{h_i} \end{bmatrix} \begin{bmatrix} y_i \\ y_{i+1} \\ s_i \\ s_{i+1} \end{bmatrix}$$

or

$$\frac{s_{i-1}}{h_{i-1}} + 2s_i \left(\frac{1}{h_{i-1}} + \frac{1}{h_i} \right) + \frac{s_{i+1}}{h_i} = \frac{3y_{i+1}}{h_i^2} + 3y_i \left(\frac{1}{h_{i-1}^2} - \frac{1}{h_i^2} \right) - \frac{3y_{i-1}}{h_{i-1}^2}, \quad i = 1, \dots, n-1 \tag{5.10}$$

Note that Eqs. (5.5)-(5.10) describe a generalized Ferguson cubic composite curve for $t_0 \le t \le t_n$. Chapter 4 discusses a particular case wherein each knot span is normalized. That is, $h_0 = h_1 = \dots = h_{n-1} = 1$ for which case Eq. (5.10) is identical to Eq. (4.17). Nevertheless, the exercise above suggests that a C^2 continuous Ferguson curve is a spline. Eqs. (5.10) are linear in $n + 1$ unknowns, $s_0, \dots, s_n$ while the number of equations are only $(n - 1)$. Thus, two additional conditions are needed to determine the unknown second derivatives s_i at each knot t_i. These can be specified using one of the three possibilities:

(i) *Free end*: Where there is no curvature at the end knots, that is, s_0 or $s_n = 0$ at t_i or t_n, respectively. This gives a *natural spline*.

(ii) *Built-in (clamped) end*: Where the first derivatives at t_0 and t_n are specified as $\Phi_0'(t_0) = g_0$ or $\Phi_{n-1}'(t_n) = g_n$, that is

$$\Phi_0'(t_0) = -s_0 h_0/2 - (y_0/h_0 - s_0 h_0/6) + (y_1/h_0 - s_1 h_0/6) = g_0$$

and $\quad \Phi_{n-1}'(t_{n-1}) = -s_{n-1}h_{n-1}/2 - (y_{n-1}/h_{n-1} - s_{n-1}h_{n-1}/6) + (y_n/h_{n-1} - s_n h_{n-1}/6) = g_n \quad$ (5.11)

(iii) *Quadratic end spans*: Where the end spans are quadratic, the end curvatures are constant, that is, $s_0 = s_1$ and $s_{n-1} = s_n$.

We may use different combinations of end conditions from the above. Note that Eq. (5.10) form a tri-diagonal system that can be solved efficiently to get the piecewise composite spline $\Phi(t) = \{\Phi_i(t), i = 0, \ldots, n-1\}$. We can set the values of $y_i, i = 0, \ldots, n$ to shape the polynomial spline as a basis function shown in Figure 5.2. As is, a polynomial spline is a two-dimensional composite curve, however, with few disadvantages. Relocation of one or more data points requires computing the entire spline again. Also, cubic polynomial splines are curvature continuous everywhere implying that it may not be possible to model real life curves with slope or curvature discontinuities.

Example 5.1. Compute a cubic polynomial spline to fit the data points $(0, 0)$, $(1, 3)$ and $(2, 0)$ with free end conditions.

The three knots $t_0 = 0$, $t_1 = 1$ and $t_2 = 2$ are uniformly placed so that $h_0 = h_1 = 1$. From Eq. (5.10), the following equation is to be solved for unknown slopes, s_0, s_1 and s_2.

$$\frac{s_0}{h_0} + 4s_1\left(\frac{1}{h_0} + \frac{1}{h_1}\right) + \frac{s_2}{h_1} = \frac{3y_2}{h_1^2} + 3y_1\left(\frac{1}{h_0^2} - \frac{1}{h_1^2}\right) - \frac{3y_0}{h_0^2}$$

For free end conditions, $s_0 = s_2 = 0$. Further using y_0, y_1 and y_2 as 0, 3 and 0 respectively,

$$4s_1 = 0 \Rightarrow s_1 = 0$$

Using Eq. (5.7)

$$\Phi_0(t) = \begin{bmatrix} 1 & t & t^2 & t^3 \end{bmatrix} \begin{bmatrix} 1 & 0 & 0 & 0 \\ 1 & 1 & 1 & 1 \\ 0 & 1 & 0 & 0 \\ 0 & 1 & 2 & 3 \end{bmatrix}^{-1} \begin{bmatrix} 0 \\ 3 \\ 0 \\ 0 \end{bmatrix} = 9t^2 - 6t^3$$

and

$$\Phi_1(t) = \begin{bmatrix} 1 & t & t^2 & t^3 \end{bmatrix} \begin{bmatrix} 1 & 1 & 1 & 1 \\ 1 & 2 & 4 & 8 \\ 0 & 1 & 2 & 3 \\ 0 & 1 & 4 & 12 \end{bmatrix}^{-1} \begin{bmatrix} 3 \\ 0 \\ 0 \\ 0 \end{bmatrix} = 6t^2 - 27t^2 + 36t - 12$$

Note that $\Phi_0(1) = \Phi_1(1) = 3$. Further, $\Phi_0'(1) = \Phi_1'(1) = 0$ and $\Phi_0''(1) = \Phi_1''(1) = -18$. A plot of the two cubic spline segments is shown in Figure 5.4.

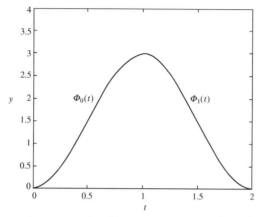

Figure 5.4 Plot of Splines $\Phi_0(t) = 9t^2 - 6t^3$, $0 \le t \le 1$ and $\Phi_1(t) = 6t^3 - 27t^2 + 36t - 12$, $1 \le t \le 2$

5.4 B-Splines (Basis-Splines)

In Section 5.3, generation of a cubic polynomial spline $\Phi(t)$ was discussed with $\Phi_i(t_i) = y_i$ and $\Phi_i(t_{i+1}) = y_{i+1}$ $i = 0, \ldots, n - 1$ with slope and curvature continuity at each knot, t. There were $n+1$ unknowns $s_0, \ldots, s_n$, the first derivatives of the spline at each knot, with $n-1$ equations. Two additional conditions were required to match the number of unknowns. In all, data points and two conditions, that is, $n+3$ conditions were needed to completely determine the spline with n knot spans. Construction of cubic polynomial spline is performed next such that its form appears like a bell-shaped basis function much like the one in Figure 5.4. Consider a cubic spline $\Phi(t)$ with $\Phi(t)$, $\Phi'(t)$ and $\Phi''(t)$ all zero at each end of the knot vector leading to 6 conditions. From above, we can observe that the number of spans n, required to determine a unique cubic spline is 3 ($n+3 = 6$). Let the four knots be denoted by t_{i-3}, t_{i-2}, t_{i-1} and t_i. The solution obtained over these knots, that is, $\Phi(t)$ $\equiv 0$ is trivial however and, therefore, we would need to increase the number of spans by 1 or introduce a new knot, say t_{i-4}. Thus, for $n = 4$ or in the knot span $t_{i-4} \le t \le t_i$, an additional condition is required. (This is because from among the required $n+3 = 7$ conditions, 6 are already known). We can specify a non-zero value of the spline at an internal knot, or alternatively, can standardize the spline. A way suggested by Cox (1972) and de Boor (1972) is

$$\int_{t_{i-4}}^{t_i} \Phi(t)\,dt = \frac{1}{m} \tag{5.12}$$

where m is the order (degree + 1) of the spline. For a cubic spline, $m = 4$. We can realize that computing the cubic spline as above is an arduous procedure. Example 5.2 provides an insight even though it is simplified for a uniform knot span.

Example 5.2. Construct a standard cubic spline over the knot span $t_i = i$, $i = 0, \ldots, 4$.
We may use the fact here that the knot placement being uniform and the boundary conditions being symmetric, the standardized spline will be symmetric about $t = 2$. It is thus required to compute the spline only in two segments, $\Phi_0(t)$ in $0 \le t \le 1$ and $\Phi_1(t)$ in $1 \le t \le 2$. Since the spline is cubic,

$$\Phi_0(t) = a_0 + a_1 t + a_2 t^2 + a_3 t^3$$

Noting that $\Phi_i(0) = \Phi_i'(0) = \Phi_i'(0) = 0$, $a_0 = a_1 = a_2 = 0$ so that $\Phi_0(t) = a_3 t^3$ where a_3 is an unknown. Next, a cubic expression for $\phi_1(t)$ may be written as

$$\Phi_1(t) = b_0 + b_1 t + b_2 t^2 + b_3 t^3$$

As the spline is continuous up to the second derivative, at $t_1 = 1$, we have $\Phi_0(1) = \Phi_1(1)$, $\Phi_0'(1)$, $\Phi_1'(1)$ and $\Phi_0''(1) = \Phi_1''(1)$. These conditions, respectively, yield

$$b_0 + b_1 + b_2 + b_3 = a_3$$

$$b_1 + 2b_2 + 3b_3 = 3a_3$$

$$2b_2 + 6b_3 = 6a_3$$

Solving the above in terms of b_3, we get

$$\Phi_1(t) = (a_3 - b_3) - 3(a_3 - b_3)t + 3(a_3 - b_3)t^2 + b_3 t^3$$

Also, since the spline is symmetric about $t = 2$, it is expected that $\phi_1''(2) = 0$ which gives

$$-3(a_3 - b_3) + 12(a_3 - b_3) + 12b_3 = 0 \quad \text{or} \quad b_3 = -3a_3$$

Thus

$$\Phi_1(t) = 4a_3 - 12a_3 t + 12a_3 t^2 - 3a_3 t^3$$

The unknown constant a_3 can be determined using the standardization integral in Eq. (5.12). Using symmetry and noting that the order of the curve is 4,

$$\int_0^4 \Phi(t)\,dt = 2\int_0^2 \Phi(t)\,dt = 2\int_0^1 \Phi_0(t)\,dt + 2\int_1^2 \Phi_1(t)\,dt$$

$$= \frac{a_3}{2} + \frac{11a_3}{2} = 6a_3 = \frac{1}{m} = \frac{1}{4} \Rightarrow a_3 = \frac{1}{24}$$

Thus, $\Phi_0(t) = \frac{1}{24} t^3$ and $\Phi_1(t) = \frac{1}{24}[4 - 12t + 12t^2 - 3t^3]$. Since $\Phi(t)$ is symmetric about $t = 2$, $\Phi(2 + \delta) = \Phi(2 - \delta)$. For $2 - \delta = t$, $2 + \delta = 4 - t$ and so $\Phi(t) = \Phi(4 - t)$. More specifically, the splines in knot spans $2 \le t \le 3$ and $3 \le t \le 4$ are $\Phi_2(t) = \frac{1}{24}[4 - 12(4 - t) + 12(4 - t)^2 - 3(4 - t)^3]$ and $\Phi_4(t) = \frac{1}{24}(4 - t)^3$. The plot of the computed spline is shown in Figure 5.5.

The general form of the standardized spline in the knot span $t_{i-4} \le t \le t_i$ is shown in Figure 5.6. Note that the spline is extended indefinitely from the end points t_{i-4} to the left and t_i to the right, respectively, on the t axis. Thus, the spline has an indefinite number of spans and is non-zero over precisely 4 spans. It is also termed as a *fundamental spline*, or the spline of *minimal support* the support being the number of spans over which the spline is non-zero. Note that this spline is of the lowest order that can be C^2 continuous for which reason, it is called the fundamental spline. We would realize later in this chapter that such standardized splines have barycentric properties similar

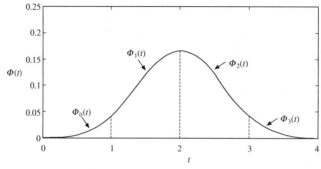

Figure 5.5 **Computed normalized cubic spline with knots $t_i = i$, $i = 0, \ldots ,4$ for Example 5.2**

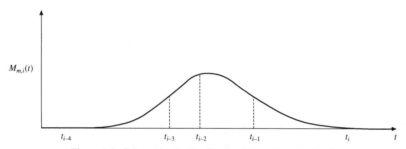

Figure 5.6 **Schematic of a B-spline basis function of order 4**

to the Bernstein polynomials, and thus can be used as weighting or *basis* functions. For this reason, standardized splines are also termed as *basis-* or *B-splines*. In general, a B-spline of order m with the last knot as t_i can be denoted as $M_{m,i}(t)$. Similar to a cubic B-spline, $M_{m,i}(t)$ may be computed with the end conditions, $M_{m,i}(t) = dM_{m,i}(t)/dt = d^2M_{m,i}(t)/dt^2 = \ldots = d^{m-2}M_{m,i}(t)/dt^{m-2} = 0$ at both ends, i.e., $2(m-1)$ conditions with continuity conditions $M_{m,i}(t), dM_{m,i}(t)/dt, d^2M_{m,i}(t)/dt^2, \ldots, d^{m-2}M_{m,i}(t)/dt^{m-2}$ continuous at interior knots. As mentioned above, this method of computing B-splines is quite tedious and requires many algebraic manipulations. Alternatively, the *divided difference* approach may be employed.

5.5 Newton's Divided Difference Method

The divided difference scheme (discussed briefly in Section 3.1) uses the following curve interpolation approach for given points (x_i, y_i), $i = 0, \ldots, n - 1$. A polynomial of degree $n - 1$ can be written as

$$y = p_{n-1}(x)$$
$$= \alpha_0 + \alpha_1 (x - x_0) + \alpha_2 (x - x_0) (x - x_1) + \ldots + \alpha_{n-1} (x - x_0) (x - x_1) \ldots (x - x_{n-2}) \quad (5.13)$$

where the unknown coefficients $\alpha_0, \alpha_1, \ldots, \alpha_{n-1}$ can be determined using the following substitutions.

$$y_0 = p_{n-1}(x_0) = \alpha_0$$
$$y_1 = p_{n-1}(x_1) = \alpha_0 + \alpha_1 (x_1 - x_0) \Rightarrow \alpha_1 = \frac{y_1 - y_0}{x_1 - x_0}$$

$$y_2 = p_{n-1}(x_2) = \alpha_0 + \alpha_1 (x_2 - x_0) + \alpha_2 (x_2 - x_0)(x_2 - x_1)$$

$$\Rightarrow \qquad \alpha_2 = \frac{1}{(x_2 - x_0)}\left(\frac{y_2 - y_1}{x_2 - x_1} - \frac{y_1 - y_0}{x_1 - x_0}\right)$$

$$\cdots$$

$$y_{n-1} = \alpha_0 + \alpha_1(x_{n-1} - x_0) + \ldots + \alpha_{n-1}(x_{n-1} - x_0)(x_{n-1} - x_1) \ldots (x_{n-1} - x_{n-2}) \tag{5.14}$$

The scheme works using forward substitutions with an advantage that if a new data point (x_n, y_n) is introduced, only one unknown α_n needs to be determined without altering the previously calculated coefficients. Note that α_0 depends only on y_0, α_1 depends on y_0 and y_1, α_2 depends on y_0, y_1 and y_2, and so on. This dependence is usually expressed as

$$\alpha_i = y[x_0, x_1, \ldots, x_i] \tag{5.15}$$

with
$$\alpha_0 = y[x_0] = y_0$$

$$\alpha_1 = y[x_0, x_1] = \frac{y_1 - y_0}{x_1 - x_0} = \frac{y[x_1] - y[x_0]}{x_1 - x_0}$$

$$\alpha_2 = y[x_0, x_1, x_2] = \frac{1}{(x_2 - x_0)}\left(\frac{y_2 - y_1}{x_2 - x_1} - \frac{y_1 - y_0}{x_1 - x_0}\right) = \frac{y[x_1, x_2] - y[x_0, x_1]}{x_2 - x_0}$$

Thus, by inspection

$$\alpha_r = y[x_0, x_1, x_2, \ldots, x_r] = \frac{y[x_1, x_2, \ldots, x_r] - y[x_0, x_1, \ldots, x_{r-1}]}{x_r - x_0}$$

It is possible to construct similar entities from any consecutive set of data points. Thus, in general

$$y[x_s, x_{s+1}, x_{s+2}, \ldots, x_r] = \frac{y[x_{s+1}, x_{s+2}, \ldots, x_r] - y[x_s, x_{s+1}, \ldots, x_{r-1}]}{x_r - x_s} \tag{5.16}$$

The expressions $y[x_s, x_{s+1}, x_{s+2}, \ldots, x_r]$ are known as *divided differences* and can be computed in the tabular form (Table 5.1).

Table 5.1 Computation of divided differences

x values	y values	1st differences	2nd differences	3rd differences
x_0	$y[x_0]$			
		$y[x_0, x_1]$		
x_1	$y[x_1]$		$y[x_0, x_1, x_2]$	
		$y[x_1, x_2]$		$y[x_0, x_1, x_2, x_3]$
x_2	$y[x_2]$		$y[x_1, x_2, x_3]$	
		$y[x_2, x_3]$		
x_3	$y[x_3]$			

Figure 5.7 gives the geometric interpretation of the divided differences. For the curve that passes through the specified points (x_i, y_i), $i = 0, \ldots, n-1$, the zeroth divided difference $y[x_s] = y_s$ represents

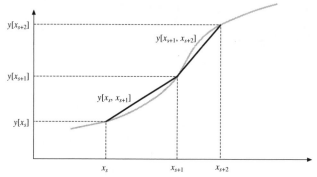

Figure 5.7   Geometric interpretation of the divided differences

some intermediate point on the curve, the first divided differences $y[x_{s+1}, x_{s+2}] = \dfrac{y_{s+2} - y_{s+1}}{x_{s+2} - x_{s+1}}$ and

$y[x_s, x_{s+1}] = \dfrac{y_{s+1} - y_s}{x_{s+1} - x_s}$ represent the slopes, the second divided difference $y[x_s, x_{s+1}, x_{s+2}] =$

$\dfrac{y[x_{s+1}, x_{s+2}] - y[x_s, x_{s+1}]}{x_{s+2} - x_s}$ represents the rate of change of slope or the second derivative of the

curve, and so on.

Thus, an $(n-1)$th divided difference is representative of the $(n-1)$th derivative of a curve. For an $(n-1)$th degree polynomial, the $(n-1)$th divided differences are equal and so the nth divided differences are zero. For instance, for a line, the first divided differences are equal (to the slope) while the second divided differences are zero (since the slope is constant).

In algebraic form, the divided difference, $y[x_j, x_{j+1}, ..., x_{j+k}]$ can be written as

$$y[x_j, x_{j+1}, \dots, x_{j+k}] = \sum_{r=0}^{k} \frac{y_{j+r}}{w'(x_{j+r})} \tag{5.17}$$

where $w(x) = (x - x_j)(x - x_{j+1}) \dots (x - x_{j+k})$ and $w'(x) = dw/dx$.

Example 5.3. Show, using examples, that the result in Eq. (5.17) holds.

For $k = 0$, $y[x_j] = \dfrac{y(x_j)}{w'(x_j)} = y_j$ since $w(x) = (x - x_j)$

For $k = 1$, $y[x_j, x_{j+1}] = \dfrac{y_j}{w'(x_j)} + \dfrac{y_{j+1}}{w'(x_{j+1})}$

Here, $w(x) = (x - x_j)(x - x_{j+1})$ so that $w'(x) = (x - x_{j+1}) + (x - x_j)$

Thus, $w'(x_j) = (x_j - x_{j+1})$ and $w'(x_{j+1}) = (x_{j+1} - x_j)$

On substitution, we get

$$y[x_j, x_{j+1}] = \frac{y_j}{(x_j - x_{j+1})} + \frac{y_{j+1}}{(x_{j+1} - x_j)} = \frac{y_{j+1} - y_j}{x_{j+1} - x_j}$$

For $k = 2$, $y[x_j, x_{j+1}, x_{j+2}] = \dfrac{y_j}{w'(x_j)} + \dfrac{y_{j+1}}{w'(x_{j+1})} + \dfrac{y_{j+2}}{w'(x_{j+2})}$

Here, $w(x) = (x - x_j)(x - x_{j+1})(x - x_{j+2})$ so that

$$w'(x) = (x - x_{j+1})(x - x_{j+2}) + (x - x_j)(x - x_{j+2}) + (x - x_j)(x - x_{j+1})$$

Thus,

$$w'(x_j) = (x_j - x_{j+1})(x_j - x_{j+2}), \; w'(x_{j+1}) = (x_{j+1} - x_j)(x_{j+1} - x_{j+2}) \text{ and } w'(x_{j+2}) = (x_{j+2} - x_j)(x_{j+2} - x_{j+1})$$

This gives

$$y[x_j, x_{j+1}, x_{j+2}]$$

$$= \frac{y_j}{(x_j - x_{j+1})(x_j - x_{j+2})} + \frac{y_{j+1}}{(x_{j+1} - x_j)(x_{j+1} - x_{j+2})} + \frac{y_{j+2}}{(x_{j+2} - x_j)(x_{j+2} - x_{j+1})}$$

$$= \frac{y_j(x_{j+2} - x_{j+1}) - y_{j+1}(x_{j+2} - x_j) + y_{j+2}(x_{j+1} - x_j)}{(x_{j+2} - x_j)(x_{j+1} - x_j)(x_{j+2} - x_{j+1})}$$

$$= \frac{y_j(x_{j+2} - x_{j+1}) - y_{j+1}(x_{j+2} - x_{j+1} + x_{j+1} - x_j) + y_{j+2}(x_{j+1} - x_j)}{(x_{j+2} - x_j)(x_{j+1} - x_j)(x_{j+2} - x_{j+1})}$$

$$= \frac{(x_{j+2} - x_{j+1})(y_j - y_{j+1}) + (x_{j+1} - x_j)(y_{j+2} - y_{j+1})}{(x_{j+2} - x_j)(x_{j+1} - x_j)(x_{j+2} - x_{j+1})}$$

$$= \frac{1}{(x_{j+2} - x_j)}\left[\frac{(y_{j+2} - y_{j+1})}{(x_{j+2} - x_{j+1})} - \frac{(y_{j+1} - y_j)}{(x_{j+1} - x_j)}\right] = \frac{y[x_{j+1}, x_{j+2}] - y[x_j, x_{j+1}]}{(x_{j+2} - x_j)}$$

5.5.1 Divided Difference Method to Compute B-Spline Basis Functions

To compute a B-spline basis function of order m using divided differences, consider a truncated power function (Figure 5.8(a))

$$f(t) = t_+^{m-1} = \begin{cases} t^{m-1}, t \geq 0 \\ 0 \quad\;, t < 0 \end{cases} \tag{5.18}$$

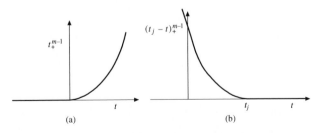

(a) (b)

Figure 5.8 Plots of truncated power functions

Note that $f(t) = f'(t) = f''(t) = \ldots = f^{m-2}(t) = 0$ at $t = 0$. However, $f^{m-1}(t = 0_+) = (m-1)!$ while $f^{m-1}(t = 0-) = 0$ implying that $f^{m-1}(t)$ is discontinuous at $t = 0$. Thus, by definition, $f(t) = t_+^{m-1}$ is a spline of order m over the entire range of t. Next, for some knot t_j, consider the function $f(t_j; t) = (t_j - t)_+^{m-1}$ which is continuous at $t = t_j$ and so are its derivatives up to $m - 2$ as above. Thus, $(t_j - t)_+^{m-1}$ is a spline of order m as well (Figure 5.8(b)). A linear combination of such splines considered over a knot span $t_0, \ldots, t_n$, that is

$$\psi(t) = \sum_{r=0}^{n} \alpha_r (t_r - t)_+^{m-1} \tag{5.19}$$

with non-zero constants α_r will be a spline of order m. This was first established by Sohenberg and Whitney in 1953. A B-spline basis function can be computed as the mth divided difference of the truncated power function $f(t_j; t) = (t_j - t)_+^{m-1}$. Considering t as constant and computing the m th divided difference for $t_j = t_{i-m}, t_{i-m+1}, \ldots, t_i$, we have

$$f[t_{i-m}, t_{i-m+1}, \ldots, t_i; t] = \sum_{r=0}^{m} \frac{(t_{i+r-m} - t)_+^{m-1}}{w'(t_{i+r-m})} = \psi(t) \tag{5.20}$$

where $w(t) = (t - t_{i-m})(t - t_{i-m+1}) \ldots (t - t_i)$ and $w'(t) = dw/dt$. That $\psi(t)$ is a linear combination of individual splines of order m and thus $\psi(t)$ by itself is a spline of the same order is established by Eq. (5.19). Further, $\psi(t)$ is a B-spline basis function $M_{m,i}(t)$ for the following reasons:

(i) $\psi(t) = 0$, for $t > t_i$ since the individual truncated functions $(t_{i+r-m} - t)_+^{m-1}$, $r = 0, \ldots, m$ are all zero.

(ii) $\psi(t) = 0$, for $t \le t_{i-m}$ since $\psi(t)$ is the m th divided difference of a pure $(m-1)$ degree polynomial in t. The mth divided difference is representative of (but not equal to) the mth derivative which is zero for a pure polynomial of degree upto $m - 1$.

(iii) We can further show that $\psi(t)$ is standardized, or

$$\int_{t_{i-m}}^{t_i} \psi(t)\,dt = \int_{t_{i-m}}^{t_i} f[t_{i-m}, t_{i-m+1}, \ldots, t_i; t]\,dt = \frac{1}{m} \tag{5.21}$$

For this, Peano's theorem for divided differences may be used.

$$(m - 1)!\, g[t_{i-m}, t_{i-m+1}, \ldots, t_i; t] = \int_{t_{i-m}}^{t_i} \psi(t) g^m(t)\,dt \tag{5.22}$$

for any $g(t)$. Choosing $g(t) = t^m$, $g^m(t) = (m)!$. Further, the mth divided difference of t^m is 1 (which can be verified using hand calculations for smaller values of m). Thus, Eq. (5.22) becomes

$$(m - 1)! = (m)! \int_{t_{i-m}}^{t_i} \psi(t)\,dt \quad \Rightarrow \quad \int_{t_{i-m}}^{t_i} \psi(t)\,dt = \frac{1}{m}$$

Example 5.3. Show, using plots, that $\psi(t) = f[t_0, t_1, t_2, t_3; t] = \sum_{r=0}^{3} \dfrac{(t_r - t)_+^2}{w'(t_r)}$ is a quadratic B-spline basis function with $w(t) = (t - t_0)(t - t_1) \ldots (t - t_3)$ and $w'(t) = dw/dt$. Assume that $t_0 < t_1 < t_2 < t_3$.

It is required to show that (a) $\psi(t)$ is a quadratic spline, (b) $\psi(t) = 0$ for $t < t_0$ and $t > t_3$, (c) $\psi(t)$ is non-negative for all t and, finally, (d) $\int_{t_0}^{t_3} \psi(t)\,dt = \frac{1}{3}$.

(a) It is known that the truncated power functions of order m (degree $m-1$) are continuous up to the $m-2$ derivatives so that $f[t_0, t_1, t_2, t_3; t]$ is a quadratic spline.

(b) Further, to show that it is a B-spline, we consider the expanded form of $f[t_0, t_1, t_2, t_3; t]$, that is,

$$\psi(t) = f[t_0, t_1, t_2, t_3; t] = \frac{(t_0 - t)_+^2}{w'(t_0)} + \frac{(t_1 - t)_+^2}{w'(t_1)} + \frac{(t_2 - t)_+^2}{w'(t_2)} + \frac{(t_3 - t)_+^2}{w'(t_3)}$$

For $t > t_3$, all truncated functions in $\psi(t)$ are zero and thus $\psi(t)$ is zero. For $t < t_0$, all truncated functions are pure quadratic functions. Ignoring the truncation (+) sign

$$\psi(t < t_0) = \frac{(t_0 - t)^2}{w'(t_0)} + \frac{(t_1 - t)^2}{w'(t_1)} + \frac{(t_2 - t)^2}{w'(t_2)} + \frac{(t_3 - t)^2}{w'(t_3)}$$

Noting that the above is the third divided difference of a quadratic polynomial $(t_j - t)^2$, the tabular form may be used to compute $\psi(t < t_0)$.

t values	$f[t_j; t]$	1st differences	2nd differences	3rd differences
t_0	$(t_0 - t)^2$			
		$[(t_1 - t)^2 - (t_0 - t)^2]/(t_1 - t_0)$ $= t_0 + t_1 - 2t$		
t_1	$(t_1 - t)^2$		$(t_2 - t_0)/(t_2 - t_0) = 1$	
		$[(t_2 - t)^2 - (t_1 - t)^2]/(t_2 - t_1)$ $= t_1 + t_2 - 2t$		$0 = \psi(t < t_0)$
t_2	$(t_2 - t)^2$		$(t_3 - t_1)/(t_3 - t_1) = 1$	
		$[(t_3 - t)^2 - (t_2 - t)^2]/(t_3 - t_2)$ $= t_2 + t_3 - 2t$		
t_3	$(t_3 - t)^2$			

Thus, $\psi(t < t_0) = 0$ which is expected by definition (Eq. (5.20)).

(c) Showing that $\psi(t)$ is non-negative is deferred until the next section though we may be convinced by referring to the plot in Figure 5.9 for $t_0 = 0$, $t_1 = 2$, $t_2 = 3$ and $t_3 = 6$.

(d) Showing that $\psi(t)$ is standardized can be done using Eq. (5.22) for $m = 3$.

5.6 Recursion Relation to Compute B-Spline Basis Functions

Eq. (5.20) provides two ways to compute the B-spline basis function of order m either by computing the divided differences in the tabular form as in the left hand side or computing it algebraically as in the right hand side. The third alternative proposed by Cox and de Boor (1972) is the *recursion relation* that can be derived from divided differences. Using Leibnitz result on divided differences of the product of two functions $h(t) = f(t)g(t)$, we have

$$h[t_0, t_1, \ldots, t_k] = \sum_{r=0}^{k} f[t_0, t_1, \ldots, t_r]g[t_r, t_{r+1}, \ldots, t_k]$$

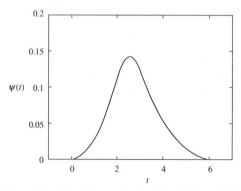

Figure 5.9   Plot of the quadratic spline in Example 5.3

$$= f[t_0]g[t_0, t_1, \ldots, t_r] + f[t_0, t_1]g[t_1, \ldots, t_k] + \ldots + f[t_0, t_1, \ldots, t_{k-1}]g[t_{k-1}, t_k]$$

$$+ f[t_0, t_1, \ldots, t_k] \, g[t_k] \tag{5.23}$$

For $h_k(t_j; t) = (t_j - t)_+^{k-1} = (t_j - t)_+^{k-2} (t_j - t)_+ = h_{k-1}(t_j; t) (t_j - t)_+$ using Eq. (5.23) yields

$$h_k[t_{i-k}, \ldots, t_i; t] = h_{k-1}[t_{i-k}, \ldots, t_{i-1}; t] + h_{k-1}[t_{i-k}, \ldots, t_i; t](t_i - t)$$

where $h_k[t_{i-k}, \ldots, t_i; t]$ is the kth divided difference of $(t_j - t)_+^{k-1}$ and hence is a B-spline $M_{k,i}(t)$. Likewise, $M_{k-1, i-1}(t) = h_{k-1}[t_{i-k}, \ldots, t_{i-1}; t]$. $h_{k-1}[t_{i-k}, \ldots, t_i; t]$ may be expressed as $\dfrac{h_{k-1}[t_{i-k+1}, \ldots, t_i; t] - h_{k-1}[t_{i-k}, \ldots, t_{i-1}; t]}{t_i - t_{i-k}}$ using Eq. (5.16).

Thus, the above relation becomes

$$M_{k,i}(t) = M_{k-1, i-1}(t) + \frac{t_i - t}{t_i - t_{i-k}} \{h_{k-1}[t_{i-k+1}, \ldots, t_i; t] - h_{k-1}[t_{i-k}, \ldots, t_{i-1}; t]\}$$

or

$$M_{k,i}(t) = M_{k-1, i-1}(t) + \frac{t_i - t}{t_i - t_{i-k}} \{M_{k-1, i}(t) - M_{k-1, i-1}(t)\}$$

or

$$M_{k,i}(t) = \frac{t - t_{i-k}}{t_i - t_{i-k}} M_{k-1, i-1}(t) + \frac{t_i - t}{t_i - t_{i-k}} M_{k-1, i}(t) \tag{5.24}$$

Eq. (5.24) is the recursion relation to compute B-spline basis functions. It appears very similar to the de Casteljau's algorithm discussed in Chapter 4 that employs repeated linear interpolation between data points to compute Bézier curves. Only here, repeated linear interpolation is performed between two consecutive splines of one order less. Like in the de Casteljau's algorithm, a table for constructing splines may also be generated as in Table 5.2.

Table 5.2 also suggests that $M_{k,i}(t)$ can be computed once splines of order 1, i.e. $M_{1, i-k+1}(t)$, $M_{1, i-k+2}(t), \ldots, M_{1,i}(t)$ are all known. As $M_{k,i}(t)$ is non-zero in the knot span $t_{i-k} \le t < t_i$ and zero elsewhere, $M_{1,i}(t)$ is non-zero only in one span $t_{i-1} \le t < t_i$ and can be computed using the standardization condition in Eq. (5.12). Note that being of degree 0, $M_{1,i}(t)$ is constant. Thus

Table 5.2   Recursion to compute B-spline basis functions

$[t_{i-k}, t_{i-k+1})$	$M_{1,i-k+1}(t)$	
		$M_{2,i-k+2}(t)$
$[t_{i-k+1}, t_{i-k+2})$	$M_{1,i-k+2}(t)$	
		$M_{2,i-k+3}(t)$
$[t_{i-k+2}, t_{i-k+3})$	$M_{1,i-k+3}(t)$	
$\vdots$	$\cdots$	
$[t_{i-3}, t_{i-2})$	$M_{1,i-2}(t)$	
		$M_{2,i-1}(t)$
$[t_{i-2}, t_{i-1})$	$M_{1,i-1}(t)$	
		$M_{2,i}(t)$
$[t_{i-1}, t_i)$	$M_{1,i}(t)$	

$M_{k-1,i-1}(t)$, $M_{k-1,i}(t)$ → $M_{k,i}(t)$

$$\int_{t_{i-1}}^{t_i} M_{1,i}(t)\,dt = M_{1,i}(t) \int_{t_{i-1}}^{t_i} dt = M_{1,i}(t)(t_i - t_{i-1}) = 1$$

or

$$M_{1,i}(t) = \frac{1}{t_i - t_{i-1}} \quad \text{for} \quad t \in [t_{i-1}, t_i); \tag{5.25}$$

$$= 0 \text{ elsewhere}$$

Combining Eqs. (5.24) and (5.25), the recursion relation for a B-spline basis function may be written as

$$M_{1,i}(t) = \frac{1}{t_i - t_{i-1}} \quad \text{for} \quad t \in [t_{i-1}, t_i);$$

$$= 0 \text{ elsewhere}$$

$$M_{k,i}(t) = \frac{t - t_{i-k}}{t_i - t_{i-k}} M_{k-1,i-1}(t) + \frac{t_i - t}{t_i - t_{i-k}} M_{k-1,i}(t) \quad \text{for} \quad t \in [t_{i-k}, t_i); \tag{5.26}$$

$$= 0 \text{ elsewhere}$$

5.6.1 Normalized B-Spline Basis Functions

The normalized B-spline weight $N_{k,i}(t)$, which are used more frequently in the design of B-spline curves may be computed as

$$N_{k,i}(t) = (t_i - t_{i-k})M_{k,i}(t) \tag{5.27}$$

From Eqs. (5.25) and (5.27), $N_{1,i}(t) = (t_i - t_{i-1})M_{1,i}(t) = 1$ for t in the range $[t_{i-1}, t_i)$ and $N_{1,i}(t) = 0$ for all other values of t. We may combine the two results as $N_{1,i}(t) = \delta_i$, where $\delta_i = 1$ for $t \in [t_{i-1}, t_i)$ and $\delta_i = 0$ elsewhere. For higher order normalized B-splines, the recursion relation can be derived using Eqs. (5.24) and (5.27). Starting with (5.24)

$$M_{k,i}(t) = \frac{t - t_{i-k}}{t_i - t_{i-k}} M_{k-1,i-1}(t) + \frac{t_i - t}{t_i - t_{i-k}} M_{k-1,i}(t)$$

$$\Rightarrow \qquad \frac{N_{k,i}(t)}{t_i - t_{i-k}} = \frac{t - t_{i-k}}{t_i - t_{i-k}} \frac{N_{k-1,i-1}(t)}{t_{i-1} - t_{i-k}} + \frac{t_i - t}{t_i - t_{i-k}} \frac{N_{k-1,i}(t)}{t_i - t_{i-k+1}}$$

$$\Rightarrow \qquad N_{k,i}(t) = \frac{t - t_{i-k}}{t_{i-1} - t_{i-k}} N_{k-1,i-1}(t) + \frac{t_i - t}{t_i - t_{i-k+1}} N_{k-1,i}(t)$$

The recursion relation to compute normalized B-splines is then

$$N_{1,i}(t) = \delta_i \text{ such that } \delta_i = 1 \text{ for } t \in [t_{i-1}, t_i)$$

$$= 0, \text{ elsewhere}$$

$$N_{k,i}(t) = \frac{t - t_{i-k}}{t_{i-1} - t_{i-k}} N_{k-1,i-1}(t) + \frac{t_i - t}{t_i - t_{i-k+1}} N_{k-1,i}(t) \qquad (5.28)$$

Normalized B-splines may be used as basis functions to generate B-spline curves as Hermite and Bernstein polynomials are used in designing Ferguson and Bézier curves, respectively (Chapter 4). In that regard, a study of the properties of B-spline basis functions becomes essential. It may be mentioned that in some publications the notation $N_{i,k}(t)$ is used instead of $N_{k,i}(t)$, where k is the degree of the polynomial in t and t_i is the first knot value.

5.7 Properties of Normalized B-Spline Basis Functions

(A) $N_{k,i}(t)$ is a degree $k–1$ polynomial in t
From Eqs. (5.20) and (5.24), $M_{k,i}(t)$ is a piecewise polynomial of degree $k–1$ in the knot span $[t_{i-k}, \ldots t_i)$ and therefore from Eq. (5.27), $N_{k,i}(t)$ is a polynomial of degree $k–1$.

(B) Non-negativity: For all i, k and t, $N_{k,i}(t)$ is non-negative
The property can be deduced by induction. In a given knot span $t_{i-k} < t_{i-k+1} < \ldots < t_i$,

$N_{1,i}(t) = 1$ for $t \in [t_{i-1}, t_i)$

$N_{1,i}(t) = 0$ elsewhere from Eq. (5.28)

Thus, $N_{1,i}(t) \geq 0$ in $[t_{i-k}, t_i)$

Similarly, $N_{1,i-1}(t) = 1$ for $t \in [t_{i-2}, t_{i-1})$ and $N_{1,i-1}(t) = 0$ elsewhere

Also, $N_{1,i-2}(t) = 1$ for $t \in [t_{i-3}, t_{i-2})$ and $N_{1,i-2}(t) = 0$ elsewhere

Thus, both $N_{1,i-1}(t) \geq 0$ and $N_{1,i-2}(t) \geq 0$ in $[t_{i-k}, t_i)$ $\qquad (5.29)$

From Eqs. (5.28) and (5.29),

$$N_{2,i}(t) = \frac{t - t_{i-2}}{t_{i-1} - t_{i-2}} N_{1,i-1}(t) + \frac{t_i - t}{t_i - t_{i-1}} N_{1,i}(t) \text{ for } t \in [t_{i-2}, t_i) \text{ and } N_{2,i}(t) = 0 \text{ elsewhere.}$$

Now, $\qquad\qquad N_{2,i}(t) = \frac{t - t_{i-2}}{t_{i-1} - t_{i-2}} \geq 0 \text{ for } t \in [t_{i-2}, t_{i-1}]$

$$= \frac{t_i - t}{t_i - t_{i-1}} \geq 0 \text{ for } t \in [t_{i-1}, t_i)$$

$$= 0 \text{ elsewhere}$$

thus, $\qquad\qquad\qquad\qquad N_{2,i}(t) \geq 0$ for t in $[t_{i-k}, t_i)$ $\qquad\qquad\qquad$ (5.30)

Likewise, $\qquad\qquad\qquad N_{2,i-1}(t) \geq 0$ for t in $[t_{i-k}, t_i)$ $\qquad\qquad$ (5.31)

Next, assume that Eqs. (5.30) and (5.31) are true for the $(k-1)$th order normalized splines, that is

$$\mathrm{N}_{k-1,i}(t) \geq 0,\ t \in\ [t_{i-k}, t_i)\ \text{and}\ \mathrm{N}_{k-1,i-1}(t) \geq 0,\ t \in\ [t_{i-k}, t_i) \qquad (5.32)$$

Then, using Eq. (5.28)

$$N_{k,i}(t) = \frac{t - t_{i-k}}{t_{i-1} - t_{i-k}}\, N_{k-1,i-1}(t) + \frac{t_i - t}{t_i - t_{i-k+1}}\, N_{k-1,i}(t)$$

From Eq. (5.32) and additionally, since $\dfrac{t - t_{i-k}}{t_{i-1} - t_{i-k}} \geq 0$ and $\dfrac{t_i - t}{t_i - t_{i-k+1}} \geq 0$ for $t \in\ [t_{i-k},\ t_i)$, $N_{k,i}(t) \geq 0$ for t in $[t_{i-k}, t_i)$.

Example 5.4. Verify using plots, the non-negativity property of $N_{4,i}(t)$ with knots $t_{i-4} = 0$, $t_{i-3} = 1$, $t_{i-2} = 2$, $t_{i-1} = 3$, $t_i = 4$.

The plot for various normalized B-splines is shown in Figure 5.10. $N_{1,j}(t)$, $j = 1, \ldots, 4$ are step functions which are equal to 1 in $[t_{j-1}, t_j)$ and are zero otherwise. $N_{2,j}(t)$, $j = 2, \ldots, 4$ are linear triangle-shaped functions, $N_{3,j}(t)$, $j = 3, 4$ are the inverted bell-shaped quadratics while $N_{4,4}(t)$ is the cubic B-spline function (thickest solid line). Note that all splines are non-zero in their domains of definition.

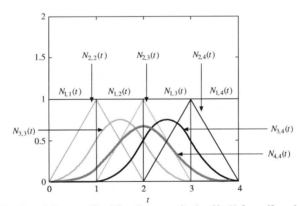

Figure 5.10   Plot of the normalized B-splines constituting $N_{4,i}(t)$ for uniform knot spacing

(C) Local support: $N_{k,i}(t)$ is a non-zero polynomial in (t_{i-k}, t_i)
From Eq. (5.20) $M_{k,i}(t) = 0$ for $t \geq t_i$. Since $M_{k,i}(t)$ is the kth divided difference in $[t_{i-k}, t_i)$ of a linear combination of pure polynomials of degree $k-1$ each, $M_{k,i}(t) = 0$ for $t \leq t_{i-k}$. Thus, from Eq. (5.27), $N_{k,i}(t) = 0$ for $t \leq t_{i-k}$ and $t \geq t_i$. From Eq. (5.21), the integral of $M_{k,i}(t)$ over the interval is $1/k$. This implies that $N_{k,i}(t)$ over $[t_{i-k}, t_i)$ is at least not zero entirely in the interval. Additionally, the non-negativity property above suggests that $N_{k,i}(t)$ does not have any root in $[t_{i-k}, t_i)$ and so is a non-zero polynomial in (t_{i-k}, t_i). Thus, in a given parent knot sequence $(t_0, t_1, \ldots, t_n)$, all B-spline functions $N_{k,i}(t)$, $i = k, \ldots, n$ $(k \leq n)$ have their subdomains in $[t_{i-k}, t_i)$ wherein they are non-zero.

(D) On any span $[t_i, t_{i+1})$, at most p order p normalized B-spline functions are non-zero

This follows from the local support property mentioned above. For any r, $N_{p,r}(t) \geq 0$ in the knot span $[t_{r-p}, t_r)$. So that $[t_i, t_{i+1})$ is contained in $[t_{r-p}, t_r)$, it must be ensured that there is at least one order p B-spline with t_i as the first knot and at least one with t_{i+1} as the last knot. Thus, $r - p = i$ and $r = i + 1$ provide the range in r, that is, $r = i + 1, \ldots, i + p$ for which $N_{p,r}(t)$ is non-zero in $[t_i, t_{i+1})$. This adds up to p B-splines. Figure 5.11 demonstrates this property for $p = 4$ (a normalized cubic B-spline). It is this property that provides local control when reshaping B-spline curves discussed later.

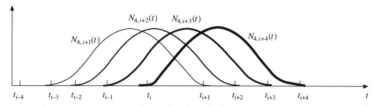

Figure 5.11 **Schematic of the normalized fourth order B-splines that are non-zero over $[t_i, t_{i+1})$**

(E) Partition of unity*: The sum of all non-zero order p B-spline functions over the span $[t_i, t_{i+1})$ is 1

Example 5.5. Demonstrate the partition of unity property for quadratic normalized B-splines.
We know from above that over $[t_i, t_{i+1})$, the quadratic normalized B-splines $N_{3,i+1}(t)$, $N_{3,i+2}(t)$ and $N_{3,i+3}(t)$ are non-zero.

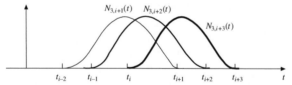

Figure 5.12 **Schematic of the normalized third order non-zero B-splines over $[t_i, t_{i+1})$**

Using Eq. (5.28), first $N_{3,i}(t)$ is computed.

$$N_{2,i-1}(t) = \frac{t - t_{i-3}}{t_{i-2} - t_{i-3}} \delta_{i-2} + \frac{t_{i-1} - t}{t_{i-1} - t_{i-2}} \delta_{i-1}$$

$$N_{2,i}(t) = \frac{t - t_{i-2}}{t_{i-1} - t_{i-2}} \delta_{i-1} + \frac{t_i - t}{t_i - t_{i-1}} \delta_i$$

$$N_{3,i}(t) = \frac{t - t_{i-3}}{t_{i-1} - t_{i-3}} N_{2,i-1}(t) + \frac{t_i - t}{t_i - t_{i-2}} N_{2,i}(t)$$

---

*This property can be proved using mathematical induction for a generic case.

$$= \frac{t - t_{i-3}}{t_{i-1} - t_{i-3}} \left[\frac{t - t_{i-3}}{t_{i-2} - t_{i-3}} \delta_{i-2} + \frac{t_{i-1} - t}{t_{i-1} - t_{i-2}} \delta_{i-1} \right] + \frac{t_i - t}{t_i - t_{i-2}} \left[\frac{t - t_{i-2}}{t_{i-1} - t_{i-2}} \delta_{i-1} + \frac{t_i - t}{t_i - t_{i-1}} \delta_i \right]$$

or

$$N_{3,i}(t) = \frac{(t - t_{i-3})^2}{(t_{i-1} - t_{i-3})(t_{i-2} - t_{i-3})} \delta_{i-2} + \left[\frac{(t - t_{i-3})(t_{i-1} - t)}{(t_{i-1} - t_{i-3})(t_{i-1} - t_{i-2})} + \frac{(t_i - t)(t - t_{i-2})}{(t_i - t_{i-2})(t_{i-1} - t_{i-2})} \right] \delta_{i-1}$$

$$+ \frac{(t_i - t)^2}{(t_i - t_{i-1})(t_i - t_{i-2})} \delta_i \qquad (5.33)$$

Thus

$$N_{3,i+1}(t) = \frac{(t - t_{i-2})^2}{(t_i - t_{i-2})(t_{i-1} - t_{i-2})} \delta_{i-1} + \left[\frac{(t - t_{i-2})(t_i - t)}{(t_i - t_{i-2})(t_i - t_{i-1})} + \frac{(t_{i+1} - t)(t - t_{i-1})}{(t_{i+1} - t_{i-1})(t_i - t_{i-1})} \right] \delta_i$$

$$+ \frac{(t_{i+1} - t)^2}{(t_{i+1} - t_i)(t_{i+1} - t_{i-1})} \delta_{i+1}$$

$$N_{3,i+2}(t) = \frac{(t - t_{i-1})^2}{(t_{i+1} - t_{i-1})(t_i - t_{i-1})} \delta_i + \left[\frac{(t - t_{i-1})(t_{i+1} - t)}{(t_{i+1} - t_{i-1})(t_{i+1} - t_i)} + \frac{(t_{i+2} - t)(t - t_i)}{(t_{i+2} - t_i)(t_{i+1} - t_i)} \right] \delta_{i+1}$$

$$+ \frac{(t_{i+2} - t)^2}{(t_{i+2} - t_{i+1})(t_{i+2} - t_i)} \delta_{i+2}$$

and

$$N_{3,i+3}(t) = \frac{(t - t_i)^2}{(t_{i+2} - t_i)(t_{i+1} - t_i)} \delta_{i+1} + \left[\frac{(t - t_i)(t_{i+2} - t)}{(t_{i+2} - t_i)(t_{i+2} - t_{i+1})} + \frac{(t_{i+3} - t)(t - t_{i+1})}{(t_{i+3} - t_{i+1})(t_{i+2} - t_{i+1})} \right] \delta_{i+2}$$

$$+ \frac{(t_{i+3} - t)^2}{(t_{i+3} - t_{i+2})(t_{i+3} - t_{i+1})} \delta_{i+3}$$

Since for $t \in [t_i, t_{i+1})$, $\delta_{i+1} = 1$ and $\delta_{i-1} = \delta_i = \delta_{i+2} = \delta_{i+3} = 0$, we have

$$N_{3,i+1}(t) + N_{3,i+2}(t) + N_{3,i+3}(t) = \frac{(t_{i+1} - t)^2}{(t_{i+1} - t_i)(t_{i+1} - t_{i-1})} + \frac{(t - t_{i-1})(t_{i+1} - t)}{(t_{i+1} - t_{i-1})(t_{i+1} - t_i)}$$

$$+ \frac{(t_{i+2} - t)(t - t_i)}{(t_{i+2} - t_i)(t_{i+1} - t_i)} + \frac{(t - t_i)^2}{(t_{i+2} - t_i)(t_{i+1} - t_i)}$$

$$= \frac{(t_{i+1} - t)(t_{i+1} - t_{i-1})}{(t_{i+1} - t_i)(t_{i+1} - t_{i-1})} + \frac{(t - t_i)(t_{i+2} - t_i)}{(t_{i+2} - t_i)(t_{i+1} - t_i)}$$

$$= \frac{(t_{i+1} - t)}{(t_{i+1} - t_i)} + \frac{(t - t_i)}{(t_{i+1} - t_i)} = \frac{(t_{i+1} - t_i)}{(t_{i+1} - t_i)} = 1$$

(F) For $m + 1$ number of knots, degree $p-1$ basis functions and $n + 1$ number of control points, $m = n + p$

For $n+1$ control points and hence basis functions of order p, this property puts a limit on the number of knots. The first normalized spline on the knot set $[t_0, t_m)$ is $N_{p,p}(t)$ while the last spline on this set is $N_{p,m}(t)$ making a total of $m - p + 1$ basis splines. Letting $n + 1 = m - p + 1$ gives the result ($m = n + p$).

(G)   Multiple knots: Some knots in a given knot span may be equal for which some knot spans may not exist

If a knot t_i appears k times (i.e. $t_{i-k+1} = t_{i-k+2} = \ldots = t_i$), where $k > 1$, t_i is called a *multiple knot* or a knot of *multiplicity k*. Otherwise, for $k = 1$, t_i is termed as a *simple knot*. Multiple knots significantly change the properties of basis functions and are very useful in the design of B-spline curves. We may note here that to ensure right continuity of $M_{k,i}(t)$ and $N_{k,i}(t)$ in case k consecutive knots coincide, one assumes $\frac{0}{0} = 0$ convention when computing B-spline basis functions. If $t_{i-1} = t_i$, then $M_{1,i}(t)$ and $N_{1,i}(t)$ are defined as *zero*. Some properties of normalized B-splines with multiple knots are as follows.

G1: At a knot i of multiplicity k, the basis function $N_{p,i}(t)$ is C^{p-1-k} continuous at that knot

Example 5.6. Verify using plots the discontinuity property above for quadratic normalized B-splines using the parent knot sequence [0 1 2 3].

The plots for $N_{3,i}(t)$ using Eq. (5.33) are shown in Figure 5.13 for the knot sequence: (a) $[t_{i-3}, t_{i-2}, t_{i-1}, t_i] \equiv [0\ 1\ 2\ 3]$, (b) [0 1 3 3] ($k = 2$ at $t_i = 3$) and (c) [0 3 3 3] ($k = 3$ at $t_i = 3$). For knots with multiplicity 1, $N_{3,i}(t)$ is expected to be C^1 continuous everywhere, especially at the knot value 3 where the slope is zero. Raising knot multiplicity by one at knot value 3 results in slope discontinuity as for $t \to 3_-$, the slope is non-zero while for $t \to 3_+$, the slope is zero. Thus, the B-spline is C^0 continuous at $t = 3$. Further increase in knot multiplicity by 1 at knot value 3 makes $N_{3,i}(t)$ position discontinuous since for $t \to 3_-$, $N_{3,i}(t) = 1$ while for $t \to 3_+$, $N_{3,i}(t) = 0$.

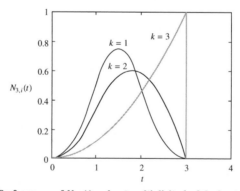

Figure 5.13   Performance of $N_{3,i}(t)$ as knot multiplicity k of the knot at 3 is increased

G2. Over each internal knot of multiplicity k, the number of non-zero order p basis functions is at most $p-k$

The property is elucidated using normalized order 4 (cubic) B-splines. Figure 5.14(a) shows three such splines, that is, $N_{4,i-3}$, $N_{4,i-2}$ and $N_{4,i-1}$ over the knot span $[t_{i-7}, t_{i-1})$ which are non-zero over a simple knot t_{i-4} concurring with the property for $p = 4$ and $k = 1$. If knot t_{i-3} is moved to t_{i-4}

(Figure 5.14b) making the latter of multiplicity $k=2$, the total number of non-zero B-splines over t_{i-4} gets reduced from three to two ($p-k = 4-2$) since $N_{4,i-3}$ gets eliminated from that set. Further, for $t_{i-4} = t_{i-3} = t_{i-2}$ raising the multiplicity of t_{i-4} to $k = 3$, $N_{4,i-2}$ gets removed from the set of non-zero splines (Figure 5.14(c)) leaving only $N_{4,i-1}$ ($p - k = 4-3$) in the set and for $t_{i-4} = t_{i-3} = t_{i-2} = t_{i-1}$ with $k = 4$ for t_{i-4} no ($p-k = 4-4$) non-zero splines over t_{i-4} exists (Figure 5.14(d)).

Note that due to **G1**, $N_{4,i-1}(t)$ would be position discontinuous at t_{i-1}.

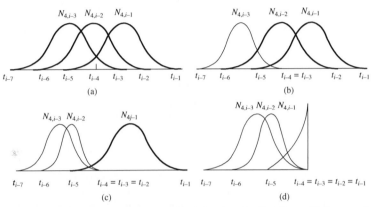

Figure 5.14   Schematic behavior of the normalized fourth order B-splines with increase in the knot multiplicity; non-zero splines are shown using thick lines

5.8   B-Spline Curves: Definition

With an insight into the properties of normalized B-splines shape functions, we may now attempt to design B-spline curves. Given $n+1$ control points $\mathbf{b}_0, \mathbf{b}_1, \ldots, \mathbf{b}_n$ and a knot vector $\mathbf{T} = \{t_0, t_1, \ldots, t_m\}$, the B-spline curve, $\boldsymbol{b}(t)$ of order p may be expressed as a weighted linear combination using the normalized B-spline functions as

$$\mathbf{b}(t) = \sum_{i=0}^{n} N_{p,p+i}(t)\mathbf{b}_i \tag{5.34}$$

This form of B-spline curve is very similar to a Bézier curve wherein the basis functions are the Bernstein polynomials. The degree of the Bernstein basis functions is one less than the number of control points for a Bézier segment. However, in case of B-spline curves, the degree of the basis functions is an independent choice specified by the user. The number of knots ($m+1$) get determined by the relation $m = n + p$ with the total number of basis functions ($n+1$) being the same as the number of control points (Eq. (5.34) and p being the order of the basis functions and hence the curve. The spline in Eq. (5.34) is called an approximating spline as the curve usually does not pass through the data points. However, a B-spline curve is more proximal to the control polyline than a Bézier segment.

Though Eq. (5.34) is valid for all t in $[-\infty, \infty]$, $\mathbf{b}(t) = 0$ for $t \leq t_0$ and $t > t_m$. Thus, restricting the parameter range in $[t_0, t_m)$ seems reasonable. A more restrictive range for t may be one in which *full support* of the basis functions is achieved, that is, over any knot span $[t_j, t_{j+1})$ in $[t_0, t_m)$, atmost p B-splines of order p are non-zero (Section 5.7D). The first such span is $[t_{p-1}, t_p)$ where p basis functions $N_{p,p}, \ldots, N_{p,2p-1}$ are non-zero (Figure 5.15) while the last span is $[t_{m-p}, t_{m-p+1})$ where basis functions $N_{p,m-p+1}, \ldots, N_{p,m}$ are non-zero. Combining the two results gives the range $[t_{p-1}, t_{m-p+1})$ wherein for any value of t, it is assured that there are always p basis splines that are non-zero. The above discussion assumes that all knots used are simple knots.

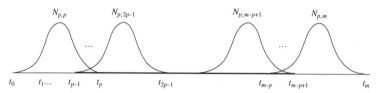

Figure 5.15   Parametric range (thick line) for B-spline curves with full support of basis functions.

Example 5.7 discusses two kinds of B-spline curves, viz. *unclamped* and *clamped*. In the former, the curve does not pass through any control point while in the latter, it passes through one or both the end points. If a spline is to be clamped at the first control point, we can enforce $p - 2$ knots to the left of t_{p-1} to be equal, that is, $t_1 = \ldots = t_{p-2} = t_{p-1}$. Likewise, for the curve to pass through the last control point, $p - 2$ knots to the right of t_{m-p+1} must be equal, to the latter that is, $t_{m-p+1} = t_{m-p+2} = \ldots = t_{m-1}$. Section 5.9(b) discusses why a B-spline curve passes through a data point using knot multiplicity. A B-spline curve clamped at both ends behaves like a Bézier curve that passes through the end points and is also tangent to the first and the last leg of the control polyline.

Example 5.7. For data points, $(0, 0)$, $(0, 1)$, $(2, 3)$, $(2.5, 6)$, $(5, 2)$, $(6, 0)$ and $(7, -3)$, design a B-spline curve using cubic normalized B-spline basis.

It is required to use the fourth order B-splines for 7 data points. Using Section 5.7 (F), the number of knots is determined as $7 + 4 = 11$. First, to generate an open spline, all knots must be simple (with multiplicity 1 each) and we choose a uniform sequence as $[0, 1, 2, \ldots, 10]$. Computing the B-splines using Eq. (5.28) and applying Eq. (5.34), we get the plot in Figure 5.16(a) for t in $[3, 7]$. The thin line shows the control polyline while the thick line shows the B-spline curve.

To clamp the curve at the first data point, the knot sequence is modified to line $[0, 3, 3, 3, 4, 5, \ldots, 10]$.

Clamping at both ends is performed using the sequence $[0, 3, 3, 3, 4, 5, 6, 7, 7, 7, 10]$. The respective plots are shown in Figure 5.16 (b) and (c).

That $\mathbf{b}(t)$ is a linear combination of $N_{p,p+i}(t)$, B-spline curves inherit all properties from those of the normalized basis functions.

5.8.1   Properties of B-spline Curves

(A) B-spline curve is a piecewise curve with each component an order p segment
This is because each basis function of $\mathbf{b}(t)$ in Eq. (5.34) by itself is a piecewise order p curve.

(B) Equality $m = n + p$ must be satisfied
Each control point requires a basis function and the number of such functions when added to the order of the B-splines provides the number of knots required.

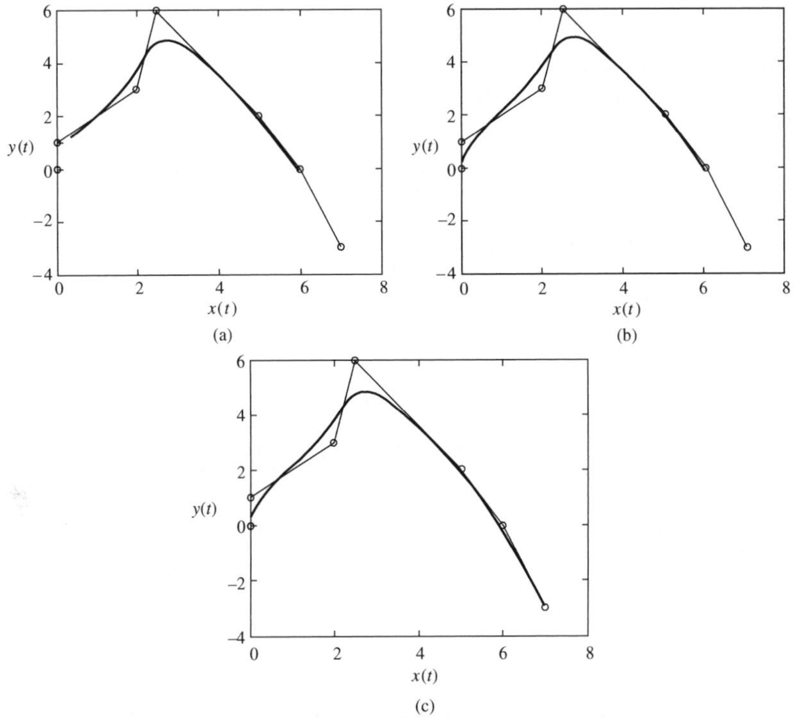

Figure 5.16   **Unclamped and clamped B-spline curves: (a) unclamped spline, (b) spline clamped at one end and (c) spline clamped at both ends**

(C) Strong convex hull property: The B-spline curve, b(t) is contained in the convex hull defined by the polyline [b$_j$, b$_{j+1}$, . . . , b$_{j+p-1}$] for t in [t_{j+p-1}, t_{j+p}). This convex hull is the subset of the parent hull [b$_0$, b$_1$, . . . , b$_n$]

For t in the knot span [t_{j+p-1}, t_{j+p}), $j+p-1 = 0$, . . . $m-1$, p basis functions, i.e. $N_{p,j+p}(t)$, $N_{p,j+p+1}(t)$ . . . , $N_{p,j+2p-1}(t)$ are non-zero from the property in Section 5.7(D). As $N_{p,p+k}(t)$ is the coefficient of b$_k$, only p control points, namely, b$_j$, b$_{j+1}$, . . . , b$_{j+p-1}$ have non-zero coefficients for $t \in$ [t_{j+p-1}, t_{j+p}). These coefficients also sum to one (property in Section 5.7(E)] making them barycentric in nature like the Bernstein polynomials. Hence their weighted average, **b**(t) must lie in the convex hull defined by p data points, b_j, b_{j+1}, . . . , b_{j+p-1}. The term *strong* implies that this convex hull is the subset of the original convex hull of $n+1$ control points. As t crosses t_{j+p}, $N_{p,j+p}(t)$ becomes zero while $N_{p,j+2p}(t)$ becomes non-zero. Consequently, **b**(t) for $t \in$ [t_{j+p}, t_{j+p+1}) lies in the new convex hull defined by [b_{j+1}, b_{j+2}, . . . , b_{j+p}] which again is the subset of the parent hull. The convex hull property is elucidated in Figure 5.17 for an open B-spline curve in Figure 5.16 (a) for Example 5.7. For $3 \leq t < 4$, **b**(t) lies in the convex hull of (0, 0), (0, 1), (2, 3) and (2.5, 6). For $4 \leq t < 5$, the new convex hull is defined by (0, 1), (2, 3), (2.5, 6) and (5, 2) and so on. For $6 \leq t < 7$, the convex hull is given by the last four data points in the set.

(D) b(t) is C^{p-k-1} continuous at a knot of multiplicity k

If $t = t_i$, a knot of multiplicity k, since $N_{p,i}(t)$ is C^{p-k-1} continuous, so is the curve $\mathbf{b}(t)$ at that knot. For any other t which is not a knot, the B-spline curve is a polynomial of order p and is infinitely differentiable.

(E) Variation diminishing property

The variation diminishing property discussed previously for Bézier segments also holds for B-spline curves. This feature along with the strong convex hull property helps predict the shape of B-spline curves better than that of Bézier segments.

(F) Local modification scheme: Relocating b_i only affects the curve b(t) in the interval $[t_i, t_{i+p})$

This follows from the local support property in Section 5.7(C) of B-spline basis functions. Let the control point $\mathbf{b}_i$ be moved to a new position $\mathbf{b}_i + \mathbf{v}$. Then, the new B-spline curve, $\mathbf{c}(t)$ from Eq. (5.34) is

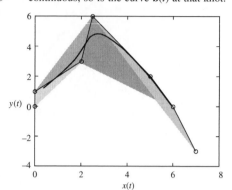

Figure 5.17   The convex hull property of B-spline curves

$$\mathbf{c}(t) = \sum_{k=0}^{i-1} N_{p,p+k}(t)\mathbf{b}_k + N_{p,p+i}(t)(\mathbf{b}_i + \mathbf{v}) + \sum_{k=i+1}^{n} N_{p,p+k}(t)\mathbf{b}_k$$

$$= \sum_{k=0}^{n} N_{p,p+k}(t)\mathbf{b}_k + N_{p,p+i}(t)\mathbf{v} = \mathbf{b}(t) + N_{p,p+i}(t)\mathbf{v} \qquad (5.35)$$

The coefficient of $\mathbf{v}$, i.e., $N_{p,i+p}(t)$ is non-zero in $[t_i, t_{i+p})$. For t not in this interval, $N_{p,i+p}(t)\mathbf{v}$ has no effect on the shape of $\mathbf{b}(t)$. However, for $t \in [t_i, t_{i+p})$, $N_{p,i+p}(t)$ is non-zero and the curve $\mathbf{b}(t)$ gets locally modified by $N_{p,i+p}(t)\mathbf{v}$. To show this, the data point $(5, 2)$ in Example 5.7 is moved to a new location $(8, 6)$ for which the local change in the open spline in Figure 5.16 (a) is shown in Figure 5.18 (dotted lines).

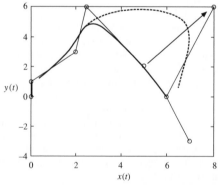

Figure 5.18   Local shape modification of B-spline curves

5.9 Design Features with B-Spline Curves

Generating unclamped and clamped B-spline curves, illustrated in Example 5.7, shows how knot multiplicity (a case of knot positioning) can be used to have a curve pass through the end points. Curve design with B-spline basis functions is more flexible than with Bernstein polynomials. B-spline curves require additional information (the control points and the order of piecewise curve segments) compared to the Bézier curves for which the two requirements are related (the number of control points is also the order of Bézier segment). More precisely, the shape of a B-spline curve is dependent on, and can be controlled using (i) the position of control points, (ii) the order of normalized basis functions and (iii) the position of knots.

(a) Shape manipulation using control points

That the shape of a B-spline curve changes only locally when a control point is moved to a different location is known from Eq. (5.35). We can employ the convex hull property to further manipulate the curve shape by relocating data points. For instance, we can *force a curve segment to become a line segment* by making any p adjacent control points collinear. Thus, if $\mathbf{b}_i, \mathbf{b}_{i+1}, \ldots, \mathbf{b}_{i+p-1}$, all are in a straight line, the curve segment that lies in their convex hull for t in $[t_{i+p-1}, t_{i+p})$ will be a straight line. For t, however, not belonging to the interval, the curve segments will not be linear. If $p-1$ of these control points are identical, say, $\mathbf{b}_i = \mathbf{b}_{i+1} = \ldots = \mathbf{b}_{i+p-2}$, the convex hull degenerates to a line segment $\mathbf{b}_i \mathbf{b}_{i+p-1}$ and the curve passes through $\mathbf{b}_i$. Further if $\mathbf{b}_{i-1}, \mathbf{b}_i = \mathbf{b}_{i+1} = \ldots \mathbf{b}_{i+p-2}$ and $\mathbf{b}_{i+p-1}$ are collinear, the line segment $\mathbf{b}_{i-1} \mathbf{b}_{i+p-1}$ is tangent to the curve at $\mathbf{b}_i$. Using the knot sequence as $[0, \ldots, 10]$, the first four data points in Example 5.7 are modified as $(0, 0)$, $(1, 1)$, $(2, 2)$ and $(3, 3)$, respectively. The resultant open B-spline is shown in Figure 5.19 (a) with a linear segment for t in $[3, 4)$. Next, the data points are modified to $(0, 0)$, $(0, 1)$, $(2, 3)$, $(2, 3)$, $(2, 3)$, $(6, 0)$ and $(7, -3)$. The curve is shown in Figure 5.19 (b) which passes through the point $(2, 3)$. Notice the slope discontinuity at this point that can also be achieved as a design feature using multiple data points. Further, the data point $(6, 0)$ is moved to a new location $(4, 5)$ so that $(0, 1)$, $(2, 3)$ and $(4, 5)$ are collinear. Figure 5.19 (c) shows that the curve not only passes through $(2, 3)$ but also is tangent to the polyline with end points $(0, 1)$ and $(4, 5)$.

Clamping of a B-spline curve discussed in Example 5.7 using knot-multiplicity can also be achieved by repeating the first and/or the last data point(s). Thus, if $\mathbf{b}_0 = \mathbf{b}_1 = \ldots \mathbf{b}_{p-2}$, the curve will pass through $\mathbf{b}_0$. This is shown in Figure 5.19(d) with the first three of the parent data points in Example 5.7 as $(0, 0)$. Further, with the last three data points set as $(7, -3)$, Figure 5.19(e) shows a spline clamped at both ends. Finally, a *closed* B-spline curve is shown in Figure 5.19(f) which is obtained using the control points $(4, -4)$, $(4, -4)$, $(4, -4)$, $(2, -4)$, $(0, 0)$, $(2, 4)$, $(4, 6)$, $(8, 0)$, $(6, -4)$, $(4, -4)$, $(4, -4)$, $(4, -4)$. These are 12 in number, and for an order 4 B-spline curve, 16 knots are required for which the sequence $[0, \ldots, 15]$ is used. Note that the first and last data points, that is, $(4, -4)$ are repeated three $(p-1)$ times each. The curve passes through $(4, -4)$ and is slope continuous at this point since points $(2, -4)$, $(4, -4)$ and $(6, -4)$ are collinear.

(b) Shape manipulation using knot modification

Knot modification may be another way to incorporate changes in the shape of a B-spline curve. This is because each piece of the B-spline curve is defined over a knot span, and modifying the position of one or more knots changes the behavior of the basis functions and thus the shape of the curve. However, since the change in shape of respective basis functions is not predictable with the change in position of the simple knots, this mode of shape control is not recommended.

Change in curve shape using multiple knots on the other hand can be predictable. Examples of

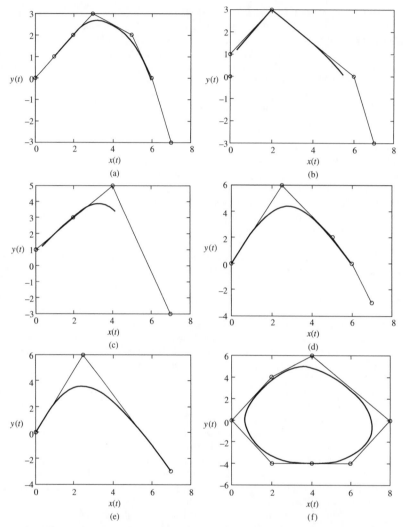

Figure 5.19 **(a) Linear segments with B-spline curves, (b) B-spline curve passing through an intermediate point, (c) curve passing through an intermediate data point and is also tangent to the line containing it, (d) curve clamped at the first data point using multiplicity $p-1$ of data points, (e) curve clamped at both ends using multiplicity $p-1$ of end points and (f) a closed spline using multiplicity of data points**

clamped B-splines have already been shown in Figure 5.16. Figure 5.20 shows a schematic of 7 cubic ($p = 4$) basis functions to explain what happens when the knot multiplicity is raised. Recall from the

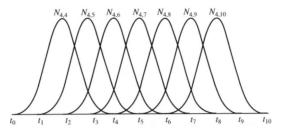

Figure 5.20 **A schematic of cubic B-spline basis functions**

barycentric property of B-spline basis functions that over the span $[t_5, t_6)$, there are four non-zero such functions, that is, $N_{4,6}$, $N_{4,7}$, $N_{4,8}$ and $N_{4,9}$ that sum to 1. At $t = t_5$, $N_{4,9} = 0$, implying that $N_{4,6}(t_5) + N_{4,7}(t_5) + N_{4,8}(t_5) = 1$. This is consistent with the knot multiplicity property (Section 5.7 G2) that over a simple knot t_5 ($k = 1$), the number of non-zero basis functions are three ($p-k = 4 - 1$). Now, if knot t_4 is moved to t_5 raising the multiplicity of the latter to 2, the function $N_{4,8}(t_4 = t_5)$ becomes zero leaving $N_{4,6}(t_5) + N_{4,7}(t_5) = 1$. Further, if $t_3 = t_4 = t_5$ so that the multiplicity of t_5 is 3, $N_{4,7}(t_3 = t_5) = 0$. This implies that $N_{4,6}(t_5) = 1$. Or, in other words, from Eq. (5.34), the B-spline curve will pass through $\mathbf{b}_2$ for $t = t_5$. In general, therefore, if $t_{i+1} = t_{i+2} =, \ldots, = t_{i+p-1}$ with t_{i+p-1} having multiplicity $p-1$, only one basis function $N_{p,p+i}$ will be non-zero over t_{i+p-1}, and from the barycentric property, $N_{p,p+i}$ will be 1, implying that the B-spline curve will pass through $\mathbf{b}_i$.

Example 5.8. Using control points, $(4, -4)$, $(4, -4)$, $(4, -4)$, $(2, -4)$, $(0, 0)$, $(2, 4)$, $(4, 6)$, $(8, 0)$, $(6, -4)$, $(4, -4)$, $(4, -4)$, $(4, -4)$ for a cubic B-spline curve in Figure 5.19 (f), use knot multiplicity to ensure that the curve passes through control points $(0, 0)$ and $(8, 0)$. Start with a uniform knot sequence $t_i = i$, $i = 0, \ldots, 15$.

The range of full support for this example is $[3, 12]$. To have the spline pass through $(0, 0)$, which is the fifth control point ($i = 4$), the knot $t_{4+4-1} = t_7 = 7$ should have multiplicity 3, that is, $t_5 = t_6 = t_7 = 7$ (say). Otherwise, to have the spline pass through $(8, 0)$ which is the eighth control point ($i = 7$), t_{10} should have multiplicity 3, that is, $t_8 = t_9 = t_{10} = 10$ (say). The two results are shown in Figure 5.21. Note the slope discontinuity at the two respective control points which is expected from

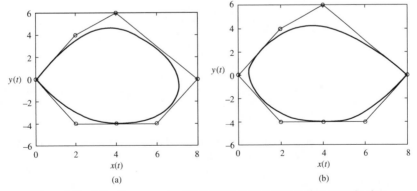

(a)　　　　　　　　　　　　　　(b)

Figure 5.21 **A B-spline curve passing through desired intermediate control points**

property in Section 5.7G1. Shape functions $N_{4,7}(t)$ and $N_{4,10}(t)$ are only C^0 (position) continuous at respective knots t_7 and t_{10}.

5.10   Parameterization

For an approximating spline in Eq. (5.34), we require $n+1$ control points $\mathbf{b}_0, \mathbf{b}_1, \ldots, \mathbf{b}_n$ and the order p of the curve as input. From the properties mentioned earlier, it is required that the number of knots, $m+1$ must satisfy the equality $m = n + p$. However, these knots are not known *a priori* and can be chosen in a number of ways. One way is to assign a parameter u_i to each control point $\mathbf{b}_i$ and then compute the knot vector from these parameters. Parameter assignment may be accomplished in any of the following ways.

Uniformly spaced method

For $n+1$ parameters to be equally spaced in $[a, b]$, we have

$$u_i = a + i\,\frac{b - a}{n}, i = 0, \ldots, n \tag{5.36}$$

This assignment scheme, though simple, does not work well when the control points are placed unevenly. In such cases, curves with unsatisfactory shapes might result.

Chord length method

To ensure that the curve shape closely follows the shape of the corresponding polyline, this method of parameterization may be employed. Herein, parameters are placed proportional to the chord lengths of control polyline, that is, if the first parameter corresponding to $\mathbf{b}_0$ is u_0, then the subsequent parameters u_i corresponding to $\mathbf{b}_i$ may be written as

$$u_i = u_0 + \sum_{k=1}^{i} |\,\mathbf{b}_k - \mathbf{b}_{k-1}|, i = 1, \ldots, n \tag{5.37}$$

We may normalize the parameterization in Eq. (5.37) by setting u_0 to 0 and dividing u_i by the total chord length of the polyline $L = \sum_{k=1}^{n} |\,\mathbf{b}_k - \mathbf{b}_{k-1}|$. Otherwise, we may choose to set this parameterization in a chosen domain $[a, b]$ for which

$$u_i = a + (b - a)\,\frac{\sum_{k=1}^{i} |\,\mathbf{b}_k - \mathbf{b}_{k-1}|}{\sum_{k=1}^{n} |\,\mathbf{b}_k - \mathbf{b}_{k-1}|}, i = 1, \ldots, n \tag{5.38}$$

The chord length method is widely used and it usually performs well. Sometimes, a longer chord may cause its curve segment to have a bulge bigger than necessary, which is a common problem with the chord length method.

Centripetal method

This is derived from a concept analogous to the centripetal acceleration when a point is traversing along a curve. The notion is that the centripetal acceleration should not be too large at sharp turns (smaller radii of curvature). For the parameters to lie in the domain $[a, b]$, the centripetal method gives the parameter values as

$$u_i = a + (b-a)\frac{\sum_{k=1}^{i} |\mathbf{b}_k - \mathbf{b}_{k-1}|^{\frac{1}{2}}}{\sum_{k=1}^{n} |\mathbf{b}_k - \mathbf{b}_{k-1}|^{\frac{1}{2}}}, \; i = 1, \ldots, n \tag{5.39}$$

with u_0 as a. Note that Eqs. (5.36), (5.38) and (5.39) can be generalized to

$$u_i = a + (b-a)\frac{\sum_{k=1}^{i} |\mathbf{b}_k - \mathbf{b}_{k-1}|^{e}}{\sum_{k=1}^{n} |\mathbf{b}_k - \mathbf{b}_{k-1}|^{e}}, \; i = 1, \ldots, n \tag{5.40}$$

for some chosen exponent $e \geq 0$. For $e = 0$, uniformly spaced parameterization is obtained while for $e = 1$ and $\frac{1}{2}$, respectively, chord length and centripetal parameterizations are achieved.

5.10.1 Knot Vector Generation

Once a set of parameters is obtained, the knot vector may be generated. The placement of knots would, however, depend on the end conditions. For unclamped splines, all $m+1$ knots are simple. Of those, $n+1$ knots $(t_p, \ldots, t_{n+p})$ may be chosen as the parameters, u_i, $i = 0, \ldots, n$, respectively from Eq. (5.40) while the remaining first p knots $(t_0, \ldots, t_{p-1})$ may be chosen freely, that is, $t_{i+p} = u_i$, $i = 0, \ldots, n$ while $t_0, t_1, \ldots, t_{p-1}$ are free choices of simple knots. In case the B-spline curve is clamped at one end, the knot corresponding to that end must be repeated at least $p-1$ times. If the spline is clamped at the first control point, then $t_1 = \ldots = t_{p-1}$ and $t_{i+p} = u_i$, $i = 0, \ldots, n$. We may still have a free choice for t_0 that can be taken as equal to t_1. Likewise, to clamp the spline at the last control point $t_i = u_i$, $i = 0, \ldots, n$ while $t_{n+1} = \ldots = t_{n+p}$ is the free choice.

For a B-spline curve to be clamped at both ends, both knots t_{p-1} and t_{m-p+1}, the two limits of the full support range, may each be repeated p times, that is, $t_0 = \ldots = t_{p-1}$ and $t_{n+1} = \ldots = t_{n+p}$. With $2p$ knots determined, the remaining $n-p+1$ internal knots $t_p, \ldots, t_n$ may be as follows:

Internal knots may be evenly spaced. The $n-p+1$ internal knots divide the chosen interval $[a, b]$ into $n-p+2$ spans. For their even spacing

$$t_0 = t_1 = \ldots = t_{p-1} = a$$

$$t_{j+p-1} = a + (b-a)\frac{j}{n-p+2}, \; j = 1, 2, \ldots, n-p+1$$

$$t_{n+1} = t_{n+2} = \ldots = t_{n+p} = b \tag{5.41}$$

The uniformly spaced knot vector does not require the knowledge of the position of control points, and is simple to generate.

Internal knots may be averaged with respect to the parameters. As suggested by de Boor

$$t_0 = t_1 = \ldots = t_{p-1} = a$$

$$t_{j+p-1} = \frac{1}{p-1}\sum_{i=j}^{j+p-2} u_i, \, j = 1, 2, \ldots, n-p+1$$

$$t_{n+1} = t_{n+2} = \ldots = t_{n+p} = b \tag{5.42}$$

Thus, the first internal knot t_p is the average of $p-1$ parameters $u_1, u_2, \ldots, u_{p-1}$; the second internal knot t_{p+1} is the average of the next $p-1$ *parameters, $u_2, u_3, \ldots, u_p$*, and so on, with u's given by Eq. (5.40).

5.11   Interpolation with B-Splines

Given $n+1$ data points $\mathbf{p}_0, \mathbf{p}_1, \ldots, \mathbf{p}_n$ it is desired to fit them with a B-spline curve of given order $p \le n$. We can select a set of parameters $u_0, u_1, \ldots, u_n$ corresponding to each data point as discussed in section 5.10. A knot vector $[t_0, t_1, \ldots, t_m]$ may then be computed so that $m = n + p$. Let the set of unknown control points be $\mathbf{b}_0, \mathbf{b}_1, \ldots, \mathbf{b}_n$. The B-spline curve may be expressed as

$$\mathbf{b}(t) = \sum_{i=0}^{n} N_{p,p+i}(t)\,\mathbf{b}_i \tag{5.43}$$

Substituting the correspondence of the data points with the parameters, we get

$$\mathbf{p}_k = \mathbf{b}(u_k) = \sum_{i=0}^{n} N_{p,p+i}(u_k)\,\mathbf{b}_i, \, k = 0, \ldots, n \tag{5.44}$$

or
$$\mathbf{P} = \begin{pmatrix} \mathbf{p}_0 \\ \mathbf{p}_1 \\ \mathbf{p}_2 \\ \ldots \\ \ldots \\ \mathbf{p}_n \end{pmatrix} = \begin{pmatrix} N_{p,p}(u_0) & N_{p,p+1}(u_0) & N_{p,p+2}(u_0) & \ldots & N_{p,n+p}(u_0) \\ N_{p,p}(u_1) & N_{p,p+1}(u_1) & N_{p,p+2}(u_1) & \ldots & N_{p,n+p}(u_1) \\ N_{p,p}(u_2) & N_{p,p+1}(u_2) & N_{p,p+2}(u_2) & \ldots & N_{p,n+p}(u_2) \\ \ldots & \ldots & \ldots & \ldots & \ldots \\ N_{p,p}(u_n) & N_{p,p+1}(u_n) & N_{p,p+2}(u_n) & \ldots & N_{p,n+p}(u_n) \end{pmatrix} \begin{pmatrix} \mathbf{b}_0 \\ \mathbf{b}_1 \\ \mathbf{b}_2 \\ \ldots \\ \ldots \\ \mathbf{b}_n \end{pmatrix} = \mathbf{NB} \tag{5.45}$$

Note that for a spatial interpolating B-spline curve, each data point $\mathbf{p}_k$ would be expressed as a triad (x_k, y_k, z_k) in Cartesian coordinates implying that both $\mathbf{B}$ and $\mathbf{P}$ would be $(n+1) \times 3$ in size. Inverting $\mathbf{N}$ and pre-multiplying the result with $\mathbf{P}$ would give the triads corresponding to the unknown control points and hence the interpolating spline. Even though the B-spline basis functions satisfy the local support property, the shape change of the curve in the interpolation method discussed above is global. If the position of a single data point is changed, even though the matrix $\mathbf{N}$ and its inverse is unchanged, the $\mathbf{P}$ matrix, and therefore the $\mathbf{B}$ matrix change, thereby changing the shape of the interpolating curve overall.

Example 5.9. Interpolate the data points, (0, 0), (0, 1), (2, 3), (2.5, 6), (5, 2), (6, 0) and (7, −3), using a B-spline curve with piecewise cubic polynomial segments.

The parameters u are first generated for given data points using the chord length method (Eq. (5.38)). First, the distances between successive data points are computed, that is, $d_1 = \sqrt{(1^2 + 0^2)} = 1$; $d_2 = \sqrt{(2^2 + 2^2)} = 2.83$; $d_3 = \sqrt{(0.5^2 + 3^2)} = 3.04$; $d_4 = \sqrt{(2.5^2 + 4^2)} = 4.72$; $d_5 = \sqrt{(1^2 + 2^2)} = 2.24$; $d_6 = \sqrt{(1^2 + 3^2)} = 3.16$. The sum L of these distances is 17. Setting $u_0 = 0$, we have

$$u_1 = u_0 + d_1/L = 0.058 \qquad u_4 = u_3 + d_4/L = 0.682$$
$$u_2 = u_1 + d_2/L = 0.225 \qquad u_5 = u_4 + d_5/L = 0.814$$
$$u_3 = u_2 + d_3/L = 0.404 \qquad u_6 = u_5 + d_6/L = 1.000$$

With 7 data points and 4 as the order of the piecewise curves, the number of knots required are 11. We can use $u_0, \ldots, u_6$ as 7 knots with 4 free choices. Let these choices be arbitrary, say –2 and –1 to the left and, 2 and 3 to the right. Then the knot vector is

$$[t_0, \ldots, t_m] \equiv [-2, -1, 0, 0.058, 0.225, 0.404, 0.682, 0.814, 1, 2, 3]$$

We can use the above knot sequence to compute the B-spline functions $N_{p,p+i}(t)$ in Eq. (5.43). Further, we can use the u values above to compute the coefficient matrix in Eq. (5.45). Substituting for the data points **P**, the unknown control points can be determined as

$$\mathbf{B} \equiv \begin{pmatrix} 0.69 & -5.90 \\ -0.20 & 1.74 \\ 2.80 & 2.38 \\ 1.92 & 8.32 \\ 4.77 & 2.34 \\ 6.14 & -0.10 \\ 8.86 & -8.46 \end{pmatrix}$$

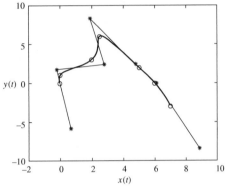

The control points '∗' and the interpolated curve (thick line) with data points '∘' are shown in Figure 5.22. The interpolating spline is plotted between 0 and 1 which are the lower and upper bounds of parameter u. Note, however, that $t \in$ [0, 1] does not correspond to the interval of full support, which is [0.058, 0.814].

Figure 5.22   Interpolation with B-spline curves.

Bézier curves are special cases of B-spline curves clamped at both ends. If the order of a B-spline curve is chosen as the number of control points (i.e., $p = n+1$), then $m+1 = 2n+2 = 2p$ knots are required of which p knots are clamped at each end and the B-spline curve reduces to a Bézier curve.

5.12   Non-Uniform Rational B-Splines (NURBS)

Rational Bézier curves are first introduced in section 4.6 wherein, in addition to specifying the data points, a user is also required to specify respective weights to gain additional design freedom. However, local shape control is still not possible with rational Bézier segments. Noting that B-spline basis functions are locally barycentric that render local shape control to B-spline curves, analogous to Eqs. (4.66) and (5.34), rational B-spline curves can be defined as

$$\mathbf{b}(t) = \frac{\sum\limits_{i=0}^{n} w_i N_{p,p+i}(t) \mathbf{b}_i}{\sum\limits_{i=0}^{n} w_i N_{p,p+i}(t)} \tag{5.46}$$

The term *non-uniform* signifies that the knots are not placed at regular intervals in a general setting. Here again, setting w_i to zero implies that the location of $\mathbf{b}_i$ does not affect the curve's shape. For larger values of w_i, the curve gets pushed towards $\mathbf{b}_i$. Because they offer a great deal of flexibility in design and also because they possess local shape control and all other properties of B-spline curves, NURBS are widely employed in freeform modeling of curves. NURBS are also capable of accurately modeling many analytic curves. Since NURBS are the generalization of B-spline curves (setting all weights in Eq. 5.46 to 1 yields the B-spline curve), discussion in this chapter pertaining to the design of B-spline curves all apply to NURBS.

Example 5.10. For data points in Example 5.7, that is, (0, 0), (0, 1), (2, 3), (2.5, 6), (5, 2), (6, 0) and (7, –3), design a rational B-spline curve with basis functions of order 4. First set all weights to 1. Increase the weight w_3 corresponding to (2.5, 6) to visualize the shape change.

The example is solved using a uniform knot vector [0, 1, 2, . . ., 10) for an open rational spline and the knot vector with multiple knots, that is, [0, 3, 3, 3, 4, 5, 6, 7, 7, 7, 10) for rational spline clamped at both ends. NURBS curves are shown in Figure 5.23. Notice that in both cases, for $w_3 = 0$, $\mathbf{P}_3 = (2.5, 6)$ is not considered a part of the control polyline and the NURBS curves lie within the convex hull of (0, 0), (0, 1), (2, 3), (5, 2), (6, 0) and (7, –3). With increase in w_3, the curves get closer to (2.5, 6).

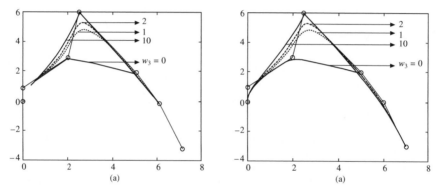

Figure 5.23 **(a) Open and (b) clamped rational B-spline curves**

EXERCISES

1. Compute a quadratic B-spline basis function as a polynomial spline like in Example 5.2. Take the knot vector as [0, 1, 2, 3].
2. Verify the result obtained above using the divided differences table for truncated power series function
$$f[t_j, t] = (t_j - t)_+^2.$$
3. A B-spline curve is defined as

$$\mathbf{b}(t) = \sum_{i=0}^{n} N_{p,p+i}(t)\,\mathbf{b}_i$$

(a)Explain and provide the *full support* interval for $\mathbf{b}(t)$ · (*b*) Demonstrate algebraically the local shape control property if $\mathbf{b}_j$ is relocated to $\mathbf{b}_j + \mathbf{v}$. For what interval of t would the curve change in shape.
4. A first order basis function is defined as, say

$$M_{1,i}(t) = \frac{C}{t_i - t_{i-1}} \text{ for } t \in [t_{i-1}, t_i]$$

or $\qquad\qquad N_{1,i}(t) = C[\text{for } t \in [t_{i-1}, t_i]$

What should be the value of C if (a) $t_{i-1} \neq t_i$ and (b) if $t_{i-1} = t_i$.

[Hint: For $t_{i-1} \neq t_i$, use normalization; for $t_{i-1} = t_i$, find C from the condition $M_{2,i}(t_i) = N_{2,i}(t_i) = 0$]. For $t \in [t_{i-1}, t_i)$, would one need to evaluate $M_{1,i}(t)$ or $N_{1,i}(t)$ at $t = t_i$. Would the right open parenthesis in $[t_{i-1}, t_i)$ serve the purpose of C.

5. The recursion relation for a normalized B-spline basis function is given as

$$N_{1,i}(t) = \delta_i \text{ such that } \delta_i = 1 \text{ for } t \in [t_{i-1}, t_i)$$

$$= 0, \text{ elsewhere}$$

$$N_{k,i}(t) = \frac{t - t_{i-k}}{t_{i-1} - t_{i-k}} N_{k-1,i-1}(t) + \frac{t_i - t}{t_i - t_{i-k+1}} N_{k-1,i}(t)$$

where k is the order (degree+1) of the spline and i the last knot over which $N_{k,i}(t)$ is defined. Show in a general case that the sum of all non-zero B-spline basis functions over a knot span $[t_j, t_{j+1})$ is 1.

6. Show that the derivative of a B-spline basis function of order k is given as

$$\frac{d}{dt} N_{k,j}(t) = N'_{k,j}(t) = \frac{k-1}{t_{j-1} - t_{j-k}} N_{k-1,j-1} - \frac{k-1}{t_j - t_{j-k+1}} N_{k-1,j}$$

and thus the derivative of a B-spline curve $\mathbf{b}(t) = \sum\limits_{i=0}^{n} N_{p,p+i}(t)\mathbf{b}_i$ is

$$\frac{d}{dt} \mathbf{b}(t) = \sum\limits_{i=0}^{n-1} N_{p-1,p+i}\mathbf{q}_i$$

where $$\mathbf{q}_i = \frac{(p-1)}{t_{p+i} - t_{i+1}} (\mathbf{b}_{i+1} - \mathbf{b}_i)$$

7. Write a generic code to compute the normalized B-spline basis function $N_{k,i}(t)$. Device ways to make the computations robust. [Hint: Note that $N_{k,i}(t) = 0$ for $t \notin [t_{i-k}, t_i)$. Further, for $t \in [t_{i-k}, t_i)$, computing $N_{m,j}(t)$ requires $N_{m-1,j-1}(t)$ and $N_{m-1,j}(t)$, $m = 1, \ldots, k$, $j = i-k+1, \ldots, i$, which form a triangular pattern shown in Table 5.2. Judge if all basis functions are needed for computations, or some are known to be zero *a priori*].

8. Explore *knot insertion* and *blossoming* as alternative methods to compute B-spline basis functions.

9. How would one get a Bernstein polynomial (Bézier) curve from a B-spline basis function (a B-spline) curve? Explain and illustrate.

10. Given the control polyline for a B-spline curve, is it possible to (graphically) obtain the Bézier control points for the same curve? If so, under what conditions? (We may want to work out an example).

11. Write a procedure to implement Eq. (5.34) for a general 2-D case.

12. In Exercise 11, provide for (a) data points to be interactively relocated and flexibility and (b) curve clamping at the ends using knot multiplicity.

13. Let $\mathbf{r}_0, \mathbf{r}_1, \mathbf{r}_2, \mathbf{r}_3, \mathbf{r}_4$ be five control points specified by the user. Show that the equations of 3 B-spline curve segments with uniform knot vectors are given by

$$\mathbf{r}_1(u) = \frac{1}{3}(3-u)^2 \mathbf{r}_0 + \frac{1}{2}[(u-1)(3-u) + (4-u)(u-2)]\mathbf{r}_1 + \frac{1}{2}(u-2)^2 \mathbf{r}_2$$

$$\mathbf{r}_2(u) = \frac{1}{2}(4-u)^2 \mathbf{r}_1 + \frac{1}{2}[(u-2)(4-u) + (5-u)(u-3)]\mathbf{r}_2 + \frac{1}{2}(u-3)^2 \mathbf{r}_3$$

$$\mathbf{r}_3(u) = \frac{1}{2}(5-u)^2 \mathbf{r}_2 + \frac{1}{2}[(u-3)(5-u) + (6-u)(u-4)]\mathbf{r}_3 + \frac{1}{2}(u-4)^2 \mathbf{r}_4$$

(a) Each of the segments can be normalized with $0 \le u < 1$ by substituting $(u+2)$ in place of u. Show that the matrix form of the curve segments is given by

$$\mathbf{r}_1(u) = \frac{1}{2} [u^2 \; u \; 1] \begin{pmatrix} 1 & -2 & 1 \\ -2 & 2 & 0 \\ 1 & 1 & 0 \end{pmatrix} \begin{pmatrix} \mathbf{r}_0 \\ \mathbf{r}_1 \\ \mathbf{r}_2 \end{pmatrix}$$

$$\mathbf{r}_2(u) = \frac{1}{2} [u^2 \; u \; 1] \begin{pmatrix} 1 & -2 & 1 \\ -2 & 2 & 0 \\ 1 & 1 & 0 \end{pmatrix} \begin{pmatrix} \mathbf{r}_1 \\ \mathbf{r}_2 \\ \mathbf{r}_3 \end{pmatrix}$$

$$\mathbf{r}_3(u) = \frac{1}{2} [u^2 \; u \; 1] \begin{pmatrix} 1 & -2 & 1 \\ -2 & 2 & 0 \\ 1 & 1 & 0 \end{pmatrix} \begin{pmatrix} \mathbf{r}_2 \\ \mathbf{r}_3 \\ \mathbf{r}_4 \end{pmatrix} \quad \text{or} \quad \mathbf{r}_i(u) = \frac{1}{2} [u^2 \; u \; 1] \begin{pmatrix} 1 & -2 & 1 \\ -2 & 2 & 0 \\ 1 & 1 & 0 \end{pmatrix} \begin{pmatrix} \mathbf{r}_{i-1} \\ \mathbf{r}_i \\ \mathbf{r}_{i+1} \end{pmatrix}$$

(b) Similarly, as above, show that the ith segment of a cubic B-spline curve is given by

$$\mathbf{r}_i(u) = \frac{1}{6} [u^3 \; u^2 \; u \; 1] \begin{pmatrix} -1 & 3 & -3 & 1 \\ 3 & -6 & 3 & 0 \\ -3 & 0 & 3 & 0 \\ 1 & 4 & 1 & 0 \end{pmatrix} \begin{pmatrix} \mathbf{r}_{i-1} \\ \mathbf{r}_i \\ \mathbf{r}_{i+1} \\ \mathbf{r}_{i+2} \end{pmatrix}$$

14. Using the above formulation, closed cubic B-spline curves can be generated. For example, let there be 7 control points $\mathbf{r}_0, \mathbf{r}_1, \mathbf{r}_2, \ldots\ldots, \mathbf{r}_6$ ($\mathbf{r}_i$, $i = 0, 1, 2, \ldots, 6$). There will be $n + 1 = 7$ curve segments each of them will be cubic B-spline, and can be written as

$$\mathbf{r}_j(u) = \frac{1}{6} [u^3 \; u^2 \; u \; 1] \begin{pmatrix} -1 & 3 & -3 & 1 \\ 3 & -6 & 3 & 0 \\ -3 & 0 & 3 & 0 \\ 1 & 4 & 1 & 0 \end{pmatrix} \begin{pmatrix} \mathbf{r}_{(j-1)\bmod(n+1)} \\ \mathbf{r}_{j\bmod(n+1)} \\ \mathbf{r}_{(j+1)\bmod(n+1)} \\ \mathbf{r}_{(j+2)\bmod(n+1)} \end{pmatrix}$$

Here, 'mod' is the "modulo" function which means that, if $j = 2$, $j \bmod 7 = 2$ (the remainder as a result of this division).

Chapter 6
Differential Geometry of Surfaces

Surfaces define the boundaries of a solid. They themselves are bounded by curves (Figure 6.1). Surface design may be regarded as an extension of curve design in two parametric dimensions. Developments in previous chapters, therefore, can all be applied in surface modeling. In curve design, emphasis is laid on the generic (non-analytical) parametric representation of low degree polynomial segments that can be composed together to model a curve. Reasons are to encompass a variety of shapes that analytical curves fail to provide, to prevent undue oscillations that may be observed in higher degree polynomial segments, and to make the representation free from singularities like vertical slopes. Also, parametric representation makes easier to compute the intersection points or plot curves easier that is not so with explicit and implicit representations. Following the above, we can treat surface modeling in a manner similar to curve design, that is, we represent *surface patches*

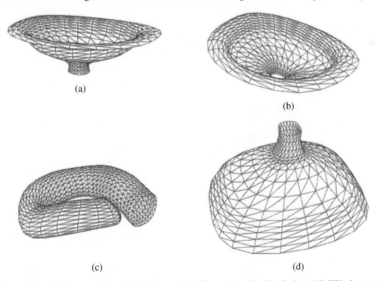

(a)

(b)

(c)   (d)

Figure 6.1   Surface models of (a) kitchen sink (b) wash basin (c) air duct (d) TV-picture tube

in parametric form using low order polynomials and then *knit* these patches to form a composite surface with continuous slopes and/or curvatures at their boundaries. Given that the parametric form of a surface patch is known, this chapter deals with determining the *differential properties of the patch* to facilitate composite fitting.

6.1 Parametric Representation of Surfaces

A surface patch (Figure 6.2) is a set of points whose position vectors are given by $\mathbf{r} = \mathbf{r}(u, v)$ for parameters u and v each varying in the interval [0, 1]. For constant $u = u_c$, $\mathbf{r}(u_c, v)$ is a parametric curve in v while for $v = v_c$, $\mathbf{r}(u, v_c)$ is a curve that varies only with u. Thus, a parametric surface $\mathbf{r}(u, v)$ may be regarded as a set of matted curves. Values of u and v determine the position of a point on the surface and thus u and v may be regarded as the *curvilinear* or *Gaussian* coordinates. For scalar functions $x(u, v)$, $y(u, v)$ and $z(u, v)$, a parametric surface may be represented in vector form as

$$\mathbf{r}(u, v) = x(u, v)\mathbf{i} + y(u, v)\mathbf{j} + z(u, v)\mathbf{k} \qquad (6.1)$$

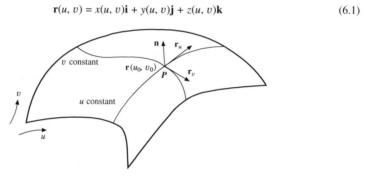

Figure 6.2   A parametric surface r (u, v)

Example 6.1. Some commonly known analytical surfaces can be represented in the parametric form.
(a) *An x-y Plane:* $\mathbf{r}(u, v) = u\mathbf{i} + v\mathbf{j}$ (in cartesian form, $u, v) = u \cos v\ \mathbf{i} + u \sin v\ \mathbf{j}$ (in polar form) ($u \neq 0$, $0 \leq v \leq 2\pi$).
(b) *A Sphere:* A point P on the sphere is given by $\mathbf{r}(u, v) = a \cos u \cos v\mathbf{i} + a\ \sin u \cos v\mathbf{j} + a \sin v\mathbf{k}$, where a is the radius, v is the angle made by the radius vector $\mathbf{r}(u, v)$ with the x-y plane and u the angle made between the x-axis and the projection of the radius vector on the x-y plane as shown in Figure 6.3. Angles u and v are called *longitude* and *latitude*, respectively. The *circles* of latitude are v = constant ($v = 0$ is the equator), whereas u = constant, $u \in [0, 2\pi)$ are called *meridians*.
(c) *A Catenoid*: Rotation of a *catenary* $y = a \cosh (x/a)$ about its *directrix* (x-axis) results in a *catenoid*. A point on the catenoid is then given by

$$\mathbf{r}(u, v) = u\mathbf{i} + a \cosh \left(\frac{u}{a}\right) \cos v\mathbf{j} + a \cosh \left(\frac{u}{a}\right) \sin v\mathbf{k},$$

where v is the angle of rotation in [0, 2π].
(d) *The Pseudosphere:* Tractrix is a planar curve having the property that the segment of its tangent between the contact point P and some fixed straight line (called its asymptote) in the plane is

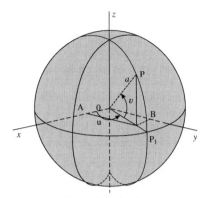

Figure 6.3   A sphere

of constant length a. Rotation of a tractrix about its asymptote results in a pseudosphere. If the asymptote is the x-axis, the equation of a tractrix for u varying between 0 and $\pi/2$ is given by

$$x(u) = a \cos u + a \ln \left| \tan \frac{u}{2} \right|, \quad y(u) = a \sin u$$

and for the rotation angle v varying from 0 to 2π, the equation of the *pseudosphere* is

$$x(u, v) = a \cos u + a \ln \left| \tan \frac{u}{2} \right|, \quad y(u, v) = a \sin u \sin v, \quad z(u, v) = a \sin u \cos v$$

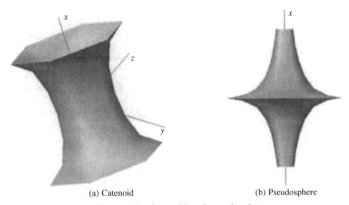

(a) Catenoid   (b) Pseudosphere

Figure 6.4   Catenoid and pseudosphere

(e) *A Helicoid:* This is the surface formed by the perpendiculars dropped from a circular helix to its axis. The parametric equation of a helicoid for v varying from 0 to 2π is represented by

$$x(u, v) = u \cos v, \quad y(u, v) = u \sin v, \quad z(u, v) = av$$

Equations in cartesian and parametric form for some known analytical surfaces are given below.

Analytic surface	Parametric form
Ellipsoid $\left(\dfrac{x^2}{a^2} + \dfrac{y^2}{b^2} + \dfrac{z^2}{c^2} = 1 \right)$	$\mathbf{r}(\alpha, \beta) \equiv [a \cos \alpha \cos \beta, \, b \cos \alpha \sin \beta, \, c \sin \alpha]$
Elliptic Hyperboloid $\left(\dfrac{x^2}{a^2} + \dfrac{y^2}{b^2} = cz \right)$	$\mathbf{r}(u, v) \equiv \left[au, \, bv, \, \dfrac{u^2 + v^2}{c} \right]$
Hyperboloid of one sheet $\left(\dfrac{x^2}{a^2} + \dfrac{y^2}{b^2} - \dfrac{z^2}{c^2} = 1 \right)$	$\mathbf{r}(\alpha, \beta) \equiv \left[a \dfrac{\cos \beta}{\cos \alpha}, \, b \dfrac{\sin \beta}{\cos \alpha}, \, c \tan \alpha \right]$
Hyperboloid of two sheets $\left(\dfrac{x^2}{a^2} - \dfrac{y^2}{b^2} - \dfrac{z^2}{c^2} = 1 \right)$	$\mathbf{r}(\alpha, \beta) \equiv \left[\dfrac{a}{\cos \alpha}, \, b \tan \alpha \cos \beta, \, c \tan \alpha \sin \beta \right]$
Cone $\left(\dfrac{x^2}{a^2} + \dfrac{y^2}{b^2} - \dfrac{z^2}{c^2} = 0 \right)$	$\mathbf{r}(u, \beta) \equiv [au \cos \beta, \, bu \sin \beta, \, cu]$
Hyperbolic paraboloid $\left(\dfrac{x^2}{a^2} - \dfrac{y^2}{b^2} = cz \right)$	$\mathbf{r}(u, v) \equiv \left[au, \, bv, \, \dfrac{u^2 - v^2}{c} \right]$
Quadric Circular Cylinder $(x - a)^2 + (y - b)^2 = c^2, \, z = h$	Quadric Circular Cylinder $\mathbf{r}(\theta, h) \equiv [a + c \cos \theta, \, b + c \sin \theta, \, h]$
Quadric Parabolic Cylinder $(y - a)^2 = bx, \, z = h$	Quadric Parabolic Cylinder $\mathbf{r}(\theta, h) \equiv \left[\sin^2 \theta, \, a + \sqrt{b} \sin \theta, \, h \right]$
Torus $x^2 + y^2 + z^2 - 2b\sqrt{a^2 - z^2} = a^2 + b^2$	Torus $x = (b + a \cos u) \cos v$ $y = (b + a \cos u) \sin v$ $z = a \sin u, \quad b > a,$ $0 \le u \le 2\pi, \quad 0 \le v \le 2\pi$

A *simple sheet* of surface $\mathbf{r}(u, v)$ is continuous and obtained from a rectangular sheet by stretching, squeezing and bending but without tearing or gluing. For instance, a cylinder is not a simple sheet, for it cannot be obtained from a rectangle without gluing at the edges. Similarly, a sphere and a cone are not simple sheets. A flat sheet with an annular hole is also not a simple sheet. However, a cylindrical surface with a cut all along or an annular sheet with an open sector, are both simple sheets. If, for points **P** on the surface, a portion of the surface containing **P** can be cut, and if that portion is a simple surface, then the entire surface is called an *ordinary surface*.

6.1.1 Singular Points and Regular Surfaces

Let $\mathbf{r}_u$ and $\mathbf{r}_v$ define the derivatives along the curvilinear coordinates u and v at a point P ($\mathbf{r}(u_0, v_0)$) on the surface (Figure 6.2), then P is called a *regular point* if

$$\mathbf{r}_u(u_0, v_0) = \left. \frac{\partial \mathbf{r}(u, v)}{\partial u} \right|_{u_0, v_0}, \; \mathbf{r}_v(u_0, v_0) = \left. \frac{\partial \mathbf{r}(u, v)}{\partial v} \right|_{u_0, v_0} \tag{6.2}$$

and $$\mathbf{r}_u(u_0, v_0) \times \mathbf{r}_v(u_0, v_0) \neq \mathbf{0}$$

that is, the slopes $\mathbf{r}_u$ and $\mathbf{r}_v$ should exist at P and that their cross product should not be *zero*. If all points on the surface are regular points, the surface is a *regular surface*. If at some point P, $\mathbf{r}_u(u_0, v_0) \times \mathbf{r}_v(u_0, v_0) = \mathbf{0}$, then P is a *singular point* where the slopes $\mathbf{r}_u$ and $\mathbf{r}_v$ are either undefined (non-unique/non-existing) or *zero* or coincident. Condition (Eq. (6.2)) requires that at least one of the Jacobians (J_1, J_2, J_3) described below is non-zero.

If $$x_u = \frac{\partial x(u, v)}{\partial u}, x_v = \frac{\partial x(u, v)}{\partial v}, y_u \frac{\partial y(u, v)}{\partial u}, y_v = \frac{\partial y(u, v)}{\partial v}, z_u = \frac{\partial z(u, v)}{\partial u}, z_v = \frac{\partial z(u, v)}{\partial v}$$

then $$\mathbf{r}_u(u_0, v_0) \times \mathbf{r}_v(u_0, v_0) = \begin{bmatrix} \mathbf{i} & \mathbf{j} & \mathbf{k} \\ x_u & y_u & z_u \\ x_v & y_v & z_v \end{bmatrix} = J_1\mathbf{i} + J_2\mathbf{j} + J_3\mathbf{k}$$

with the Jacobians

$$J_1 = \begin{vmatrix} y_u & z_u \\ y_v & z_v \end{vmatrix}, J_2 = \begin{vmatrix} z_u & x_u \\ z_v & x_v \end{vmatrix}, J_3 = \begin{vmatrix} x_u & y_u \\ x_v & y_v \end{vmatrix}, \tag{6.3}$$

Figure 6.5 shows some examples of singular points or lines on the surface. In Figure 6.5 (a) and (b), slope $\mathbf{r}_u$ is not uniquely defined while in (c), P represents the tip of a cone where the slope again is non-unique.

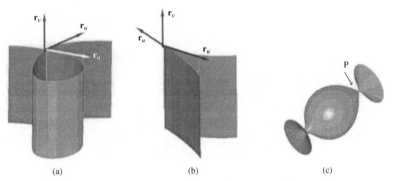

(a)   (b)   (c)

Figure 6.5   Singular points and lines on surfaces

6.1.2   Tangent Plane and Normal Vector on a Surface

Referring to Eq. (6.2) for tangents $\mathbf{r}_u(u, v)$ and $\mathbf{r}_v(u, v)$ at $\mathbf{P}(u, v)$ on the surface, the normal at $\mathbf{P}$ is a vector perpendicular to the plane containing $\mathbf{r}_u(u, v)$ and $\mathbf{r}_v(u, v)$. The normal $\mathbf{N}(u, v)$ and the unit normal $\mathbf{n}(u, v)$ are given by

$$\mathbf{N}(u, v) = \mathbf{r}_u(u, v) \times \mathbf{r}_v(u, v), \quad \mathbf{n}(u, v) = \frac{\mathbf{r}_u(u, v) \times \mathbf{r}_v(u, v)}{|\mathbf{r}_u(u, v) \times \mathbf{r}_v(u, v)|} \tag{6.4}$$

From the foregoing discussion, we may realize that at a regular point, the normal to the suface is well-defined.

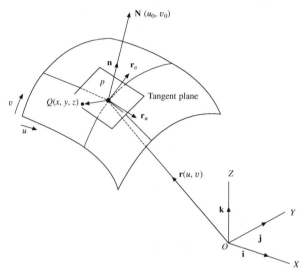

Figure 6.6 Normal and tangent plane

For a surface in implicit form, that is, $f(x, y, z) = 0$, the normal $\mathbf{N}$ and unit normal $\mathbf{n}$ at a point can be obtained from

$$\mathbf{N} = \frac{\partial f}{\partial x}\mathbf{i} + \frac{\partial f}{\partial y}\mathbf{j} + \frac{\partial f}{\partial z}\mathbf{k}, \quad \mathbf{n} = \frac{\mathbf{N}}{|\mathbf{N}|} \tag{6.5}$$

The plane containing the tangent vectors $\mathbf{r}_u(u, v)$ and $\mathbf{r}_v(u, v)$ at $\mathbf{P}(u_0, v_0) = \mathbf{P}(x_0, y_0, z_0)$ on the surface is called the *tangent plane*. To determine its equation, we can select any generic point $Q(x, y, z)$ on the tangent plane, different from P. Since the normal $\mathbf{N}(u_0, v_0)$ and the vector $\mathbf{PQ}$ are perpendicular to each other, their scalar product is zero. With

$$\mathbf{PQ} = (x - x_0)\mathbf{i} + (y - y_0)\mathbf{j} + (z - z_0)\mathbf{k} \quad \text{and} \quad \mathbf{PQ} \cdot \mathbf{N} = 0$$

we have

$$\mathbf{PQ} \cdot (\mathbf{r}_u(u, v) \times \mathbf{r}_v(u, v)) = \begin{vmatrix} x - x_0 & y - y_0 & z - z_0 \\ x_u & y_u & z_u \\ x_v & y_v & z_v \end{vmatrix} = 0 \tag{6.6}$$

where (x_u, y_u, z_u) and (x_v, y_v, z_v) are defined in Eq. (6.3) and are evaluated at (u_0, v_0). Following the expression of the normal in Eq. (6.5), for a surface in the form $f(x, y, z) = 0$, the tangent plane is given by

$$(x - x_0)\frac{\partial f}{\partial x} + (y - y_0)\frac{\partial f}{\partial y} + (z - z_0)\frac{\partial f}{\partial z} = 0 \tag{6.7}$$

with the derivatives evaluated at (x_0, y_0, z_0).

6.2 Curves on a Surface

A curve c on a parametric surface $\mathbf{r}(u, v)$ may be expressed in terms of an additional parameter t as $c(t) = [u(t)\ v(t)]^T$ by letting the parameters u and v as functions of t (Figure 6.7).

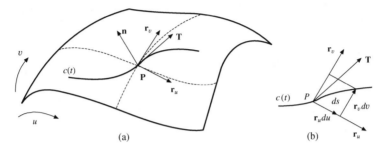

Figure 6.7 A curve $c(t)$ on a parametric surface

The tangent to the curve is given by

$$
\mathbf{T} = \frac{d\mathbf{c}(u, v)}{dt} = \frac{\partial \mathbf{r}(u, v)}{\partial u}\frac{du}{dt} + \frac{\partial \mathbf{r}(u, v)}{\partial v}\frac{dv}{dt} = [\mathbf{r}_u \quad \mathbf{r}_v]\begin{bmatrix} \dfrac{du}{dt} \\ \dfrac{dv}{dt} \end{bmatrix} = \mathbf{A}\begin{bmatrix} \dfrac{du}{dt} \\ \dfrac{dv}{dt} \end{bmatrix} = \begin{bmatrix} \dfrac{\partial x}{\partial u} & \dfrac{\partial x}{\partial v} \\ \dfrac{\partial y}{\partial u} & \dfrac{\partial y}{\partial v} \\ \dfrac{\partial z}{\partial u} & \dfrac{\partial z}{\partial v} \end{bmatrix}\begin{bmatrix} \dfrac{du}{dt} \\ \dfrac{dv}{dt} \end{bmatrix}
$$

$$(6.8a)$$

The differential arc length ds of the curve is given by

$$
ds = \left| \frac{d\mathbf{c}(u, v)}{dt} \right| dt = \left| \mathbf{r}_u \frac{du}{dt} + \mathbf{r}_v \frac{dv}{dt} \right| dt = \sqrt{\left(\mathbf{r}_u \frac{du}{dt} + \mathbf{r}_v \frac{dv}{dt} \right) \cdot \left(\mathbf{r}_u \frac{du}{dt} + \mathbf{r}_v \frac{dv}{dt} \right)}\, dt
$$

$$
= \sqrt{\begin{bmatrix} \dfrac{du}{dt} & \dfrac{dv}{dt} \end{bmatrix}\begin{bmatrix} \mathbf{r}_u \\ \mathbf{r}_v \end{bmatrix} \cdot [\mathbf{r}_u \ \mathbf{r}_v]\begin{bmatrix} \dfrac{du}{dt} \\ \dfrac{dv}{dt} \end{bmatrix}}\, dt = \sqrt{\begin{bmatrix} \dfrac{du}{dt} & \dfrac{dv}{dt} \end{bmatrix}\begin{bmatrix} \mathbf{r}_u \cdot \mathbf{r}_u & \mathbf{r}_u \cdot \mathbf{r}_v \\ \mathbf{r}_v \cdot \mathbf{r}_u & \mathbf{r}_v \cdot \mathbf{r}_v \end{bmatrix}\begin{bmatrix} \dfrac{du}{dt} \\ \dfrac{dv}{dt} \end{bmatrix}}\, dt
$$

$$
= \sqrt{\begin{bmatrix} \dfrac{du}{dt} & \dfrac{dv}{dt} \end{bmatrix}\mathbf{G}\begin{bmatrix} \dfrac{du}{dt} \\ \dfrac{dv}{dt} \end{bmatrix}}\, dt
$$

$$(6.8b)$$

where

$$
\mathbf{G} = \begin{bmatrix} \mathbf{r}_u \cdot \mathbf{r}_u & \mathbf{r}_u \cdot \mathbf{r}_v \\ \mathbf{r}_v \cdot \mathbf{r}_u & \mathbf{r}_v \cdot \mathbf{r}_v \end{bmatrix} = \mathbf{A}^T\mathbf{A} = \begin{bmatrix} G_{11} & G_{12} \\ G_{21} & G_{22} \end{bmatrix} = \begin{bmatrix} E & F \\ F & G \end{bmatrix}
$$

The symmetric matrix $\mathbf{G}$ is termed as the *first fundamental matrix* of the surface. In literature, the components of $\mathbf{G}$ matrix are also written as

$$G_{11} = E = \mathbf{r}_u \cdot \mathbf{r}_u, \ G_{12} = \mathbf{r}_u \cdot \mathbf{r}_v, \ G_{21} = \mathbf{r}_u \cdot \mathbf{r}_v, \ G_{22} = \mathbf{r}_v \cdot \mathbf{r}_v \tag{6.9a}$$

G_{11}, G_{12} and G_{22} are called the *first fundamental form coefficients*. The length of the curve $\mathbf{c}(t)$ lying on the surface is given by

$$ds^2 = \left\{ G_{11} \left(\frac{du}{dt} \right)^2 + 2G_{12} \left(\frac{du}{dt} \right) \left(\frac{dv}{dt} \right) + G_{22} \left(\frac{dv}{dt} \right)^2 \right\} (dt)^2 \tag{6.9b}$$

This equation can also be expressed as

$$ds^2 = G_{11}du^2 + 2G_{12}dudv + G_{22}dv^2 \tag{6.9c}$$

The *first fundamental form* for a surface is given by

$$I = \left(\frac{1}{G_{11}} \right) \{(G_{11}du + G_{12}dv)^2 + (G_{11}G_{22} - G_{12}^2)dv^2\} \tag{6.9d}$$

The unit tangent $\mathbf{t}$ to the curve is given by

$$\mathbf{t} = \frac{\dfrac{\partial \mathbf{r}(u, v)}{\partial u} \dfrac{du}{dt} + \dfrac{\partial \mathbf{r}(u, v)}{\partial v} \dfrac{dv}{dt}}{\left| \dfrac{\partial \mathbf{r}(u, v)}{\partial u} \dfrac{du}{dt} + \dfrac{\partial \mathbf{r}(u, v)}{\partial v} \dfrac{dv}{dt} \right|} = \frac{\dfrac{\partial \mathbf{r}(u, v)}{\partial u} \dfrac{du}{dt} + \dfrac{\partial \mathbf{r}(u, v)}{\partial v} \dfrac{dv}{dt}}{\sqrt{\begin{bmatrix} \dfrac{du}{dt} & \dfrac{dv}{dt} \end{bmatrix} \mathbf{G} \begin{bmatrix} \dfrac{du}{dt} \\ \dfrac{dv}{dt} \end{bmatrix}}} \tag{6.10}$$

Thus, for $\mathbf{t}$ to exist, $\mathbf{G}$ should always be positive definite. For any 2×2 matrix $\mathbf{M}$, the condition for positive definiteness is that (a) $\mathbf{M}_{11} > 0$ and (b) $\mathbf{M}_{11}\mathbf{M}_{22} - \mathbf{M}_{12}\mathbf{M}_{21} > 0$, where $\mathbf{M}_{ij}$ is the entry in the *i*th row and *j*th column of $\mathbf{M}$. Now $G_{11} = \mathbf{r}_u \cdot \mathbf{r}_u > 0$ and also $G_{11}G_{22} - G_{12}G_{21} = (\mathbf{r}_u \cdot \mathbf{r}_u)(\mathbf{r}_v \cdot \mathbf{r}_v) - (\mathbf{r}_u \cdot \mathbf{r}_v)^2 = (\mathbf{r}_u \times \mathbf{r}_v) \cdot (\mathbf{r}_u \times \mathbf{r}_v) > 0$ and so $\mathbf{G}$ is always positive definite. The length of the curve segment in $t_0 \leq t \leq t_1$ can be computed using Eq. (6.8b) as

$$s = \int_{t_0}^{t_1} ds = \int_{t_0}^{t_1} \sqrt{\begin{bmatrix} \dfrac{du}{dt} & \dfrac{dv}{dt} \end{bmatrix} \mathbf{G} \begin{bmatrix} \dfrac{du}{dt} \\ \dfrac{dv}{dt} \end{bmatrix}} \, dt \tag{6.11}$$

If $\mathbf{c}(t_1)$ and $\mathbf{c}(t_2)$ are two curves on the surface $\mathbf{r}(u, v)$ that intersect, the angle of intersection θ can be computed using

$$\mathbf{t}_1 \cdot \mathbf{t}_2 = \frac{\dfrac{\partial \mathbf{r}(u, v)}{\partial u} \dfrac{du}{dt_1} + \dfrac{\partial \mathbf{r}(u, v)}{\partial v} \dfrac{dv}{dt_1}}{\sqrt{\begin{bmatrix} \dfrac{du}{dt_1} & \dfrac{dv}{dt_1} \end{bmatrix} \mathbf{G} \begin{bmatrix} \dfrac{du}{dt_1} \\ \dfrac{dv}{dt_1} \end{bmatrix}}} \cdot \frac{\dfrac{\partial \mathbf{r}(u, v)}{\partial u} \dfrac{du}{dt_2} + \dfrac{\partial \mathbf{r}(u,v)}{\partial v} \dfrac{dv}{dt_2}}{\sqrt{\begin{bmatrix} \dfrac{du}{dt_2} & \dfrac{dv}{dt_2} \end{bmatrix} \mathbf{G} \begin{bmatrix} \dfrac{du}{dt_2} \\ \dfrac{dv}{dt_2} \end{bmatrix}}} = \cos \theta \tag{6.12}$$

The two curves are orthogonal to each other if

$$\left(\frac{\partial \mathbf{r}(u, v)}{\partial u} \frac{du}{dt_1} + \frac{\partial \mathbf{r}(u, v)}{\partial v} \frac{dv}{dt_1} \right) \cdot \left(\frac{\partial \mathbf{r}(u, v)}{\partial u} \frac{du}{dt_2} + \frac{\partial \mathbf{r}(u, v)}{\partial v} \frac{dv}{dt_2} \right) = 0$$

or

$$G_{11} \frac{du}{dt_1} \frac{du}{dt_2} + G_{12} \left(\frac{du}{dt_1} \frac{dv}{dt_2} + \frac{du}{dt_2} \frac{dv}{dt_1} \right) + G_{22} \frac{dv}{dt_1} \frac{dv}{dt_2} = 0 \qquad (6.13)$$

If the two curves coincide, respectively, with u and v iso-parametric curves of the surface, then we may regard $u \equiv t_1$ and $v \equiv t_2$ for which the dot product in Eq. (6.12) becomes

$$\mathbf{t}_1 \cdot \mathbf{t}_2 = \cos \theta = \frac{\dfrac{\partial \mathbf{r}(u, v)}{\partial u}}{\sqrt{G_{11}}} \cdot \frac{\dfrac{\partial \mathbf{r}(u, v)}{\partial v}}{\sqrt{G_{22}}} = \frac{G_{12}}{\sqrt{G_{11} G_{22}}} = \frac{\mathbf{r}_u \cdot \mathbf{r}_v}{|\mathbf{r}_u| |\mathbf{r}_v|} = \frac{\mathbf{r}_u \cdot \mathbf{r}_v}{\sqrt{\mathbf{r}_u \cdot \mathbf{r}_u} \sqrt{\mathbf{r}_v \cdot \mathbf{r}_v}} \qquad (6.14)$$

Thus, the iso-parametric curves are orthogonal if $G_{12} = 0$. To compute the area of a surface patch $\mathbf{r}(u, v)$, let a small patch on the surface be formed by the curves between $u = u_0$, $u = u_0 + du$, $v = v_0$ and $v = v_0 + dv$. The four corners of this patch are $\mathbf{r}(u_0, v_0)$, $\mathbf{r}(u_0 + du, v_0)$, $\mathbf{r}(u_0, v_0 + dv)$ and $\mathbf{r}(u_0 + du, v_0 + dv)$ as shown in Fig. 6.8. The infinitesimal area dA is approximated by

$$dA = |\mathbf{r}_u du \times \mathbf{r}_v dv| = |\mathbf{r}_u \times \mathbf{r}_v| du dv = \sqrt{G_{11} G_{22} - G_{12}^2} \, du dv = \sqrt{|\mathbf{G}|} \, du dv$$

Therefore, the area of the patch is given by

$$A = \int_{\text{Domain}} \sqrt{|\mathbf{G}|} \, du dv \qquad (6.15)$$

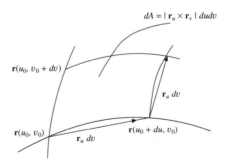

Figure 6.8   Infinitesimal area on the surface

6.3   Deviation of the Surface from the Tangent Plane: Second Fundamental Matrix

In Figure 6.9, let R($\mathbf{r}(u_0 + du, v_0 + dv)$) be a point on the surface a small distance away from P($\mathbf{r}(u_0, v_0)$). The deviation d of R from P along the normal $\mathbf{n}$ at P may be written as

$$d = [\mathbf{r}(u_0 + du, v_0 + dv) - \mathbf{r}(u_0, v_0)] \cdot \mathbf{n}$$

which using Taylor series expansion is

$$d \cong \left[\frac{\partial \mathbf{r}}{\partial u}du + \frac{\partial \mathbf{r}}{\partial v}dv + \frac{\partial^2 \mathbf{r}}{\partial u \partial v}dudv + \frac{1}{2}\frac{\partial^2 \mathbf{r}}{\partial u^2}(du)^2 + \frac{1}{2}\frac{\partial^2 \mathbf{r}}{\partial v^2}(dv)^2\right] \cdot \mathbf{n}$$

$$= \mathbf{r}_u \cdot \mathbf{n}du + \mathbf{r}_v \cdot \mathbf{n}dv + \left(\frac{\partial^2 \mathbf{r}}{\partial u \partial v}\right) \cdot \mathbf{n}dudv + \frac{1}{2}\frac{\partial^2 \mathbf{r}}{\partial u^2} \cdot \mathbf{n}(du)^2 + \frac{1}{2}\frac{\partial^2 \mathbf{r}}{\partial v^2} \cdot \mathbf{n}(dv)^2$$

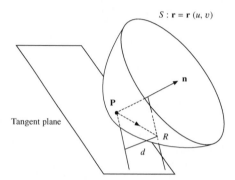

$S : \mathbf{r} = \mathbf{r}(u, v)$

Figure 6.9   Deviation of *R* from the tangent plane

Since $\mathbf{n}$ is perpendicular to the tangent plane, $\mathbf{r}_u \cdot \mathbf{n} = \mathbf{r}_v \cdot \mathbf{n} = 0$, hence

$$d = \left(\frac{\partial^2 \mathbf{r}}{\partial u \partial v}\right) \cdot \mathbf{n}dudv + \frac{1}{2}\frac{\partial^2 \mathbf{r}}{\partial u^2} \cdot \mathbf{n}(du)^2 + \frac{1}{2}\frac{\partial^2 \mathbf{r}}{\partial v^2} \cdot \mathbf{n}(dv)^2$$

$$= \frac{1}{2}[du \ dv]\begin{bmatrix} \mathbf{r}_{uu} \cdot \mathbf{n} & \mathbf{r}_{uv} \cdot \mathbf{n} \\ \mathbf{r}_{uv} \cdot \mathbf{n} & \mathbf{r}_{vv} \cdot \mathbf{n} \end{bmatrix}\begin{bmatrix} du \\ dv \end{bmatrix} \qquad (6.16)$$

The matrix $\mathbf{D} = \begin{bmatrix} \mathbf{r}_{uu} \cdot \mathbf{n} & \mathbf{r}_{uv} \cdot \mathbf{n} \\ \mathbf{r}_{uv} \cdot \mathbf{n} & \mathbf{r}_{vv} \cdot \mathbf{n} \end{bmatrix}$ is called the *second fundamental matrix* of the surface. Using $G_{11}G_{22} - G_{12}G_{21} = (\mathbf{r}_u \times \mathbf{r}_v) \cdot (\mathbf{r}_u \times \mathbf{r}_v)$ in Eq. (6.4), we get

$$\mathbf{n} = \frac{\mathbf{r}_u \times \mathbf{r}_v}{|\ \mathbf{r}_u \times \mathbf{r}_v\ |} = \frac{\mathbf{r}_u \times \mathbf{r}_v}{\sqrt{G_{11}G_{22} - G_{12}^2}} \qquad (6.17)$$

From Eq. (6.16)

$$2d = [du \ dv]\begin{bmatrix} \mathbf{r}_{uu} \cdot \mathbf{n} & \mathbf{r}_{uv} \cdot \mathbf{n} \\ \mathbf{r}_{uv} \cdot \mathbf{n} & \mathbf{r}_{vv} \cdot \mathbf{n} \end{bmatrix}\begin{bmatrix} du \\ dv \end{bmatrix} = \mathbf{r}_{uu} \cdot \mathbf{n}(du)^2 + 2\mathbf{r}_{uv} \cdot \mathbf{n}dudv + \mathbf{r}_{vv} \cdot \mathbf{n}(dv)^2$$

or

$$2d = L(du)^2 + 2Mdudv + N(dv)^2$$

where

$$L = \mathbf{r}_{uu} \cdot \mathbf{n}, \quad M = \mathbf{r}_{uv} \cdot \mathbf{n}, \quad N = \mathbf{r}_{vv} \cdot \mathbf{n} \qquad (6.18)$$

Here L, M and N are the *second fundamental form coefficients*. The second fundamental matrix **D** can then be expressed as

$$\mathbf{D} = \begin{bmatrix} L & M \\ M & N \end{bmatrix} \tag{6.19}$$

Using Eqs. (6.17), (6.18) and (6.19), we have

$$\mathbf{D} = \begin{bmatrix} \mathbf{r}_{uu} \cdot \mathbf{n} & \mathbf{r}_{uv} \cdot \mathbf{n} \\ \mathbf{r}_{uv} \cdot \mathbf{n} & \mathbf{r}_{vv} \cdot \mathbf{n} \end{bmatrix} = \frac{1}{\sqrt{G_{11} G_{22} - G_{12}^2}} \begin{bmatrix} \mathbf{r}_{uu} \cdot (\mathbf{r}_u \times \mathbf{r}_v) & \mathbf{r}_{uv} \cdot (\mathbf{r}_u \times \mathbf{r}_v) \\ \mathbf{r}_{uv} \cdot (\mathbf{r}_u \times \mathbf{r}_v) & \mathbf{r}_{vv} \cdot (\mathbf{r}_u \times \mathbf{r}_v) \end{bmatrix} \tag{6.20}$$

From Eq. (6.1) we have

$$\mathbf{r}_{uu} = x_{uu}\mathbf{i} + y_{uu}\mathbf{j} + z_{uu}\mathbf{k}$$

$$\mathbf{r}_{uv} = x_{uv}\mathbf{i} + y_{uv}\mathbf{j} + z_{uv}\mathbf{k}$$

$$\mathbf{r}_{vv} = x_{vv}\mathbf{i} + y_{vv}\mathbf{j} + z_{vv}\mathbf{k} \tag{6.21a}$$

and

$$\mathbf{r}_u \times \mathbf{r}_v = \begin{vmatrix} \mathbf{i} & \mathbf{j} & \mathbf{k} \\ x_u & y_u & z_u \\ x_v & y_v & z_v \end{vmatrix}$$

we get from Eq. (6.20)

$$\tag{6.21b}$$

$$\mathbf{D} = \frac{1}{\sqrt{A^2 + B^2 + C^2}} \begin{bmatrix} D_{11} & D_{12} \\ D_{21} & D_{22} \end{bmatrix}$$

where

$$A = y_u z_v - y_v z_u, \quad B = z_u x_v - z_v x_u, \quad C = x_u y_v - x_v y_u$$

and

$$D_{11} = \begin{vmatrix} x_{uu} & y_{uu} & z_{uu} \\ x_u & y_u & z_u \\ x_v & y_v & z_v \end{vmatrix}, \quad D_{12} = D_{21} = \begin{vmatrix} x_{uv} & y_{uv} & z_{uv} \\ x_u & y_u & z_u \\ x_v & y_v & z_v \end{vmatrix}, \quad D_{22} = \begin{vmatrix} x_{vv} & y_{vv} & z_{vv} \\ x_u & y_u & z_u \\ x_v & y_v & z_v \end{vmatrix}$$

$$\tag{6.22}$$

6.4 Classification of Points on a Surface

From Eq. (6.18), we observe that deviation d of a point R on the surface from a point P along the normal **n** through P is given by

$$d = \frac{1}{2}(L du^2 + 2M du dv + N dv^2) \tag{6.23}$$

To realize on which side of the tangent plane R lies, we can determine whether d is positive, negative or zero. The tangent plane at point P will intersect the surface at all points where $d = 0$, that is

$$L du^2 + 2M du dv + N dv^2 = 0 \Rightarrow du = \frac{-M \pm \sqrt{M^2 - LN}}{L} dv \tag{6.24}$$

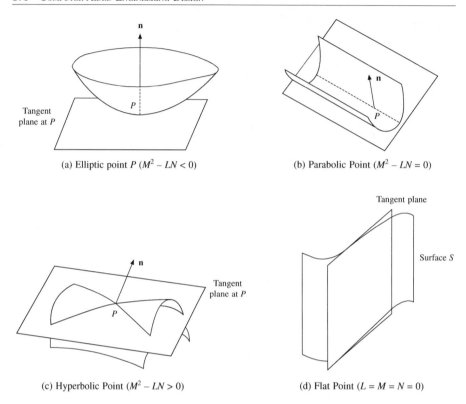

(a) Elliptic point P ($M^2 - LN < 0$)

(b) Parabolic Point ($M^2 - LN = 0$)

(c) Hyperbolic Point ($M^2 - LN > 0$)

(d) Flat Point ($L = M = N = 0$)

Figure 6.10   Classification of points on a surface

Case 1. $M^2 - LN < 0$: For any departure dv from point P, there is no real value of du. This implies that the tangent plane at P does not intersect the surface at any other point, or P is the only point common between the tangent plane and the surface and that the *surface lies only on one side of the tangent plane*. Point P is called the *elliptic point* of the surface. All points on an ellipsoid, a sphere, an elliptic paraboloid and hyperboloid of two sheets are *elliptic points*.

Case 2. $M^2 - LN = 0$, $L^2 + M^2 + N^2 > 0$: There are no double roots and $du = -(M/L)dv$. For P (u_0, v_0) and R (u, v), the result implies $u - u_0 = -(M/L)(v - v_0)$ which is the equation of a straight line in u and v. Thus, the tangent plane intersects the surface along the aforementioned straight line. P is then called a *parabolic point*. A cylinder or a truncated (frustrum) cone consists entirely of parabolic points. All regular points of any *developable surface* (covered later) are parabolic points.

Case 3. $M^2 - LN > 0$: There exist two real roots, and the tangent plane at P intersects the surface along two lines passing through P. For $du = u - u_0$ and $dv = v - v_0$

$$du = \frac{-M + \sqrt{M^2 - LN}}{L} dv \quad \text{and} \quad du = \frac{-M - \sqrt{M^2 - LN}}{L} dv$$

here P is called a *hyperbolic point* on the surface. All regular points of a *pseudosphere* and all points of a *catenoid* are hyperbolic (Figure 6.4). All points of the hyperbolic paraboloid and hyperboloid of one sheet, or a non-developable ruled surface (discussed later) are also hyperbolic points.

Case 4. $L = M = N = 0$: The tangent plane is not only tangent to the surface at P, but also has a contact of higher order with the surface. In this case, point P on the surface is called a *flat point,* because points in the small neighborhood of P are also the points of tangency to the same tangent plane. A monkey saddle $z = x(x^2 - 3y^2)$ has a flat point at $(0, 0, 0)$.

Example 6.2. Show that the lines $u = \pm \frac{1}{2} \pi$ are parabolic lines of the torus $x = (b + a \cos u) \cos v$, $y = (b + a \cos u) \sin v$, $z = a \sin u$ with $b > a$. These lines divide the torus into two domains. Show that the exterior domain $-\pi/2 < u < \pi/2$ consists of elliptic points and the interior domain $\pi/2 < u < 3\pi/2$ consists of hyperbolic points.

The expression for $M^2 - LN$ is computed using Eqs. (6.19) and (6.22) to obtain

$$x(u, v) = (b + a \cos u) \cos v, \qquad y(u, v) = (b + a \cos u) \sin v, \qquad z(u, v) = a \sin u$$

$$x_u = -a \sin u \cos v, \qquad y_u = -a \sin u \sin v, \qquad z_u = a \cos u$$

$$x_v = -(b + a \cos u) \sin v, \qquad y_v = (b + a \cos u) \cos v, \qquad z_v = 0$$

$$x_{uu} = -a \cos u \cos v, \qquad y_{uu} = -a \cos u \sin v, \qquad z_{uu} = -a \sin u$$

$$x_{uv} = a \sin u \sin v, \qquad y_{uv} = -a \cos v \sin u, \qquad z_{uv} = 0$$

$$x_{vv} = -(b + a \cos u) \cos v, \qquad y_{vv} = -(b + a \cos u) \sin v, \qquad z_{vv} = 0$$

$$D_{11} = \begin{vmatrix} x_{uu} & y_{uu} & z_{uu} \\ x_u & y_u & z_u \\ x_v & y_v & z_v \end{vmatrix}, D_{12} = \begin{vmatrix} x_{uv} & y_{uv} & z_{uv} \\ x_u & y_u & z_u \\ x_v & y_v & z_v \end{vmatrix}, D_{22} = \begin{vmatrix} x_{vv} & y_{vv} & z_{vv} \\ x_u & y_u & z_u \\ x_v & y_v & z_v \end{vmatrix}$$

Substitution and simplification of the determinants yields

$$D_{11} = a^2 (b + a \cos u), \quad D_{12} = 0, \quad D_{22} = a(b + a \cos u)^2 \cos u$$

$$M^2 - LN = -\frac{a^3 \cos u(b + a \cos u)^3}{\sqrt{A^2 + B^2 + C^2}}$$

where A, B and C are defined in Eq. (6.21b). For $u = \pm \frac{1}{2} \pi$, $\cos(u) = 0$ and hence $M^2 - LN = 0$. This corresponds to Case 2 above for all values of the parameter v and hence $u = \pm \frac{1}{2} \pi$ are parabolic points on the surface of the torus.

For $-\frac{\pi}{2} < u < \frac{\pi}{2}$, $\cos(u) > 0$, since a and $(b + a \cos u)$ are both greater than 0, $M^2 - LN < 0$ for all values of v. This shows (Case 1) that the exterior part of the torus has all elliptic points. For $\frac{\pi}{2} < u < \frac{3\pi}{2}$, $\cos u < 0$, with a and $(b + a \cos u) > 0$. Thus, $M^2 - LN > 0$ corresponds to Case 3 above. Hence, the surface patch corresponding to $\frac{\pi}{2} < u < \frac{3\pi}{2}$ has all hyperbolic points.

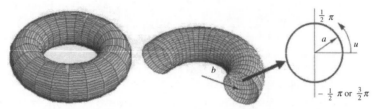

Figure 6.11 Parabolic, elliptic and hyperbolic points on the torus.

6.5 Curvature of a Surface: Gaussian and Mean Curvature

The curve C in Figure 6.12 lies on a surface passing through a point P. Let π denote the tangent plane containing vectors $\mathbf{r}_u$ and $\mathbf{r}_v$ and $\mathbf{n}$ be the unit normal to the surface at P. The unit tangent vector $\mathbf{t}$ to the curve at P also lies on π. Let κ denote the curvature and $\mathbf{n}_c$ be the unit normal to curve at P. The curvature vector $\mathbf{k}$ is in the direction of $\mathbf{n}_c$ and can be decomposed into two components: (a) $\mathbf{k}_n$ in the direction of $\mathbf{n}$ and (b) $\mathbf{k}_g$ in the plane π but perpendicular to $\mathbf{t}$. Now

$$\mathbf{k} = \kappa \mathbf{n}_c = \frac{d\mathbf{t}}{ds} = \mathbf{k}_n + \mathbf{k}_g \quad \text{with} \quad \mathbf{k}_n = \kappa_n \mathbf{n} \tag{6.25}$$

Here $\mathbf{k}_n$ and $\mathbf{k}_g$ are called the vectors of *normal curvature* and *geodesic curvature*, respectively. κ_n is called the *normal curvature of the surface* at P. Since $\mathbf{n}$ and $\mathbf{t}$ are mutually perpendicular, $\mathbf{n} \cdot \mathbf{t} = 0$.

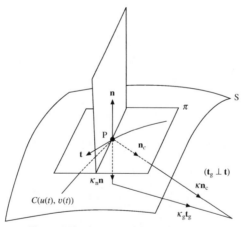

Figure 6.12 Curvature of a surface

Thus

$$\frac{d\mathbf{n}}{ds} \cdot \mathbf{t} + \mathbf{n} \cdot \frac{d\mathbf{t}}{ds} = 0 \tag{6.26}$$

where s is the arc length parameter of the curve C. From Eq. (6.25), since $\mathbf{k}_g$ and $\mathbf{n}$ are perpendicular, $\mathbf{k}_g \cdot \mathbf{n} = 0$ using which

$$\frac{d\mathbf{t}}{ds} \cdot \mathbf{n} = (\mathbf{k}_n + \mathbf{k}_g) \cdot \mathbf{n} = \mathbf{k}_n \cdot \mathbf{n} = \kappa_n \mathbf{n} \cdot \mathbf{n} = \kappa_n = -\mathbf{t} \cdot \frac{d\mathbf{n}}{ds} \tag{6.27}$$

Therefore, $\qquad \kappa_n = -\dfrac{d\mathbf{r}}{ds} \cdot \dfrac{d\mathbf{n}}{ds} = -\dfrac{d\mathbf{r} \cdot d\mathbf{n}}{d\mathbf{r} \cdot d\mathbf{r}}$ with $ds^2 \approx d\mathbf{r} \cdot d\mathbf{r}$ $\qquad$ (6.28)

We can simplify Eq. (6.28) by decomposing $d\mathbf{r}$ and $d\mathbf{n}$ along parametric lengths du and dv, that is

$$d\mathbf{r} = \mathbf{r}_u du + \mathbf{r}_v dv, \quad d\mathbf{n} = \mathbf{n}_u du + \mathbf{n}_v dv$$

$$\Rightarrow \qquad d\mathbf{r} \cdot d\mathbf{n} = \mathbf{r}_u \cdot \mathbf{n}_u (du)^2 + (\mathbf{r}_u \cdot \mathbf{n}_v + \mathbf{r}_v \cdot \mathbf{n}_u) du dv + \mathbf{r}_v \cdot \mathbf{n}_v (dv)^2$$

$$d\mathbf{r} \cdot d\mathbf{r} = \mathbf{r}_u \cdot \mathbf{r}_u (du)^2 + (\mathbf{r}_u \cdot \mathbf{r}_v + \mathbf{r}_v \cdot \mathbf{r}_u) du dv + \mathbf{r}_v \cdot \mathbf{r}_v (dv)^2$$

$$= G_{11} du^2 + 2 G_{12}\, du dv + G_{22} dv^2 \qquad (6.29)$$

Since $\mathbf{r}_u$ and $\mathbf{r}_v$ are both perpendicular to $\mathbf{n}$, using Eq. (6.18), we get

$$\mathbf{r}_u \cdot \mathbf{n} = 0 \Rightarrow \mathbf{r}_{uu} \cdot \mathbf{n} + \mathbf{r}_u \cdot \mathbf{n}_u = 0 \Rightarrow \mathbf{r}_u \cdot \mathbf{n}_u = -\mathbf{r}_{uu} \cdot \mathbf{n} = -L \qquad (6.30a)$$

$$\mathbf{r}_v \cdot \mathbf{n} = 0 \Rightarrow \mathbf{r}_{vv} \cdot \mathbf{n} + \mathbf{r}_v \cdot \mathbf{n}_v = 0 \Rightarrow \mathbf{r}_v \cdot \mathbf{n}_v = -\mathbf{r}_{vv} \cdot \mathbf{n} = -N$$

$$\mathbf{r}_u \cdot \mathbf{n} = 0 \Rightarrow \mathbf{r}_{uv} \cdot \mathbf{n} + \mathbf{r}_u \cdot \mathbf{n}_v = 0 \Rightarrow \mathbf{r}_u \cdot \mathbf{n}_v = -\mathbf{r}_{uv} \cdot \mathbf{n} = -M \qquad (6.30b)$$

$$\mathbf{r}_v \cdot \mathbf{n} = 0 \Rightarrow \mathbf{r}_{vu} \cdot \mathbf{n} + \mathbf{r}_v \cdot \mathbf{n}_u = 0 \Rightarrow \mathbf{r}_v \cdot \mathbf{n}_u = -\mathbf{r}_{vu} \cdot \mathbf{n} = -M$$

Using Eqs. (6.28), (6.29) with (6.30), the expression for the normal curvature

$$\kappa_n = \frac{L du^2 + 2 M du dv + N dv^2}{G_{11} du^2 + 2 G_{12}\, du dv + G_{22} dv^2} = \frac{L + 2 M \mu + N \mu^2}{G_{11} + 2 G_{12} \mu + G_{22} \mu^2} \qquad (6.31)$$

where $\mu = \dfrac{dv}{du}$.

Equation (6.31) can be rewritten as

$$(G_{11} + 2 G_{12} \mu + G_{22} \mu^2) \kappa_n = L + 2 M \mu + N \mu^2 \qquad (6.32)$$

For an optimum value of the normal curvature $\dfrac{d\kappa_n}{d\mu} = 0$. Differentiating Eq. (6.32) yields

$$(G_{11} + 2 G_{12} \mu + G_{22} \mu^2) \frac{d\kappa_n}{d\mu} + 2(G_{12} + G_{22} \mu) \kappa_n = 2(M + N \mu) \qquad (6.33a)$$

$$\Rightarrow \qquad (G_{12} + G_{22} \mu) \kappa_n = (M + N \mu) \qquad (6.33b)$$

Equating Eqs. (6.31) and (6.33(b)), we get

$$\kappa_n = \frac{M + N \mu}{G_{12} + G_{22} \mu} = \frac{L + 2 M \mu + N \mu^2}{G_{11} + 2 G_{12} \mu + G_{22} \mu^2} = \frac{(L + M \mu) + \mu(M + N \mu)}{(G_{11} + G_{12} \mu) + \mu(G_{12} + G_{22} \mu)}$$

which can be simplified as

$$\kappa_n = \frac{M + N \mu}{G_{12} + G_{22} \mu} = \frac{L + M \mu}{G_{11} + G_{12} \mu}$$

$$\Rightarrow \qquad (M - G_{12} \kappa_n) + (N - G_{22} \kappa_n) \mu = 0$$

$$(L - G_{11}\kappa_n) + (M - G_{12}\kappa_n)\mu = 0$$

$$\Rightarrow \quad \begin{bmatrix} (M - G_{12}\kappa_n) & (N - G_{22}\kappa_n) \\ (L - G_{11}\kappa_n) & (M - G_{12}\kappa_n) \end{bmatrix} \begin{bmatrix} 1 \\ \mu \end{bmatrix} = \begin{bmatrix} 0 \\ 0 \end{bmatrix} \tag{6.34}$$

For a non-trivial solution of μ, the determinant of the coefficient matrix must be zero. Therefore

$$\begin{vmatrix} (M - G_{12}\kappa_n) & (N - G_{22}\kappa_n) \\ (L - G_{11}\kappa_n) & (M - G_{12}\kappa_n) \end{vmatrix} = 0$$

or $\quad (G_{11}G_{22} - G_{12}^2)\,\kappa_n^2 - (G_{11}N + G_{22}L - 2G_{12}M)\kappa_n + (LN - M^2) = 0 \tag{6.35}$

The above can be further simplified to find the two optimal values of the normal curvature. Thus

$$\kappa_n^2 - \frac{G_{11}N + G_{22}L - 2G_{12}M}{G_{11}G_{22} - G_{12}^2}\,\kappa_n + \frac{LN - M^2}{G_{11}G_{22} - G_{12}^2} = 0$$

$$\Rightarrow \quad \kappa_n^2 - 2H\kappa_n + K = 0$$

with $\quad H = \dfrac{G_{11}N + G_{22}L - 2G_{12}M}{2(G_{11}G_{22} - G_{12}^2)}$ and $K = \dfrac{LN - M^2}{G_{11}G_{22} - G_{12}^2}$

$$\Rightarrow \kappa_n = (H \pm \sqrt{H^2 - K})$$

Thus $\quad (\kappa_n)_{max} = \kappa_{max} = (H + \sqrt{H^2 - K})$

$$(\kappa_n)_{min} = \kappa_{min} = (H - \sqrt{H^2 - K}) \tag{6.36}$$

The maximum and minimum normal curvatures (κ_{max} and κ_{min}) at a point on a surface can be calculated as above. K and H are called the *Gaussian* and *mean curvatures,* respectively. It can be shown that the discriminant ($H^2 - K$) is either positive or zero. When the discriminant is 0, the surface point is called an *umbilical point* for which $\kappa_n = H$. When $K = H = 0$, the point is a *flat* or *planar point*. We note that

$$K = \kappa_{max}\kappa_{min}, \quad H = \frac{\kappa_{max} + \kappa_{min}}{2} \tag{6.37}$$

Example 6.3. The parametric equation of a monkey saddle surface (Figure 6.13a) is given by

$$\mathbf{r}(u, v) = u\mathbf{i} + v\mathbf{j} + (u^3 - 3uv^2)\mathbf{k} = (u, v, u^3 - 3uv^2)$$

Compute the Gaussian and mean curvatures.

We can compute

$$\mathbf{r}_u = (1, 0, 3u^2 - 3v^2), \quad \mathbf{r}_v = (0, 1, -6uv),$$

$$\mathbf{r}_{uu} = (0, 0, 6u), \quad \mathbf{r}_{uv} = (0, 0, -6v), \mathbf{r}_{vv} = (0, 0, -6u)$$

Therefore, from Eqs. (6.9) and (6.16), we can calculate the coefficients of the first and second fundamental forms of the monkey saddle as

$$G_{11} = \mathbf{r}_u \cdot \mathbf{r}_u = 1 + (3u^2 - 3v^2)^2, \quad G_{12} = \mathbf{r}_u \cdot \mathbf{r}_v = -6uv\,(3u^2 - 3v^2),$$

$$G_{22} = \mathbf{r}_v \cdot \mathbf{r}_v = 1 + 36u^2v^2$$

The surface normal is determined as

$$\mathbf{n} = \frac{\mathbf{r}_u \times \mathbf{r}_v}{|\,\mathbf{r}_u \times \mathbf{r}_v\,|} = \frac{(-3u^2 + 3v^2, 6uv, 1)}{\sqrt{1 + 9u^4 + 9v^4 + 18u^2v^2}}$$

Therefore, $L = \mathbf{n} \cdot \mathbf{r}_{uu} = \dfrac{6u}{\sqrt{1 + 9u^4 + 9v^4 + 18u^2v^2}}, \quad M = \mathbf{n} \cdot \mathbf{r}_{uv} = \dfrac{-6\,v}{\sqrt{1 + 9u^4 + 9v^4 + 18u^2v^2}}$

$$N = \mathbf{n} \cdot \mathbf{r}_{vv} = \frac{-6u}{\sqrt{1 + 9u^4 + 9v^4 + 18u^2v^2}}$$

From Eq. (6.36) we have the Gaussian and mean curvatures

$$K = \frac{LN - M^2}{G_{11}G_{22} - G_{12}^2} = \frac{-36(u^2 + v^2)}{(1 + 9u^4 + 9v^4 + 18u^2v^2)^2}$$

$$H = \frac{G_{11}N + G_{22}L - 2G_{12}M}{2(G_{11}G_{22} - G_{12}^2)} = \frac{-27u^5 + 54u^3v^2 + 81uv^4}{(1 + 9u^4 + 9v^4 + 18u^2v)^{3/2}}$$

The monkey saddle and its curvatures are shown in Figures 6.13. Maximum and minimum normal curvatures can be determined using Eqs. (6.36) with expressions of Gaussian and mean curvatures derived above.

Foregoing discussion dealt with the differential properties of surfaces that included the tangent plane and normal at a point, the first and second fundamental matrices, and principal (maximum and minimum normal), Gaussian and mean curvatures. Such properties are mainly studied for composite fitting of surface patches at their common boundaries to achieve slope and/or curvature continuity. Before generic design of surface patches is ventured into in Chapter 7, a notion about some specific patches is provided below. A user may not need to specify the slope or data point information directly for their design. However, the input sought would be indirect, more like in terms of a pair of curves (for instance Ferguson, Bézier or spline curves), or a curve and a straight line.

6.6   Developable and Ruled Surfaces

Developable surfaces can be unfolded or developed onto a plane without stretching or tearing (Figure 6.14a). Such surfaces are useful in sheet metal industry for making drums, conical funnels, convergent or divergent nozzles, ducts for air conditioning, shoes, tailoring shirts and pants, automobile upholstery, door panels, windshield, shipbuilding, fiber reinforced plastic (FRP) panels for aircraft wings, and many other applications. Hence, the design of developable surfaces cannot be ignored. Cylindrical and conic patches are well-known examples. An interesting note about a developable surface is that at every point on the surface, the *Gaussian curvature K is zero*. Thus, they are known as *singly curved surfaces,* since one of their principal curvatures is zero.

From Eq. (6.37), the Gaussian curvature K is zero if either κ_{min} or κ_{max}, or both are zero. Since $G_{11}G_{22} - G_{12}^2 > 0$ as shown earlier, $K = 0$ implies that $LN - M^2 = 0$. As discussed in Section 6.4, this

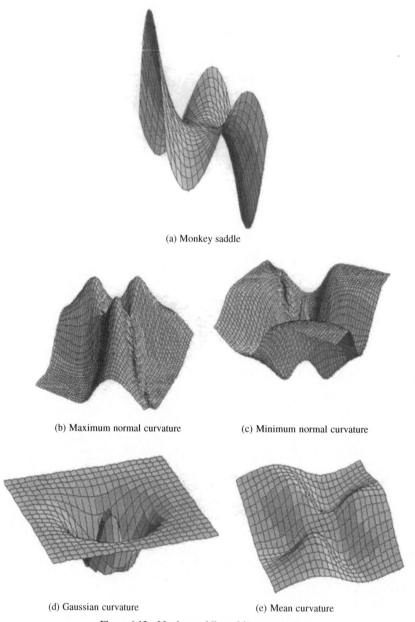

(a) Monkey saddle

(b) Maximum normal curvature   (c) Minimum normal curvature

(d) Gaussian curvature   (e) Mean curvature

Figure 6.13   Monkey saddle and its curvatures

condition conveys that the tangent plane touches the surface along a straight line (in u and v) at *parabolic points*. If both the curvatures κ_{min} and κ_{max} are nonzero, the surface is called a *doubly curved surface*.

Consider two curves $\mathbf{p}(u)$ and $\mathbf{q}(u)$ in Figure 6.14(b). If a straight line moves uniformly such that its one end is always on the curve $\mathbf{p}(u)$ and the other is always on $\mathbf{q}(u)$, a *ruled surface* is generated. The equation of the resulting surface is

$$\mathbf{r}(u, v) = (1 - v)\mathbf{p}(u) + v\mathbf{q}(u) = \mathbf{p}(u) + v[\mathbf{q}(u) - \mathbf{p}(u)] \qquad (6.38a)$$

or
$$\mathbf{r}(u, v) = \mathbf{p}(u) + v\mathbf{d}(u) \qquad (6.38b)$$

with $\mathbf{d}(u) = \mathbf{q}(u) - \mathbf{p}(u)$. For ruled surfaces $\mathbf{r}_{vv} = \mathbf{0}$ which makes $N = 0$ and the Gaussian curvature in Eq. (6.36) becomes

$$K = \frac{-M^2}{G_{11}G_{22} - G_{12}^2} \qquad (6.39a)$$

Further, for a ruled surface to be developable, it is required that K and thus M is zero, that is, $\mathbf{r}_{uv} \cdot \mathbf{n} = 0$ at every point on the surface. The mean curvature for developable ruled surfaces is then given by

$$H = \frac{G_{22}L}{2(G_{11}G_{22} - G_{12}^2)} \qquad (6.39b)$$

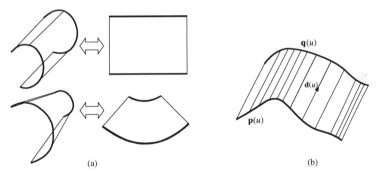

(a)   (b)

Figure 6.14   (a) Developable and (b) ruled surfaces

Example 6.4. A toothpaste tube is a ruled surface. For $\mathbf{p}(u) = a \cos u\mathbf{i} + a \sin u\,\mathbf{j} + 0\mathbf{k}$ with a as the radius of the circle and $\mathbf{q}(u) = (a - 2au/\pi)\mathbf{i} + 0\mathbf{j} + b\mathbf{k}$, the straight line parallel to the plane of the circle and at a distance b from it, the half section $\mathbf{r}(u, v)$ of the tube with u in $[0, \pi]$ is given by

$$\mathbf{r}(u, v) = [(1 - v)\, a \cos u + v(a - 2au/\pi)]\mathbf{i} + a \sin u(1 - v)\mathbf{j} + vb\mathbf{k}$$

To show that the toothpaste tube is not developable, the Gaussian curvature is determined.

$$\mathbf{r}_u = [-(1 - v)\, a \sin u - 2av/\pi]\mathbf{i} + a \cos u\, (1 - v)\mathbf{j}$$
$$\mathbf{r}_v = [-a \cos u + (a - 2\, au/\pi)]\mathbf{i} - a \sin u\mathbf{j} + b\mathbf{k}$$

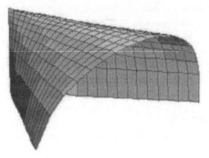

Figure 6.15 A symmetric half of the toothpaste tube

$$\mathbf{r}_{uu} = [-(1-v)\,a\cos u]\mathbf{i} - a\sin u\,(1-v)\mathbf{j}$$

$$\mathbf{r}_{uv} = [a\sin u - 2a/\pi]\mathbf{i} - a\cos u\,\mathbf{j}$$

$$\mathbf{r}_{vv} = 0$$

All we need to show then is $M \neq 0$. From Eqs. (6.19) and (6.22)

$$D_{12} = \begin{vmatrix} x_{uv} & y_{uv} & z_{uv} \\ x_u & y_u & z_u \\ x_v & y_v & z_v \end{vmatrix} = \begin{vmatrix} a\sin u - \dfrac{2a}{\pi} & -a\cos u & 0 \\ -(1-v)a\sin(u) - \dfrac{2av}{\pi} & (1-v)a\cos u & 0 \\ -a\cos u + \left(a - \dfrac{2au}{\pi}\right) & -a\sin u & b \end{vmatrix}$$

$$= b\left\{(1-v)a\cos u\left(a\sin u - \frac{2a}{\pi}\right) - a\cos u\left((1-v)a\sin(u) + \frac{2av}{\pi}\right)\right\}$$

$$= b\left\{ \begin{array}{l} a^2\cos u\sin u - \dfrac{2a^2}{\pi}\cos u - va^2\cos u\sin u + \dfrac{2a^2}{\pi}v\cos u - a^2\cos u\sin u \\[2mm] + va^2\cos u\sin u - \dfrac{2a^2v}{\pi}\cos u \end{array} \right\}$$

$$= b\left\{-\frac{2a^2}{\pi}\cos u\right\}$$

which is zero only when $\cos u = 0$ or when $u = \frac{1}{2}\pi$. Since D_{12} and hence $M \neq 0$ for other values of u, the Gaussian curvature is not zero at all points on the surface. The toothpaste tube, therefore, is non-developable and cannot be flattened without tearing or stretching.

Some other examples of ruled but non-developable surfaces are those of Plucker polar and hyperbolic paraboloid surfaces shown in Figure 6.16. The respective equations are

$$\mathbf{r}(u, v) = u\cos v\mathbf{i} + u\sin v\mathbf{j} + \sin nv\mathbf{k} \quad (n \text{ is an integer} \geq 2)$$

$$\mathbf{r}(u, v) = u\mathbf{i} + v\mathbf{j} + uv\mathbf{k}$$

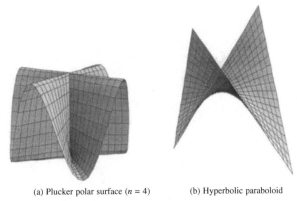

(a) Plucker polar surface ($n = 4$)   (b) Hyperbolic paraboloid

Figure 6.16   Other ruled but non-developable surfaces

6.7   Parallel Surfaces

Creation of *parallel surfaces* is useful in design and manufacture. Making of *dies* for *forging* and *castings* require modeling of parallel surfaces. Enhancing or reducing the size of free-from surfaces requires calculation of curvature and other properties of the new surface, which is parallel to the original surface.

Let S: $\mathbf{r}(u, v)$ define a surface patch with parametric curves

$$\mathbf{r}(u, v)|_{v=v_0} = \mathbf{r}(u, v_0), \mathbf{r}(u, v)|_{u=u_0} = \mathbf{r}(u_0, v) \tag{6.41}$$

Parametric curves, $\mathbf{r}(u, v_0)$, $\mathbf{r}(u_0, v)$ lie entirely on the surface S and intersect at a point P ($\mathbf{r}(u_0, v_0)$). For simplicity, let these *parametric curves* be also the *lines of curvature* of S. The tangents to these curves at P are given by

$$\mathbf{r}_u = \mathbf{r}_u(u, v_0) = \frac{\partial \mathbf{r}(u, v_0)}{\partial u}, \mathbf{r}_v = \mathbf{r}_v(u_0, v) = \frac{\partial \mathbf{r}(u_0, v)}{\partial v} \tag{6.42}$$

These tangents are the two *principal direction vectors* and hence they are *orthogonal*. From Eq. (6.14), this will mean that the angle between the two tangent vectors $\mathbf{r}_u(u, v_0)$, $\mathbf{r}_v(u_0, v)$ is $\theta = 90°$,

$$\cos \theta = \frac{G_{12}}{\sqrt{G_{11} G_{22}}} = \frac{\mathbf{r}_u \cdot \mathbf{r}_v}{\sqrt{\mathbf{r}_u \cdot \mathbf{r}_u} \sqrt{\mathbf{r}_v \cdot \mathbf{r}_v}} = 0 \Rightarrow G_{12} = 0 \tag{6.43}$$

From Eq. (6.34), the normal curvature κ_n satisfies the equations

$$(M - G_{12}\kappa_n) \, du + (N - G_{22}\kappa_n) \, dv = 0 \Rightarrow Mdu + (N - G_{22}\kappa_n) \, dv = 0$$

$$(L - G_{11}\kappa_n) \, du + (M - G_{12}\kappa_n) \, dv = 0 \Rightarrow (L - G_{11}\kappa_n) \, du + Mdv = 0 \tag{6.44}$$

These equations are true for any arbitrary values of du and dv, because u and v form an orthogonal net of lines on the surface. This implies that

$$(N - G_{22}\kappa_n) = 0 \Rightarrow \kappa_1 = \frac{N}{G_{22}}, (L - \kappa_n G_{11}) = 0 \Rightarrow \kappa_2 = \frac{L}{G_{11}}$$

$$M = 0 \Rightarrow K = \kappa_1\kappa_2 = \frac{LN}{G_{11}G_{22}}, 2H = (\kappa_1 + \kappa_2) = \frac{LG_{22} + NG_{11}}{G_{11}G_{22}} \quad (6.45)$$

Create a surface S* parallel to S by shifting each point P on S through a distance a along the unit normal $\mathbf{n} = \dfrac{\mathbf{r}_u \times \mathbf{r}_v}{|\,\mathbf{r}_u \times \mathbf{r}_v\,|}$ on S at P. From Figure 6.17, the point P* on the parallel surface is given by $\mathbf{r}^*(u, v)$, where $\mathbf{r}^*(u, v) = \mathbf{r}(u, v) + a\mathbf{n}$. The tangents on the parallel surface S* are given by

$$\mathbf{r}_u^* = \mathbf{r}_u + a\mathbf{n}_u, \mathbf{r}_v^* = \mathbf{r}_v + a\mathbf{n}_v \quad (6.46)$$

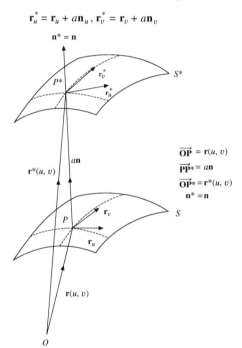

$\overrightarrow{OP} = \mathbf{r}(u, v)$

$\overrightarrow{PP^*} = a\mathbf{n}$

$\overrightarrow{OP^*} = \mathbf{r}^*(u, v)$

$\mathbf{n}^* = \mathbf{n}$

Figure 6.17   Parallel surfaces

To find the Gaussian and mean curvatures, K^* and H^* (of the parallel surface S*), one needs the coefficients $G_{11}^*, G_{12}^*, G_{22}^*$ and L^*, M^*, N^* of the first and second fundamental forms in terms of G_{11}, G_{12}, G_{22} and L, M, N of surface S.

It can be shown that $\mathbf{n}_u$ and $\mathbf{n}_v$ are normal to $\mathbf{n}$ and therefore, lie in the tangent plane at point P on surface S

$\mathbf{n} \cdot \mathbf{n} = 1 \Rightarrow \mathbf{n}_u \cdot \mathbf{n} + \mathbf{n} \cdot \mathbf{n}_u = 0 \Rightarrow \mathbf{n}_u \cdot \mathbf{n} = 0 \Rightarrow \mathbf{n}_u \perp \mathbf{n}$. Similarly, $\mathbf{n}_v \perp \mathbf{n}$.

Since, $\mathbf{r}_u$ and $\mathbf{r}_v$ are orthogonal vectors through P and lie on the tangent plane, they can be used as orthogonal basis for $\mathbf{n}_u$ and $\mathbf{n}_v$. Thus, $\mathbf{n}_u$ and $\mathbf{n}_v$ can be expressed as linear combinations of $\mathbf{r}_u$ and $\mathbf{r}_v$

$\mathbf{n}_u = a_1\mathbf{r}_u + b_1\mathbf{r}_v \Rightarrow \mathbf{n}_u \cdot \mathbf{r}_u = a_1\mathbf{r}_u \cdot \mathbf{r}_u + b_1\mathbf{r}_v \cdot \mathbf{r}_u \Rightarrow -L = a_1G_{11} + b_1G_{12} \Rightarrow a_1 = -\dfrac{L}{G_{11}} \; (\because G_{12} = 0)$

$$\mathbf{n}_u \cdot \mathbf{r}_v = a_1 \mathbf{r}_u \cdot \mathbf{r}_v + b_1 \mathbf{r}_v \cdot \mathbf{r}_v \Rightarrow -M = a_1 G_{12} + b_1 G_{22} \Rightarrow b_1 = 0 \ (\because M = 0, \ G_{12} = 0)$$

Therefore,
$$\mathbf{n}_u = -\frac{L}{G_{11}} \mathbf{r}_u \qquad (6.47)$$

Similarly,
$$\mathbf{n}_v = a_2 \mathbf{r}_u + b_2 \mathbf{r}_v \Rightarrow \mathbf{n}_v \cdot \mathbf{r}_u = a_2 \mathbf{r}_u \cdot \mathbf{r}_u + b_1 \mathbf{r}_v \cdot \mathbf{r}_u \Rightarrow -M = a_2 G_{11} + b_2 G_{12}$$

$$\Rightarrow a_2 = 0 \ (\because G_{12} = 0, \ M = 0)$$

$$\mathbf{n}_v \cdot \mathbf{r}_v = a_2 \mathbf{r}_u \cdot \mathbf{r}_v + b_2 \mathbf{r}_v \cdot \mathbf{r}_v \Rightarrow -N = b_2 G_{22} \Rightarrow b_2 = -\frac{N}{G_{22}} \ (\because a_2 = 0)$$

Therefore,
$$\mathbf{n}_v = -\frac{N}{G_{22}} \mathbf{r}_v \qquad (6.48)$$

Using these expressions for $\mathbf{n}_u$ and $\mathbf{n}_v$, K^* and H^* can be determined as follows:

$$\mathbf{r}_u^* = \mathbf{r}_u - a \frac{L}{G_{11}} \mathbf{r}_u = \left(1 - \frac{aL}{G_{11}}\right) \mathbf{r}_u, \ \mathbf{r}_v^* = \mathbf{r}_v - a \frac{N}{G_{22}} \mathbf{r}_v = \left(1 - \frac{aN}{G_{22}}\right) \mathbf{r}_v \qquad (6.49a)$$

$$G_{11}^* = \mathbf{r}_u^* \cdot \mathbf{r}_u^* = \left(1 - \frac{aL}{G_{11}}\right)^2 \mathbf{r}_u \cdot \mathbf{r}_u = \left(1 - \frac{aL}{G_{11}}\right)^2 G_{11} \qquad (6.49b)$$

$$G_{12}^* = \mathbf{r}_u^* \cdot \mathbf{r}_v^* = \left(1 - \frac{aL}{G_{11}}\right)\left(1 - \frac{aN}{G_{22}}\right) \mathbf{r}_v \cdot \mathbf{r}_u = \left(1 - \frac{aL}{G_{11}}\right)\left(1 - \frac{aN}{G_{22}}\right) G_{12} = 0 \qquad (6.49c)$$

$$G_{22}^* = \mathbf{r}_v^* \cdot \mathbf{r}_v^* = \left(1 - \frac{aN}{G_{22}}\right)^2 \mathbf{r}_v \cdot \mathbf{r}_v = \left(1 - \frac{aN}{G_{22}}\right)^2 G_{22} \qquad (6.49d)$$

$$L^* = -\mathbf{r}_u^* \cdot \mathbf{n}_u^* = -\left(1 - \frac{aL}{G_{11}}\right) \mathbf{r}_u \cdot \mathbf{n}_u = L\left(1 - \frac{aL}{G_{11}}\right), \ M^* = -\mathbf{r}_u^* \cdot \mathbf{r}_v^* = -\left(1 - \frac{aL}{G_{11}}\right) \mathbf{r}_u \cdot \mathbf{n}_v = 0$$
$$(6.49e)$$

$$N^* = -\mathbf{r}_v^* \cdot \mathbf{n}_v^* = -\left(1 - \frac{aN}{G_{22}}\right) \mathbf{r}_v \cdot \mathbf{n}_v = \left(1 - \frac{aN}{G_{22}}\right) N \qquad (6.49f)$$

The principal curvatures at point P* on S* can now be determined as follows:

$$\kappa_1^* = \frac{L^*}{G_{11}^*} = \frac{L}{\left(1 - \frac{aL}{G_{11}}\right) G_{11}}, \ \kappa_2^* = \frac{N^*}{G_{22}^*} = \frac{N}{\left(1 - \frac{aN}{G_{22}}\right) G_{22}} \qquad (6.50)$$

The Gaussian and mean curvatures of S* are given by

$$K^* = \kappa_2^* \kappa_2^* = \frac{NL}{G_{22} G_{11}\left(1 - \frac{aL}{G_{11}}\right)\left(1 - \frac{aN}{G_{22}}\right)}$$

$$= \frac{K}{1 - a\left(\frac{L}{G_{11}} + \frac{N}{G_{22}}\right) + a^2 \frac{NL}{G_{11}G_{22}}} = \frac{K}{1 - 2aH + a^2 K} \qquad (6.51)$$

$$H^* = \frac{1}{2}\left(\kappa_1^* + \kappa_2^*\right) = \frac{1}{2}\left(\frac{L}{G_{11}\left(1 - \frac{aL}{G_{11}}\right)} + \frac{N}{G_{22}\left(1 - \frac{aN}{G_{22}}\right)}\right) = \frac{H + Ka}{1 - 2aH + a^2 K} \qquad (6.52)$$

Example 6.5. Find the equation of surfaces parallel to the sphere

$$x(u, v) = a \cos u \cos v, \quad y(u, v) = a \cos u \sin v, \quad z(u, v) = a \sin u$$

and the catenoid

$$\mathbf{r}(u, v) = a \cos u \cosh (v/a)\mathbf{i} + a \sin u \cosh (v/a)\mathbf{j} + v\mathbf{k}$$

A surface S* parallel to the surface S is obtained from

$$\mathbf{r}^*(u, v, t) = \mathbf{r}(u, v) + t\,\mathbf{n}(u, v)$$

where *t* is the separation of the parallel surface and $\mathbf{n}(u, v)$ is the unit normal to S at the corresponding point with $\mathbf{n}(u, v) = \dfrac{\mathbf{r}_u \times \mathbf{r}_v}{|\,\mathbf{r}_u \times \mathbf{r}_v\,|}$.

For the sphere,

$$\mathbf{r}_u = - a \sin u \cos v\mathbf{i} - a \sin u \sin v\mathbf{j} + a \cos u\mathbf{k}$$

$$\mathbf{r}_v = -a \cos u \sin v\mathbf{i} + a \cos u \cos v\mathbf{j} + 0\mathbf{k}$$

$$\mathbf{r}_u \times \mathbf{r}_v = -a^2 \cos^2 u \cos v\mathbf{i} - a^2 \cos^2 u \sin v\mathbf{j} - a^2 \cos u \sin u\,\mathbf{k}$$

$$|\,\mathbf{r}_u \times \mathbf{r}_v\,| = a^2 \cos u$$

$$\mathbf{r}^*(u, v, t) = \mathbf{r}(u, v) + t\mathbf{n}(u, v) = \cos u \cos v(a - t)\,\mathbf{i} + \cos u \sin v(\,a - t)\mathbf{j} + \sin u(a - t)\mathbf{k}$$

Two parallel spheres are shown with $t = 0$, the original surface S is between Figure 6.18(a). In the example, $a = 1$, $t = -\frac{1}{2}$, and $t = \frac{1}{2}$. Using the above method, we can determine the equation of the surface parallel to the catenoid separated by a distance *t* as

$$\mathbf{r}^*(u, v, t) = [a \cos u \cosh (v/a) + t \cos u \operatorname{sech} (v/a)]\mathbf{i}$$
$$+ [a \sin u \cosh (v/a) + t \sin u \operatorname{sech} (v/a)]\mathbf{j} + [v - t \tanh (v/a)]\mathbf{k}$$

Parallel surfaces for the catenoid have been generated for $a = 1$, $0 < u < 3\pi/2$, $-1.5 < v\ 1.5$ and three values of *t* at $-\frac{1}{2}$, 0 and $\frac{1}{2}$ in Figure 6.18 (b).

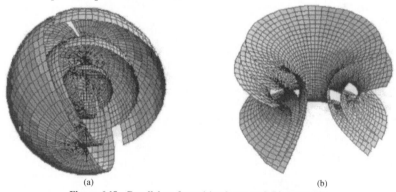

(a)   (b)

Figure 6.18   Parallel surfaces (a) spheres and (b) catenoids

6.8   Surfaces of Revolution

A large number of common objects such as cans and bottles, funnels, wine glasses, pitchers, football, legs of furniture, torus, ellipsoid, paraboloid and sphere are all surfaces of revolution.

For a curve on a plane, we can form a surface of revolution by rotating it about a given line (axis). A surface of revolution is expressed as

$$\mathbf{r}(u, v) = \lambda(v) \cos u\mathbf{i} + \lambda(v) \sin u\mathbf{j} + \mu(v)\mathbf{k} \equiv [\lambda(v) \cos u, \lambda(v) \sin u, \mu(v)] \qquad (6.53)$$

for a curve, called the *profile curve,* lying on the *x-z* plane given by

$$\mathbf{r}(v) = \lambda(v)\,\mathbf{i} + \mu(v)\mathbf{k} = (\lambda(v), 0, \mu(v)) \qquad (6.54)$$

This profile curve when rotated about the z-axis through an angle u, gives the equation of the surface patch. Various positions of the profile curve around the axis are called *meridians.* Each point on this curve creates a circle called *parallels.* The tangents to the surface of revolution are given by

$$\mathbf{r}_v = \left[\frac{d\lambda}{dv} \cos u, \quad \frac{d\lambda}{dv} \sin u, \quad \frac{d\mu}{dv} \right] = [\dot{\lambda} \cos u, \quad \dot{\lambda} \sin u, \quad \dot{\mu}] \qquad (6.55)$$

$$\mathbf{r}_u = [-\lambda \sin u, \quad \lambda \cos u, \quad 0]$$

The normal at a point and the coefficients (G_{11}, G_{12}, G_{22}) and (L, M, N) of the first and second fundamental forms of the surface can be determined as

$$\mathbf{N} = \mathbf{r}_u \times \mathbf{r}_v = [\lambda\dot{\mu} \cos u, \quad \lambda\dot{\mu} \sin u, \quad -\lambda\dot{\lambda}] = \lambda[\dot{\mu} \cos u, \quad \dot{\mu} \sin u, \quad -\dot{\lambda}]$$

$$\mathbf{n} = \frac{\mathbf{r}_u \times \mathbf{r}_v}{|\mathbf{r}_u \times \mathbf{r}_v|} = \frac{[\dot{\mu} \cos u, \quad \dot{\mu} \sin u, \quad -\dot{\lambda}]}{\sqrt{\dot{\mu}^2 + \dot{\lambda}^2}}$$

$$G_{11} = \mathbf{r}_u \cdot \mathbf{r}_u = \lambda^2, \quad G_{12} = \mathbf{r}_u \cdot \mathbf{r}_v = 0, \quad G_{22} = \mathbf{r}_v \cdot \mathbf{r}_v = \dot{\mu}^2 + \dot{\lambda}^2 \qquad (6.56)$$

$$L = \mathbf{r}_{uu} \cdot \mathbf{n} = \frac{-\lambda\dot{\mu}}{\sqrt{\dot{\mu}^2 + \dot{\lambda}^2}}, \quad M = \mathbf{r}_{uv} \cdot \mathbf{n} = 0, \quad N = \mathbf{r}_{vv} \cdot \mathbf{n} = \frac{\ddot{\lambda}\dot{\mu} + \dot{\lambda}\ddot{\mu}}{\sqrt{\dot{\mu}^2 + \dot{\lambda}^2}}$$

Since $G_{12} = 0$, and from Eq. (6.14) $\cos \theta = G_{12}/\sqrt{(G_{11}G_{22})} = 0$, the meridians and parallels are orthogonal as the angle θ between the tangents $\mathbf{r}_u$ and $\mathbf{r}_v$ is 90°. Since both $G_{12} = 0$ and $M = 0$, the conditions for the meridians and parallels to be the lines of curvature are also met. The Gaussian and mean curvatures (K and H, respectively) in Eq. (6.36) for a surface of revoution are given by

$$K = \frac{LN - M^2}{G_{11}G_{22} - G_{12}^2} = \frac{-(\ddot{\lambda}\dot{\mu}^2 + \dot{\lambda}\dot{\mu}\ddot{\mu})}{\lambda(\dot{\mu}^2 + \dot{\lambda}^2)^2}$$

$$H = \frac{G_{11}N + G_{22}L - 2G_{12}M}{2(G_{11}G_{22} - G_{12}^2)} = \frac{(\lambda\ddot{\lambda} - \dot{\lambda}^2)\dot{\mu} + \lambda\dot{\lambda}\ddot{\mu} - \dot{\mu}^3}{2\lambda(\dot{\mu}^2 + \dot{\lambda}^2)^{3/2}} \qquad (6.57)$$

Example 6.6. A curve $\mathbf{r}(u) = u\,\mathbf{i} + \log u\mathbf{k}$ lies in the *x-z* plane. It is rotated about the *z*-axis through an angle v. Find the properties of the surface.

The equation of the surface, tangents, normal and coefficients of the first and second fundamental forms are given by

$$\mathbf{r}(u, v) = u \cos v\mathbf{i} + u \sin v\mathbf{j} + \log u\mathbf{k} = (u \cos v, u \sin v, \log u)$$

$$\Rightarrow \qquad \mathbf{r}_u = \left(\cos v, \sin v, \frac{1}{u} \right), \mathbf{r}_v = (-u \sin v, u \cos v, 0), \mathbf{r}_{uu} = \left(-\sin v, \cos v, -\frac{1}{u^2} \right)$$

$$\mathbf{r}_{vv} = (-u \cos v, -u \sin v, 0), \ \mathbf{r}_{uv} = (-\sin v, \cos v, 0), \ \mathbf{N} = \mathbf{r}_u \times \mathbf{r}_v = (-\cos v, \ -\sin v, u)$$

$$\mathbf{n} = \frac{(-\cos v, \ -\sin v, u)}{\sqrt{1 + u^2}}, \ G_{11} = \mathbf{r}_u \cdot \mathbf{r}_u = \frac{1 + u^2}{u^2}, \ G_{22} = \mathbf{r}_v \cdot \mathbf{r}_v = u^2, \ G_{12} = \mathbf{r}_u \cdot \mathbf{r}_v = 0$$

$$L = \mathbf{r}_{uu} \cdot \mathbf{n} = -\frac{1}{u\sqrt{1 + u^2}}, \ M = \mathbf{r}_{uv} \cdot \mathbf{n} = 0, \ N = r_{vv} \cdot \mathbf{n} = \frac{u}{\sqrt{1 + u^2}}$$

From the above, Gaussion and mean curvatures can be calculated:

$$K = \frac{LN}{G_{11}G_{22}} = -\frac{1}{(1 + u^2)^2}, \ H = \frac{G_{11}N + G_{22}L}{G_{11}G_{22}} = \frac{1}{u(1 + u^2)^{\frac{3}{2}}}$$

The surface of revolution is shown as a funnel in Figure 6.19. The parallels are the circles with $u = u_0$, a constant, while meridians are the curves for $v = v_0$, a constant.

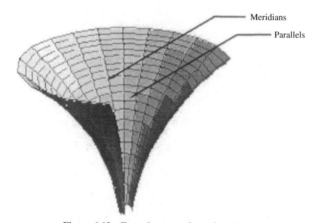

Meridians

Parallels

Figure 6.19   Funnel as a surface of revolution

6.9   Sweep Surfaces

A large number of objects created by engineers are designed with sweep surfaces. Common examples are *wash-basin, volute of a hydraulic pump, aircondition ducting, helical pipe, corrugated sheets* and many more. A sweep surface consists of "cross section curves" swept along a directrix curve or cross section curves with Hermite or B-spline blending.

A cylinder may be regarded as a sweep surface. If one considers the elliptical cross section curve lying on the *x-y* plane swept linearly along the *z*-axis, it will form a cylinder. The equation of the cross section curve is given in homogeneous coordinates by

$$\mathbf{C}(u) = [a \cos u, \ b \sin u, \ 0, \ 1]^T$$

Sweeping the curve along the *z*-axis through a distance *v* will mean applying a transformation matrix

$$\mathbf{T}(v) = \begin{bmatrix} 1 & 0 & 0 & 0 \\ 0 & 1 & 0 & 0 \\ 0 & 0 & 1 & v \\ 0 & 0 & 0 & 1 \end{bmatrix}$$

Thus, the equation of the cylinder (in *homogeneous co-ordinates*) produced by this sweep is given by

$$\mathbf{r}(u, v) = \mathbf{T}(v)\mathbf{C}(u) = [a \cos u, \, b \sin u, \, v, \, 1]^T$$

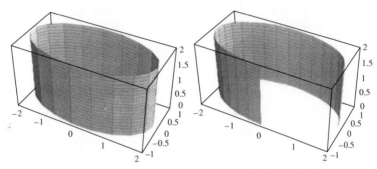

Figure 6.20   Cylinders (a) $0 \leq u \leq 2\pi$ and (b) $(0 \leq u \leq 3\pi/2)$

Following examples illustrate some other sweep surfaces:

Example 6.7

(a) Helical tube: The surface of a helical tube is produced if we sweep a circular cross section of radius r along a helix $\gamma(t) = (a \cos t, \, a \sin t, \, bt)$. The unit tangent $\mathbf{t}(t)$, binormal $\mathbf{b}(t)$, and normal $\mathbf{n}(t)$ to the helical curve are given by

$$\mathbf{t} = \frac{d\gamma/dt}{|\, d\gamma/dt \,|} = \frac{(-a \sin t, \, a \cos t, \, b)}{\sqrt{a^2 + b^2}},$$

$$\mathbf{b} = \frac{(-a \sin t, \, a \cos t, \, b) \times (-a \cos t, \, -a \sin t, \, 0)}{(-a \sin t, \, a \cos t, \, b) \times (-a \cos t, \, -a \sin t, \, 0)} = \frac{(b \sin t, \, -b \cos t, \, a)}{\sqrt{a^2 + b^2}}$$

$$\mathbf{n} = \mathbf{b} \times \mathbf{t} = (-\cos t, \, -\sin t, \, 0)$$

The equation of the tube thus formed is given by

$$\mathbf{r}(t, \theta) = \gamma(t) + r[-\mathbf{n} \cos \theta + \mathbf{b} \sin \theta]$$

$$\mathbf{r}(t, \theta) = \gamma(t) + r[-\cos \theta(-\cos t, \, -\sin t, \, 0) + \frac{\sin \theta}{\sqrt{a^2 + b^2}} (b \sin t, \, -b \cos t, \, a)]$$

(b) Seashell: A seashell is a helical tube but the radius r of the tube increases as the circular cross section sweeps along the helical backbone curve. One may use the following model for the surface:

$$\mathbf{r}(t, \theta) = \boldsymbol{\gamma}(t) + rt[-\cos \theta(-\cos t, -\sin t, 0) + \frac{\sin \theta}{\sqrt{a^2 + b^2}} (b \sin t, -b \cos t, a)]$$

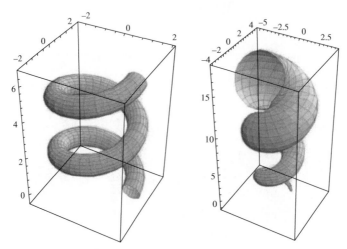

Figure 6.21   (a) Helix tube ($a = 2$, b, $r = 0.5$) (b) Seashell ($a = 2$, $b = 1$, $r = 0.2$)

(c) Corrugated sheet: A sinusoidal curve on the x-z plane along x-axis is given by

$$\boldsymbol{\gamma}(t) = (t, 0, a \sin t)$$

Sweeping this curve along y-axis by a distance v gives

$$\mathbf{r}(t, v) = \begin{bmatrix} 1 & 0 & 0 & 0 \\ 0 & 1 & 0 & v \\ 0 & 0 & 1 & 0 \\ 0 & 0 & 0 & 1 \end{bmatrix} \begin{bmatrix} t \\ 0 \\ a \sin t \\ 1 \end{bmatrix} = \begin{bmatrix} t \\ v \\ a \sin t \\ 1 \end{bmatrix}$$

Note the use of *homogeneous co-ordinate system*. A plot of the corrugated surface is shown here with $a = 1$ and $v = 25$ and t ranging from 0 to 8π.

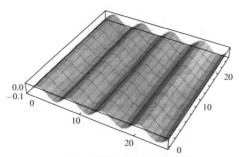

Figure 6.22   Corrugated sheet

6.10   Curve of Intersection between Two Surfaces

In engineering design, one has to deal with situations where two surfaces are made to intersect. Examples can be that of design of the outer shell of an automobile where surfaces such as the glass

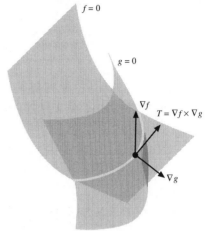

window, the roof top and the bonnet are made to intersect along various curves of intersection. In case of air-conditioning ducts as well, various cylindrical and spherical surfaces intersect in curves. These curves of intersection are important from the manufacturing viewpoint as they define the boundaries of various surfaces to be assembled or conformed. In many cases, it is often difficult to determine the curve of intersection in explicit form. It is useful, therefore, to know the properties of such a curve, like torsion and curvature, using which we can numerically integrate to determine the curve of intersection.

Let $f(x, y, z) = 0$ and $g(x, y, z) = 0$ be two surfaces intersecting in a curve of intersection whose equation cannot be determined in a simple form. The curvature and torsion of the curve may still be determined in the following manner. Let the curve of intersection be given by $\mathbf{r} = \mathbf{r}(s)$. The unit tangent vector $\mathbf{t}$ at any point on this curve is

Figure 6.23   Curve of intersection between two surfaces

orthogonal to the surface normals at that point on each of the surfaces. The surface normals are given by

$$\nabla f = \frac{\partial f}{\partial x} \mathbf{i} + \frac{\partial f}{\partial y} \mathbf{j} + \frac{\partial f}{\partial z} \mathbf{k}, \quad \nabla g = \frac{\partial g}{\partial x} \mathbf{i} + \frac{\partial g}{\partial y} \mathbf{j} + \frac{\partial g}{\partial z} \mathbf{k} \tag{6.53}$$

Thus, $\mathbf{t}$ will be proportional to $\mathbf{p} = \nabla f \times \nabla g$. For a scalar λ which is a function of s

$$\lambda \mathbf{t} = \nabla f \times \nabla g = \mathbf{p} \Rightarrow (\lambda \mathbf{t}) \cdot (\lambda \mathbf{t}) = (\nabla f \times \nabla g) \cdot (\nabla f \times \nabla g) \Rightarrow \lambda^2 = (\nabla f \times \nabla g)^2 \tag{6.54}$$

$$\lambda \mathbf{t} = \lambda \frac{d\mathbf{r}}{ds} = \nabla f \times \nabla g \xrightarrow{\text{Define oparator}} \lambda \frac{d}{ds} \equiv \left(h_1 \frac{\partial}{\partial x} + h_2 \frac{\partial}{\partial y} + h_3 \frac{\partial}{\partial x} \right) \equiv \Delta \tag{6.55}$$

where, $\quad h_1 = \lambda \dfrac{dx}{ds}, \quad h_2 = \lambda \dfrac{dy}{ds}, \quad h_3 = \lambda \dfrac{dz}{ds}.$

Using the operator Δ on $\lambda \mathbf{t}$ results in

$$\lambda \frac{d(\lambda \mathbf{t})}{ds} = \lambda^2 \frac{d\mathbf{t}}{ds} + \lambda \frac{d\lambda}{ds} \mathbf{t} = \Delta \mathbf{P} = \lambda^2 \kappa \mathbf{n} + \lambda \frac{d\lambda}{ds} \mathbf{t} \tag{6.56}$$

Where κ is the curvature and $\mathbf{t}$ the unit normal to the curve of intersection.

Taking the cross product with $\lambda \mathbf{t} = \mathbf{p}$

$$\lambda \mathbf{t} \times \left(\lambda^2 \frac{d\mathbf{t}}{ds} + \lambda \frac{d\lambda}{ds} \mathbf{t} \right) = \mathbf{p} \times \Delta \mathbf{p} \cdot \text{ Here } \mathbf{t} \times \frac{d\mathbf{t}}{ds} = \kappa \mathbf{b}, \quad \mathbf{t} \times \mathbf{t} = 0 \text{ with } \mathbf{b} \text{ being the unit binormal.}$$

Hence,
$$\lambda^3 \kappa \mathbf{b} = \mathbf{p} \times \varDelta \mathbf{p} = \mathbf{m} \text{ (say)} \tag{6.57}$$

Therefore,
$$|\mathbf{m}| = \lambda^3 \kappa \tag{6.58}$$

Using the operator $\varDelta$ on (6.57),

$$\lambda \frac{d}{ds} = (\lambda^3 \kappa \mathbf{b}) = \varDelta \mathbf{m} \Rightarrow \lambda \frac{d}{ds}(\lambda^3 \kappa)\mathbf{b} + \lambda^4 \kappa \frac{d\mathbf{b}}{ds} = \varDelta \mathbf{m} \tag{6.59}$$

$$\frac{d\mathbf{b}}{ds} = -\tau \mathbf{n}$$

where τ is the torsion of the curve of intersection (Chapter 3).

Taking the scalar product with (6.56) results in

$$\left(\lambda \frac{d}{ds}(\lambda^3 \kappa)\mathbf{b} - \lambda^4 \kappa \tau \mathbf{n} \right) \cdot \left(\lambda^2 \kappa \mathbf{n} + \lambda \frac{d\lambda}{ds}\mathbf{t} \right) = \varDelta \mathbf{m} \cdot \varDelta \mathbf{p} \tag{6.60}$$

$$\Rightarrow \qquad \lambda^6 \kappa^2 \tau = -\varDelta \mathbf{m} \cdot \varDelta \mathbf{p}$$

From Eqs. (6.58) and (6.60) we can determine the curvature and torsion at any point on the curve of intersection.

Example 6.8. Determine the torsion and curvature of the curve of intersection between a plane and a sphere given by

$$x = 2, \quad x^2 + y^2 + z^2 = 9$$

The plane intersects the sphere in a circle of radius $\sqrt{5}$, having its center on the x-axis at $(2, 0, 0)$. Let

$$f = \frac{1}{2}(x^2 + y^2 + z^2 - 9), \quad g = (x - 2), \quad \nabla = \left(\mathbf{i}\frac{\partial}{\partial x} + \mathbf{j}\frac{\partial}{\partial y} + \mathbf{k}\frac{\partial}{\partial z} \right)$$

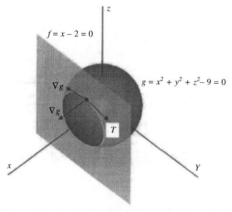

Figure 6.24   Intersection between a sphere and a plane in Example 6.8

Then

$$\nabla f = (x\mathbf{i} + y\mathbf{j} + z\mathbf{k}) = (x, y, z) = \mathbf{N}_f, \quad \nabla g = (\mathbf{i} + 0\mathbf{j} + 0\mathbf{k}) = (1, 0, 0) = \mathbf{N}_g$$

Here, $\mathbf{N}_f$ and $\mathbf{N}_g$ are vectors normal to the sphere f and the plane g, respectively.

$$\nabla f \times \nabla g = \begin{vmatrix} \mathbf{i} & \mathbf{j} & \mathbf{k} \\ x & y & z \\ 1 & 0 & 0 \end{vmatrix} = (0, z, -y) = \lambda \mathbf{t} = \mathbf{p}$$

Since $\mathbf{t}$ is the unit tangent vector to the curve of intersection

$$(\lambda \mathbf{t}) \cdot (\lambda \mathbf{t}) = \lambda^2 = (-y)^2 + z^2 \Rightarrow \lambda = \pm (y^2 + z^2)^{\frac{1}{2}}$$

It can be observed that at $y = 0$ on the $x = 2$ plane, the point on the circle of intersection is given by $(2, 0, \sqrt{5})$ and $\lambda \mathbf{t} = 0\mathbf{i} + z\mathbf{j} + 0\mathbf{k} = \sqrt{5}\mathbf{j}$. This shows that the unit tangent $\mathbf{t}$ is along the $\mathbf{j}$-direction (Oy-axis). Using Eqs. (6.55) and (6.57),

$$\mathbf{p} = 0\mathbf{i} + z\mathbf{j} - y\mathbf{k} \text{ and } \Delta \equiv \left(0\frac{\partial}{\partial x} + z\frac{\partial}{\partial y} - y\frac{\partial}{\partial z} \right) \Rightarrow \Delta \mathbf{p} = 0\mathbf{i} - y\mathbf{j} - z\mathbf{k}$$

Therefore,
$$\mathbf{p} \times \Delta \mathbf{p} = \begin{vmatrix} \mathbf{i} & \mathbf{j} & \mathbf{k} \\ 0 & z & -y \\ 0 & -y & -z \end{vmatrix} = -(y^2 + z^2)\mathbf{i} = -\lambda^2 \mathbf{i}$$

Now, from equation (5),

$$\lambda^3 \kappa \mathbf{b} = \mathbf{p} \times \Delta \mathbf{p} = -\lambda^2 \mathbf{i} \Rightarrow \lambda \kappa \mathbf{b} = -1\mathbf{i}$$

This shows that the bi-normal $\mathbf{b}$ is along the negative x-axis and the curvature $\kappa = \dfrac{1}{(y^2 + z^2)^{\frac{1}{2}}}$. The

radius of the circle is $\sqrt{5}$.

Again, from equation (6.57) $\Rightarrow \mathbf{p} \times \Delta \mathbf{p} = \mathbf{m}$.

Therefore, $\Delta \mathbf{m} = \left(z\frac{\partial}{\partial y} - y\frac{\partial}{\partial z} \right)\{-(y^2 + z^2)\mathbf{i} + 0\mathbf{j} + 0\mathbf{k}\} = \{-(2zy - 2yz)\mathbf{i} + 0\mathbf{j} + 0\mathbf{k}\}$

$$\Rightarrow \Delta \mathbf{m} = (0,0,0)$$

From equation (6.60),

$$-\Delta \mathbf{m} \cdot \Delta \mathbf{p} = -(0, 0, 0) \cdot (0, -y, -z) = 0 = \lambda^6 \kappa^2 \tau$$

Hence, the torsion of the curve of intersection $\tau = 0$. This is true, because the curve is a circle of radius $\sqrt{5}$ lying on the plane $x = 2$.

Example 6.9. Intersection between a sphere and a cylinder, or a cylinder and another cylinder is quite common in mechanical design. In some cases, it may be possible to get a parametric representation of the curve discussed as follows.

Intersection of a Cylinder and a Sphere: Viviani Curve (1692)

Let the equations of the sphere and the cylinder be given by

$$x^2 + y^2 + z^2 = 4a^2$$

$$(x - a)^2 + y^2 = a^2$$

The parametric equation of the cylinder can be written as

$$x = a(1 + \cos u),\ y = a \sin u,\ z$$

The parametric equation of the curve common to the cylinder and the sphere can be written as

$$x = a(1 + \cos u),\ y = a \sin u,\ z = 2a \sin (u/2)$$

This curve is known as the Viviani curve as shown in Figure 6.25.

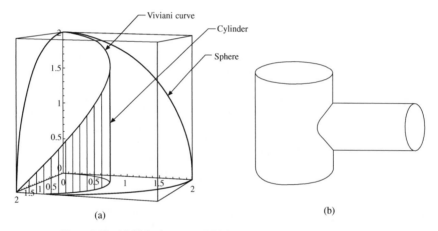

Figure 6.25 (a) Viviani curve and (b) intersection between two cylinders

With
$$\mathbf{r}(u) = \{a(1 + \cos u),\ a \sin u,\ 2a \sin (u/2)\}$$

$\Rightarrow$
$$\dot{\mathbf{r}}(u) = \{-a \sin u,\quad a \cos u,\quad a \cos (u/2)\}$$

$$\ddot{\mathbf{r}}(u) = \{-a \cos u,\quad -a \sin u,\quad -(a/2) \sin (u/2)\}$$

$$\dddot{\mathbf{r}}(u) = \{a \sin u,\quad -a \cos u,\quad -(a/4) \cos (u/2)\}$$

$\Rightarrow$
$$\kappa = \frac{|\dot{\mathbf{r}} \times \ddot{\mathbf{r}}|}{|\dot{\mathbf{r}}|^3} = \frac{(13 + 3 \cos u)^{\frac{1}{2}}}{a(3 + \cos u)^{\frac{3}{2}}}, \quad \tau = \frac{(\dot{\mathbf{r}} \times \ddot{\mathbf{r}}) \cdot \dddot{\mathbf{r}}}{|\dot{\mathbf{r}} \times \ddot{\mathbf{r}}|^2} = \frac{6 \cos (u/2)}{a(13 + 3 \cos u)}$$

Curve of Intersection of Two Perpendicular Cylinders

The equation of two cylinders can be written as

$$x^2 + y^2 = a^2 \text{ along } z\text{-axis} \Rightarrow (a \cos u, \quad a \sin u, \quad z) \Rightarrow x = \sqrt{a^2 - y^2}$$

$$y^2 + z^2 = b^2 \text{ along } x\text{-axis} \Rightarrow (x, b \cos u, \quad b \sin u) \Rightarrow \pm\sqrt{a^2 - b^2 \cos^2 u}, \ b \cos u, \ b \sin u)$$

The last equation gives the curves of intersection along the x-axis. For $a = 1.1$, $b = 1$, the curves of intersection appears as shown in Figure 6.26.

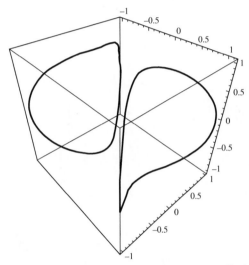

Figure 6.26   Curves of intersection along x-axis between two perpendicular cylinders

EXERCISES

1. For the surfaces shown in Figure P 6.1, determine the tangents, normal, coefficients of the first and second fundamental forms, Gaussian curvature, mean curvature and surface area (use numerical integration if closed form integration is not possible). Evaluate the same at [$u = 0.5$, $v = 0.5$] (in case the parametric range is not [0, 1], then evaluate at the middle of the parametric range, e.g. if the range is [0, 2π] then evaluate at π):

    The equations of the surfaces are given by

    (a): $\mathbf{r}(u, v) = (u^3 - 13u^2 + 6)\mathbf{i} + (-7u^3 + 8u^2 + 5u)\mathbf{j} + 6v\mathbf{k};$ $u \in [0, 1], v \in [0, 1]$

    (b): $\mathbf{r}(u, v) = u \cos v\mathbf{i} + u \sin v\mathbf{j} + u^2\mathbf{k};$ $u \in [0, 2], v \in [0, 2\pi]$

    (c): $\mathbf{r}(u, v) = \{(2 + 0.5 \sin 2u) \cos v, (2 + 0.5 \sin 2u) \sin v, u\};$ $u \in [0, \pi], v \in [0, 2\pi]$

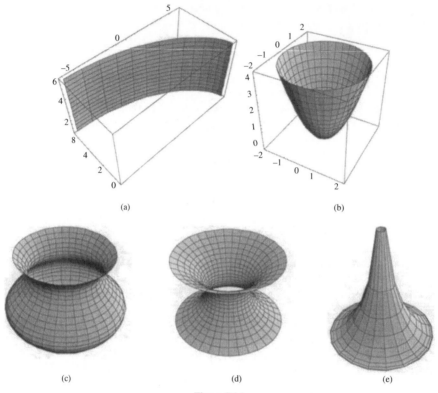

Figure P6.1

(d): $\mathbf{r}(u, v) = \left\{\left(2 \cosh \dfrac{1}{2} u\right) \cos v, \left(2 \cosh \dfrac{1}{2} u\right) \sin v, u\right\}; \quad u \in [-3, 3], v \in [0, 2\pi]$

(e): $\mathbf{r}(u, v) = \left\{2 \sin u \cos v, 2 \sin u \sin v, 2 \cos u + 2 \ln\left(\tan \dfrac{u}{2}\right)\right\}; \quad u \in \left[\dfrac{\pi}{2}, \dfrac{3\pi}{2}\right], v \in [0, 2\pi]$

2. Figure P6.2 shows a Mobius strip. Find the tangents and normal for the surface. Show that the normal at $(u, 0)$ has two different values at the same point, that is, $\lim\limits_{u \to \pi} \mathbf{n}(u, 0) = (0, 0, -1)$ and $\lim\limits_{u \to -\pi} \mathbf{n}(u, 0) = (0, 0, 1)$ depending upon whether we move along $v = 0$ in the CCW direction or clockwise direction. The equation of the surface is

$$\mathbf{r}(u, v) = \left\{\cos u + v \sin\left(\dfrac{u}{2}\right) \cos u, \sin u + v \sin\left(\dfrac{u}{2}\right) \sin u, v \cos\left(\dfrac{u}{2}\right)\right\}, \, u \in [-\pi, \pi], v \in [-0.5, 0.5]$$

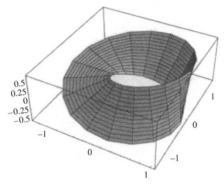

Figure P6.2   Mobius strip

3. Develop a procedure for viewing surface geometry. Your program should display a parametric surface by drawing iso-parametric curves and surface geometry moving along any such curve picked by the user. The following geometric entities should be displayable: (i) two partial derivatives, (ii) unit surface normal and (iii) tangent plane at the point.

4. Given a bi-cubic patch

$$\mathbf{r}(u, v) = [u^3 \ u^2 \ u^1 \ 1] \begin{bmatrix} [1 \ 2 \ 0] & [1 \ 4 \ 4] & [0 \ 2 \ 4] & [0 \ 2 \ 4] \\ [4 \ 6 \ 2] & [4 \ 8 \ 6] & [0 \ 2 \ 4] & [0 \ 2 \ 4] \\ [5 \ 2 \ -2] & [5 \ 2 \ -2] & 0 & 0 \\ [3 \ 4 \ 2] & [3 \ 4 \ 2] & 0 & 0 \end{bmatrix} \begin{bmatrix} v^3 \\ v^2 \\ v \\ 1 \end{bmatrix}$$

Determine whether it is a developable surface.

5. A bi-cubic patch is given by

$$\mathbf{r}(u, v) = [u^3 \ u^2 \ u \ 1] \begin{bmatrix} [0 \ 0 \ 10] & [10 \ 0 \ 0] & [16 \ 0 \ 0] & [0 \ 0 \ -16] \\ [0 \ 10 \ 8] & [18 \ 10 \ 0] & [24 \ 0 \ 0] & [0 \ 0 \ -14] \\ [0 \ 10 \ -2] & [8 \ 10 \ 0] & [8 \ 0 \ 0] & [0 \ 0 \ 2] \\ [0 \ 10 \ -2] & [8 \ 10 \ 0] & [8 \ 0 \ 0] & [0 \ 0 \ 2] \end{bmatrix} \begin{bmatrix} v^3 \\ v^2 \\ v \\ 1 \end{bmatrix}$$

Determine the
(a) coordinates on the surface at **P** (0.5, 0.5)
(b) unit normal at **P**
(c) unit tangent at **P**
(d) equation of the tangent plane at **P**
(e) Gaussian quadrature at **P**

6. Prove the Weingarten relations

$$H^2\mathbf{n}_u = (G_{12}M - G_{22}L)\mathbf{r}_u + (G_{12}L - G_{11}M)\mathbf{r}_v$$

$$H^2\mathbf{n}_v = (G_{12}N - G_{22}M)\mathbf{r}_u + (G_{12}M - G_{11}N)\mathbf{r}_v$$

and show that

$$H(\mathbf{n}_u \times \mathbf{n}_v) = (LN - M^2)\mathbf{n}$$

Hint: One may express $\mathbf{n}_u$ and $\mathbf{n}_v$ as respective linear combinations of $\mathbf{r}_u$ and $\mathbf{r}_v$, that is,

$$\mathbf{n}_u = c_1\mathbf{r}_u + c_2\mathbf{r}_v$$

$$\mathbf{n}_v = d_1\mathbf{r}_u + d_2\mathbf{r}_v$$

where c_1, c_2, d_1 and d_2 are scalars. Taking dot product of the above with $\mathbf{r}_u$ and $\mathbf{r}_v$ would yield the values of c_1, c_2, d_1 and d_2 in terms of L, M and N. Elimination of the scalars leads to the Weingarten equations. To get the third relation, consider the vector product of Weingarten relations and simplify.

7. Show that $(\mathbf{n}_u \times \mathbf{n}_v) = K\mathbf{r}_u \times \mathbf{r}_v$ where K is the Gaussian curvature.
8. Show that the following surfaces are not developable.

$$\mathbf{r}(u, v) = u \cos v \, \mathbf{i} + u \sin v \, \mathbf{j} + \sin nv \, \mathbf{k}$$

$$\mathbf{r}(u, v) = u\mathbf{i} + v\mathbf{j} + uv \, \mathbf{k}$$

Chapter 7

Design of Surfaces

A closed, connected composite surface represents the shape of a solid. This surface, in turn, is composed of *surface patches*, much like how a composite curve is a collection of juxtaposed segments. Surface design may be influenced by a variety of factors such as aesthetics, aerodynamics, fluid flow (for turbine blades, flow passages in a gas turbine, ship hull), ergonomics and many others. A free-formed surface like the aircraft wings and fuselage, car body and its doors, seats and windshields are all designed by combining surface patches at their boundaries.

Surface patches, similar to curve segments, can be modeled mathematically in *parametric* form using two parameters u and v:

$$f(u, v) = [x(u, v) \ y(u, v) \ z(u, v)], \quad u \in [0, 1], v \in [0, 1] \tag{7.1}$$

where $x(u, v)$, $y(u, v)$ and $z(u, v)$ are scalar polynomials in parameters (u, v). Note that a surface patch is bounded by the curves $f(u, 0), f(u, 1), f(0, v)$ and $f(1, v)$.

With an implicit representation $\phi(x, y, z) = 0$, many analytical surfaces such as a sphere, an ellipsoid, a paraboloid and others can be represented accurately. In parametric form, surface patches can be constructed (approximated) to closely represent the analytical counterparts. In a reverse engineering application, a surface may also be required to *fit* a *point cloud* or a given set of large number of points in space, usually obtained when scanning an existing surface by laser beam or the *faro arm*[1].

Though a set of data points and boundary curves may be known, the shape of the surface patch is to be designed based on designers' intuition and some qualitative data such as smoothness, flatness, bumps and change in curvature. Design of the car rooftop as a surface and the manifold for coolant circulation in the engine may require different methods of surface generation. We need to create a surface without knowing a large number of points, because the analytic form of the desired surface may not be known at the time of conception. A designer may often be required to change the shape *interactively* to achieve the desired shape. In design, it is more convenient to deal with surface patches and create a composite surface by *stitching* the patches ensuring C^n ($n = 0$ for position, 1 for slope and 2 for curvature) continuity. Boundary surfaces (and hence solids) are to be eventually manufactured using automated machine tools, press dies as in sheet metal forming, casting, molding and other processes. The mathematical description of a surface must be eventually transformable via the CAD/CAM interface to generate the manufacturing data and tool path generation codes.

---

[1] A robot arm that has an end sensor to locate the coordinates of points on a physical surface.

In Chapter 6, differential properties of surface patches are discussed using analytical surfaces. They include the plane, ruled or lofted patches, surfaces of revolution, and sweep patches. These patches are described in parametric form with examples of the mathematical background required for the design of synthetic surfaces. In surface design, free form surfaces are created using surface patches. The following gives a broad classification of surface patches

(a) Parametric polynomial patches (or tensor product surfaces)
(b) Boundary interpolating surfaces
(c) Sweep (linear or rotational) surface patches
(d) Quadric surface patches

Surface patches are *bi-parametric*, and the curve models developed in Chapters 4 and 5 are directly extendible to their design, that is, Hermite, Bézier or B-spline surface patches can be created using the basis functions for the respective curves described in these chapters.

7.1 Tensor Product Surface Patch

Given Φ and Ψ as two sets of univariate functions such that

$$\Phi = \left\{\varphi_i(u)\right\}_{i=0}^{m}, \ \Psi = \left\{\psi_j(v)\right\}_{j=0}^{n} \tag{7.2}$$

with interval domains $u \in U$ and $v \in V$, a surface

$$\mathbf{r}(u, v) = \sum_{j=0}^{n} \sum_{i=0}^{m} \mathbf{C}_{ij}\varphi_i(u)\psi_j(v) \tag{7.3}$$

is called a *tensor product surface* with domain $U \times V$. The surface is *bi-quadratic* for $m = n = 2$ and *bi-cubic* for $m = n = 3$.

Example 7.1. Consider the first and second order Bézier basis functions

$$\Phi(u) = \{\varphi_0(u) \ \varphi_1(u)\} = \{(1 - u) \ u\},$$

$$\Psi(v) = \{\psi_0(v) \ \psi_1(v) \ \psi_2(v)\} = \{(1 - v)^2 \ 2v(1 - v) \ v^2\}$$

The equation of the tensor product surface is given by

$$\mathbf{r}(u, v) = \mathbf{C}_{00}\varphi_0\psi_0 + \mathbf{C}_{01}\varphi_0\psi_1 + \mathbf{C}_{02}\varphi_0\psi_2 + \mathbf{C}_{10}\varphi_1\psi_0 + \mathbf{C}_{11}\varphi_1\psi_1 + \mathbf{C}_{12}\varphi_1\psi_2$$

Given $\mathbf{C}_{ij}$ as $\mathbf{C}_{00} = [0\ 0\ 0]$, $\mathbf{C}_{10} = [1\ 2\ 0]$, $\mathbf{C}_{01} = [0\ 2\ 4]$, $\mathbf{C}_{11} = [1\ 2\ 4]$ $\mathbf{C}_{02} = [0\ -1\ 3]$, $\mathbf{C}_{12} = [1\ -1\ 3]$, the equation of the tensor product surface may be written in the following form where the ordered triple $[x\ y\ z]$ is a function of parameters u and v.

$$\mathbf{r}(u, v) = [x\ y\ z] = [(1 - u)\ u]\begin{bmatrix}(0, 0, 0) & (0, 2, 4) & (0, -1, 3) \\ (1, 2, 0) & (1, 2, 4) & (1, -1, 3)\end{bmatrix}\begin{bmatrix}(1 - v)^2 \\ 2v(1 - v) \\ v^2\end{bmatrix} \tag{7.4}$$

The surface generated is shown in Figure 7.1. The thick lines represent v = constant values and the thick curve represents u = constant values on the surface.

We can generalize the form for a tensor product surface as

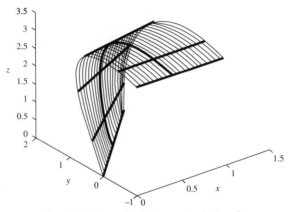

Figure 7.1 **Example of a tensor product surface**

$$\mathbf{r}(u, v) = \sum_{j=0}^{n} \sum_{i=0}^{m} \mathbf{D}_{ij} u^i v^j = [u^m \ u^{m-1} \ \dots \ 1] \begin{bmatrix} \mathbf{D}_{mn} & \mathbf{D}_{m(n-1)} & \cdot\cdot & \mathbf{D}_{m0} \\ \mathbf{D}_{(m-1)n} & \mathbf{D}_{(m-1)(n-1)} & \cdot\cdot & \mathbf{D}_{(m-1)} \\ \cdot & \cdot & \cdot\cdot & \cdot \\ \cdot & \cdot & \cdot\cdot & \cdot \\ \mathbf{D}_{0n} & \mathbf{D}_{0(n-1)} & \cdot\cdot & \mathbf{D}_{00} \end{bmatrix} \begin{bmatrix} v^n \\ v^{n-1} \\ \cdot \\ \cdot \\ 1 \end{bmatrix} \quad (7.5)$$

where m and n are user-chosen degrees in parameters u and v. For a bi-cubic surface patch, we need to specify 16 sets of data as control points and/or slopes. Though we can model patches with degrees in u and v greater than 3 and can as well choose the degrees unequal ($m \neq n$), like in Example 7.1, for most applications, use of bi-cubic surface patches seems adequate. Cubic curve models developed in previous chapters can now be extended to fit in the schema given in Eq. (7.5).

7.1.1 Ferguson's Bi-cubic Surface Patch

From Eq. 4.7, a point $\mathbf{r}(u)$ on the Hermite-Ferguson curve is given by

$$\mathbf{r}(u) = \varphi_0 \mathbf{r}(0) + \varphi_1 \mathbf{r}(1) + \varphi_2 \mathbf{r}_u(0) + \varphi_3 \mathbf{r}_u(1)$$

Here, $\mathbf{r}(0)$ and $\mathbf{r}(1)$ are two end points of the curve and $\mathbf{r}_u(0)$, $\mathbf{r}_u(1)$ are the end tangents. The Hermite blending functions $\Phi_i(u)$, ($i = 0, 1, 2, 3$) are given below.

$$\varphi_0 = (2u^3 - 3u^2 + 1), \quad \varphi_1 = (-2u^3 + 3u^2), \quad \varphi_2 = (u^3 - 2u^2 + u), \quad \varphi_3 = (u^3 - u^2)$$

In matrix form, the equation for the above curve is written as

$$\mathbf{r}(u) = [\varphi_0 \ \ \varphi_1 \ \ \varphi_2 \ \ \varphi_3] \begin{bmatrix} \mathbf{r}(0) \\ \mathbf{r}(1) \\ \mathbf{r}_u(0) \\ \mathbf{r}_u(1) \end{bmatrix} = [u^3 \ \ u^2 \ \ u \ \ 1] \begin{bmatrix} 2 & -2 & 1 & 1 \\ -3 & 3 & -2 & -1 \\ 0 & 0 & 1 & 0 \\ 1 & 0 & 0 & 0 \end{bmatrix} \begin{bmatrix} \mathbf{r}(0) \\ \mathbf{r}(1) \\ \mathbf{r}_u(0) \\ \mathbf{r}_u(1) \end{bmatrix} \quad (7.6)$$

Construction of a *tensor product surface patch* using Hermite blending functions can be similarly accomplished. We have to consider two parameters u and v and correspondingly, the two Hermite blending functions $\Phi_i(u)$, $(i = 0, 1, 2, 3)$ and $\Phi_j(v)$, $(j = 0, 1, 2, 3)$. The equation of the surface (or the position vector of any general point P on the surface) is given by

$$\mathbf{r}(u, v) = [\varphi_0(u) \quad \varphi_1(u) \quad \varphi_2(u) \quad \varphi_3(u)] \begin{bmatrix} \mathbf{C}_{00} & \mathbf{C}_{01} & \mathbf{C}_{02} & \mathbf{C}_{03} \\ \mathbf{C}_{10} & \mathbf{C}_{11} & \mathbf{C}_{12} & \mathbf{C}_{13} \\ \mathbf{C}_{20} & \mathbf{C}_{21} & \mathbf{C}_{22} & \mathbf{C}_{23} \\ \mathbf{C}_{30} & \mathbf{C}_{31} & \mathbf{C}_{32} & \mathbf{C}_{33} \end{bmatrix} \begin{bmatrix} \varphi_0(v) \\ \varphi_1(v) \\ \varphi_2(v) \\ \varphi_3(v) \end{bmatrix} \quad (7.7)$$

Each $\mathbf{C}_{ij}$ has 3 components and there are 16 of them. Thus there are (16×3) 48 unknowns to be determined for constructing the Hermite tensor product surface. These can be determined from the following data:

(a) four *corner points* $\mathbf{r}(0, 0)$, $\mathbf{r}(0, 1)$, $\mathbf{r}(1, 0)$ and $\mathbf{r}(1, 1)$ of the surface patch,
(b) eight *tangents* along the boundary curves $\mathbf{r}(0, v)$, $\mathbf{r}(1, v)$, $\mathbf{r}(u, 0)$, $\mathbf{r}(u, 1)$ with two at each corner point. These slopes are given as

$$\left.\frac{d\mathbf{r}(0, v)}{dv}\right|_{v=0} = \mathbf{r}_v(0, 0) \qquad \left.\frac{d\mathbf{r}(0, v)}{dv}\right|_{v=1} = \mathbf{r}_v(0, 1) \qquad \left.\frac{d\mathbf{r}(u, 0)}{du}\right|_{u=0} = \mathbf{r}_u(0, 0)$$

$$\left.\frac{d\mathbf{r}(u, 0)}{du}\right|_{u=1} = \mathbf{r}_u(1, 0) \qquad \left.\frac{d\mathbf{r}(1, v)}{dv}\right|_{v=0} = \mathbf{r}_v(1, 0) \qquad \left.\frac{d\mathbf{r}(1, v)}{dv}\right|_{v=1} = \mathbf{r}_v(1, 1) \qquad (7.8)$$

$$\left.\frac{d\mathbf{r}(u, 1)}{du}\right|_{u=0} = \mathbf{r}_u(0, 1) \qquad \left.\frac{d\mathbf{r}(u, 1)}{du}\right|_{u=1} = \mathbf{r}_u(1, 1)$$

(c) four *twist vectors* at the corners

$$\left.\frac{\partial^2 \mathbf{r}(u, v)}{\partial u \partial v}\right|_{u=0,v=0} = \mathbf{r}_{uv}(0, 0) \qquad \left.\frac{\partial^2 \mathbf{r}(u, v)}{\partial u \partial v}\right|_{u=0,v=1} = \mathbf{r}_{uv}(0, 1)$$

$$\left.\frac{\partial^2 \mathbf{r}(u, v)}{\partial u \partial v}\right|_{u=1,v=0} = \mathbf{r}_{uv}(1, 0) \qquad \left.\frac{\partial^2 \mathbf{r}(u, v)}{\partial u \partial v}\right|_{u=1,v=1} = \mathbf{r}_{uv}(1, 1) \qquad (7.9)$$

At any $u = u_j$, there is a curve $\mathbf{r}(u_j, v)$ and a tangent $\mathbf{r}_u(u_j, v)$. As we move along $\mathbf{r}(u_j, v)$ by varying v, we get different points $\mathbf{r}(u_j, v_i)$ on the surface as well as different tangents $\mathbf{r}_u(u_j, v_i)$, which vary both in direction and magnitude. Twist vectors $\mathbf{r}_{uv}(u_j, v_i)$ represent the rate of change of the tangent vector $\mathbf{r}_u(u_j, v)$ with respect to v at $\mathbf{r}(u_j, v_i)$. Function $\mathbf{r}(u, v)$ is such that the twist vectors $\mathbf{r}_{uv}(u, v) = \mathbf{r}_{vu}(u, v)$, that is, the partial mixed derivatives are symmetric with respect to u and v at every point on the surface.

Expanding the right hand side of Eq. (7.7) and using the Hermite blending functions φ and derivatives, we can evaluate $\mathbf{C}_{ij}$ as

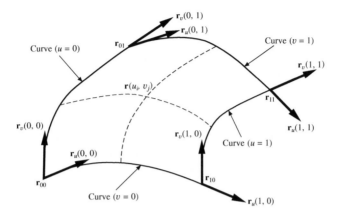

Figure 7.2   Hermite-Ferguson patch

$$
\begin{array}{llll}
\mathbf{r}(0,0) = \mathbf{C}_{00}; & \mathbf{r}(1,0) = \mathbf{C}_{10}; & \mathbf{r}_u(0,0) = \mathbf{C}_{20}; & \mathbf{r}_u(1,0) = \mathbf{C}_{30} \\[4pt]
\mathbf{r}(0,1) = \mathbf{C}_{01}; & \mathbf{r}(1,1) = \mathbf{C}_{11}; & \mathbf{r}_u(0,1) = \mathbf{C}_{21}; & \mathbf{r}_u(1,1) = \mathbf{C}_{31} \\[4pt]
\mathbf{r}_v(0,0) = \mathbf{C}_{02}; & \mathbf{r}_v(1,0) = \mathbf{C}_{12}; & \mathbf{r}_{uv}(0,0) = \mathbf{C}_{22}; & \mathbf{r}_{uv}(1,0) = \mathbf{C}_{32} \\[4pt]
\mathbf{r}_v(0,1) = \mathbf{C}_{03}; & \mathbf{r}_v(1,1) = \mathbf{C}_{13}; & \mathbf{r}_{uv}(0,1) = \mathbf{C}_{23}; & \mathbf{r}_{uv}(1,1) = \mathbf{C}_{33}
\end{array}
\tag{7.10}
$$

so that Eq. (7.7) can now be written in the form

$$
\mathbf{r}(u,v) = [\varphi_0(u)\ \varphi_1(u)\ \varphi_2(u)\ \varphi_3(u)]
\begin{bmatrix}
\mathbf{r}(0,0) & \mathbf{r}(0,1) & \mathbf{r}_v(0,0) & \mathbf{r}_v(0,1) \\
\mathbf{r}(1,0) & \mathbf{r}(1,1) & \mathbf{r}_v(1,0) & \mathbf{r}_v(1,1) \\
\mathbf{r}_u(0,0) & \mathbf{r}_u(0,1) & \mathbf{r}_{uv}(0,0) & \mathbf{r}_{uv}(0,1) \\
\mathbf{r}_u(1,0) & \mathbf{r}_u(1,1) & \mathbf{r}_{uv}(1,0) & \mathbf{r}_{uv}(1,1)
\end{bmatrix}
\begin{bmatrix}
\varphi_0(v) \\
\varphi_1(v) \\
\varphi_2(v) \\
\varphi_3(v)
\end{bmatrix}
\tag{7.11}
$$

Knowing that $[\varphi_0(u)\ \varphi_1(u)\ \varphi_2(u)\ \varphi_3(u)] = \mathbf{UM}$ in Eq. (4.7), Eq. (7.11) can be written as

$$
\mathbf{r}(u,v) = \mathbf{UMGM}^{\mathrm{T}}\mathbf{V}^{\mathrm{T}}
\tag{7.12}
$$

with $\mathbf{M}$ as the Ferguson's coefficient matrix in Eqs. ((4.7), (7.6)), $\mathbf{V}$ as $[v^3\ v^2\ v\ 1]$ and $\mathbf{G}$, the geometric matrix as

$$
\mathbf{G} =
\begin{bmatrix}
\mathbf{r}(0,0) & \mathbf{r}(0,1) & \mathbf{r}_v(0,0) & \mathbf{r}_v(0,1) \\
\mathbf{r}(1,0) & \mathbf{r}(1,1) & \mathbf{r}_v(1,0) & \mathbf{r}_v(1,1) \\
\mathbf{r}_u(0,0) & \mathbf{r}_u(0,1) & \mathbf{r}_{uv}(0,0) & \mathbf{r}_{uv}(0,1) \\
\mathbf{r}_u(1,0) & \mathbf{r}_u(1,1) & \mathbf{r}_{uv}(1,0) & \mathbf{r}_{uv}(1,1)
\end{bmatrix}
\tag{7.13}
$$

It is convenient to express $\mathbf{G}$ in the partitioned matrix form given in Eq. (7.14). The top left entries are corner points, the bottom left are the corner tangents (with respect to u) to the boundary curves $v = 0$ and $v = 1$, and top right are corner tangents to the boundary curves at $u = 0$ and $u = 1$. The bottom right entries indicate the twist vectors at the corners of the surface patch.

$$
G = \begin{bmatrix}
\mathbf{r}(0,0) & \mathbf{r}(0,1) & | & \mathbf{r}_v(0,0) & \mathbf{r}_v(0,1) \\
\mathbf{r}(1,0) & \mathbf{r}(1,1) & | & \mathbf{r}_v(1,0) & \mathbf{r}_v(1,1) \\
- & - & | & - & - \\
\mathbf{r}_u(0,0) & \mathbf{r}_u(0,1) & | & \mathbf{r}_{uv}(0,0) & \mathbf{r}_{uv}(0,1) \\
\mathbf{r}_u(1,0) & \mathbf{r}_u(1,1) & | & \mathbf{r}_{uv}(1,0) & \mathbf{r}_{uv}(1,1)
\end{bmatrix}
\tag{7.14}
$$

The algebraic form in Eq. (7.5) and the geometric form in Eq. (7.12) are equivalent. For $m = n = 3$,

$$
\mathbf{r}(u,v) = \sum_{j=0}^{3} \sum_{i=0}^{3} \mathbf{D}_{ij} u^i v^j = [u^3 \quad u^2 \quad u \quad 1]
\begin{bmatrix}
\mathbf{D}_{33} & \mathbf{D}_{32} & \mathbf{D}_{31} & \mathbf{D}_{30} \\
\mathbf{D}_{23} & \mathbf{D}_{22} & \mathbf{D}_{21} & \mathbf{D}_{20} \\
\mathbf{D}_{13} & \mathbf{D}_{12} & \mathbf{D}_{11} & \mathbf{D}_{10} \\
\mathbf{D}_{03} & \mathbf{D}_{02} & \mathbf{D}_{01} & \mathbf{D}_{00}
\end{bmatrix}
\begin{bmatrix}
v^3 \\ v^2 \\ v \\ 1
\end{bmatrix}
$$

$$
= \mathbf{U}\mathbf{D}\mathbf{V}^T = \mathbf{U}\mathbf{M}\mathbf{G}\mathbf{M}^T\mathbf{V}^T \quad \Rightarrow \quad \mathbf{D} = \mathbf{M}\mathbf{G}\mathbf{M}^T \quad \text{or} \quad \mathbf{G} = \mathbf{M}^{-1}\mathbf{D}(\mathbf{M}^T)^{-1}
$$

implying that the algebraic coefficients $\mathbf{D}_{ij}$ and geometric coefficients $\mathbf{G}_{ij}$ can be obtained from each other, each having three components (3-tuple).

A simple Ferguson's patch can be expressed using the following geometric matrix

$$
G = \begin{bmatrix}
\mathbf{r}(0,0) & \mathbf{r}(0,1) & | & \mathbf{r}_v(0,0) & \mathbf{r}_v(0,1) \\
\mathbf{r}(1,0) & \mathbf{r}(1,1) & | & \mathbf{r}_v(1,0) & \mathbf{r}_v(1,1) \\
- & - & | & - & - \\
\mathbf{r}_u(0,0) & \mathbf{r}_u(0,1) & | & 0 & 0 \\
\mathbf{r}_u(1,0) & \mathbf{r}_u(1,1) & | & 0 & 0
\end{bmatrix}
\tag{7.15a}
$$

This has been found to be convenient because one can have an intuitive feeling for the corner points and for the tangents, to a certain extent. *It is quite difficult to have any intuitive feel about the twist vectors.*

In general, the tangents and twist vectors in Eq. (7.13) can be expressed as unit vectors ($\mathbf{t}$) in given directions along with magnitude values. Using short notation such as $\mathbf{r}_u(a, b) = \alpha_{ab}\, \mathbf{t}_{ab}^u$, the geometric matrix can be expressed as follows

$$
G = \begin{bmatrix}
\mathbf{r}_{00} & \mathbf{r}_{01} & \beta_{00}\mathbf{t}_{00}^v & \beta_{01}\mathbf{t}_{01}^v \\
\mathbf{r}_{10} & \mathbf{r}_{11} & \beta_{10}\mathbf{t}_{10}^v & \beta_{11}\mathbf{t}_{11}^v \\
\alpha_{00}\mathbf{t}_{00}^u & \alpha_{01}\mathbf{t}_{01}^u & \gamma_{00}\mathbf{t}_{00}^{uv} & \gamma_{01}\mathbf{t}_{01}^{uv} \\
\alpha_{10}\mathbf{t}_{10}^u & \alpha_{11}\mathbf{t}_{11}^u & \gamma_{10}\mathbf{t}_{10}^{uv} & \gamma_{11}\mathbf{t}_{11}^{uv}
\end{bmatrix}
\tag{7.15b}
$$

The 12 coefficients α, β and γ can be selectively changed to get a desired surface (recall the change in shape of the Hermite PC curve by selecting the values of the tangent magnitudes, in Chapter 4).

7.1.2 Shape Interrogation

Shape interrogation is to extract the differential properties like curvatures, normal and tangents (discussed in Chapter 6) for a surface patch. The unit normal is given by Eq. (6.4) as

$$\mathbf{n}(u, v) = \frac{\mathbf{r}_u(u, v) \times \mathbf{r}_v(u, v)}{|\mathbf{r}_u(u, v) \times \mathbf{r}_v(u, v)|} \tag{7.16}$$

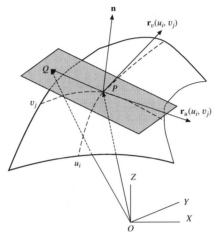

Figure 7.3   Tangent plane on the surface patch.

The equation of the *tangent plane* at P = $\mathbf{r}(u_i, v_j)$ with Q(x, y, z) as any general point on the plane may be obtained using Eqs. (6.6) and (6.7) as

$$\mathbf{r}(u_i, v_j) = x(u_i, v_j)\mathbf{i} + y(u_i, v_j)\mathbf{j} + z(u_i, v_j)\mathbf{k} = x_{ij}\mathbf{i} + y_{ij}\mathbf{j} + z_{ij}\mathbf{k}$$

$$\mathbf{r}_u(u_i, v_j) = x_u(u_i, v_j)\mathbf{i} + y_u(u_i, v_j)\mathbf{j} + z_u(u_i, v_j)\mathbf{k} = x_{ij}^u\mathbf{i} + y_{ij}^u\mathbf{j} + z_{ij}^u\mathbf{k}$$

$$\mathbf{r}_v(u_i, v_j) = x_v(u_i, v_j)\mathbf{i} + y_v(u_i, v_j)\mathbf{j} + z_v(u_i, v_j)\mathbf{k} = x_{ij}^v\mathbf{i} + y_{ij}^v\mathbf{j} + z_{ij}^v\mathbf{k} \tag{7.17}$$

$$\begin{vmatrix} x - x_{ij} & x_{ij}^u & x_{ij}^v \\ y - y_{ij} & y_{ij}^u & y_{ij}^v \\ z - z_{ij} & z_{ij}^u & z_{ij}^v \end{vmatrix} = 0$$

The Gaussian curvature K and the mean curvature H may also be obtained using Eq. (6.36) after computing the expressions for G_{11}, G_{12}, G_{22} and L, M, N depending on the derivatives of $\mathbf{r}(u, v)$, and are detailed in Chapter 6.

$$G_{11} = \mathbf{r}_u(u, v) \cdot \mathbf{r}_u(u, v) \qquad G_{12} = \mathbf{r}_u(u, v) \cdot \mathbf{r}_v(u, v) \qquad G_{22} = \mathbf{r}_v(u, v) \cdot \mathbf{r}_v(u, v)$$
$$L = \mathbf{r}_{uu}(u, v) \cdot \mathbf{n}(u, v) \qquad M = \mathbf{r}_{uv}(u, v) \cdot \mathbf{n}(u, v) \qquad N = \mathbf{r}_{vv}(u, v) \cdot \mathbf{n}(u, v) \tag{7.18}$$

From these the principal curvatures can be determined. Issues such as developability and point classification (whether the given point on the surface is elliptic, hyperbolic, or parabolic) can also be answered by examining whether the function $LN - M^2 = 0$ (parabolic point), $LN - M^2 > 0$ (elliptic point) or $LN - M^2 < 0$ (hyperbolic point). The surface area of the patch can be determined using

$$S = \iint_{u,v} |\mathbf{r}_u(u, v) \times \mathbf{r}_v(u, v)| \, du \, dv \tag{7.19}$$

For Ferguson's bi-cubic patch, from Eq. (7.12)

$$\mathbf{r}_u(u, v) = \mathbf{U} \mathbf{M}^u \mathbf{G} \mathbf{M}^T \mathbf{V}^T \qquad \mathbf{r}_v(u, v) = \mathbf{U} \mathbf{M} \mathbf{G} (\mathbf{M}^v)^T \mathbf{V}^T$$

$$\mathbf{r}_{uu}(u, v) = \mathbf{U} \mathbf{M}^{uu} \mathbf{M}^T \mathbf{V}^T \qquad \mathbf{r}_{uv}(u, v) = \mathbf{U} \mathbf{M}^u \mathbf{G} (\mathbf{M}^v)^T \mathbf{V}^T \tag{7.20}$$

$$\mathbf{r}_{vv}(u, v) = \mathbf{U} \mathbf{M} \mathbf{G} (\mathbf{M}^{vv})^T \mathbf{V}^T$$

where $\mathbf{M}^u$ and $\mathbf{M}^{uu}$ are $\mathbf{M}_1$ and $\mathbf{M}_2$, respectively in Eq. (4.9). $\mathbf{M}^{uu}$ and $\mathbf{M}^{vv}$ are identical with the difference that they are used with their respective parameter matrices $\mathbf{U}$ and $\mathbf{V}$.

$$\mathbf{M}^u = \mathbf{M}^v = \begin{bmatrix} 0 & 0 & 0 & 0 \\ 6 & -6 & 3 & 3 \\ -6 & 6 & -4 & -2 \\ 0 & 0 & 1 & 0 \end{bmatrix}, \mathbf{M}^{uu} = \mathbf{M}^{vv} = \begin{bmatrix} 0 & 0 & 0 & 0 \\ 0 & 0 & 0 & 0 \\ 12 & -12 & 6 & 6 \\ -6 & 6 & -4 & -2 \end{bmatrix} \tag{7.21}$$

Example 7.2. A Ferguson surface patch has the following geometric coefficients:

$$\mathbf{G} = \begin{bmatrix} (6, 0, 0) & (6, 0, 6) & (0, 0, 6) & (0, 0, 6) \\ (0, 6, 0) & (0, 6, 6) & (0, 0, 6) & (0, 0, 6) \\ (0, 5, 0) & (0, 5, 0) & 0 & 0 \\ (-5, 0, 0) & (-5, 0, 0) & 0 & 0 \end{bmatrix}$$

Determine the tangents, normal, Gaussian curvature, mean curvature, principal curvatures, and equation of the tangent plane at $\mathbf{r}(u = 0.5, v = 0.5)$. Also determine whether the surface is developable and find the surface area of the patch.

The patch is given by $\mathbf{r}(u, v) = \mathbf{U} \mathbf{M} \mathbf{G} \mathbf{M}^T \mathbf{V}^T = [x \ y \ z] = [(7u^3 - 13u^2 + 6) \ (-7u^3 + 8u^2 + 5u) \ (6v)]$ whose plot is shown in Figure 7.3. At $(u = 0.5, v = 0.5)$, the co-ordinates are $(3.625, 3.625, 3)$. Using Eq. (7.20), the slopes at any point on the surface are given by

$$\mathbf{r}_u(u, v) = [(21u^2 - 26u) \ (-21u^2 + 16u + 5) \ 0]$$

$$\mathbf{r}_v(u, v) = [0 \ 0 \ 6]$$

In particular, at $(u = 0.5, v = 0.5)$,

$$\mathbf{r}_u(0.5, 0.5) = [-7.75 \ 7.75 \ 0]$$

$$\mathbf{r}_v(0.5, 0.5) = [0 \ 0 \ 6]$$

The unit normal can be determined using

$$\begin{vmatrix} \mathbf{i} & \mathbf{j} & \mathbf{k} \\ -7.75 & 7.75 & 0 \\ 0 & 0 & 6 \end{vmatrix} = 46.5\mathbf{i} + 46.5\mathbf{j}$$

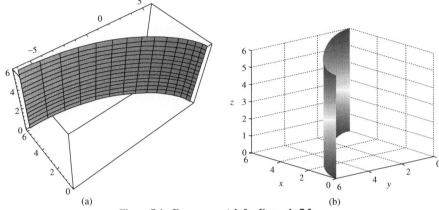

Figure 7.4   Ferguson patch for Example 7.2.

$$\mathbf{n}(0.5, 0.5) = \frac{46.5\mathbf{i} + 46.5\mathbf{j}}{\sqrt{46.5^2 + 46.5^2}} = \frac{1}{\sqrt{2}}(\mathbf{i} + \mathbf{j})$$

Now

$$\mathbf{r}_{uu}(u, v) = \mathbf{U}\mathbf{M}^{uu}\mathbf{G}\mathbf{M}^T\mathbf{V}^T = [(42u - 26)\ (-42u + 16)\ 0]$$

$$\mathbf{r}_{uv}(u, v) = \mathbf{U}\mathbf{M}^u\mathbf{G}\mathbf{M}^{vT}\mathbf{V}^T = [0\ \ 0\ \ 0]$$

$$\mathbf{r}_{vv}(u, v) = \mathbf{U}\mathbf{M}\mathbf{G}\mathbf{M}^{vvT}\mathbf{V}^T = [0\ \ 0\ \ 0]$$

using which

$$L = \mathbf{r}_{uu}(u, v) \cdot \mathbf{n}(u, v) = \frac{-1260u^2 + 1260u - 780}{\sqrt{(-126u^2 + 96u + 30)^2 + (-126u^2 + 156u)^2}}$$

$$M = \mathbf{r}_{uv}(u, v) \cdot \mathbf{n}(u, v) = 0$$

$$N = \mathbf{r}_{vv}(u, v) \cdot \mathbf{n}(u, v) = 0$$

It can be seen that $LN - M^2 = 0$ for which the Gaussian curvature $K = \dfrac{LN - M^2}{G_{11}G_{22} - G_{12}^2} = 0$ at all points on the surface. Hence the surface is developable. Further, G_{11}, G_{12}, G_{22} are given from Eq. (7.18) by

$$G_{11} = \mathbf{r}_u(u, v) \cdot \mathbf{r}_u(u, v) = (21u^2 - 26u)^2 + (-21u^2 + 16u + 5)^2$$

$$G_{12} = \mathbf{r}_u(u, v) \cdot \mathbf{r}_v(u, v) = 0$$

$$G_{22} = \mathbf{r}_v(u, v) \cdot \mathbf{r}_v(u, v) = 36$$

$$= 36\ [(21u^2 - 26u)^2 + (-21u^2 + 16u + 5)^2]$$

$$H = \frac{G_{11}N + G_{22}L - 2G_{12}M}{2[G_{11}G_{22} - G_{12}^2]} = \frac{L}{2G_{11}} = \frac{\mathbf{r}_{uu}(0.5, 0.5) \cdot \mathbf{n}(0.5, 0.5)}{2G_{11}(0.5, 0.5)}$$

Here $\mathbf{r}_{uu}(0.5, 0.5) = (-5\mathbf{i} - 5\mathbf{j})$, $\mathbf{n}(0.5, 0.5) = (\mathbf{i} + \mathbf{j})/\sqrt{2}$, for which $H = -0.0294$. The equation of the tangent plane at $(u = 0.5, v = 0.5)$ is given using Eq. (7.17) as

$$\begin{vmatrix} (x - 3.625) & -7.75 & 0 \\ (y - 3.625) & 7.75 & 0 \\ (z - 3) & 0 & 6 \end{vmatrix} = 0$$

$$46.5 \, (x - 3.625) + 46.5(y - 3.625) = 0, \text{ that is } x + y = 7.25$$

The surface area of the patch can be obtained by Eq. (7.19) as

$$\mathbf{r}_u(u, v) \times \mathbf{r}_v(u, v) = 6 \, (-21u^2 + 16u + 5)\mathbf{i} - 6(21u^2 - 26u)\mathbf{j}$$

$$\mid \mathbf{r}_u(u, v) \times \mathbf{r}_v(u, v) \mid = 6[(-21u^2 + 16u + 5)^2 + (21u^2 - 26u)^2]^{\frac{1}{2}} = f(u)$$

$$S = \int_{u=0}^{1} \int_{v=0}^{1} f(u) \, du \, dv = 54.64$$

7.1.3 Sixteen Point Form Surface Patch

For 16 uniformly spaced points on a surface patch, to fit a bi-cubic tensor product surface of the form given in Eq. (7.5) with $m = n = 3$,

$$\mathbf{r}(u, v) = \sum_{j=0}^{3} \sum_{i=0}^{3} \mathbf{D}_{ij} u^i v^j = [u^3 \quad u^2 \quad u \quad 1] \begin{bmatrix} \mathbf{D}_{33} & \mathbf{D}_{32} & \mathbf{D}_{31} & \mathbf{D}_{30} \\ \mathbf{D}_{23} & \mathbf{D}_{22} & \mathbf{D}_{21} & \mathbf{D}_{20} \\ \mathbf{D}_{13} & \mathbf{D}_{12} & \mathbf{D}_{11} & \mathbf{D}_{10} \\ \mathbf{D}_{03} & \mathbf{D}_{02} & \mathbf{D}_{01} & \mathbf{D}_{00} \end{bmatrix} \begin{bmatrix} v^3 \\ v^2 \\ v \\ 1 \end{bmatrix} \quad (7.22)$$

has 16 unknowns $\mathbf{D}_{ij}$ to be determined. Let $u \in [0, 1]$, $v \in [0, 1]$ and each interval be subdivided as [0, 1/3, 2/3, 1]. The given sixteen points on the surface are (each $\mathbf{r}_{ij}$ is a triplet) such that

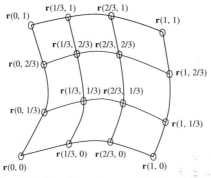

Figure 7.5 Bi-cubic surface patch in 16 point form

$$
\begin{bmatrix}
\mathbf{r}_{33} & \mathbf{r}_{32} & \mathbf{r}_{31} & \mathbf{r}_{30} \\
\mathbf{r}_{23} & \mathbf{r}_{22} & \mathbf{r}_{21} & \mathbf{r}_{20} \\
\mathbf{r}_{13} & \mathbf{r}_{12} & \mathbf{r}_{11} & \mathbf{r}_{10} \\
\mathbf{r}_{03} & \mathbf{r}_{02} & \mathbf{r}_{01} & \mathbf{r}_{00}
\end{bmatrix}
=
\begin{bmatrix}
\mathbf{r}(1,1) & \mathbf{r}\left(1,\frac{2}{3}\right) & \mathbf{r}\left(1,\frac{1}{3}\right) & \mathbf{r}(1,0) \\
\mathbf{r}\left(\frac{2}{3},1\right) & \mathbf{r}\left(\frac{2}{3},\frac{2}{3}\right) & \mathbf{r}\left(\frac{2}{3},\frac{1}{3}\right) & \mathbf{r}\left(\frac{2}{3},0\right) \\
\mathbf{r}\left(\frac{1}{3},1\right) & \mathbf{r}\left(\frac{1}{3},\frac{2}{3}\right) & \mathbf{r}\left(\frac{1}{3},\frac{1}{3}\right) & \mathbf{r}\left(\frac{1}{3},0\right) \\
\mathbf{r}(0,1) & \mathbf{r}\left(0,\frac{2}{3}\right) & \mathbf{r}\left(0,\frac{1}{3}\right) & \mathbf{r}(0,0)
\end{bmatrix}
\tag{7.23}
$$

From Eqs. (7.22) and (7.23)

$$
\begin{bmatrix}
\mathbf{r}_{33} & \mathbf{r}_{32} & \mathbf{r}_{31} & \mathbf{r}_{30} \\
\mathbf{r}_{23} & \mathbf{r}_{22} & \mathbf{r}_{21} & \mathbf{r}_{20} \\
\mathbf{r}_{13} & \mathbf{r}_{12} & \mathbf{r}_{11} & \mathbf{r}_{10} \\
\mathbf{r}_{03} & \mathbf{r}_{02} & \mathbf{r}_{01} & \mathbf{r}_{00}
\end{bmatrix}
=
\begin{bmatrix}
1 & 1 & 1 & 1 \\
\frac{8}{27} & \frac{4}{9} & \frac{2}{3} & 1 \\
\frac{1}{27} & \frac{1}{9} & \frac{1}{3} & 1 \\
0 & 0 & 0 & 1
\end{bmatrix}
\begin{bmatrix}
\mathbf{D}_{33} & \mathbf{D}_{32} & \mathbf{D}_{31} & \mathbf{D}_{30} \\
\mathbf{D}_{23} & \mathbf{D}_{22} & \mathbf{D}_{21} & \mathbf{D}_{20} \\
\mathbf{D}_{13} & \mathbf{D}_{12} & \mathbf{D}_{11} & \mathbf{D}_{10} \\
\mathbf{D}_{03} & \mathbf{D}_{02} & \mathbf{D}_{01} & \mathbf{D}_{00}
\end{bmatrix}
\begin{bmatrix}
1 & \frac{8}{27} & \frac{1}{27} & 0 \\
1 & \frac{4}{9} & \frac{1}{9} & 0 \\
1 & \frac{2}{3} & \frac{1}{3} & 0 \\
1 & 1 & 1 & 1
\end{bmatrix}
\tag{7.24}
$$

using which

$$
\begin{bmatrix}
\mathbf{D}_{33} & \mathbf{D}_{32} & \mathbf{D}_{31} & \mathbf{D}_{30} \\
\mathbf{D}_{23} & \mathbf{D}_{22} & \mathbf{D}_{21} & \mathbf{D}_{20} \\
\mathbf{D}_{13} & \mathbf{D}_{12} & \mathbf{D}_{11} & \mathbf{D}_{10} \\
\mathbf{D}_{03} & \mathbf{D}_{02} & \mathbf{D}_{01} & \mathbf{D}_{00}
\end{bmatrix}
= \mathbf{M}_{16}
\begin{bmatrix}
\mathbf{r}_{33} & \mathbf{r}_{32} & \mathbf{r}_{31} & \mathbf{r}_{30} \\
\mathbf{r}_{23} & \mathbf{r}_{22} & \mathbf{r}_{21} & \mathbf{r}_{20} \\
\mathbf{r}_{13} & \mathbf{r}_{12} & \mathbf{r}_{11} & \mathbf{r}_{10} \\
\mathbf{r}_{03} & \mathbf{r}_{02} & \mathbf{r}_{01} & \mathbf{r}_{00}
\end{bmatrix}
\mathbf{M}_{16}^T, \text{ where } \mathbf{M}_{16} =
\begin{bmatrix}
1 & 1 & 1 & 1 \\
\frac{8}{27} & \frac{4}{9} & \frac{2}{3} & 1 \\
\frac{1}{27} & \frac{1}{9} & \frac{1}{3} & 1 \\
0 & 0 & 0 & 1
\end{bmatrix}^{-1}
$$

The expression for a surface patch interpolating 16 uniformly spaced points is then

$$
\mathbf{r}(u,v) = \begin{bmatrix} u^3 & u^2 & u & 1 \end{bmatrix} \mathbf{M}_{16}
\begin{bmatrix}
\mathbf{r}_{33} & \mathbf{r}_{32} & \mathbf{r}_{31} & \mathbf{r}_{30} \\
\mathbf{r}_{23} & \mathbf{r}_{22} & \mathbf{r}_{21} & \mathbf{r}_{20} \\
\mathbf{r}_{13} & \mathbf{r}_{12} & \mathbf{r}_{11} & \mathbf{r}_{10} \\
\mathbf{r}_{03} & \mathbf{r}_{02} & \mathbf{r}_{01} & \mathbf{r}_{00}
\end{bmatrix}
\mathbf{M}_{16}^T
\begin{bmatrix} v^3 \\ v^2 \\ v \\ 1 \end{bmatrix}
\tag{7.25}
$$

7.1.4 Bézier Surface Patches
Similar to Bézier curves employing Bernstein polynomials as weight functions with control points (Chapter 4), a tensor product Bézier surface patch is given by

$$
\mathbf{r}(u,v) = \sum_{i=0}^{m} \sum_{j=0}^{n} \mathbf{r}_{ij} B_i^m(u) B_j^n(v)
\tag{7.26}
$$

where $\mathbf{r}_{ij}$, $i = 0, \ldots, m$, $j = 0, \ldots, n$ are the control points and $B_i^m(u)$ and $B_j^n(v)$ are Bernstein polynomials in parameters u and v. The control points form the *control polyhedron* or *control polynet* of the surface (Figure 7.6). For any $u = u_0$, $\mathbf{r}(u_0, v)$ is a Bézier curve of degree n. Likewise, for any $v = v_0$, $\mathbf{r}(u, v_0)$ is a Bézier curve of degree m. Eq. (7.26) may be written in the form

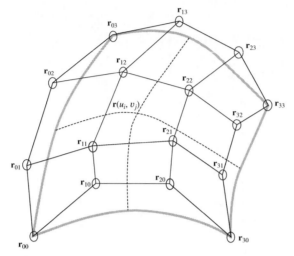

Figure 7.6   Schematic of a bi-cubic Bézier patch with its control polynet

$$
[(1-u)^m \quad mu(1-u)^{m-1} \quad \ldots \quad u^m]
\begin{bmatrix}
\mathbf{r}_{00} & \mathbf{r}_{01} & \cdot\cdot & \mathbf{r}_{0(n-1)} & \mathbf{r}_{0n} \\
\mathbf{r}_{10} & \mathbf{r}_{11} & \cdot\cdot & \mathbf{r}_{1(n-1)} & \mathbf{r}_{1n} \\
\cdot & \cdot & \cdot\cdot & \cdot & \cdot \\
\cdot & \cdot & \cdot\cdot & \cdot & \cdot \\
\mathbf{r}_{(m-1)0} & \mathbf{r}_{(m-1)1} & \cdot\cdot & \mathbf{r}_{(m-1)(n-1)} & \mathbf{r}_{(m-1)n} \\
\mathbf{r}_{m0} & \mathbf{r}_{m1} & \cdot\cdot & \mathbf{r}_{m(n-1)} & \mathbf{r}_{mn}
\end{bmatrix}
\begin{bmatrix}
(1-v)^n \\
n(1-v)^{n-1}v \\
\vdots \\
v^n
\end{bmatrix}
\tag{7.27}
$$

For a bi-cubic Bézier surface patch, for instance, the equation above becomes

$$
\mathbf{r}(u,v) = [(1-u)^3 \quad 3u(1-u)^2 \quad 3u^2(1-u) \quad u^3]
\begin{bmatrix}
\mathbf{r}_{00} & \mathbf{r}_{01} & \mathbf{r}_{02} & \mathbf{r}_{03} \\
\mathbf{r}_{10} & \mathbf{r}_{11} & \mathbf{r}_{12} & \mathbf{r}_{13} \\
\mathbf{r}_{20} & \mathbf{r}_{21} & \mathbf{r}_{22} & \mathbf{r}_{23} \\
\mathbf{r}_{30} & \mathbf{r}_{31} & \mathbf{r}_{32} & \mathbf{r}_{33}
\end{bmatrix}
\begin{bmatrix}
(1-v)^3 \\
3v(1-v)^2 \\
3v^2(1-v) \\
v^3
\end{bmatrix}
$$

$$
= [u^3 \quad u^2 \quad u \quad 1]
\begin{bmatrix}
-1 & 3 & -3 & 1 \\
3 & -6 & 3 & 0 \\
-3 & 3 & 0 & 0 \\
1 & 0 & 0 & 0
\end{bmatrix}
\begin{bmatrix}
\mathbf{r}_{00} & \mathbf{r}_{01} & \mathbf{r}_{02} & \mathbf{r}_{03} \\
\mathbf{r}_{10} & \mathbf{r}_{11} & \mathbf{r}_{12} & \mathbf{r}_{13} \\
\mathbf{r}_{20} & \mathbf{r}_{21} & \mathbf{r}_{22} & \mathbf{r}_{23} \\
\mathbf{r}_{30} & \mathbf{r}_{31} & \mathbf{r}_{32} & \mathbf{r}_{33}
\end{bmatrix}
\begin{bmatrix}
-1 & 3 & -3 & 1 \\
3 & -6 & 3 & 0 \\
-3 & 3 & 0 & 0 \\
1 & 0 & 0 & 0
\end{bmatrix}^T
\begin{bmatrix}
v^3 \\
v^2 \\
v \\
1
\end{bmatrix}
$$

$$= [u^3 \quad u^2 \quad u \quad 1] \mathbf{M}_B \begin{bmatrix} \mathbf{r}_{00} & \mathbf{r}_{01} & \mathbf{r}_{02} & \mathbf{r}_{03} \\ \mathbf{r}_{10} & \mathbf{r}_{11} & \mathbf{r}_{12} & \mathbf{r}_{13} \\ \mathbf{r}_{20} & \mathbf{r}_{21} & \mathbf{r}_{22} & \mathbf{r}_{23} \\ \mathbf{r}_{30} & \mathbf{r}_{31} & \mathbf{r}_{32} & \mathbf{r}_{33} \end{bmatrix} \mathbf{M}_B^T \begin{bmatrix} v^3 \\ v^2 \\ v \\ 1 \end{bmatrix} \tag{7.28}$$

where $\mathbf{M}_B$ is the Bézier coefficient matrix defined in Eq. (4.40).

The properties of Bézier curves are inherited by Bézier patches, some notable properties being: (a) the four corner points of the patch are the respective corner points in the control polyhedron, (b) boundary curves are tangent to the polyhedron edges at corner points and (c) the patch is contained within the convex hull of the polyhedron. Most solid modeling packages use bi-quintic ($m = n = 5$) or bi-septic ($m = n = 7$) patches to provide more flexibility to a user when designing a composite surface. We know from Chapter 4 that when designing composite Bézier curves, 3 control points are constrained to be collinear when requiring C^1 continuity at the junction point and, in addition, 2 more control points (a total of five) are required to be coplanar for curvature continuity.

Example 7.3. The control points for a quadratic-cubic Bézier patch are given as

$$\begin{bmatrix} \mathbf{r}_{00} & \mathbf{r}_{01} & \mathbf{r}_{02} & \mathbf{r}_{03} \\ \mathbf{r}_{10} & \mathbf{r}_{11} & \mathbf{r}_{12} & \mathbf{r}_{13} \\ \mathbf{r}_{20} & \mathbf{r}_{21} & \mathbf{r}_{22} & \mathbf{r}_{23} \end{bmatrix} = \begin{bmatrix} (0,0,0) & (1,0,1) & (2,0,1) & (3,0,0) \\ (0,1,0) & (1,1,1) & (2,1,1) & (3,1,0) \\ (0,2,0) & (1,2,0) & (2,2,1) & (3,2,0) \end{bmatrix}$$

Plot the Bézier's patch. Determine the unit normal, equation of the tangent plane and curvature at ($u = 0.5$, $v = 0.5$) on the surface.

Equation (7.27) defines the surface with $m = 2$ and $n = 3$, two opposite boundaries are quadratic Bézier curves and the remaining is a pair of cubic Bézier curves. The expression for the surface patch is

$$\mathbf{r}(u,v) = [(1-u)^2 \quad 2u(1-u) \quad u^2] \begin{bmatrix} \mathbf{r}_{00} & \mathbf{r}_{01} & \mathbf{r}_{02} & \mathbf{r}_{03} \\ \mathbf{r}_{10} & \mathbf{r}_{11} & \mathbf{r}_{12} & \mathbf{r}_{13} \\ \mathbf{r}_{20} & \mathbf{r}_{21} & \mathbf{r}_{22} & \mathbf{r}_{23} \end{bmatrix} \begin{bmatrix} (1-v)^3 \\ 3v(1-v)^2 \\ 3v^2(1-v) \\ v^3 \end{bmatrix}$$

$$= [(1-u^2) \quad 2u(1-u) \quad u^2] \begin{bmatrix} (0,0,0) & (1,0,1) & (2,0,1) & (3,0,0) \\ (0,1,0) & (1,1,1) & (2,1,1) & (3,1,0) \\ (0,2,0) & (1,2,0) & (2,2,1) & (3,2,0) \end{bmatrix} \begin{bmatrix} (1-v)^3 \\ 3v(1-v)^2 \\ 3v^2(1-v) \\ v^3 \end{bmatrix}$$

$$= [3v, 2u, 3v(1-v)]$$

which is plotted in Figure 7.7 (a).

Unit normal to the surface is given by

$$\mathbf{n}(u,v) = \frac{\mathbf{r}_u(u,v) \times \mathbf{r}_v(u,v)}{|\mathbf{r}_u(u,v) \times \mathbf{r}_v(u,v)|} = \frac{\{6 - 12v, 0, -6\}}{\sqrt{(6-12v)^2 + 36}} \quad \Rightarrow \quad \mathbf{n}(0.5, 0.5) = \{0, 0, -1\}$$

Equation of the tangent plane at $\mathbf{r}(0.5, 0.5) \equiv \{1.5, 1, 0.75\}$ is

$$\begin{vmatrix} x - x_p & x_{up} & x_{xp} \\ y - y_p & y_{up} & y_{vp} \\ z - z_p & z_{up} & z_{vp} \end{vmatrix} = 0 \Rightarrow \begin{vmatrix} x - 1.5 & 0 & 3 \\ y - 1 & 2 & 0 \\ z - 0.75 & 0 & 0 \end{vmatrix} = 0 \Rightarrow z = 0.75$$

Also, at $u = 0.5$, $v = 0.5$

$$\mathbf{r}_u = \{0, 2, 0\}; \qquad \mathbf{r}_v = \{3, 0, 0\}$$

$$\mathbf{r}_{uu} = \{0, 0, 0\} \qquad \mathbf{r}_{uv} = \{0, 0, 0\} \qquad \mathbf{r}_{vv} = \{0, 0, -6\}$$

$$\Rightarrow \qquad G_{11} = 4,\ G_{12} = 0,\ G_{22} = 9;\ L = M = 0,\ N = 6$$

$$K = \kappa_1 \kappa_2 = \frac{LN - M^2}{G_{11}G_{22} - G_{12}^2} = 0 \qquad H = \frac{1}{2}(\kappa_1 + \kappa_2) = \frac{G_{11}N + G_{22}L - 2G_{12}M}{2(G_{11}G_{22} - G_{12}^2)} = \frac{1}{3}$$

$$\Rightarrow \qquad \kappa_1 = \frac{2}{3},\ \kappa_2 = 0$$

Hence, the radius of curvature (at $\mathbf{r}_p$) $= 1/\kappa_1 = 1.5$ and since the Gaussian curvature $K = 0$, the surface is developable.

For control points $\mathbf{r}_{10}$ and $\mathbf{r}_{13}$ changed as $(0, 1, 0.5)$ and $(3, 1, 0.5)$ respectively, lifting them up by 0.5 units each along the z-direction, the new equation of the surface is

$$\mathbf{r}(u, v) = [(1 - u)^2 \quad 2u(1 - u) \quad u^2] \begin{bmatrix} (0, 0, 0) & (1, 0, 1) & (2, 0, 1) & (3, 0, 0) \\ (0, 1, 0.5) & (1, 1, 1) & (2, 1, 1) & (3, 1, 0.5) \\ (0, 2, 0) & (1, 2, 1) & (2, 2, 1) & (3, 2, 0) \end{bmatrix} \begin{bmatrix} (1 - v)^3 \\ 3v(1 - v)^2 \\ 3v^2(1 - v) \\ v^3 \end{bmatrix}$$

The new shape of the surface is shown in Figure 7.7(b).

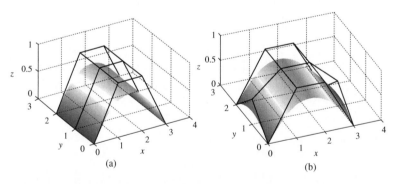

(a)   (b)

Figure 7.7 (a) Example of a quadratic-cubic Bézier surface with control polynet and (b) change in patch's shape with relocation of two control points

Example 7.4. The control points for a bi-cubic Bézier surface are given by

$$\mathbf{r}_{00} = \{0, 0, 0\} \quad \mathbf{r}_{10} = \{1, 0, 1\} \quad \mathbf{r}_{20} = \{2, 0, 1\} \quad \mathbf{r}_{30} = \{3, 0, 0\}$$
$$\mathbf{r}_{01} = \{0, 1, 1\} \quad \mathbf{r}_{11} = \{1, 1, 2\} \quad \mathbf{r}_{21} = \{2, 1, 2\} \quad \mathbf{r}_{31} = \{3, 1, 1\}$$
$$\mathbf{r}_{02} = \{0, 2, 1\} \quad \mathbf{r}_{12} = \{1, 2, 2\} \quad \mathbf{r}_{22} = \{2, 2, 2\} \quad \mathbf{r}_{32} = \{3, 2, 1\}$$
$$\mathbf{r}_{03} = \{0, 3, 0\} \quad \mathbf{r}_{13} = \{1, 3, 1\} \quad \mathbf{r}_{23} = \{2, 3, 1\} \quad \mathbf{r}_{33} = \{3, 3, 0\}$$

Plot the bi-cubic surface.

The equation for the bi-cubic Bézier surface patch is given in (7.28) and the parent surface is shown in Figure 7.8 (a). As an exercise, we may determine the tangents, normal, and Gaussian and mean curvatures at $u = 0.5$, $v = 0.7$. The effect of relocating control points is shown in Figures 7.8 (b and c). For new control points $\mathbf{r}_{11} = \{2, 2, 6\}$ and $\mathbf{r}_{21} = \{4, 2, 6\}$ we get Figure 7.8(b) and for $\mathbf{r}_{12} = \{4, 6, 4\}$ and $\mathbf{r}_{22} = \{4, 4, 4\}$, Figure 7.8(c) is obtained.

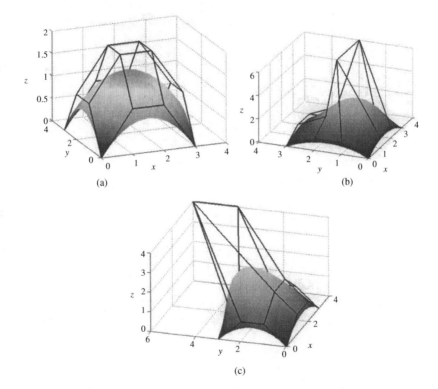

(a)   (b)

(c)

Figure 7.8   (a) Bézier bi-cubic patch in Example 7.4, (b) and (c) patches with data points relocated.

Example 7.5. Triangular Bi-Cubic Bézier Patch. Collapsing the data points for any boundary curve can create a triangular bi-cubic Bézier surface patch. Create the surface patch with the following control points:

$$\mathbf{P}_{00} = \{0, 0, 3\}; \quad \mathbf{P}_{10} = \{1, 1, 3\}; \; \mathbf{P}_{20} = \{1, 2, 3\}; \; \mathbf{P}_{30} = \{0, 3, 3\};$$

$$\mathbf{P}_{01} = \{0, -1, 2\}; \; \mathbf{P}_{11} = \{2, 1, 2\}; \; \mathbf{P}_{21} = \{2, 2, 2\}; \; \mathbf{P}_{31} = \{1, 3, 2\};$$

$$\mathbf{P}_{02} = \{0, -1, 1\}; \; \mathbf{P}_{12} = \{1, 1, 1\}; \; \mathbf{P}_{22} = \{2, 2, 1\}; \; \mathbf{P}_{32} = \{1, 3, 1\};$$

$$\mathbf{P}_{03} = \{0, 0, 0\}; \quad \mathbf{P}_{13} = \{0, 0, 0\}; \; \mathbf{P}_{23} = \{0, 0, 0\}; \; \mathbf{P}_{33} = \{0, 0, 0\}$$

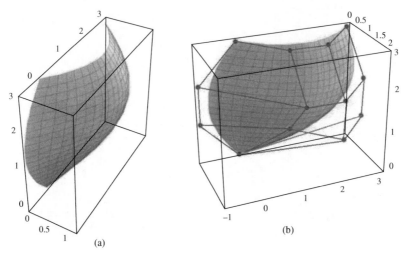

Figure 7.9   A triangular Bézier patch (a) without control points and (b) with control points

7.1.5   Triangular Surface Patch

Usually, a surface is created or modeled as a set of triangular or rectangular patches with continuity conditions satisfied across the boundaries of adjoining patches. A way to generate a triangular patch is described in Example 7.5.

Another way to model a triangular patch is to use three parameters u, v and w such that they are constrained to sum to 1. With three parameters and a constraint, the patch still remains bi-parametric. The triangular patch is defined by a set of control points $\mathbf{r}_{ijk}$ arranged in a triangular manner (Figure 7.10). Each control point is three dimensional and the indices i, j, k are such that $0 \le i, j, k \le n$,

Figure 7.10   Schematic of the placement of data points for a triangular patch

$i + j + k = n$. The value of n is user's choice. A large n will carry finer details of the patch but at increased computational cost. The number of control points used are $\frac{1}{2}(n + 1)(n + 2)$.

Index $i = 0$ corresponds to the left side of the triangle, $j = 0$ to the base and $k = 0$ to the right side of the triangular table of control points. There are $n + 1$ points on each side of the triangle. The surface patch is defined by

$$\mathbf{r}(u, v, w) = \sum_{i+j+k=n} \mathbf{r}_{ijk}\, \frac{n!}{i!\,j!\,k!}\, u^i v^j w^k, \quad u + v + w = 1,\ i + j + k = n \tag{7.29}$$

The three boundary curves are given by $\{u = 0,\ v,\ w = (1 - v)\}$, $\{u = (1 - w),\ v = 0,\ w\}$ and $\{u,\ v = (1 - u),\ w = 0\}$. Thus,

$$\mathbf{r}(v) = \sum_{i+k=n} \mathbf{r}_{0,j,k}\, \frac{n!}{j!\,k!}\, v^j (1 - v)^k = \sum_{j=0}^{n} \mathbf{r}_{0,j,n-j}\, \frac{n!}{j!(n - j)!}\, v^j (1 - v)^{n-j}$$

$$\mathbf{r}(u) = \sum_{i+j=n} \mathbf{r}_{i,j,0}\, \frac{n!}{i!\,j!}\, u^i (1 - u)^j = \sum_{i=0}^{n} \mathbf{r}_{i,(n-i),0}\, \frac{n!}{i!(n - i)!}\, u^i (1 - u)^{n-i} \tag{7.30}$$

$$\mathbf{r}(w) = \sum_{k+i=n} \mathbf{r}_{i,0,k}\, \frac{n!}{i!\,k!}\, w^k (1 - w)^i = \sum_{k=0}^{n} \mathbf{r}_{(n-k),0,k}\, \frac{n!}{k!(n - k)!}\, w^k (1 - w)^{n-k}$$

Example 7.6. Generate a triangular Bézier patch with $n = 2$ and the following 6 control points:

$$\mathbf{r}_{020} = (1, 3, 1)$$

$$\mathbf{r}_{011} = (0.5, 1, 0); \ \mathbf{r}_{110} = (1.5, 1, 0)$$

$$\mathbf{r}_{002} = (0, 0, 0); \ \mathbf{r}_{101} = (1, 0, -1); \ \mathbf{r}_{200} = (2, 0, 0)$$

The patch generated using Eq. (7.30) is shown in Figure 7.11.

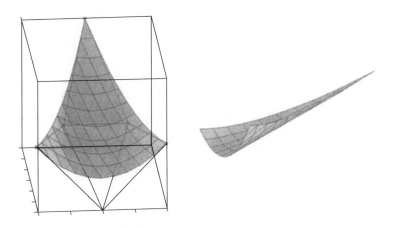

Figure 7.11 Triangular surface patch.

7.2 Boundary Interpolation Surfaces

Ruled and lofted patches are some examples of boundary interpolation surfaces. Given two parametric curves, $\mathbf{r}_1(u)$ and $\mathbf{r}_2(u)$, u in [0, 1], a linear blending of curves in parameter v provides a *ruled* patch. Discussed in section 6.6, the result is

$$\mathbf{r}(u, v) = (1 - v)\mathbf{r}_1(u) + v\mathbf{r}_2(u) = \mathbf{r}_1(u) + v[\mathbf{r}_2(u) - \mathbf{r}_1(u)] \tag{7.31}$$

Here $\mathbf{r}_2(u) - \mathbf{r}_1(u)$ is the direction vector along the straight line rulings. If, in addition, the *cross boundary tangents* $\mathbf{t}_1(u)$ and $\mathbf{t}_2(u)$ are also provided with respective curves $\mathbf{r}_1(u)$ and $\mathbf{r}_2(u)$, then the Hermite blending of four conditions (positions and slopes) along v for every u can be performed as in Eq. (7.32) with Hermite functions in Eq. (7.6). The resultant patch is called a *lofted* surface. Figure 7.12 differentiates between ruled and lofted patches for two boundary curves.

$$\mathbf{r}(u, v) = \varphi_0(v)\mathbf{r}_1(u) + \varphi_1(v)\mathbf{r}_2(u) + \varphi_2(v)\mathbf{t}_1(u) + \varphi_3(v)\mathbf{t}_2(u) \tag{7.32}$$

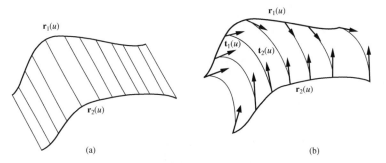

(a) (b)

Figure 7.12   (a) A ruled patch and (b) a lofted patch

Example 7.7: Let

$$\mathbf{r}(u, 0) = \{\cos[\pi(1 - u)], \sin[-\pi u], 0\}$$

$$\mathbf{r}(u, 1) = \{(2u - 1), -2u(1 - u), 1\}$$

be two given boundary curves of the ruled surface, with $u \in [0, 1]$. The equation of the surface for $v \in [0, 1]$ is given by

$$\mathbf{r}(u, v) = \{(1 - v)\cos[\pi(1 - u)] + v(2u - 1), \quad -(-v\sin[\pi u] - 2u)(1 - u)v, v\}$$

The surface is shown in Figure 7.13.

It can be verified that the tangent vector and unit normal vector at $(u = 0.5, v = 0.5)$ are

$$\mathbf{r}_u(0.5, 0.5) = \left\{\frac{\pi}{2} + 1, 0, 0\right\}; \mathbf{r}_v(0.5, 0.5) = \left\{0, \frac{1}{2}, 1\right\}$$

$$\mathbf{r}_u \times \mathbf{r}_v = \left\{0, \frac{\pi}{2} + 1, \frac{1}{2}\left(\frac{\pi}{2} + 1\right)\right\} \Rightarrow \text{unit normal } \mathbf{n} = \{0, 0.895, 0.448\}$$

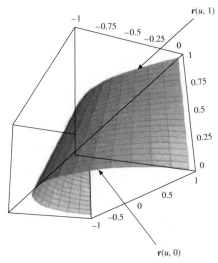

Figure 7.13   Ruled surface in Example 7.7

7.2.1   Coon's Patches

Coon's patches can use either linear or Hermite blending in surface approximation using four boundary curves. Given those curves as $\mathbf{a}_0(v)$, $\mathbf{a}_1(v)$, $\mathbf{b}_0(u)$ and $\mathbf{b}_1(u)$ that intersect at four corner points $\mathbf{P}_{00}$, $\mathbf{P}_{01}$, $\mathbf{P}_{10}$ and $\mathbf{P}_{11}$ as shown in Figure 7.14(a), ruled surfaces can be obtained by combining any two pairs of opposite curves

$$\begin{aligned} \mathbf{r}_1\,(u, v) &= (1 - v)\,\mathbf{b}_0\,(u) + v\,\mathbf{b}_1\,(u) \\ \mathbf{r}_2\,(u, v) &= (1 - u)\,\mathbf{a}_0\,(v) + u\,\mathbf{a}_1\,(u) \end{aligned} \qquad (7.33)$$

A linear Coon's patch $\mathbf{r}(u, v)$ is the sum of the two surfaces above, and a surface $\mathbf{r}_3(u, v)$ is subtracted as the *correction surface* so that the boundary conditions are met. The patch may be expressed as

$$\mathbf{r}(u, v) = \mathbf{r}_1(u, v) + \mathbf{r}_2(u, v) - \mathbf{r}_3(u, v) \qquad (7.34)$$

Note that

$$\begin{aligned} \mathbf{r}(u, 0) = \mathbf{b}_0(u) &= \mathbf{r}_1(u, 0) + \mathbf{r}_2(u, 0) - \mathbf{r}_3(u, 0) \\ &= \mathbf{b}_0(u) + (1 - u)\,\mathbf{a}_0(0) + u\mathbf{a}_1(0) - \mathbf{r}_3(u, 0) = \mathbf{b}_0(u) + (1 - u)\mathbf{P}_{00} + u\mathbf{P}_{10} - \mathbf{r}_3(u, 0) \end{aligned}$$

which implies

$$\mathbf{r}_3(u, 0) = (1 - u)\mathbf{P}_{00} + u\mathbf{P}_{10} \qquad (7.35a)$$

Similarly,

$$\begin{aligned} \mathbf{r}(u, 1) = \mathbf{b}_1(u) &= \mathbf{r}_1(u, 1) + \mathbf{r}_2(u, 1) - \mathbf{r}_3(u, 1) \\ &= \mathbf{b}_1(u) + (1 - u)\,\mathbf{a}_0(1) + u\,\mathbf{a}_1(1) - \mathbf{r}_3(u, 1) = \mathbf{b}_1(u) + (1 - u)\mathbf{P}_{01} + u\mathbf{P}_{11} - \mathbf{r}_3(u, 1) \end{aligned}$$

which gives

$$\mathbf{r}_3(u, 1) = (1 - u)\mathbf{P}_{01} + u\mathbf{P}_{11} \qquad (7.35b)$$

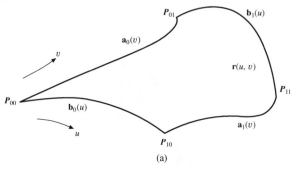

(a)

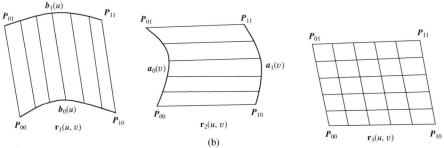

(b)

Figure 7.14   (a) Bi-linear Coon's patch and (b) constituents of the Coon's patch

From Eqs. (7.35a) and (7.35b), we realize that the two boundary curves for $\mathbf{r}_3(u, v)$, $\mathbf{r}_3(u, 0)$ and $\mathbf{r}_3$ $(u, 1)$, are available that can be linearly blended along parameter v. Or

$$\mathbf{r}_3(u, v) = (1 - v)\mathbf{r}_3(u, 0) + v\mathbf{r}_3(u, 1)$$

$$= (1 - v)[(1 - u)\mathbf{P}_{00} + u\mathbf{P}_{10}] + v[(1 - u)\mathbf{P}_{01} + u\mathbf{P}_{11}]$$

$$= (1 - v)(1 - u)\mathbf{P}_{00} + u(1 - v)\mathbf{P}_{10} + (1 - u)v\mathbf{P}_{01} + uv\mathbf{P}_{11} \qquad (7.36)$$

We may as well attempt to meet boundary conditions using $\mathbf{r}(0, v) = \mathbf{a}_0(v)$ and $\mathbf{r}(1, v) = \mathbf{a}_1(v)$ to get

$$\mathbf{r}(0, v) = (1 - v)\mathbf{b}_0(0) + v\mathbf{b}_1(0) + \mathbf{a}_0(v) - \mathbf{r}_3(0, v)$$

$$= (1 - v)\mathbf{P}_{00} + v\mathbf{P}_{01} + \mathbf{a}_0(v) - \mathbf{r}_3(0, v) = \mathbf{a}_0(v)$$

$$\Rightarrow \qquad \mathbf{r}_3(0, v) = (1 - v)\mathbf{P}_{00} + v\mathbf{P}_{01} \qquad (7.37a)$$

and

$$\mathbf{r}(1, v) = (1 - v)\mathbf{b}_0(1) + v\mathbf{b}_1(1) + \mathbf{a}_1(v) - \mathbf{r}_3(1, v)$$

$$= (1 - v)\mathbf{P}_{10} + v\mathbf{P}_{11} + \mathbf{a}_1(v) - \mathbf{r}_3(1, v) = \mathbf{a}_1(v)$$

$$\Rightarrow \qquad \mathbf{r}_3(1, v) = (1 - v)\mathbf{P}_{10} + v\mathbf{P}_{11} \qquad (7.37b)$$

and thereafter linearly blend $\mathbf{r}_3(0, v)$ and $\mathbf{r}_3(1, v)$ with respect to u. However, observe from Eq. (7.36) that the conditions in Eq. (7.37) are satisfied. In other words, using Eqs. (7.37) to determine the correction surface would yield the same result as in Eq. (7.36). In matrix form, the linear Coon's patch can be expressed as

$$\mathbf{r}(u, v) = [(1-u) \quad u] \begin{bmatrix} \mathbf{a}_0(v) \\ \mathbf{a}_1(v) \end{bmatrix} + [(1-v) \quad v] \begin{bmatrix} \mathbf{b}_0(u) \\ \mathbf{b}_1(u) \end{bmatrix} - [(1-u) \quad u] \begin{bmatrix} \mathbf{P}_{00} & \mathbf{P}_{01} \\ \mathbf{P}_{10} & \mathbf{P}_{11} \end{bmatrix} \begin{bmatrix} (1-v) \\ v \end{bmatrix} \quad (7.38a)$$

It is clear that $[(1-u) \; u]$ and $[(1-v) \; v]$ are the blending functions for the Coon's Patch, and they are *barycentric*.

Any other set of functions $\varphi_0(u)$, $\varphi_1(u)$; $\psi_0(v)$, $\psi_1(v)$ may also qualify as a set of blending functions so long as they satisfy the following properties:
- Barycentric property: $\varphi_0(u) + \varphi_1(u) = 1$; $\psi_0(v) + \psi_1(v) = 1$
- Corner conditions:

$$\varphi_0(0) = 1; \varphi_1(0) = 0; \psi_0(0) = 1, \psi_1(0) = 0$$

$$\varphi_0(1) = 0; \varphi_1(1) = 1; \psi_0(1) = 0; \psi_1(1) = 1$$

The Coon's patch, in general, will be given by

$$\mathbf{r}(u, v) = [\varphi_0(u) \quad \varphi_1(u)] \begin{bmatrix} \mathbf{a}_0(v) \\ \mathbf{a}_1(v) \end{bmatrix} + [\psi_0(v) \quad \psi_1(v)] \begin{bmatrix} \mathbf{b}_0(u) \\ \mathbf{b}_1(u) \end{bmatrix} - [\varphi_0(u) \quad \varphi_1(u)] \begin{bmatrix} \mathbf{p}_{00} & \mathbf{p}_{01} \\ \mathbf{p}_{10} & \mathbf{p}_{11} \end{bmatrix} \begin{bmatrix} \psi_0(v) \\ \psi_1(v) \end{bmatrix}$$

$$(7.38b)$$

If, in addition to the four boundary curves and corner points, the respective cross boundary tangents $\mathbf{s}_0(v)$, $\mathbf{s}_1(v)$, $\mathbf{t}_0(u)$ and $\mathbf{t}_1(u)$ are also given as shown in Figure 7.15, a Hermite or bi-cubic Coon's patch can be created in a similar manner as discussed above.

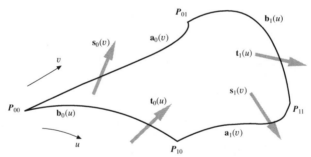

Figure 7.15 Schematic of a bi-cubic Coon's patch

Blending boundary curves $\mathbf{b}_0(u)$ and $\mathbf{b}_1(u)$ using cross boundary tangents, $\mathbf{t}_0(u)$ and $\mathbf{t}_1(u)$ gives

$$\mathbf{r}_1(u, v) = \varphi_0(v) \, \mathbf{b}_0(u) + \varphi_1(v) \, \mathbf{b}_1(u) + \varphi_2(v) \, \mathbf{t}_0(u) + \varphi_3(v) \, \mathbf{t}_1(u) \quad (7.39a)$$

Likewise, bi-cubic blending of $\mathbf{a}_0(v)$ and $\mathbf{a}_1(v)$ using cross boundary tangents, $\mathbf{s}_0(v)$ and $\mathbf{s}_1(v)$ gives

$$\mathbf{r}_2(u, v) = \varphi_0(u) \, \mathbf{a}_0(v) + \varphi_1(u) \, \mathbf{a}_1(v) + \varphi_2(u) \, \mathbf{s}_0(v) + \varphi_3(u) \, \mathbf{s}_1(v) \quad (7.39b)$$

The bi-cubic Coon's patch is expressed as in Eq. (7.34) with $\mathbf{r}_3(u, v)$ as the correction surface so that the boundary conditions are met. Now

$$\mathbf{r}(u, 0) = \varphi_0(0) \, \mathbf{b}_0(u) + \varphi_1(0) \, \mathbf{b}_1(u) + \varphi_2(0) \, \mathbf{t}_0(u) + \varphi_3(0) \, \mathbf{t}_1(u)$$

$$+ \varphi_0(u) \, \mathbf{a}_0(0) + \varphi_1(u) \, \mathbf{a}_1(0) + \varphi_2(u) \, \mathbf{s}_0(0) + \varphi_3(u) \, \mathbf{s}_1(0) - \mathbf{r}_3(u, 0) = \mathbf{b}_0(u)$$

$\Rightarrow \qquad \mathbf{r}_3(u, 0) = \varphi_0(u) \, \mathbf{P}_{00} + \varphi_1(u) \, \mathbf{P}_{10} + \varphi_2(u) \, \mathbf{s}_0(0) + \varphi_3(u) \, \mathbf{s}_1(0)$ \hfill (7.40a)

$\mathbf{r}(u, 1) = \varphi_0(1) \, \mathbf{b}_0(u) + \varphi_1(1) \, \mathbf{b}_1(u) + \varphi_2(1) \, \mathbf{t}_0(u) + \varphi_3(1) \, \mathbf{t}_1(u)$

$\qquad\qquad + \varphi_0(u) \, \mathbf{a}_0(1) + \varphi_1(u) \, \mathbf{a}_1(1) + \varphi_2(u) \, \mathbf{s}_0(1) + \varphi_3(u) \, \mathbf{s}_1(1) - \mathbf{r}_3(u, 1) = \mathbf{b}_1(u)$

$\Rightarrow \qquad \mathbf{r}_3(u, 1) = \varphi_0(u) \, \mathbf{P}_{01} + \varphi_1(u) \, \mathbf{P}_{11} + \varphi_2(u) \, \mathbf{s}_0(1) + \varphi_3(u) \, \mathbf{s}_1(1)$ \hfill (7.40b)

Now,

$$\frac{\partial}{\partial v} \, \mathbf{r}(u, v) = \frac{\partial}{\partial v} \, \mathbf{r}_1(u, v) + \frac{\partial}{\partial v} \, \mathbf{r}_2(u, v) - \frac{\partial}{\partial v} \, \mathbf{r}_3(u, v)$$

$$= \frac{\partial}{\partial v} \, \varphi_0(v) \, \mathbf{b}_0(u) + \frac{\partial}{\partial v} \, \varphi_1(v) \, \mathbf{b}_1(u) + \frac{\partial}{\partial v} \, \varphi_2(v) \, \mathbf{t}_0(u) + \frac{\partial}{\partial v} \, \varphi_3(v) \, \mathbf{t}_1(u)$$

$$+ \varphi_0(u) \, \frac{\partial}{\partial v} \, \mathbf{a}_0(v) + \varphi_1(u) \, \frac{\partial}{\partial v} \, \mathbf{a}_1(v) + \varphi_2(u) \, \frac{\partial}{\partial v} \, \mathbf{s}_0(v)$$

$$+ \varphi_3(u) \, \frac{\partial}{\partial v} \, \mathbf{s}_1(v) - \frac{\partial}{\partial v} \, \mathbf{r}_3(u, v)$$ \hfill (7.41)

The twist vectors $\boldsymbol{\chi}_{ij}$, initially introduced in section 7.1.1, are the mixed derivatives defined as

$$\boldsymbol{\chi}_{ij} = \frac{\partial^2}{\partial u \partial v} \, \mathbf{r}(u, v) \Big|_{u=i, v=j} = \frac{\partial}{\partial v} \, \mathbf{s}_i(v) \Big|_{v=j} = \frac{\partial}{\partial u} \, \mathbf{t}_j(u) \Big|_{u=i}, \; i = 0, 1; j = 0, 1 \qquad (7.42)$$

Thus, for $v = 0$, realizing from Figure 7.15 that $\dfrac{\partial}{\partial v} \, \mathbf{a}_0(0) = \mathbf{t}(0)$ and $\dfrac{\partial}{\partial v} \, \mathbf{a}_1(0) = \mathbf{t}_0(1)$, Eq. (7.41) becomes

$$\frac{\partial}{\partial v} \, \mathbf{r}(u, 0) = \mathbf{t}_0(u) + \varphi_0(u) \, \mathbf{t}_0(0) + \varphi_1(u) \, \mathbf{t}_0(1) + \varphi_2(u) \, \boldsymbol{\chi}_{00} + \varphi_3(u) \, \boldsymbol{\chi}_{10} - \frac{\partial}{\partial v} \, \mathbf{r}_3(u, 0)$$

$$= \mathbf{t}_0(u)$$

$\Rightarrow \qquad \dfrac{\partial}{\partial v} \, \mathbf{r}_3(u, 0) = \varphi_0(u) \, \mathbf{t}_0(0) + \varphi_1(u)\mathbf{t}_0(1) + \varphi_2(u)\boldsymbol{\chi}_{00} + \varphi_3(u) \, \boldsymbol{\chi}_{10}$ \hfill (7.43a)

Similarly, for $v = 1$, noting that $\dfrac{\partial}{\partial v} \, \mathbf{a}_0(1) = \mathbf{t}_1(0)$ and $\dfrac{\partial}{\partial v} \, \mathbf{a}_1(1) = \mathbf{t}_1(1)$,

$$\frac{\partial}{\partial v} \, \mathbf{r}(u, 1) = \mathbf{t}_1(u) + \varphi_0(u) \, \mathbf{t}_1(0) + \varphi_1(u) \, \mathbf{t}_1(1) + \varphi_2(u) \, \boldsymbol{\chi}_{01} + \varphi_3(u) \, \boldsymbol{\chi}_{11} - \frac{\partial}{\partial v} \, \mathbf{r}_3(u, 1)$$

$$= \mathbf{t}_1(u)$$

$\Rightarrow \qquad \dfrac{\partial}{\partial v} \, \mathbf{r}_3(u, 1) = \varphi_0(u) \, \mathbf{t}_1(0) + \varphi_1(u) \, \mathbf{t}_1(1) + \varphi_2(u)\boldsymbol{\chi}_{01} + \varphi_3(u) \, \boldsymbol{\chi}_{11}$ \hfill (7.43b)

From Eqs. (7.40) and (7.43), we can use bi-cubic lofting with respect to v to get the corrected surface, that is

$$\mathbf{r}_3(u, v) = \varphi_0(v) \, \mathbf{r}_3(u, 0) + \varphi_1(v) \, \mathbf{r}_3(u, 1) + \varphi_2(v) \, \frac{\partial}{\partial v} \, \mathbf{r}_3(u, 0) + \varphi_3(v) \, \frac{\partial}{\partial v} \, \mathbf{r}_3(u, 1)$$

or $\qquad \mathbf{r}_3(u, v) = \varphi_0(v) \, [\varphi_0(u) \, \mathbf{P}_{00} + \varphi_1(u) \mathbf{P}_{10} + \varphi_2(u) \, \mathbf{s}_0(0) + \varphi_3(u) \, \mathbf{s}_1(0)]$

$$+ \varphi_1(v) \, [\varphi_0(u) \, \mathbf{P}_{01} + \varphi_1(u) \, \mathbf{P}_{11} + \varphi_2(u) \, \mathbf{s}_0(1) + \varphi_3(u) \, \mathbf{s}_1(1)]$$

$$+ \varphi_2(v) \, [\varphi_0(u) \, \mathbf{t}_0(0) + \varphi_1(u) \, \mathbf{t}_0(1) + \varphi_2(u) \, \boldsymbol{\chi}_{00} + \varphi_3(u) \, \boldsymbol{\chi}_{10}]$$

$$+ \varphi_3(v) \, [\varphi_0(u) \, \mathbf{t}_1(0) + \varphi_1(u) \, \mathbf{t}_1(1) + \varphi_2(u) \, \boldsymbol{\chi}_{01} + \varphi_3(u) \, \boldsymbol{\chi}_{11}]$$

or in matrix form

$$
\mathbf{r}_3(u, v) = [\varphi_0(v) \ \varphi_1(v) \ \varphi_2(v) \ \varphi_3(v)]
\begin{bmatrix}
\mathbf{P}_{00} & \mathbf{P}_{10} & \mathbf{s}_0(0) & \mathbf{s}_1(0) \\
\mathbf{P}_{01} & \mathbf{P}_{11} & \mathbf{s}_0(1) & \mathbf{s}_1(1) \\
\mathbf{t}_0(0) & \mathbf{t}_0(1) & \chi_{00} & \chi_{10} \\
\mathbf{t}_1(0) & \mathbf{t}_1(1) & \chi_{01} & \chi_{11}
\end{bmatrix}
\begin{bmatrix}
\varphi_0(u) \\
\varphi_1(u) \\
\varphi_2(u) \\
\varphi_3(u)
\end{bmatrix}
\tag{7.44}
$$

The overall bi-cubic Coon's patch is given by

$$
\mathbf{r}(u, v) = [\varphi_0(v) \ \ \varphi_1(v) \ \ \varphi_2(v) \ \ \varphi_3(v)] \, [\mathbf{b}_0(u) \ \ \mathbf{b}_1(u) \ \ \mathbf{t}_0(u) \ \ \mathbf{t}_1(u)]^\mathrm{T}
$$

$$
+ [\mathbf{a}_0(v) \ \ \mathbf{a}_1(v) \ \ \mathbf{s}_0(v) \ \ \mathbf{s}_1(v)]
\begin{bmatrix}
\varphi_0(u) \\
\varphi_1(u) \\
\varphi_2(u) \\
\varphi_3(u)
\end{bmatrix}
$$

$$
- [\varphi_0(v) \ \ \varphi_1(v) \ \ \varphi_2(v) \ \ \varphi_3(v)]
\begin{bmatrix}
\mathbf{P}_{00} & \mathbf{P}_{10} & \mathbf{s}_0(0) & \mathbf{s}_1(0) \\
\mathbf{P}_{01} & \mathbf{P}_{11} & \mathbf{s}_0(1) & \mathbf{s}_1(1) \\
\mathbf{t}_0(0) & \mathbf{t}_0(1) & \chi_{00} & \chi_{10} \\
\mathbf{t}_1(0) & \mathbf{t}_1(1) & \chi_{01} & \chi_{11}
\end{bmatrix}
\begin{bmatrix}
\varphi_0(u) \\
\varphi_1(u) \\
\varphi_2(u) \\
\varphi_3(u)
\end{bmatrix}
\tag{7.45}
$$

To verify from above that the other boundary conditions are met, we see that

$$
\mathbf{r}(0, v) = [\varphi_0(v) \ \ \varphi_1(v) \ \ \varphi_2(v) \ \ \varphi_3(v)] \, [\mathbf{b}_0(0) \ \ \mathbf{b}_1(0) \ \ \mathbf{t}_0(0) \ \ \mathbf{t}_1(0)]^\mathrm{T}
$$

$$
+ \mathbf{a}_0(v) - [\varphi_0(v) \ \ \varphi_1(v) \ \ \varphi_2(v) \ \ \varphi_3(v)]
\begin{bmatrix}
\mathbf{P}_{00} \\
\mathbf{P}_{01} \\
\mathbf{t}_0(0) \\
\mathbf{t}_1(0)
\end{bmatrix}
= \mathbf{a}_0(v)
$$

$$
\mathbf{r}(1, v) = [\varphi_0(v) \ \ \varphi_1(v) \ \ \varphi_2(v) \ \ \varphi_3(v)] \, [\mathbf{b}_0(1) \ \ \mathbf{b}_1(1) \ \ \mathbf{t}_0(1) \ \ \mathbf{t}_1(1)]^\mathrm{T}
$$

$$
+ \mathbf{a}_1(v) - [\varphi_0(v) \ \ \varphi_1(v) \ \ \varphi_2(v) \ \ \varphi_3(v)]
\begin{bmatrix}
\mathbf{P}_{10} \\
\mathbf{P}_{11} \\
\mathbf{t}_0(1) \\
\mathbf{t}_1(1)
\end{bmatrix}
= \mathbf{a}_1(v)
$$

$$
\frac{\partial}{\partial u} \mathbf{r}(u, v) = [\varphi_0(v) \ \ \varphi_1(v) \ \ \varphi_2(v) \ \ \varphi_3(v)] \left[\frac{\partial}{\partial u} \mathbf{b}_0(u) \ \ \frac{\partial}{\partial u} \mathbf{b}_1(u) \ \ \frac{\partial}{\partial u} \mathbf{t}_0(u) \ \ \frac{\partial}{\partial u} \mathbf{t}_1(u) \right]^\mathrm{T}
$$

$$+[\mathbf{a}_0(v) \quad \mathbf{a}_1(v) \quad \mathbf{s}_0(v) \quad \mathbf{s}_1(v)] \begin{bmatrix} \dfrac{\partial}{\partial u}\varphi_0(u) \\[2mm] \dfrac{\partial}{\partial u}\varphi_1(u) \\[2mm] \dfrac{\partial}{\partial u}\varphi_2(u) \\[2mm] \dfrac{\partial}{\partial u}\varphi_3(u) \end{bmatrix}$$

$$-[\varphi_0(v) \quad \varphi_1(v) \quad \varphi_2(v) \quad \varphi_3(v)] \begin{bmatrix} \mathbf{P}_{00} & \mathbf{P}_{10} & \mathbf{s}_0(0) & \mathbf{s}_1(0) \\ \mathbf{P}_{01} & \mathbf{P}_{11} & \mathbf{s}_0(1) & \mathbf{s}_1(1) \\ \mathbf{t}_0(0) & \mathbf{t}_0(1) & \boldsymbol{\chi}_{00} & \boldsymbol{\chi}_{10} \\ \mathbf{t}_1(0) & \mathbf{t}_1(1) & \boldsymbol{\chi}_{01} & \boldsymbol{\chi}_{11} \end{bmatrix} \begin{bmatrix} \dfrac{\partial}{\partial u}\varphi_0(u) \\[2mm] \dfrac{\partial}{\partial u}\varphi_1(u) \\[2mm] \dfrac{\partial}{\partial u}\varphi_2(u) \\[2mm] \dfrac{\partial}{\partial u}\varphi_3(u) \end{bmatrix}$$

so that
$$\frac{\partial}{\partial u}\mathbf{r}(0, v) = [\varphi_0(v) \quad \varphi_1(v) \quad \varphi_2(v) \quad \varphi_3(v)] [\mathbf{s}_0(0) \quad \mathbf{s}_0(1) \quad \boldsymbol{\chi}_{00} \quad \boldsymbol{\chi}_{01}]^{\mathrm{T}} + \mathbf{s}_0(v)$$

$$-[\varphi_0(v) \quad \varphi_1(v) \quad \varphi_2(v) \quad \varphi_3(v)] \begin{bmatrix} \mathbf{s}_0(0) \\ \mathbf{s}_0(1) \\ \boldsymbol{\chi}_{00} \\ \boldsymbol{\chi}_{01} \end{bmatrix} = \mathbf{s}_0(v)$$

and
$$\frac{\partial}{\partial u}\mathbf{r}(1, v) = [\varphi_0(v) \quad \varphi_1(v) \quad \varphi_2(v) \quad \varphi_3(v)] [\mathbf{s}_1(0) \quad \mathbf{s}_1(1) \quad \boldsymbol{\chi}_{10} \quad \boldsymbol{\chi}_{11}]^{\mathrm{T}}$$

$$+\mathbf{s}_1(v) - [\varphi_0(v) \quad \varphi_1(v) \quad \varphi_2(v) \quad \varphi_3(v)] \begin{bmatrix} \mathbf{s}_0(0) \\ \mathbf{s}_0(1) \\ \boldsymbol{\chi}_{10} \\ \boldsymbol{\chi}_{11} \end{bmatrix} = \mathbf{s}_1(v)$$

Example 7.8. The boundary curves of a Coon's patch consist of four cubic Bézier curves with the following control points

$\mathbf{a}_0(v) \equiv \mathbf{r}_{00} = (0, 0, 0)$ $\quad \mathbf{r}_{01} = \left(\dfrac{1}{3\sqrt{2}}, \dfrac{1}{3\sqrt{2}}, \dfrac{1}{3}\right)$ $\quad \mathbf{r}_{02} = \left(\sqrt{2} - \dfrac{1}{3\sqrt{2}}, \dfrac{1}{3\sqrt{2}}, \dfrac{1}{3}\right)$ $\quad \mathbf{r}_{03} = (\sqrt{2}, 0, 0)$

$\mathbf{b}_1(u) \equiv \mathbf{r}_{10} = \mathbf{r}_{03}$ $\quad \mathbf{r}_{11} = \mathbf{r}_{02}$ $\quad \mathbf{r}_{12} = \left(\sqrt{2} - \dfrac{1}{3\sqrt{2}}, \sqrt{2} - \dfrac{1}{3\sqrt{2}}, \dfrac{1}{3}\right)$ $\quad \mathbf{r}_{13} = (\sqrt{2}, \sqrt{2}, 0)$

$\mathbf{a}_1(v) \equiv \mathbf{r}_{20} = \mathbf{r}_{13}$ $\quad \mathbf{r}_{21} = \mathbf{r}_{12}$ $\quad \mathbf{r}_{22} = \left(\dfrac{1}{3\sqrt{2}}, \sqrt{2} - \dfrac{1}{3\sqrt{2}}, \dfrac{1}{3}\right)$ $\quad \mathbf{r}_{23} = (0, \sqrt{2}, 0)$

$\mathbf{b}_0(u) \equiv \mathbf{r}_{30} = \mathbf{r}_{23}$ $\quad \mathbf{r}_{31} = \mathbf{r}_{22}$ $\quad \mathbf{r}_{32} = \mathbf{r}_{01}$ $\quad \mathbf{r}_{33} = \mathbf{r}_{00}$

To construct a bi-linear Coon's patch, the four boundary curves are given by

$$
\mathbf{a}_0(v) = [v^3 \quad v^2 \quad v \quad 1]\begin{bmatrix} -1 & 3 & -3 & 1 \\ 3 & -6 & 3 & 0 \\ -3 & 3 & 0 & 0 \\ 1 & 0 & 0 & 0 \end{bmatrix}\begin{bmatrix} \mathbf{r}_{00} \\ \mathbf{r}_{01} \\ \mathbf{r}_{02} \\ \mathbf{r}_{03} \end{bmatrix}, \mathbf{a}_1(v) = [v^3 \quad v^2 \quad v \quad 1]\begin{bmatrix} -1 & 3 & -3 & 1 \\ 3 & -6 & 3 & 0 \\ -3 & 3 & 0 & 0 \\ 1 & 0 & 0 & 0 \end{bmatrix}\begin{bmatrix} \mathbf{r}_{20} \\ \mathbf{r}_{21} \\ \mathbf{r}_{22} \\ \mathbf{r}_{23} \end{bmatrix}
$$

$$
\mathbf{b}_0(u) = [u^3 \quad u^2 \quad u \quad 1]\begin{bmatrix} -1 & 3 & -3 & 1 \\ 3 & -6 & 3 & 0 \\ -3 & 3 & 0 & 0 \\ 1 & 0 & 0 & 0 \end{bmatrix}\begin{bmatrix} \mathbf{r}_{30} \\ \mathbf{r}_{31} \\ \mathbf{r}_{32} \\ \mathbf{r}_{33} \end{bmatrix}, \mathbf{b}_1(u) = [u^3 \quad u^2 \quad u \quad 1]\begin{bmatrix} -1 & 3 & -3 & 1 \\ 3 & -6 & 3 & 0 \\ -3 & 3 & 0 & 0 \\ 1 & 0 & 0 & 0 \end{bmatrix}\begin{bmatrix} \mathbf{r}_{10} \\ \mathbf{r}_{11} \\ \mathbf{r}_{12} \\ \mathbf{r}_{13} \end{bmatrix}
$$

which can be used directly with Eq. (7.38). Stepwise results are shown in Figure 7.16.

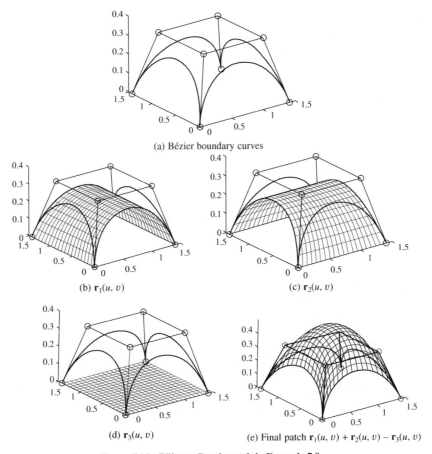

(a) Bézier boundary curves

(b) $\mathbf{r}_1(u, v)$

(c) $\mathbf{r}_2(u, v)$

(d) $\mathbf{r}_3(u, v)$

(e) Final patch $\mathbf{r}_1(u, v) + \mathbf{r}_2(u, v) - \mathbf{r}_3(u, v)$

Figure 7.16 Bilinear Coon's patch in Example 7.8

7.3   Composite Surfaces

Surface patches, in small units, need to be joined (stitched) together to form a larger surface. We can observe this in surfaces such as car roof-tops, doors, side panals, engine-hood and also in aircraft fuselage and wing-panels. In general, common boundary curves should match exactly (without any gap) and the joint should not leave any wrinkles.

Similar treatment is performed when attempting to stitch two patches together at their common boundaries as is the case with composite curves. Care is taken to maintain position (C^0), slope (C^1) and/or curvature (C^2) continuity at the boundary curves. Position continuity is obtained only when the boundary curves of two adjoining patches coincide in which case, the slope along the boundary curves is also continuous. A step further is to ensure a unique normal at any point on the common boundary. This is accomplished by coinciding the tangent planes of the two adjacent patches at that point. This section, discusses composite surfaces with Ferguson, Bézier and Coon's patches.

7.3.1   Composite Ferguson's Surface

An advantage with Ferguson's bi-cubic patch is that at least the position (C^0) continuity is ensured across patch boundaries because the corner points and slopes (and thus the boundary curves) are the same for two adjacent patches. Consider the common boundary for patches I and II, for instance, in Figure 7.17 which is a cubic curve in parameter v ($u = 1$ for patch I and $u = 0$ for patch II). It is apparent that the slope $\mathbf{r}_v$ is continuous along this common boundary. For patches I and III, the same can be stated about the continuity of the slope $\mathbf{r}_u$ along their common boundary. In addition to position continuity, therefore, the slopes along the patch boundaries are also continuous for Ferguson's patches.

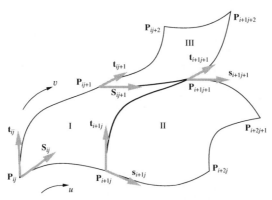

Figure 7.17   Position and slope continuity across Ferguson's patch boundaries

Note, however, that Eqs. (7.12) and (7.13) seem demanding from the user's viewpoint as they require higher order input (slopes and twist vectors) as a part of geometric information to be specified. One way to avoid is: Given a set of data points $\mathbf{P}_{ij}$, $i = 0, \ldots, m$ and $j = 0, \ldots, n$ over which it is required to fit a composite Ferguson surface (Figure 7.15), intermediate slopes $\mathbf{s}_{ij}$ (along u) and $\mathbf{t}_{ij}$ (along v) can be estimated as

$$\mathbf{s}_{ij} = C_i \frac{\mathbf{P}_{i+1j} - \mathbf{P}_{i-1j}}{|\mathbf{P}_{i+1j} - \mathbf{P}_{i-1j}|}, \text{ where } C_i = \min \left(|\mathbf{P}_{ij} - \mathbf{P}_{i-1j}|, |\mathbf{P}_{i+1j} - \mathbf{P}_{ij}| \right)$$

$$\mathbf{t}_{ij} = D_i \frac{\mathbf{P}_{ij+1} - \mathbf{P}_{ij-1}}{|\,\mathbf{P}_{ij+1} - \mathbf{P}_{ij-1}\,|}, \text{ where } D_i = \min\,(|\,\mathbf{P}_{ij} - \mathbf{P}_{ij-1}\,|, |\,\mathbf{P}_{ij+1} - \mathbf{P}_{ij}\,|) \tag{7.46}$$

The twist vectors can be assumed to be zero. The geometric matrix for this Ferguson's patch would then be

$$\mathbf{G} = \begin{bmatrix} \mathbf{P}_{ij} & \mathbf{P}_{ij+1} & | & \mathbf{t}_{ij} & \mathbf{t}_{ij+1} \\ \mathbf{P}_{i+1\,j} & \mathbf{P}_{i+1\,j+1} & | & \mathbf{t}_{i+1\,j} & \mathbf{t}_{i+1\,j+1} \\ - & - & | & - & - \\ \mathbf{s}_{ij} & \mathbf{s}_{ij+1} & | & 0 & 0 \\ \mathbf{s}_{i+1\,j} & \mathbf{s}_{i+1\,j+1} & | & 0 & 0 \end{bmatrix} \tag{7.47}$$

Note that for $i = 0$ or $i = m$, $\mathbf{P}_{-1j}$ and $\mathbf{P}_{m+1j}$, respectively, are not known and so the user will have to specify $\mathbf{s}_{0j}$ and $\mathbf{s}_{mj}$ for all $j = 0, \ldots, n$. Similarly, slopes $\mathbf{t}_{i0}$ and $\mathbf{t}_{in}$ for $i = 0, \ldots, m$ will also need to be specified. In other words, slopes along u and v are to be specified on the boundaries of the composite surface. The so-called FMILL method to generate a composite surface using Ferguson's patches described above works well for evenly spaced data points. However, local flatness or bulging for unevenly spaced data points producing unnatural surface normals is often seen. This is primarily due to the assumption of *zero* twist vectors.

Example 7.9. For given control points

$$\mathbf{P}_{00} = \{0, 0, 0\}, \mathbf{P}_{10} = \{1, 0, 0\}, \mathbf{P}_{20} = \{2, 0, 0\}$$

$$\mathbf{P}_{01} = \{0, 1, 0\}, \mathbf{P}_{11} = \{1, 1, 0\}, \mathbf{P}_{21} = \{2, 1, 0\}$$

$$\mathbf{P}_{02} = \{0, 1, 2\}, \mathbf{P}_{12} = \{1, 1, 2\}, \mathbf{P}_{22} = \{2, 1, 4\}$$

determine Ferguson's patches using the FMILL method to get the composite surface. Take the end slopes $\mathbf{s}_{0j} = \mathbf{s}_{mj} = \mathbf{t}_{i0} = \mathbf{t}_{in} = \{0, 0, 0\}$ for all $i = 0, \ldots, 2; j = 0, \ldots, 2$.

The intermediate slopes $\mathbf{s}_{ij}$ can be computed using Eq. (7.46) as

$$\mathbf{s}_{10} = [\min\,(|\,\mathbf{P}_{10} - \mathbf{P}_{00}\,|, |\,\mathbf{P}_{20} - \mathbf{P}_{10}\,|)]\,\frac{\mathbf{P}_{20} - \mathbf{P}_{00}}{|\,\mathbf{P}_{20} - \mathbf{P}_{00}\,|}$$

$$= [\min\,(|\,(1, 0, 0) - (0, 0, 0)\,|, |\,(2, 0, 0) - (1, 0, 0)\,|)\,\frac{(2, 0, 0) - (0, 0, 0)}{|\,(2, 0, 0) - (0, 0, 0)\,|} = (1, 0, 0)$$

$$\mathbf{s}_{11} = [\min\,(|\,\mathbf{P}_{11} - \mathbf{P}_{01}\,|, |\,\mathbf{P}_{21} - \mathbf{P}_{11}\,|)]\,\frac{\mathbf{P}_{21} - \mathbf{P}_{01}}{|\,\mathbf{P}_{21} - \mathbf{P}_{01}\,|}$$

$$= [\min\,(|\,(1, 1, 0) - (0, 1, 0)\,|, |\,(2, 1, 0) - (1, 1, 0)\,|)\,\frac{(2, 1, 0) - (0, 1, 0)}{|\,(2, 1, 0) - (0, 1, 0)\,|} = (1, 0, 0)$$

$$\mathbf{s}_{12} = [\min\,(|\,\mathbf{P}_{12} - \mathbf{P}_{02}\,|, |\,\mathbf{P}_{22} - \mathbf{P}_{12}\,|)]\,\frac{\mathbf{P}_{22} - \mathbf{P}_{02}}{|\,\mathbf{P}_{22} - \mathbf{P}_{02}\,|}$$

$$= [\min\,(|\,(1, 1, 2) - (0, 1, 2)\,|, |\,(2, 1, 4) - (1, 1, 2)\,|)\,\frac{(2, 1, 4) - (0, 1, 2)}{|\,(2, 1, 4) - (0, 1, 2)\,|} = \left(\frac{1}{\sqrt{2}}, 0, \frac{1}{\sqrt{2}}\right)$$

For slopes $\mathbf{t}_{ij}$

$$\mathbf{t}_{01} = [\min (|\,\mathbf{P}_{01} - \mathbf{P}_{00}\,|,\,|\,\mathbf{P}_{02} - \mathbf{P}_{01}\,|)]\,\frac{\mathbf{P}_{02} - \mathbf{P}_{00}}{|\,\mathbf{P}_{02} - \mathbf{P}_{00}\,|}$$

$$= [\min (|\,(0, 1, 0) - (0, 0, 0)\,|,\,|\,(0, 1, 2) - (0, 1, 0)\,|)]\,\frac{(0, 1, 2) - (0, 0, 0)}{|\,(0, 1, 2) - (0, 0, 0)\,|} = \left(0, \frac{1}{\sqrt{5}}, \frac{2}{\sqrt{5}}\right)$$

$$\mathbf{t}_{11} = [\min (|\,\mathbf{P}_{11} - \mathbf{P}_{10}\,|,\,|\,\mathbf{P}_{12} - \mathbf{P}_{11}\,|)]\,\frac{\mathbf{P}_{12} - \mathbf{P}_{10}}{|\,\mathbf{P}_{12} - \mathbf{P}_{10}\,|}$$

$$= [\min (|\,(1, 1, 0) - (1, 0, 0)\,|,\,|\,(1, 1, 2) - (1, 1, 0)\,|)]\,\frac{(1, 1, 2) - (1, 0, 0)}{|\,(1, 1, 2) - (1, 0, 0)\,|} = \left(0, \frac{1}{\sqrt{5}}, \frac{2}{\sqrt{5}}\right)$$

$$\mathbf{t}_{21} = [\min (|\,\mathbf{P}_{21} - \mathbf{P}_{20}\,|,\,|\,\mathbf{P}_{22} - \mathbf{P}_{21}\,|)]\,\frac{\mathbf{P}_{22} - \mathbf{P}_{20}}{|\,\mathbf{P}_{22} - \mathbf{P}_{20}\,|}$$

$$= [\min (|\,(2, 1, 0) - (2, 0, 0)\,|,\,|\,(2, 1, 4) - (2, 1, 0)\,|)]\,\frac{(2, 1, 4) - (2, 0, 0)}{|\,(2, 1, 4) - (2, 0, 0)\,|} = \left(0, \frac{1}{\sqrt{17}}, \frac{4}{\sqrt{17}}\right)$$

Repeated application of Eq. (7.12) with the geometric matrix in Eq. (7.47) results in the following composite surface with four patches shown in Figure 7.18.

To avoid local flatness or bulging, we can compute the twist vectors from the data given instead of specifying them as zero. Computations are done by imposing the C^2 continuity condition at patch boundaries. For patch I in Figure 7.17, from Eq. (7.12), we have

$$\mathbf{r}^{\mathrm{I}}(u, v) = \mathbf{U}\mathbf{M}\mathbf{G}^{\mathrm{I}}\mathbf{M}^{\mathrm{T}}\mathbf{V}^{\mathrm{T}} \qquad (7.48\mathrm{a})$$

with $\mathbf{G}^{\mathrm{I}}$ defined as

$$\mathbf{G}^{\mathrm{I}} = \begin{bmatrix} \mathbf{P}_{ij} & \mathbf{P}_{ij+1} & | & \mathbf{t}_{ij} & \mathbf{t}_{ij+1} \\ \mathbf{P}_{i+1j} & \mathbf{P}_{i+1j+1} & | & \mathbf{t}_{i+1j} & \mathbf{t}_{i+1j+1} \\ - & - & | & - & - \\ \mathbf{s}_{ij} & \mathbf{s}_{ij+1} & | & \boldsymbol{\chi}_{ij} & \boldsymbol{\chi}_{ij+1} \\ \mathbf{s}_{i+1j} & \mathbf{s}_{i+1j+1} & | & \boldsymbol{\chi}_{i+1j} & \boldsymbol{\chi}_{i+1j+1} \end{bmatrix} \qquad (7.48\mathrm{b})$$

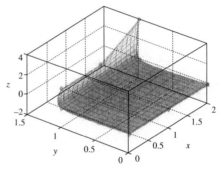

Figure 7.18   A composite Ferguson patch using the FMILL method

The unknown slopes and twist vectors can be computed as follows:
For C^2 continuity along the common boundary between patches I and II

$$\frac{\partial^2}{\partial u^2}\,\mathbf{r}^{\mathrm{I}}(1, v) = \frac{\partial^2}{\partial u^2}\,\mathbf{r}^{\mathrm{II}}(0, v)$$

$$\Rightarrow\quad [6\ 2\ 0\ 0\,]\,\mathbf{M}\mathbf{G}^{\mathrm{I}}\mathbf{M}^{\mathrm{T}}\mathbf{V}^{\mathrm{T}} = [0\ 2\ 0\ 0]\,\mathbf{M}\mathbf{G}^{\mathrm{II}}\mathbf{M}^{\mathrm{T}}\mathbf{V}^{\mathrm{T}}$$

or $\qquad [6\ {-}6\ 2\ 4]\,\mathbf{G}^{\mathrm{I}}\mathbf{M}^{\mathrm{T}}\mathbf{V}^{\mathrm{T}} = [{-}6\ 6\ {-}4\ {-}2]\,\mathbf{G}^{\mathrm{II}}\mathbf{M}^{\mathrm{T}}\mathbf{V}^{\mathrm{T}}$

or $\qquad [6\ {-}6\ 2\ 4]\,\mathbf{G}^{\mathrm{I}} = [{-}6\ 6\ {-}4\ {-}2]\,\mathbf{G}^{\mathrm{II}}$

Solving yields, four relations which can be summarized to the following two

$$\text{(A)} \quad \mathbf{s}_{ij} + 4\mathbf{s}_{i+1j} + \mathbf{s}_{i+2j} = 3(\mathbf{P}_{i+2j} - \mathbf{P}_{ij})$$

$$\text{(B)} \quad \boldsymbol{\chi}_{ij} + 4\boldsymbol{\chi}_{i+1j} + \boldsymbol{\chi}_{i+2j} = 3(\mathbf{t}_{i+2j} - \mathbf{t}_{ij}) \quad i = 0, \ldots, m-2 \text{ for fixed } j \qquad (7.49a)$$

Similarly, for the boundary between patches I and III

$$\frac{\partial^2}{\partial v^2} \mathbf{r}^{\mathrm{I}}(u, 1) = \frac{\partial^2}{\partial v^2} \mathbf{r}^{\mathrm{III}}(u, 0)$$

$$\mathbf{UMG}^{\mathrm{I}} \mathbf{M}^{\mathrm{T}} [6 \quad 2 \quad 0 \quad 0]^{\mathrm{T}} = \mathbf{UMG}^{\mathrm{III}} \mathbf{M}^{\mathrm{T}} [0 \quad 2 \quad 0 \quad 0]^{\mathrm{T}}$$

or
$$\mathbf{UMG}^{\mathrm{I}} [6 \quad -6 \quad 2 \quad 4]^{\mathrm{T}} = \mathbf{UMG}^{\mathrm{III}} [-6 \quad 6 \quad -4 \quad -2]^{\mathrm{T}}$$

or
$$\mathbf{G}^{\mathrm{I}} [6 \quad -6 \quad 2 \quad 4]^{\mathrm{T}} = \mathbf{G}^{\mathrm{III}} [-6 \quad 6 \quad -4 \quad -2]^{\mathrm{T}}$$

which yields the two relations as

$$\text{(C)} \quad \mathbf{t}_{ij} + 4\mathbf{t}_{ij+1} + \mathbf{t}_{ij+2} = 3(\mathbf{P}_{ij+2} - \mathbf{P}_{ij})$$

$$\text{(D)} \quad \boldsymbol{\chi}_{ij} + 4\boldsymbol{\chi}_{ij+1} + \boldsymbol{\chi}_{ij+2} = 3(\mathbf{s}_{ij+2} - \mathbf{s}_{ij}) \; j = 0, \ldots, n-2 \text{ for fixed } i \qquad (7.49b)$$

Thus, for given information

Data points: $\qquad \mathbf{P}_{ij}, \; i = 0, \ldots, m \text{ and } j = 0, \ldots, n$

Boundary slopes: $\qquad \mathbf{s}_{0j}, \; \mathbf{s}_{mj} \text{ for all } j = 0, \ldots, n$

$\qquad\qquad\qquad\quad \mathbf{t}_{i0}, \; \mathbf{t}_{in} \text{ for all } i = 0, \ldots, m$

Twist vectors: $\qquad \boldsymbol{\chi}_{0j}, \; \boldsymbol{\chi}_{mj} \text{ for all } j = 0, \ldots, n$

$\qquad\qquad\qquad\quad \boldsymbol{\chi}_{i0}, \; \boldsymbol{\chi}_{in} \text{ for all } i = 0, \ldots, m$

We need to solve for

$\qquad\qquad \mathbf{s}_{ij} \text{ with (A)}$

$\qquad\qquad \mathbf{t}_{ij} \text{ with (C)}$

$\qquad\qquad \boldsymbol{\chi}_{ij} \text{ using (B) and (D) for } i = 1, \ldots, m-1, j = 1, \ldots, n-1$

Eqs. (7.49) are all tri-diagonal and can be solved efficiently with algorithms available to get the Ferguson's geometric matrix for each patch. We can realize that the higher order slopes and twist vectors are still needed to be specified which is a drawback with Ferguson's composite patches.

7.3.2 Composite Bézier Surface

Both bi-cubic Ferguson and Bézier patches being tensor products, their equivalence is stated by the relation

$$\mathbf{r}(u, v) = \mathbf{UM}_{\mathrm{F}} \mathbf{G}_{\mathrm{F}} \mathbf{M}_{\mathrm{F}}^{\mathrm{T}} \mathbf{V}^{\mathrm{T}} = \mathbf{UM}_{\mathrm{B}} \mathbf{G}_{\mathrm{B}} \mathbf{M}_{\mathrm{B}}^{\mathrm{T}} \mathbf{V}^{\mathrm{T}}$$

or
$$\mathbf{M}_{\mathrm{F}} \mathbf{G}_{\mathrm{F}} \mathbf{M}_{\mathrm{F}}^{\mathrm{T}} = \mathbf{M}_{\mathrm{B}} \mathbf{G}_{\mathrm{B}} \mathbf{M}_{\mathrm{B}}^{\mathrm{T}}$$

or
$$\mathbf{G}_{\mathrm{F}} = (\mathbf{M}_{\mathrm{F}}^{-1} \mathbf{M}_{\mathrm{B}}) \, \mathbf{G}_{\mathrm{B}} (\mathbf{M}_{\mathrm{F}}^{-1} \mathbf{M}_{\mathrm{B}})^{\mathrm{T}} \qquad (7.50)$$

where the subscript F refers to the Ferguson's patch and B relates to the Bézier's patch. Using Eq. (7.6) for $\mathbf{M}_{\mathrm{F}}$ and Eq. (7.28) for $\mathbf{M}_{\mathrm{B}}$ and $\mathbf{G}_{\mathrm{B}}$, we realize that

$$\mathbf{G}_F = \begin{bmatrix} \mathbf{r}_{00} & \mathbf{r}_{03} & 3(\mathbf{r}_{01} - \mathbf{r}_{00}) & 3(\mathbf{r}_{03} - \mathbf{r}_{02}) \\ \mathbf{r}_{30} & \mathbf{r}_{33} & 3(\mathbf{r}_{31} - \mathbf{r}_{30}) & 3(\mathbf{r}_{33} - \mathbf{r}_{32}) \\ 3(\mathbf{r}_{10} - \mathbf{r}_{00}) & 3(\mathbf{r}_{13} - \mathbf{r}_{03}) & 9(\mathbf{r}_{00} - \mathbf{r}_{10} - \mathbf{r}_{01} + \mathbf{r}_{11}) & 9(\mathbf{r}_{02} - \mathbf{r}_{12} - \mathbf{r}_{03} + \mathbf{r}_{13}) \\ 3(\mathbf{r}_{30} - \mathbf{r}_{20}) & 3(\mathbf{r}_{33} - \mathbf{r}_{23}) & 9(\mathbf{r}_{20} - \mathbf{r}_{30} - \mathbf{r}_{21} + \mathbf{r}_{31}) & 9(\mathbf{r}_{22} - \mathbf{r}_{32} - \mathbf{r}_{23} + \mathbf{r}_{33}) \end{bmatrix} \qquad (7.51)$$

which implies that the gradients and twist vectors at patch corners can be expressed in terms of the characteristic Bézier polyhedron and thus the user may not need to specify higher order information. Instead, we would be more comfortable maneuvering data points to implicitly control the slope and mixed derivatives as opposed to specifying them. Figure 7.19 shows two adjacent bi-cubic Bézier patches with corner points of their control polyhedra that lie on the respective patches. From Eq. (7.28), the patches can be formulated as

$$\mathbf{r}^{\mathrm{I}}(u, v) = \mathbf{U}\mathbf{M}_{\mathrm{B}}\,\mathbf{G}_{\mathrm{B}}^{\mathrm{I}}\,\mathbf{M}_{\mathrm{B}}^{\mathrm{T}}\,\mathbf{V}^{\mathrm{T}}$$

$$\mathbf{r}^{\mathrm{II}}(u, v) = \mathbf{U}\mathbf{M}_{\mathrm{B}}\,\mathbf{G}_{\mathrm{B}}^{\mathrm{II}}\,\mathbf{M}_{\mathrm{B}}^{\mathrm{T}}\,\mathbf{V}^{\mathrm{T}} \qquad (7.52)$$

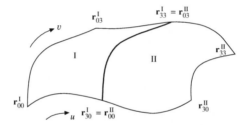

Figure 7.19   Adjacent Bézier patch boundaries

For positional continuity across the common boundary, it is required that $\mathbf{r}^{\mathrm{I}}(1, v) = \mathbf{r}^{\mathrm{II}}(0, v)$ for all values of v, that is $[1\ 1\ 1\ 1]\,\mathbf{M}_{\mathrm{B}}\,\mathbf{G}_{\mathrm{B}}^{\mathrm{I}} = [0\ 0\ 0\ 1]\mathbf{M}_{\mathrm{B}}\,\mathbf{G}_{\mathrm{B}}^{\mathrm{II}}$ or

$$[0\ 0\ 0\ 1]\begin{bmatrix} \mathbf{r}_{00} & \mathbf{r}_{01} & \mathbf{r}_{02} & \mathbf{r}_{03} \\ \mathbf{r}_{10} & \mathbf{r}_{11} & \mathbf{r}_{12} & \mathbf{r}_{13} \\ \mathbf{r}_{20} & \mathbf{r}_{21} & \mathbf{r}_{22} & \mathbf{r}_{23} \\ \mathbf{r}_{30} & \mathbf{r}_{31} & \mathbf{r}_{32} & \mathbf{r}_{33} \end{bmatrix}^{\mathrm{I}} = \begin{bmatrix} \mathbf{r}_{30} \\ \mathbf{r}_{31} \\ \mathbf{r}_{32} \\ \mathbf{r}_{33} \end{bmatrix}^{\mathrm{I}} = [1\ 0\ 0\ 0]\begin{bmatrix} \mathbf{r}_{00} & \mathbf{r}_{01} & \mathbf{r}_{02} & \mathbf{r}_{03} \\ \mathbf{r}_{10} & \mathbf{r}_{11} & \mathbf{r}_{12} & \mathbf{r}_{13} \\ \mathbf{r}_{20} & \mathbf{r}_{21} & \mathbf{r}_{22} & \mathbf{r}_{23} \\ \mathbf{r}_{30} & \mathbf{r}_{31} & \mathbf{r}_{32} & \mathbf{r}_{33} \end{bmatrix}^{\mathrm{II}} = \begin{bmatrix} \mathbf{r}_{00} \\ \mathbf{r}_{01} \\ \mathbf{r}_{02} \\ \mathbf{r}_{03} \end{bmatrix}^{\mathrm{II}}$$

implying that $\mathbf{r}^{\mathrm{I}}{}_{3j} = \mathbf{r}^{\mathrm{II}}{}_{0j}$, $j = 0, \ldots, 3$, or in other words, the boundary polygon must be common between the two patches.

Example 7.10. For blending two quadratic Bézier surfaces, a quadratic Bézier surface S_p has the following control points:

$$\begin{bmatrix} \mathbf{p}_{00} = \{0, 0, 0\} & \mathbf{p}_{01} = \{0, 1, 2\} & \mathbf{p}_{02} = \{0, 2, 2\} \\ \mathbf{p}_{10} = \{1, 0, 1\} & \mathbf{p}_{11} = \{1, 1, 3\} & \mathbf{p}_{12} = \{1, 3, 3\} \\ \mathbf{p}_{20} = \{2, 0, 0\} & \mathbf{p}_{21} = \{2, 1, 2\} & \mathbf{p}_{22} = \{2, 2, 2\} \end{bmatrix}$$

Another quadratic Bézier surface S_q has the control points

$$\begin{bmatrix} \mathbf{q}_{00} = \{0,2,2\} & \mathbf{q}_{01} = \{0,3,2\} & \mathbf{q}_{02} = \{0,4,4\} \\ \mathbf{q}_{10} = \{1,3,3\} & \mathbf{q}_{11} = \{1,4,3\} & \mathbf{q}_{12} = \{1,5,5\} \\ \mathbf{q}_{20} = \{2,2,2\} & \mathbf{q}_{21} = \{2,3,2\} & \mathbf{q}_{22} = \{2,4,4\} \end{bmatrix}$$

The individual surfaces are shown separately in Figure 7.20 (a) and (b), respectively. The composite surface joined at the common quadratic curve $\{\mathbf{p}_{02}, \mathbf{p}_{12}, \mathbf{p}_{22}\} = \{\mathbf{q}_{00}, \mathbf{q}_{10}, \mathbf{q}_{20}\}$ is shown in Figure 7.21.

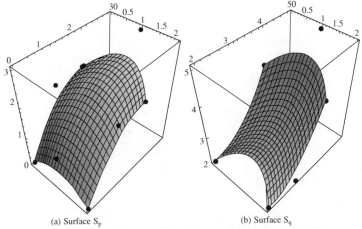

(a) Surface S_p (b) Surface S_q

Figure 7.20   Two bi-quadratic Bézier surfaces in Example 7.10

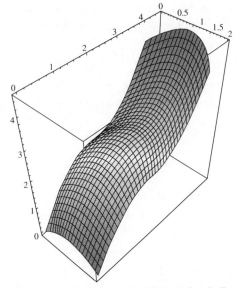

Figure 7.21   Composite bi-quadratic Bézier surface in Example 7.10

Further, for S_p remaining unaltered while S_q changed to S_{q1} having different control points, say

$$\begin{bmatrix} \mathbf{q}_{00} = \{0,2,2\} & \mathbf{q}_{01} = \{0,3,2\} & \mathbf{q}_{02} = \{0,4,1\} \\ \mathbf{q}_{10} = \{1,3,3\} & \mathbf{q}_{11} = \{1,4,2\} & \mathbf{q}_{12} = \{1,5,2\} \\ \mathbf{q}_{20} = \{2,2,2\} & \mathbf{q}_{21} = \{2,3,2\} & \mathbf{q}_{22} = \{2,4,1\} \end{bmatrix}$$

the surface S_{q1} and the resulting composite surface is shown in Figure 7.22.

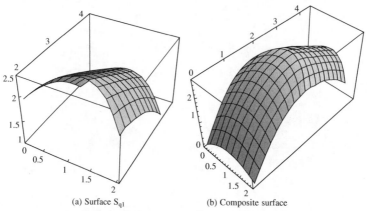

(a) Surface S_{q1} (b) Composite surface

Figure 7.22   **(a) Surface S_{q1} and (b) the composite surface with patches S_p and S_{q1}**

In Figure 7.23, for gradient continuity, the tangent plane of patch I at $u = 1$ must coincide with that of patch II at $u = 0$ for $v \in [0, 1]$. This implies that the direction of surface normal at the boundary must be unique. Or,

$$\frac{\partial}{\partial u} \mathbf{r}^{\mathrm{II}}(0, v) \times \frac{\partial}{\partial v} \mathbf{r}^{\mathrm{II}}(0, v) = \lambda(v) \frac{\partial}{\partial u} \mathbf{r}^{\mathrm{I}}(1, v) \times \frac{\partial}{\partial v} \mathbf{r}^{\mathrm{I}}(1, v) \tag{7.53}$$

where $\lambda(v)$ is a scalar function that takes into account the discontinuities in magnitudes of the surface normals. Since the positional continuity ensures that $\frac{\partial}{\partial v} \mathbf{r}^{\mathrm{II}}(0, v) = \frac{\partial}{\partial v} \mathbf{r}^{\mathrm{I}}(0, v)$, a solution for Eq. (7.53) is to have:

Case I

$$\frac{\partial}{\partial u} \mathbf{r}^{\mathrm{II}}(0, v) = \lambda(v) \frac{\partial}{\partial u} \mathbf{r}^{\mathrm{I}}(1, v) \tag{7.54}$$

$$\Rightarrow \quad [0 \quad 0 \quad 1 \quad 0] \, \mathbf{M}_B \, \mathbf{G}_B^{\mathrm{II}} \, \mathbf{M}_B^{\mathrm{T}} \mathbf{V}^{\mathrm{T}} = \lambda(v)[3 \quad 2 \quad 1 \quad 0] \, \mathbf{M}_B \, \mathbf{G}_B^{\mathrm{I}} \, \mathbf{M}_B^{\mathrm{T}} \mathbf{V}^{\mathrm{T}}$$

Since the left hand side is cubic in v, the right hand side should be such that $\lambda(v) = \lambda$, a constant to match the degree in v. Further, equating coefficients of $\mathbf{V}$ and post-multiplying with $\mathbf{M}_B^{-\mathrm{T}}$ results in

$$\mathbf{r}_{1i}^{\mathrm{II}} - \mathbf{r}_{0i}^{\mathrm{II}} = \lambda(\mathbf{r}_{3i}^{\mathrm{I}} - \mathbf{r}_{2i}^{\mathrm{I}}), i = 0, 1, 2, 3 \tag{7.55}$$

which means that the four pairs of polyhedron edges meeting at the boundary must be collinear as shown in Figure 7.23 (thick lines).

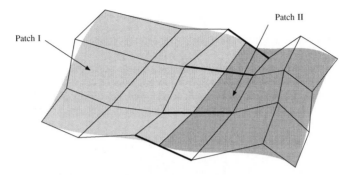

Figure 7.23   Arrangement of polyhedral edges for gradient continuity (Case I)

Once the 16 data points for a bi-cubic Bézier patch are chosen, in choosing the data points for adjacent patches, restrictions are strict. The 16 data points for the first patch, say A (Figure 7.24) can be chosen freely. For an adjacent patch B, four data points get constrained from the positional continuity requirement. For $\lambda = \lambda_1$ in Eq. (7.55), four out of the remaining 12 are further constrained to maintain the tangent plane continuity at the common boundary. Thus, only 8 points for patch B can be freely chosen. For patch C also adjacent to patch A, similar is the case in that 8 points can be freely chosen. For patch D adjacent to both B and C, only 4 of the 16 data points can be freely chosen.

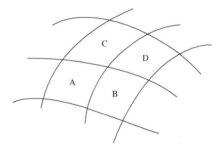

Figure 7.24   Four adjacent Bézier patches

An alternative gradient continuity condition, and a solution to Eq. (7.53) can be:

Case II

$$\frac{\partial}{\partial u} \mathbf{r}^{\mathrm{II}}(0, v) = \lambda(v) \frac{\partial}{\partial u} \mathbf{r}^{\mathrm{I}}(1, v) + \mu(v) \frac{\partial}{\partial v} \mathbf{r}^{\mathrm{I}}(1, v) \tag{7.56}$$

where $\mu(v)$ is another scalar function of v. Note that Eq. (7.56) satisfies the requirement in Eq. (7.53) and is a more general solution than that in Eq. (7.54). Eq. (7.56) suggests that $\frac{\partial}{\partial u} \mathbf{r}^{\mathrm{II}}(0, v)$ or $\mathbf{r}_u^{\mathrm{II}}(0, v)$ lies in the same plane as $\mathbf{r}_u^{\mathrm{I}}(1, v)$ and $\mathbf{r}_v^{\mathrm{I}}(1, v)$, i.e., the tangent plane of patch I at the boundary point concerned.

In matrix form, Eq. (7.56) can be written as

$$[0\ \ 0\ \ 1\ \ 0]\,\mathbf{M}_B \mathbf{G}_B^{\mathrm{II}} \mathbf{M}_B^{\mathrm{T}} \mathbf{V}^{\mathrm{T}} = \lambda(v)\,[3\ \ 2\ \ 1\ \ 0]\,\mathbf{M}_B \mathbf{G}_B^{\mathrm{I}} \mathbf{M}_B^{\mathrm{T}} \mathbf{V}^{\mathrm{T}} + \mu(v)\,[1\ \ 1\ \ 1\ \ 1]\,\mathbf{M}_B \mathbf{G}_B^{\mathrm{I}} \mathbf{M}_B^{\mathrm{T}} \begin{bmatrix} 3v^2 \\ 2v \\ 1 \\ 0 \end{bmatrix}$$

$$\tag{7.57}$$

To match the degree in v, $\lambda(v) = \lambda$, a constant, while $\mu(v) = \mu_0 + \mu_1 v$, a linear function in v. With this condition, cross boundary tangents are *discontinuous* across patch boundaries. Composite patch boundaries would have positional continuity but gradient discontinuity at all patch corners. However, the tangent directions of all four patch boundaries meeting at the intersection are coplanar. Resulting conditions impose the constraints shown in Figure 7.25 on the two adjacent polyhedra. For polyhedron II, for instance, the number of data points to be freely chosen becomes 10 as opposed to 8 in Case I. For patch C (Figure 7.24) using this scheme, the number of freely chosen data points is 10 while for patch D, they are 8. Case II is, therefore, less restrictive than Case I in terms of freely specifying the control points.

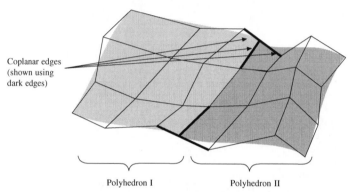

Coplanar edges
(shown using
dark edges)

Polyhedron I   Polyhedron II

Figure 7.25   Arrangement of polyhedral edges for gradient continuity (Case II)

Example 7.11. To blend two bi-cubic Bézier patches, the control points for a bi-cubic Bézier patch $\mathbf{r}^I(u, v)$ are given as

$$\begin{bmatrix} (0,0,0) & (1,0,0) & (2,0,0) & (3,0,0) \\ (0,1,0) & (1,1,1) & (2,1,1) & (3,1,0) \\ (0,2,0) & (1,2,2) & (2,2,2) & (3,2,0) \\ (0,3,0) & (1,3,3) & (2,3,3) & (3,3,0) \end{bmatrix}$$

while for an adjacent patch are given as

$$\begin{bmatrix} \mathbf{r}_{00} & \mathbf{r}_{01} & \mathbf{r}_{02} & \mathbf{r}_{03} \\ \mathbf{r}_{10} & \mathbf{r}_{11} & \mathbf{r}_{12} & \mathbf{r}_{13} \\ (0,6,0) & (1,6,5) & (2,6,5) & (3,6,0) \\ (0,7,0) & (1,8,0) & (2,8,0) & (3,7,0) \end{bmatrix}^{II}$$

where $[\mathbf{r}_{00}, \mathbf{r}_{01}, \mathbf{r}_{02}, \mathbf{r}_{03}, \mathbf{r}_{10}, \mathbf{r}_{11}, \mathbf{r}_{12}, \mathbf{r}_{13}]^{II}$ are to be determined to achieve position and slope continuity at the common boundary. Determine the unknown control points and show the composite surfaces.

The position continuity requires that the boundary polygon must be common between the two patches. Thus

$$[\mathbf{r}_{00} \quad \mathbf{r}_{01} \quad \mathbf{r}_{02} \quad \mathbf{r}_{03}]^{\mathrm{II}} = [(0, 3, 0) \quad (1, 3, 3) \quad (2, 3, 3) \quad (3, 3, 0)]$$

For slope continuity, Case I yields

$$\mathbf{r}_{1i}^{\mathrm{II}} = \mathbf{r}_{0i}^{\mathrm{II}} + \lambda\,(\mathbf{r}_{3i}^{\mathrm{I}} - \mathbf{r}_{2i}^{\mathrm{I}}), \; i = 0, 1, 2, 3$$

choosing $\lambda = 2$ results in

$$[\mathbf{r}_{10}\,\mathbf{r}_{11}\,\mathbf{r}_{12}\,\mathbf{r}_{13}]^{\mathrm{II}} = [(0, 3, 0)\ (1, 3, 3)\ (2, 3, 3)\ (3, 3, 0)] + 2[(0, 1, 0)\ (0, 1, 1)\ (0, 1, 1)\ (0, 1, 0)]$$

$$= [(0, 5, 0) \quad (1, 5, 5) \quad (2, 5, 5) \quad (3, 5, 0)]$$

The resultant composite surface is shown in Figure 7.26 with polyhedron I shown with thick linear lines while polyhedron II is shown with thin lines.

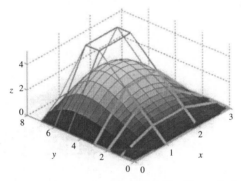

Figure 7.26   A composite Bézier surface (Example 7.11) with gradient continuity (Case I)

With Case II, from Eq. (7.57), we have

$$[0 \ \ 0 \ \ 1 \ \ 0]\,\mathbf{M}_{\mathrm{B}}\mathbf{G}_{\mathrm{B}}^{\mathrm{II}}\mathbf{M}_{\mathrm{B}}^{\mathrm{T}}\mathbf{V}^{\mathrm{T}} = \lambda[3 \ \ 2 \ \ 1 \ \ 0]\,\mathbf{M}_{\mathrm{B}}\mathbf{G}_{\mathrm{B}}^{\mathrm{I}}\mathbf{M}_{\mathrm{B}}^{\mathrm{T}}\mathbf{V}^{\mathrm{T}} + (\mu_0 + \mu_1\,v)\,[1 \ \ 1 \ \ 1 \ \ 1]\,\mathbf{M}_{\mathrm{B}}\mathbf{G}_{\mathrm{B}}^{\mathrm{I}}\mathbf{M}_{\mathrm{B}}^{\mathrm{T}} \begin{bmatrix} 3v^2 \\ 2v \\ 1 \\ 0 \end{bmatrix}$$

Comparing the coefficients of $\mathbf{V}$ results in

$$[-3 \ \ 3 \ \ 0 \ \ 0]\,\mathbf{G}_{\mathrm{B}}^{\mathrm{II}} = \lambda[3 \ \ 2 \ \ 1 \ \ 0]\,\mathbf{M}_{\mathrm{B}}\mathbf{G}_{\mathrm{B}}^{\mathrm{I}} + [1 \ \ 1 \ \ 1 \ \ 1]\,\mathbf{M}_{\mathrm{B}}\mathbf{G}_{\mathrm{B}}^{\mathrm{I}}\mathbf{M}_{\mathrm{B}}^{\mathrm{T}} \begin{bmatrix} 3\mu_1 & 3\mu_0 & 0 & 0 \\ 0 & 2\mu_1 & 2\mu_0 & 0 \\ 0 & 0 & \mu_1 & \mu_0 \\ 0 & 0 & 0 & 0 \end{bmatrix} \mathbf{M}_{\mathrm{B}}^{-\mathrm{T}}$$

Choosing $\lambda = \mu_0 = \mu_1 = 1$, which completely determines the right hand side, gives

$$3[(\mathbf{r}_{10} - \mathbf{r}_{00})\ (\mathbf{r}_{11} - \mathbf{r}_{01})\ (\mathbf{r}_{12} - \mathbf{r}_{02})\ (\mathbf{r}_{13} - \mathbf{r}_{03})\,]^{\mathrm{II}} = [(3, 3, 9)\ (4, 3, 9)\ (5, 3, 0)\ (6, 3, -18)]$$

or

$$\mathbf{r}_{10}^{II} = (0, 3, 0) + \frac{1}{3}(3, 3, 9) = (1, 4, 3)$$

$$\mathbf{r}_{11}^{II} = (1, 3, 3) + \frac{1}{3}(4, 3, 9) = \left(\frac{7}{3}, 4, 6\right)$$

$$\mathbf{r}_{12}^{II} = (2, 3, 3) + \frac{1}{3}(5, 3, 0) = \left(\frac{11}{3}, 4, 3\right)$$

$$\mathbf{r}_{13}^{II} = (3, 3, 0) + \frac{1}{3}(6, 3, -18) = (5, 4, -6)$$

The resulting composite surface is shown in Figure 7.27. To verify the coplanarity condition suggested in Eq. (7.57), we consider the arrangement of data points around the common polygon.

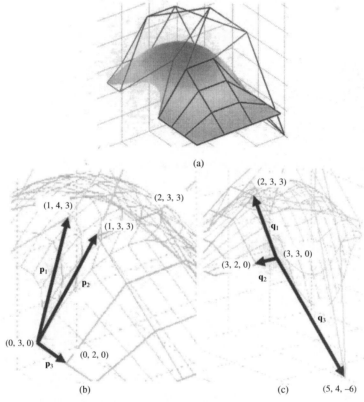

Figure 7.27   (a) A composite Bézier surface (Example 7.11) with gradient continuity (Case II), (b) vectors of the control polyhedra at a corner point in the common polygon and (c) vectors of the control polyhedra at the other corner point

Considering the triple scalar product for the vectors on the left gives

$$\mathbf{p}_1 \cdot (\mathbf{p}_2 \times \mathbf{p}_3) = \begin{vmatrix} 1 & 1 & 3 \\ 1 & 0 & 3 \\ 0 & -1 & 0 \end{vmatrix} = 1(3) + 1(0) + 3(-1) = 0$$

while for those on the right gives

$$\mathbf{q}_1 \cdot (\mathbf{q}_2 \times \mathbf{q}_3) = \begin{vmatrix} -1 & 0 & 3 \\ 0 & -1 & 0 \\ 2 & 1 & -6 \end{vmatrix} = -1(6) + 0(0) + 3(2) = 0$$

Example 7.12 (A Composite Surface with Coons Patches). Often, sharp corners in a machine component is not desirable and need to be replaced by curved surfaces at the corners and edges. An example of a composite surface is illustrated in Figure 7.28 for which the surface is divided into four patches: A is the top flat patch, B and D are the adjacent ruled patches and C is the triangular Coon's patch.

(i)   The top flat patch A is rectangular, parallel to horizontal (*x-y*) plane at a height '1' along the *z*-axes. The four corner points are (0, 0.5, 1), (0, 0.5, 1), (1, 1.5, 1) and (1, 1.5, 1). The equation of the bilinear Coon's patch is given by

$$\mathbf{r}_A(u, v) = [1 - u \quad u] \begin{bmatrix} \mathbf{P}_{00} & \mathbf{P}_{01} \\ \mathbf{P}_{10} & \mathbf{P}_{11} \end{bmatrix} \begin{bmatrix} 1 - v \\ v \end{bmatrix}$$

$$+ [1 - v \quad v] \begin{bmatrix} \mathbf{P}_{00} & \mathbf{P}_{01} \\ \mathbf{P}_{10} & \mathbf{P}_{11} \end{bmatrix} \begin{bmatrix} 1 - u \\ u \end{bmatrix} - [1 - u \quad u] \begin{bmatrix} \mathbf{P}_{00} & \mathbf{P}_{01} \\ \mathbf{P}_{10} & \mathbf{P}_{11} \end{bmatrix} \begin{bmatrix} 1 - v \\ v \end{bmatrix}$$

or $$\mathbf{r}_A(u, v) = [(1 - u) \quad u] \begin{bmatrix} (0, 0.5, 1) & (1, 0.5, 1) \\ (0, 1.5, 1) & (1, 1.5, 1) \end{bmatrix} \begin{bmatrix} (1 - v) \\ v \end{bmatrix} = \{v, (0.5 + v), 1\}$$

(ii)   Patch B has the top boundary a line $\mathbf{r}_B(1, v)$, common with A, given by the end points (0, 0.5, 1) and (1, 0.5, 1). The bottom boundary $\mathbf{r}_B(0, v)$ is also a straight line with end points (0, 0, 0) and (1, 0, 0). Let $\mathbf{r}_B(u, 0)$ and $\mathbf{r}_B(u, 1)$ represent the remaining two boundaries in terms of Hermite curves. The end points of the left boundary curve $\mathbf{r}_B(u, 0)$ are (0, 0, 0) and (0, 0.5, 1), and the end tangents are unit vectors $\mathbf{r}_{Bu}(0, 0) = \mathbf{k} = (0, 0, 1)$ and $\mathbf{r}_{Bu}(0, 1) = \mathbf{j} = (0, 1, 0)$. Similarly, the end points of the right boundary curve $\mathbf{r}_B(u, 1)$ are (1, 0, 0), (1, 0.5, 1), and the end tangents are unit vectors $\mathbf{r}_{Bu}(1, 0) = \mathbf{k} = (0, 0, 1)$ and $\mathbf{r}_{Bu}(1, 1) = \mathbf{j} = (0, 1, 0)$. Thus, the boundary curves are given by

$$\mathbf{r}_B(u, 0) = [u^3 \quad u^2 \quad u \quad 1] \begin{bmatrix} 2 & -2 & 1 & 1 \\ -3 & 3 & -2 & -1 \\ 0 & 0 & 1 & 0 \\ 1 & 0 & 0 & 0 \end{bmatrix} \begin{bmatrix} (0, 0, 0) \\ (0, 0.5, 1) \\ (0, 0, 1) \\ (0, 1, 0) \end{bmatrix}$$

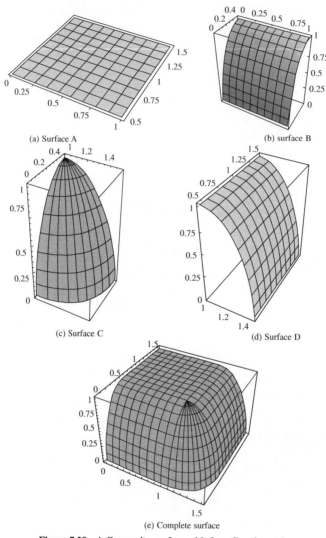

(a) Surface A

(b) surface B

(c) Surface C

(d) Surface D

(e) Complete surface

Figure 7.28   A Composite surface with four Coon's patches

$$\mathbf{r}_B(u, 1) = [u^3 \quad u^2 \quad u \quad 1] \begin{bmatrix} 2 & -2 & 1 & 1 \\ -3 & 3 & -2 & -1 \\ 0 & 0 & 1 & 0 \\ 1 & 0 & 0 & 0 \end{bmatrix} \begin{bmatrix} (1, 0, 0) \\ (1, 0.5, 1) \\ (0, 0, 1) \\ (0, 1, 0) \end{bmatrix}$$

Equation of the ruled surface B using two boundary curves is given by

$$\mathbf{r}_B(u, v) = (1 - v)\mathbf{r}_B(u, 0) + v\mathbf{r}_B(u, 1)$$

Since the boundary curves $\mathbf{r}_B(0, v)$ and $\mathbf{r}_B(1, v)$ are straight line edges of surface B, formulating $\mathbf{r}_2(u, v)$ and the correction Coon's patch $\mathbf{r}_3(u, v)$ as in Eq. (7.34) would not alter the above equation since $\mathbf{r}_2(u, v)$ and $\mathbf{r}_3(u, v)$ would cancel each other due to rectilinear nature of $\mathbf{r}_B(0, v)$ and $\mathbf{r}_B(1, v)$. Though $\mathbf{r}_B(u, v)$ is computed as a ruled surface, it is actually a bilinear Coon's patch.

(iii) Patch D is similarly obtained. The corner points are (1.5, 0.5, 0), (1, 0.5, 1), (1.5, 1.5, 0), (1, 1.5, 1) and the end tangents are {(0, 0, 1),(−1, 0, 0)}, {(0, 0, 1),(−1, 0, 0)}. The top boundary of patch D coincides with patch A {(1, 0.5, 1), (1, 1.5, 1)} and the bottom boundary is a straight line with end points {(1.5, 0.5, 0), (1.5, 1.5, 0)}. The boundary curves are

(iv)

$$\mathbf{r}_D(u, 0) = [u^3 \ u^2 \ u \ 1] \begin{bmatrix} 2 & -2 & 1 & 1 \\ -3 & 3 & -2 & -1 \\ 0 & 0 & 1 & 0 \\ 1 & 0 & 0 & 0 \end{bmatrix} \begin{bmatrix} (1.5, 0.5, 0) \\ (1, 0.5, 1) \\ (0, 0, 1) \\ (-1, 0, 0) \end{bmatrix}$$

$$\mathbf{r}_D(u, 1) = [u^3 \ u^2 \ u \ 1] \begin{bmatrix} 2 & -2 & 1 & 1 \\ -3 & 3 & -2 & -1 \\ 0 & 0 & 1 & 0 \\ 1 & 0 & 0 & 0 \end{bmatrix} \begin{bmatrix} (1.5, 1.5, 0) \\ (1, 1.5, 1) \\ (0, 0, 1) \\ (-1, 0, 0) \end{bmatrix}$$

Equation of the ruled surface D is given by

$$\mathbf{r}_D(u, v) = (1 - v) \ \mathbf{r}_D(u, 0) + v\mathbf{r}_D(u, 1)$$

(v) The triangular Coon's patch C has the corner points{(1, 0, 0), (1.5, 0.5, 0) and (1, 0.5, 1)}. Two of its boundaries are coincident with right and left boundaries of patch B and D respectively. The top boundary curve $\mathbf{r}_C(u, 1)$ is a multiple point (1, 0.5, 1). Two side boundary curves are given by

$$\mathbf{r}_C(0, v) = [u^3 \ u^2 \ u \ 1] \begin{bmatrix} 2 & -2 & 1 & 1 \\ -3 & 3 & -2 & -1 \\ 0 & 0 & 1 & 0 \\ 1 & 0 & 0 & 0 \end{bmatrix} \begin{bmatrix} (1, 0, 0) \\ (1.5, 0.5, 1) \\ (0, 0, 1) \\ (0, 1, 0) \end{bmatrix}$$

$$\mathbf{r}_C(1, v) = [u^3 \ u^2 \ u \ 1] \begin{bmatrix} 2 & -2 & 1 & 1 \\ -3 & 3 & -2 & -1 \\ 0 & 0 & 1 & 0 \\ 1 & 0 & 0 & 0 \end{bmatrix} \begin{bmatrix} (1.5, 0.5, 0) \\ (1, 0.5, 1) \\ (0, 0, 1) \\ (-1, 0, 0) \end{bmatrix}$$

The bottom boundary is given by

$$\mathbf{r}_C(u, 0) = [u^3 \ u^2 \ u \ 1] \begin{bmatrix} 2 & -2 & 1 & 1 \\ -3 & 3 & -2 & -1 \\ 0 & 0 & 1 & 0 \\ 1 & 0 & 0 & 0 \end{bmatrix} \begin{bmatrix} (1, 0, 0) \\ (1, 5, 0.5, 0) \\ (1, 0, 0) \\ (0, 1, 0) \end{bmatrix}$$

Equation of the Coon's patch C (incorporating the correction surface) is given by

$$\mathbf{r}_C(u, v) = (1 - u)\mathbf{r}_C(0, v) + u\,\mathbf{r}_C(1, v) + (1 - v)\mathbf{r}_C(u, 0) + v\mathbf{r}_C(u, 1)$$
$$- (1 - u)(1 - v)(1, 0, 0) - u(1 - v)(1.5, 0.5, 0) - v(1 - u)(1, 0.5, 1) - uv(1, 0.5, 1)$$

Plots of patches A, B, C, D and the composite coon's patch are shown in Figure 7.29.

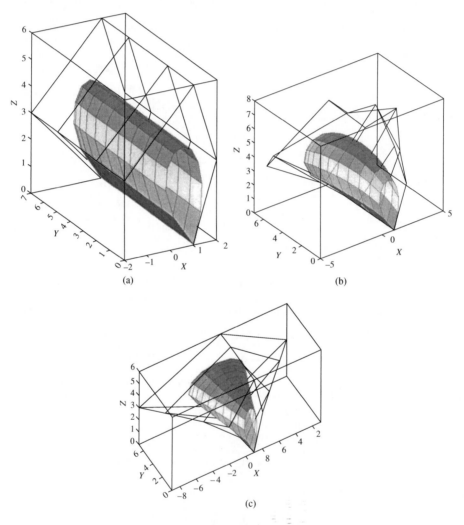

Figure 7.29   Closed Bézier surfaces

Example 7.13 (Closed Bézier Surface). Closing the polyhedron formed by the control points can create closed Bézier surfaces. It is required to create a closed tubular Bézier surface by using 5 control points (last control point being the same as the first) at each of the sections, the control point net created by using 5 such sections. The data set is given below.

(a) P1 = [1, 0, 0; –2, 0, 3; 1, 0, 6; 2, 0, 3; 1, 0, 0], P2 = [1, 2, 0; –3, 2, 4; 1, 2, 8; 3, 2, 4; 1, 2, 0],
P3 = [1, 5, 0; –5, 5, 5; 1, 5, 10; 5, 5, 5; 1, 5, 0], P4 = [1, 7, 0; –4, 7, 3; 1, 7, 6; 4, 7, 3; 1, 7, 0],
P5 = [1, 9, 0; –2, 9, 2; 1, 9, 6; 1, 9, 2; 1, 9, 0]

In the following parts (b) and (c), the control points are changed. Show the effect of the changed control points on the shape of the surface.

(b) P1 = [1, 0, 0; –2, 0, 3; 1, 0, 6; 2, 0, 3; 1, 0, 0], P2 = [1, 2, 0; –4, 2, 3; 1, 2, 6; 4, 2, 3; 1, 2, 0],
P3 = [1, 5, 0; –7, 5, 3; 1, 5, 6; 7, 5, 3; 1, 5, 0], P4 = [1, 7, 0; –9, 7, 3; 1, 7, 6; 9, 7, 3; 1, 7, 0],
P5 = [1, 9, 0; –11, 9, 3; 1, 9, 6; 11, 9, 3; 1, 9, 0]

(c) P1 = [1, 0, 0; –2, 0, 3; 1, 0, 6; 2, 0, 3; 1, 0, 0], P2 = [1, 2, 0; –2, 2, 3; 1, 2, 6; 2, 2, 3; 1, 2, 0],
P3 = [1, 5, 0; –2, 5, 3; 1, 5, 6; 2, 5, 3; 1, 5, 0], P4 = [1, 7, 0; –2, 7, 3; 1, 7, 6; 2, 7, 3; 1, 7, 0],
P5 = [1, 9, 0; –2, 9, 3; 1, 9, 6; 2, 9, 3; 1, 9, 0]

7.4 B-Spline Surface Patch

A B-spline surface patch can be created using the above definitions of the B-spline curves made in Chapter 5 and the tensor product definition of the surface.

If uniform (periodic) knot vector is used the surfaces can be easily constructed using the following equations (refer to the problems in Exercises, Chapter 5).

Uniform quadratic B-spline surface

$$\mathbf{r}(u, v) = \left(\frac{1}{2}\right)^2 [u^2 \; u \; 1] \begin{bmatrix} 1 & -2 & 1 \\ -2 & 2 & 0 \\ 1 & 1 & 0 \end{bmatrix} \begin{bmatrix} \mathbf{r}_{00} & \mathbf{r}_{01} & \mathbf{r}_{02} \\ \mathbf{r}_{10} & \mathbf{r}_{11} & \mathbf{r}_{12} \\ \mathbf{r}_{20} & \mathbf{r}_{21} & \mathbf{r}_{22} \end{bmatrix} \begin{bmatrix} 1 & -2 & 1 \\ -2 & 2 & 0 \\ 1 & 1 & 0 \end{bmatrix}^T \begin{bmatrix} v^2 \\ v \\ 1 \end{bmatrix} \tag{7.58}$$

Uniform cubic B-spline surface

$$r(u, v) = \left(\frac{1}{6}\right)^2 [u^3 \; u^2 \; u \; 1] \begin{bmatrix} -1 & 3 & -3 & 1 \\ 3 & -6 & 3 & 0 \\ -3 & 0 & 3 & 0 \\ 1 & 4 & 1 & 0 \end{bmatrix} \begin{bmatrix} \mathbf{r}_{00} & \mathbf{r}_{01} & \mathbf{r}_{02} & \mathbf{r}_{03} \\ \mathbf{r}_{10} & \mathbf{r}_{11} & \mathbf{r}_{12} & \mathbf{r}_{13} \\ \mathbf{r}_{20} & \mathbf{r}_{21} & \mathbf{r}_{22} & \mathbf{r}_{23} \\ \mathbf{r}_{30} & \mathbf{r}_{31} & \mathbf{r}_{32} & \mathbf{r}_{33} \end{bmatrix} \begin{bmatrix} -1 & 3 & -3 & 1 \\ 3 & -6 & 3 & 0 \\ -3 & 0 & 3 & 0 \\ 1 & 4 & 1 & 0 \end{bmatrix}^T \begin{bmatrix} v^3 \\ v^2 \\ v \\ 1 \end{bmatrix}$$

$$\tag{7.59}$$

Example 7.14. (Uniform Quadratic B-spline Surface). (a) A surface patch is to be constructed using the matrix formulation for uniform quadratic B-spline surface patch given by

$$\mathbf{r}(u, v) = \left(\frac{1}{2}\right)^2 [u^2 \; u \; 1] \begin{bmatrix} 1 & -2 & 1 \\ -2 & 2 & 0 \\ 1 & 1 & 0 \end{bmatrix} \begin{bmatrix} \mathbf{r}_{00} & \mathbf{r}_{01} & \mathbf{r}_{02} \\ \mathbf{r}_{10} & \mathbf{r}_{11} & \mathbf{r}_{12} \\ \mathbf{r}_{20} & \mathbf{r}_{21} & \mathbf{r}_{22} \end{bmatrix} \begin{bmatrix} 1 & -2 & 1 \\ -2 & 2 & 0 \\ 1 & 1 & 0 \end{bmatrix}^T \begin{bmatrix} v^2 \\ v \\ 1 \end{bmatrix}$$

Here,
$$\begin{bmatrix} \mathbf{r}_{00} & \mathbf{r}_{01} & \mathbf{r}_{02} \\ \mathbf{r}_{10} & \mathbf{r}_{11} & \mathbf{r}_{12} \\ \mathbf{r}_{20} & \mathbf{r}_{21} & \mathbf{r}_{22} \end{bmatrix} = \begin{bmatrix} \{0,0,0\} & \{0,1,0\} & (0,2,0\} \\ \{1,0,0\} & \{1,1,1\} & \{1,2,0\} \\ \{2,0,0\} & \{2,1,0\} & \{2,2,0\} \end{bmatrix}$$

With some care, the equation of surface can be manually derived. The surface is shown in Figure 7.30(a).

(b) A (quadratic-quadratic) B-spline surface patch is to be constructed using the following control points, without use of the matrix formulation (using the original definition of the B-spline blending functions and uniform knot vector)

$$\begin{bmatrix} \mathbf{r}_{00} & \mathbf{r}_{01} & \mathbf{r}_{02} & \mathbf{r}_{03} \\ \mathbf{r}_{10} & \mathbf{r}_{11} & \mathbf{r}_{12} & \mathbf{r}_{13} \\ \mathbf{r}_{20} & \mathbf{r}_{21} & \mathbf{r}_{22} & \mathbf{r}_{23} \\ \mathbf{r}_{30} & \mathbf{r}_{31} & \mathbf{r}_{32} & \mathbf{r}_{33} \\ \mathbf{r}_{40} & \mathbf{r}_{41} & \mathbf{r}_{42} & \mathbf{r}_{43} \end{bmatrix} = \begin{bmatrix} \{0,0,0\} & \{0,1,1\} & \{0,2,1\} & \{0,2,0\} \\ \{1,0,0\} & \{1,1,2\} & 1,2,1\} & \{1,3,2\} \\ \{2,0,0\} & \{2,1,3\} & \{2,2,2\} & \{2,3,3\} \\ \{3,0,0\} & \{3,1,2\} & \{3,2,1\} & \{3,3,2\} \\ \{4,0,0\} & \{4,1,1\} & \{4,2,1\} & \{4,2,0\} \end{bmatrix}$$

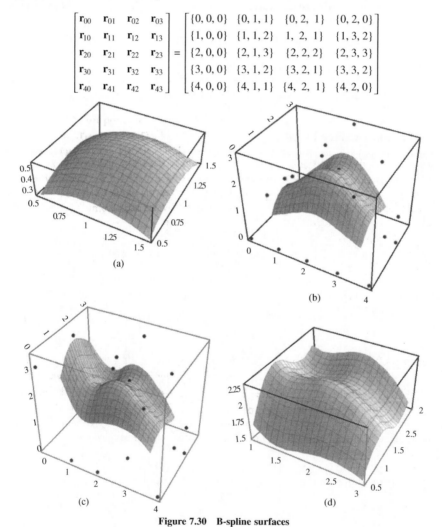

(a)

(b)

(c)

(d)

Figure 7.30   B-spline surfaces

The uniform knot vector is computed as $[t_i, i = 0, \ldots, 7] = [0, 1, 2, 3, 4, 5, 6, 7]$. A manual calculation will be obviously lengthy. It can be conveniently done using any of the programming languages or software with adequate graphics user interface. The surface is shown below in Figure 7.30(b).

(c) If some of the control points (e.g. the first row) are changed as given below, the change in shape of the surface is shown in Figure 7.30(c).

$$
\begin{bmatrix}
\mathbf{r}_{00} & \mathbf{r}_{01} & \mathbf{r}_{02} & \mathbf{r}_{03} \\
\mathbf{r}_{10} & \mathbf{r}_{11} & \mathbf{r}_{12} & \mathbf{r}_{13} \\
\mathbf{r}_{20} & \mathbf{r}_{21} & \mathbf{r}_{22} & \mathbf{r}_{23} \\
\mathbf{r}_{30} & \mathbf{r}_{31} & \mathbf{r}_{32} & \mathbf{r}_{33} \\
\mathbf{r}_{40} & \mathbf{r}_{41} & \mathbf{r}_{42} & \mathbf{r}_{43}
\end{bmatrix} =
\begin{bmatrix}
\{0, 0, 3\} & \{0, 1, 5\} & \{0, 2, 5\} & \{0, 2, 3\} \\
\{1, 0, 0\} & \{1, 1, 2\} & \{1, 2, 1\} & \{1, 3, 2\} \\
\{2, 0, 0\} & \{2, 1, 3\} & \{2, 2, 2\} & \{2, 3, 3\} \\
\{3, 0, 0\} & \{3, 1, 2\} & \{3, 2, 1\} & \{3, 3, 2\} \\
\{4, 0, 0\} & \{4, 1, 1\} & \{4, 2, 1\} & \{4, 2, 0\}
\end{bmatrix}
$$

(d) In another trial, a quadratic-cubic B-spline is created. Observe the change in shape Figure 7.30(d), for this quadratic-cubic B-spline surface for the same control points as in Figure 7.30(c).

7.5 Closed B-Spline Surface

Closed B-spline surface with uniform knot vectors can be created in a similar manner as described in Exercises, Chapter 15 for creating closed B-spline curves.

Let the control points be $\mathbf{r}_{ij}$ $(i = 0, \ldots, n)$ and $(j = 0, \ldots, m)$. Let us restrict our discussion, for simplicity, to a closed cubic uniform B-spline surface using 45 control points $\mathbf{r}_{00}, \mathbf{r}_{01}, \mathbf{r}_{02}, \mathbf{r}_{03}, \mathbf{r}_{04}, \mathbf{r}_{05}, \mathbf{r}_{06}, \mathbf{r}_{07}, \mathbf{r}_{00}; \mathbf{r}_{10}, \mathbf{r}_{11}, \mathbf{r}_{12}, \mathbf{r}_{13}, \mathbf{r}_{14}, \mathbf{r}_{15}, \mathbf{r}_{16}, \mathbf{r}_{17}, \mathbf{r}_{10}; \mathbf{r}_{20}, \mathbf{r}_{21}, \mathbf{r}_{22}, \mathbf{r}_{23}, \mathbf{r}_{24}, \mathbf{r}_{25}, \mathbf{r}_{26}, \mathbf{r}_{27}, \mathbf{r}_{20}; \mathbf{r}_{30}, \mathbf{r}_{31}, \mathbf{r}_{32}, \mathbf{r}_{33}, \mathbf{r}_{34}, \mathbf{r}_{35}, \mathbf{r}_{36}, \mathbf{r}_{37}, \mathbf{r}_{30}; \mathbf{r}_{40}, \mathbf{r}_{41}, \mathbf{r}_{42}, \mathbf{r}_{43}, \mathbf{r}_{44}, \mathbf{r}_{45}, \mathbf{r}_{46}, \mathbf{r}_{47}, \mathbf{r}_{40}$. Observe that the last control point in each segment is the same as the first. Uniform knot vector $[0\ 1\ 2\ 3\ 4\ 5\ 6\ 7\ 8\ 9\ 10\ 11\ 12\ 13\ 14]$ is to be used.

For a cubic B-spline surface, the equations of the surface segments can be written as

$$
\mathbf{r}_{1,j+1}(u, v) = \left(\frac{1}{6}\right)^2 \mathbf{UM}
\begin{bmatrix}
\mathbf{r}_{0,(j\bmod 8)+1} & \mathbf{r}_{0,((j+1)\bmod 8)+1} & \mathbf{r}_{0,((j+2)\bmod 8)+1} & \mathbf{r}_{0,((j+3)\bmod 8)+1} \\
\mathbf{r}_{1,(j\bmod 8)+1} & \mathbf{r}_{1,((j+1)\bmod 8)+1} & \mathbf{r}_{1,((j+2)\bmod 8)+1} & \mathbf{r}_{1,((j+3)\bmod 8)+1} \\
\mathbf{r}_{2,(j\bmod 8)+1} & \mathbf{r}_{2,((j+1)\bmod 8)+1} & \mathbf{r}_{2,((j+2)\bmod 8)+1} & \mathbf{r}_{2,((j+3)\bmod 8)+1} \\
\mathbf{r}_{3,(j\bmod 8)+1} & \mathbf{r}_{3,((j+1)\bmod 8)+1} & \mathbf{r}_{3,((j+2)\bmod 8)+1} & \mathbf{r}_{3,((j+3)\bmod 8)+1}
\end{bmatrix} \mathbf{M}^T \mathbf{V}
$$

$$
\mathbf{r}_{2,j+1}(u, v) = \left(\frac{1}{6}\right)^2 \mathbf{UM}
\begin{bmatrix}
\mathbf{r}_{1,(j\bmod 8)+1} & \mathbf{r}_{1,((j+1)\bmod 8)+1} & \mathbf{r}_{1,((j+2)\bmod 8)+1} & \mathbf{r}_{1,((j+3)\bmod 8)+1} \\
\mathbf{r}_{2,(j\bmod 8)+1} & \mathbf{r}_{2,((j+1)\bmod 8)+1} & \mathbf{r}_{2,((j+2)\bmod 8)+1} & \mathbf{r}_{2,((j+3)\bmod 8)+1} \\
\mathbf{r}_{3,(j\bmod 8)+1} & \mathbf{r}_{3,((j+1)\bmod 8)+1} & \mathbf{r}_{3,((j+2)\bmod 8)+1} & \mathbf{r}_{3,((j+3)\bmod 8)+1} \\
\mathbf{r}_{4,(j\bmod 8)+1} & \mathbf{r}_{4,((j+1)\bmod 8)+1} & \mathbf{r}_{4,((j+2)\bmod 8)+1} & \mathbf{r}_{4,((j+3)\bmod 8)+1}
\end{bmatrix} \mathbf{M}^T \mathbf{V}
$$

(7.60)

Here, $j \in [0, \ldots, 7]$. The surface will be closed in v and open in u direction. This will be constituted by (2×8) or 16 sliding surface segments. Matrices $\mathbf{U}$, $\mathbf{M}$ and $\mathbf{V}$ are given by

$$
\mathbf{U} = \begin{bmatrix} u^3 \\ u^2 \\ u \\ 1 \end{bmatrix}^T, \quad
\mathbf{M} = \begin{bmatrix}
-1 & 3 & -3 & 1 \\
3 & -6 & 3 & 0 \\
-3 & 0 & 3 & 0 \\
1 & 4 & 1 & 0
\end{bmatrix}, \quad
\mathbf{V} = \begin{bmatrix} v^3 \\ v^2 \\ v \\ 1 \end{bmatrix}
$$

Example 7.15. A closed cubic B-spline surface is created using the following control points created by selecting 5 sections ($i = 0, \ldots, 4$) parallel to z-axis. Eight control points combined with the first few points to close the circle create each circular cross section. The control points are given in Tables 7.1 to 7.3.

Table 7.1

i	$\mathbf{r}_{i0}$	$\mathbf{r}_{i1}$	$\mathbf{r}_{i2}$	$\mathbf{r}_{i3}$	$\mathbf{r}_{i4}$	$\mathbf{r}_{i5}$	$\mathbf{r}_{i6}$	$\mathbf{r}_{i7}$
0	(0 0 0)	(–2 2 0)	(–2 4 0)	(0 6 0)	(2 6 0)	(4 4 0)	(4 2 0)	(2 0 0)
1	(0 0 1)	(–2 2 1)	(–2 4 1)	(0 6 1)	(2 6 1)	(4 4 1)	(4 2 1)	(2 0 1)
2	(0 0 2)	(–2 2 2)	(–2 4 20)	(0 6 2)	(2 6 2)	(4 4 2)	(4 2 2)	(2 0 2)
3	(0 0 3)	(–2 2 3)	(–2 4 3)	(0 6 3)	(2 6 3)	(4 4 3)	(4 2 30)	(2 0 3)
4	(0 0 4)	(–2 2 4)	(–2 4 4)	(0 6 4)	(2 6 4)	(4 4 4)	(4 2 4)	(2 0 4)

Table 7.2

i	$\mathbf{r}_{i0}$	$\mathbf{r}_{i1}$	$\mathbf{r}_{i2}$	$\mathbf{r}_{i3}$	$\mathbf{r}_{i4}$	$\mathbf{r}_{i5}$	$\mathbf{r}_{i6}$	$\mathbf{r}_{i7}$
0	(0 0 0)	(–2 2 0)	(–2 4 0)	(0 6 0)	(2 6 0)	(4 4 0)	(4 2 0)	(2 0 0)
1	(0 0 1)	(–2 2 1)	(–2 4 1)	(0 6 1)	(2 6 1)	(4 4 1)	(4 2 1)	(2 0 1)
2	(0 0 2)	(–1 1 2)	(–1 2 2)	(0 3 2)	(2 3 2)	(2 2 2)	(2 1 2)	(1 0 2)
3	(0 0 3)	(–2 2 3)	(–2 4 3)	(0 6 3)	(2 6 3)	(4 4 3)	(4 2 3)	(2 0 3)
4	(0 0 4)	(–2 2 4)	(–2 4 4)	(0 6 4)	(2 6 4)	(4 4 4)	(4 2 4)	(2 0 4)

Table 7.3

i	$\mathbf{r}_{i0}$	$\mathbf{r}_{i1}$	$\mathbf{r}_{i2}$	$\mathbf{r}_{i3}$	$\mathbf{r}_{i4}$	$\mathbf{r}_{i5}$	$\mathbf{r}_{i6}$	$\mathbf{r}_{i7}$
0	(0 0 0)	(–2 2 0)	(–2 4 0)	(0 6 0)	(2 6 0)	(4 4 0)	(4 2 0)	(2 0 0)
1	(0 0 1)	(–2 2 1)	(–2 4 1)	(0 6 1)	(2 6 1)	(4 4 1)	(4 2 1)	(2 0 1)
2	(0 0 2)	(–2 2 2)	(–2 4 2)	(0 6 2)	(2 6 2)	(8 8 2)	(4 2 2)	(2 0 2)
3	(0 0 3)	(–2 2 3)	(–2 4 3)	(0 6 3)	(2 6 3)	(4 4 3)	(4 2 3)	(2 0 3)
4	(0 0 4)	(–2 2 4)	(–2 4 4)	(0 6 4)	(2 6 4)	(4 4 4)	(4 2 4)	(2 0 4)

Data given in Table 7.1 generates the cylindrical B-spline surface shown in Figure 7.31(a). Data given in Table 7.2 is generated by changing the control points (the row with $i = 2$) and corresponds to Figure 7.31(b). Similarly, by changing the control point $\mathbf{r}_{25}$ (Table 7.3), Figure 7.31(c) is generated. Observe the local changes in the surface due to change in control points.

7.6 Rational B-spline Patches (NURBS)

Analogous to Eq. (7.26), a B-spline surface patch can be defined as

$$\mathbf{r}(u, v) = \sum_{i=0}^{m} \sum_{j=0}^{n} N_{p,p+i}(u) N_{q,q+j}(v) \mathbf{P}_{ij} \tag{7.61}$$

for an array $\{\mathbf{P}_{ij}, i = 0, \ldots, m; j = 0, \ldots, n\}$ of control points. p and q are the orders of B-spline curves along the u and v directions, respectively. From property 5.8.1B, the number of knots required in the u direction is $m + 1 + p$ while that along the v direction is $n + 1 + q$. Knot parameterization may be performed along the two parametric directions using methods discussed in Section 5.10. More generally, a rational B-spline patch may be computed as

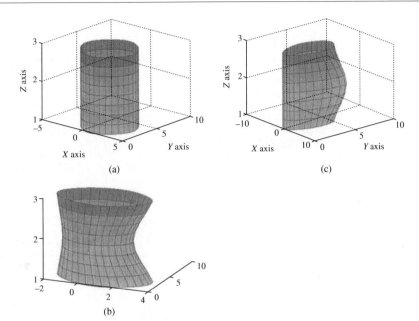

Figure 7.31   Closed B-spline surfaces

$$\mathbf{r}(u, v) = \dfrac{\displaystyle\sum_{i=0}^{m} \sum_{j=0}^{n} w_{ij} N_{p,p+i}(u) N_{q,q+j}(v) \mathbf{P}_{ij}}{\displaystyle\sum_{i=0}^{m} \sum_{j=0}^{n} w_{ij} N_{p,p+i}(u) N_{q,q+j}(v)} \qquad (7.62)$$

with user chosen weights w_{ij}.

EXERCISES

1. A bi-linear surface $\mathbf{r}(u, v)$ is defined by the points $\mathbf{r}(0, 0) = \{0, 0, 1\}$, $\mathbf{r}(0, 1) = \{1, 1, 1\}$, $\mathbf{r}(1, 0) = \{1, 0, 0\}$ and $\mathbf{r}(1, 1) = \{0, 1, 0\}$. Show the plot of the surface. Determine the unit normal to the surface at $(u = 0.5, v = 0.5)$.

2. A bi-cubic Ferguson patch is defined by the following:
   Corner points $\mathbf{r}(0, 0) = \{-100, 0, 100\}$, $\mathbf{r}(0, 1) = \{100, -100, 100\}$, $\mathbf{r}(1, 1) = \{-100, 0, -100\}$, $\mathbf{r}(1, 0) = \{-100, -100, -10\}$, u-tangent vectors $\mathbf{r}_u(0, 0) = \{10, 10, 0\}$, $\mathbf{r}_u(0, 1) = \{-1, -1, 0\}$, $\mathbf{r}_u(1, 1) = \{-1, 1, 0\}$, $\mathbf{r}_u(1, 0) = \{1, -1, 0\}$; v-tangent vectors $\mathbf{r}_v(0, 0) = \{0, -10, -10\}$, $\mathbf{r}_v(0, 1) = \{0, 1, -1\}$, $\mathbf{r}_v(1, 1) = \{0, 1, 1\}$, $\mathbf{r}_v(1, 0) = \{0, 1, 1\}$; twist vectors $\mathbf{r}_{uv}(0, 0) = \{0, 0, 0\}$, $\mathbf{r}_{uv}(0, 1) = \{0.1, 0.1, 0.1\}$, $\mathbf{r}_{uv}(1, 1) = \{0, 0, 0\}$, $\mathbf{r}_{uv}(1, 0) = \{-0.1, -0.1, -0.1\}$.
   Generate the surface and find tangents, normal and curvatures for the surface at $(0.5, 0.5)$.

3. A Coon's patch is generated using quadratic Bézier curves $\varphi_0(u)$, $\varphi_1(u)$ and $\psi_0(v)$, $\psi_1(v)$ having control points $[\{0, 0, 0\}, \{1, 0, 3\}, \{3, 0, 2\}]$; $[\{0, 3, 0\}, \{1, 3, 3\}, \{3, 3, 2\}]$ and $[\{0, 0, 0\}, \{0, 1, 3\}, \{0, 3, 2\}]$; $[\{3, 0, 2\}, \{3, 2, 3\}, \{3, 3, 2\}]$. Work out the complete analysis of individual patches and the final Coon's patch.

4. It is desired to create a closed tubular Bézier surface by using 5 control points (last control point being the same as the first) at each section, and the control point net created by using 5 such sections. Develop the program and demonstrate using an example.

5. Write a procedure to compute the coordinates of a point on a Bézier surface patch. Use this to compute a rectangular array of points to display a Bézier patch. The program should be generic and not restricted to cubic.

6. Develop and discuss the conditions required for C^0 and C^1 continuity between two Bézier patches along a common boundary.

7. Write a procedure to compute the coordinates of a point on a B-Spline surface patch. Display the surface using the code developed. Note that a B-spline patch is a tensor product surface defined as $\mathbf{r}(u, v) =$
$$\sum_{j=0}^{n} \sum_{i=0}^{n} N_{p,p+i}(u)\, N_{q,q+j}(v)\, \mathbf{r}_{ij}.$$

8. Use the code developed to compute an approximate solution to the minimum distance between two given parametric *B-spline* surfaces. First, calculate a rectangular array of points for chosen interval steps for u and v on both the patches and then proceed.

9. Write a procedure to compute the intersection between a straight line and a bi-cubic patch. Simplify your solution by first performing a transformation on both line and the patch so that the line is collinear with the z-axis. Find the intersection and perform inverse transformation.

10. (a) Generate a closed tubular surface patch using closed *B-Splines*. The fundamental aspect is in first having experience in creating a closed *B-spline* curve by taking the vertices (for 8 unique control points) $\mathbf{P}_1, \mathbf{P}_2, \mathbf{P}_3, \mathbf{P}_4, \mathbf{P}_5, \mathbf{P}_6, \mathbf{P}_7, \mathbf{P}_8, \mathbf{P}_1, \mathbf{P}_2, \mathbf{P}_3$. Uniform knot vector [0 1 2 3 4 5 6 7 8 9 10 11 12 13 14] is to be used. For a fourth order ($k = 4$) closed *B-spline* curve defined by the above polyline and for $0 \le u \le 1$, a point on the curve is calculated from the matrix formulation

$$\mathbf{r}_{j+1}(u) = \frac{1}{6}\begin{bmatrix} u^3 & u^2 & u & 1 \end{bmatrix} \begin{bmatrix} -1 & 3 & -3 & 1 \\ 3 & -6 & 3 & 0 \\ -3 & 0 & 3 & 0 \\ 1 & 4 & 1 & 0 \end{bmatrix} \begin{bmatrix} \mathbf{P}_{(j\bmod 8)+1} \\ \mathbf{P}_{((j+1)\bmod 8)+1} \\ \mathbf{P}_{((j+2)\bmod 8)+1} \\ \mathbf{P}_{((j+3)\bmod 8)+1} \end{bmatrix}$$

where $j \bmod 8$ is the remainder when j is divided by 8 (for example, $10 \bmod 8 = 2$). Let us take the case of a tubular surface above with 5 axial cross sections, each cross section having the same number of control points mentioned above. Show the effect of changing the size of different cross sections and also the effect of relocating any intermediate control point.

11. Write generic codes for the following:
   (a) C^1 continuous composite Ferguson's surface.
   (b) C^2 continuous composite Ferguson's surface.
   (c) Composite Bézier surface with tangent plane continuity. Let all the control points be freely chosen for each patch so that at least position continuity is addressed. Later, implement interactive relocation of the control points so that the tangent plane continuity is met.

12. Write a generic code for the NURBS surface patch.

Chapter 8

Solid Modeling

Solids represent a large variety of objects we see and handle. The chapters on curves and surfaces treated earlier are intended to form the basis for solid or volumetric modeling. Solid modeling techniques have been developed since early 1970's using wireframe, surface models, boundary representation (b-rep), constructive solid geometry (CSG), spatial occupancy and enumeration. A solid model not only requires surface and boundary geometry definition, but it also requires topological information such as, interior, connectivity, holes and pockets. Wire-frame and surface models cannot describe these properties adequately. Further, in design, one needs to combine and connect solids to create composite models for which spatial addressability of every point on and in the solid is required. This needs to be done in a manner that it does not become computationally intractable.

Manufacturing and Rapid Prototyping (RP) both require computationally efficient and robust solid modelers. Other usage of solid modelers is in Finite Element Analyses (as pre- and post-processing), mass-property calculations, computer aided process planning (CAPP), interference analysis for robotics and automation, tool path generation for NC machine tools, shading and rendering for realism and many others.

8.1 Solids

The treatment in previous chapters on curve and surface design is purely geometric. However, we would realize that it takes more than geometry alone to interpret solids. Solids are omnipresent, in all possible forms, shapes and sizes. From the representation viewpoint, we may mathematically regard a solid V to be a contiguous set of points in the three-dimensional Euclidean space E^3 satisfying the following attributes:

(a) *Boundedness:* A set V of points must occupy a finite volume or space in E^3. The possibility of solids with infinite volume thus gets eliminated.

(b) *Boundary and interior:* Let $b(V)$ and $I(V)$ be two subsets of V such that $b(V) \cup I(V) = V$, wherein $b(V)$ comprises boundary points and $I(V)$ is the set of interior points. A point $p \in V$ is an interior point ($p \in I(V)$) if there exists an *open ball* enclosing p that consists of points in V only. Thus, if p_0 is the center of the ball B of arbitrary small radius r, and if p_i, $p \in B \subset V$, then $| p_i - p_0 | < r$. A point p is a boundary point if $p \in V$ and $p \notin I(V)$ (Figure 8.1). Note that for p as a boundary point, if an open ball B_1 of radius r_1 is drawn around p, B_1 shall contain points belonging to $E^3 - V$ as well.

(c) *Boundary determinism:* Jordan's theorem[1] for a two-dimensional Euclidean plane E^2 states that a simple (non self intersecting) closed (Jordan) curve divides E^2 into regions interior and exterior to the curve. Formally, for C as a continuous simple closed curve in E^2, $E^2\backslash C$ (complement of C in E^2) has precisely two connected[2] components. Equivalently in E^3, a simple and orientable closed surface b (V) divides a solid V into the interior I (V) and exterior $E^3 - V$ spaces. In other words, if the closed surface b (V) of a solid V is known, the interior I (V) of the solid V is unambiguously determined.

(d) *Homogenous three-dimensionality:* A solid set V must not have disconnected or dangling subsets as shown in Figure 8.2 as such sets defy boundary determinism above.

(e) *Rigidity:* The relative positions between any two points p_1 and p_2 in V must be invariant to re-positioning or re-orientation of the solid in E^3.

(f) *Closure:* Any set operation (union, intersection and subtraction) when applied to solids V_1 and V_2 must yield a solid V_3 satisfying all the aforementioned properties.

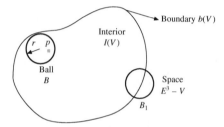

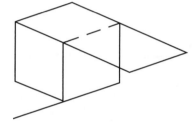

Figure 8.1   Interior and boundary points of a solid   **Figure 8.2   Example of dangling plane and line with a cube**

The discussion above suggests two ways in which a generic solid may be represented. The first, most general representation is through a set of contiguous points in 3-space. A solid object may be represented as a set of adjacent cells using a three-dimensional array. The cell size is usually the maximum resolution of the display. Space arrays have two advantages as a representation; (a) *spatial addressing* wherein it is easy to determine whether a point belongs to a solid or not and (b) *spatial uniqueness* that assures that two solids cannot occupy the same space. In contrast, it has two grave disadvantages; (a) this representation lacks *object coherence* in that there is no explicit relation between cells occupying the solid. Note that in most space arrays, only the occupancy state of a cell is stored. (b) Also, the representation is very expensive in terms of storage space.

A cell in the interior of a solid in a space array representation has the same occupancy state as its adjacent cells. However, at the object's boundaries, this is not so. Thus, a more concise approach to represent a solid is through the boundary points (or the bounding surface) that partitions the points internal and external to the body, as suggested by the boundary determinism property. In this *polyhedral representation*, usually, the bounding surface is subdivided into faces that may be planar or curved. Each face may be identified by a perimeter ring of edges that again may be planar or curved. A face may have one or more internal rings to define voids or holes. Lastly, adjacent edges intersect at

---

[1]The theorem seems geometrically plausible though its proof requires concepts from topology.
[2]For a pair of non-empty subsets U and V of E^2, U and V are connected if $U \cap V = \{\}$ and $U \cup V = E^2$.

vertices. The polyhedral representation has advantages of object coherence and compact storage which outweighs its disadvantage of spatial addressing (or *point membership classification* discussed in Chapter 9) wherein an involved algorithm is used to determine whether a point is inside the object.

The polyhedral representation is still quite broad to encompass generic (freeform) definitions of solids and boundary determinism is a strong property suggesting that one may not consider points interior to a solid, rather only the simple closed connected surface $b(V)$ would seen sufficient to represent a solid in E^3. In polyhedral representation, the surface information pertaining to faces, edges and vertices is stored in two parts. The first is *geometry* wherein physical dimensions and locations in space of individual components are specified. The other is *topology* describing the connectivity between components. It is the topology that renders the object coherence property to this representation, and that the geometry alone is inadequate. Topology regards two points as vertices that bounds a line to define an edge. Likewise, a closed ring of edges bounds a surface to define a face. Both geometry and topology are essential for a complete shape description.

The study of topology ignores the dimensions (lengths and angles) from the geometry and studies the latter for the notations of continuity and closeness. Topology studies the patterns in geometric figures for relative positions without regard to size. Topology is sometimes referred to as the *rubber sheet geometry* since a figure can be changed into an equivalent figure by bending, twisting, stretching and other such transformations, but not by cutting, tearing and gluing. Previously known as *analysis situs*, topology is thought to be initiated by Euler when the solution to the Königsberg bridge problem (Figure 8.3) was provided in 1736. The problem was to determine if the seven bridges (edges) in the city of Königsberg across four land patches (nodes) can all be traversed in a single trip without doubling

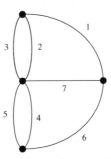

Figure 8.3 Königsberg bridge circuit (nodes are the land patches and edges are bridges)

any bridge back to which the answer was in the negative. Topology, as a subject by itself, is very broad though from the viewpoint of solid/volumetric modeling, we can restrict ourselves to the understanding of topological properties of surfaces as suggested by boundary determinism.

8.2 Topology and Homeomorphism

The aim in topology is to identify a set of rules or procedures to recognize geometrical figures. Two figures would belong to the same topical class if they have the same basic, overall structure even though differing much in details. Consider a cube, for instance, in Figure 8.4(a). So long as the internal angles are all 90° and the edge length a is the same, the form remains a cube irrespective of the edge size specified. So is true for a sphere whose form is independent of the radius specified. A cube is a special case of a block, wherein although all internal angles are 90° each, dimensions of three mutually orthogonal edges a, b and c (Figure 8.4b) can be different. If we let the internal angles to have values other than 90°, and also the edge lengths to be different, we get a form shown in Figure 8.4(c). What is common in these figures, though they are of different shapes, is that they are all hexahedrons (of six sides). Thus, from geometry of a solid, if we ignore the intricacies of size (lengths and angles), we address the topology of that solid. The illustrations in Figure 8.4 (a-c) are topologically identical. So is the illustration in Figure 8.4(d) wherein some of the internal angles are zero.

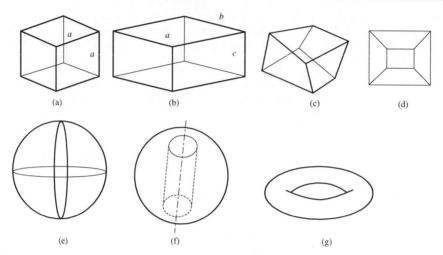

Figure 8.4   **Various shapes of a hexahedral topology (a-d), all homeoporphic to a sphere in (e). A sphere with a through hole (f) is homeomorphic to a torus (g)**

In this regard, a square, a parallelogram, and a rectangle are topologically identical as well. Since we are ignorant about edge lengths, we might as well have a side of a quadrilateral of zero length (Figure 8.5b). The result, a triangle, would still be identical in topology to a quadrilateral. We may further re-shape or re-morph the straight edges of a quadrilateral to have bends. A polygon (thick lines in Figure 8.5c), and in case the number of bends approaches infinity a simple closed curve with no self-intersection (thin lines in the figure), both, would be of the same topology as the parent quadrilateral. It thus implies that a closed polygon is topologically equivalent to a circle and this equivalence is termed as *homeomorphism*. Likewise, the hexahedrals in Figure 8.4 (a-d) are homeomorphic to themselves and to a sphere in Figure 8.4(e) since we can deform (bend, stretch or twist) the faces and edges of a hexahedral to blend with the surface of the sphere. If a cylindrical void is cut through the sphere (Figure 8.4f), the resultant topology is not homeomorphic to a sphere (without void) since no amount of bending, stretching or twisting would transform it to a sphere and vice versa. However, a sphere with a through void is homeomorphic to a torus or doughnut in Figure 8.4 (g) which, in turn, is homeomorphic to a coffee mug.

An interesting aspect to realize is the direction of motion along the boundary when viewing from a point P inside the shapes in Figure 8.5. Starting from any point Q on a closed curve, the traverse along the curve with the aim of getting back to Q would be undirectional, either anticlockwise or clockwise, and moreover, it would be continuous. Topologically, since lengths are of no importance, a line would result by fusing any two vertices of the triangle in Figure 8.5 (b) in a manner similar to how the triangle was obtained from the quadrilateral. The resulting line, however, would not be the same in topology to any of the closed curves. This is because the sense of direction of motion from a point on the line to the same point would not remain unidirectional anymore. This non-homeomorphism between a line and a closed curve can be explained alternatively. There is a *cut* involved, anywhere on the closed curve with its two ends stretched, to obtain a line.

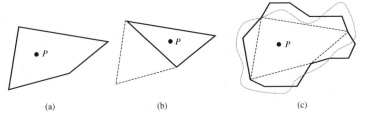

Figure 8.5 Various simply closed planar shapes of identical topology

8.3 Topology of Surfaces

To understand the polyhedral representation of solids better, study of the topological properties of surfaces becomes essential. Surfaces are compact and connected topological objects on which each point has a neighborhood (a closed curve around a point on the surface) homeomorphic to either a plane $\mathbf{R}^2$ or a half plane $\mathbf{H}^2$. Points of the first type are interior points (point P in Figure 8.6) while of the second type are the boundary points (Q and R in the figure). A set of all boundary points constitutes the boundary of the surface. The boundary can comprise one or more components, each of which is homeomorphic to a circle. In Figure 8.6, there are two boundary components, the exterior and the interior, each of which can be morphed into respective circles.

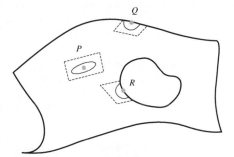

Figure 8.6 Boundary and interior points on the surface

The bounding surface in polyhedral representation of a solid must satisfy certain properties so that the solid is well-defined. A valid solid consists of a complete set of spatial points occupied by an object. Solids may vary in form with different applications. However, in general, they are *bounded* and *connected*. A *bounded* solid is defined within a finite space. If *connected*, there exists a (continuous) path, totally interior to the solid, that connects any pair of points belonging to it. Note that this is true even when a solid may have multiple cavities. For bounded and connected solids, the bounding surface must be (a) *closed*, (b) *orientable*, (c) *connected* and (d) *nonself-intersecting*. Nonself-intersection is essential for otherwise, a bounding surface may enclose two or more domains or volumes defying the Jordon's curve theorem.

A closed surface is one having no boundary. For instance, a sphere and a torus in Figure 8.4 (e) and (g). A sphere and a cube, both with cylindrical through holes, being homeomorphic to a torus are also closed surfaces. A disc has one boundary curve, a circle, and is topologically the same as a hemisphere (Figure 8.7a). A cylinder (Figure 8.7b), open at both ends (discs removed from both ends), has two boundary curves. However, a cylindrical surface (Figure 8.7c) has only one boundary curve.

8.3.1 Closed-up Surfaces

A generic surface (as in Figure 8.6) can be thought to be composed of boundary components, which

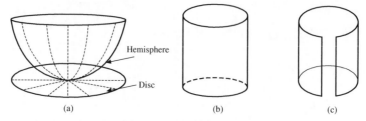

Figure 8.7   (a) Homeomorphism between a disc and a hemisphere (b) an open ended cylinder having two boundaries (c) a cylindrical surface having one boundary

are all homeomorphic to a circle. If we attach a disc to each boundary component of a surface S, the resulting surface $\hat{S}$ will be a closed one. This closing-up operation preserves homeomorphism types, that is, S_1 and S_2 are homeomorphic to each other ($S_1 \approx S_2$) if and only if $\hat{S}_1 \approx \hat{S}_2$. Thus, we can divide surfaces into classes, where two surfaces are in the same class if they have the same homeomorphic closed-up surfaces. Given $\hat{S}_1$ and $\hat{S}_2$ as two closed surfaces, we can cut out a disc from each one and attach the resultants along their cut boundaries. The result is a closed surface $\hat{S}_1 \# \hat{S}_2$ called the *connected sum* of two surfaces. As an example, in Figure 8.8, discs are cut from two spheres and the resulting surfaces are joined at the boundaries to get a double sphere which is a closed surface. The connected sum of any two surfaces does not depend not the choice of discs cut from each surface and that the connected sum operation respects homeomorphism. Thus, if $S_1 \approx S_1'$ and $S_2 \approx S_2'$, then $S_1 \# S_2 \approx S_1' \# S_2'$.

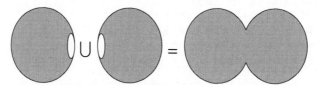

Figure 8.8   Connected sum of two spheres

8.3.2   Some Basic Surfaces and Classification

A sphere and torus introduced above are some examples of basic closed surfaces in three-dimensions. We can build a torus (doughnut) from a rectangular piece of paper (Figure 8.9a) by gluing together the edges with corresponding arrows shown. Note that as an intermediate step, an open ended cylinder is obtained with two boundaries. A Möbius strip is obtained by gluing two opposite ends of a rectangular strip (Figure 8.9b) with a twist. We may find that it has only one boundary. The construction of a torus and Möbius strip can be combined (Figure 8.9c) in a manner that we get an open cylinder in an intermediate step to twist one end by 180° and then glue the two ends. The resulting surface is a Klein bottle which cannot be built in a three-dimensional space without self intersection. Figure 8.9 (c) shows two projections of the Klein bottle which is a closed surface (zero boundaries). Finally, we can glue the opposite sides of a rectangular strip such that there is a twist about both the horizontal and vertical axes as suggested in Figure 8.9(d). The resultant surface is a real projective plane or a cross cap which, again, cannot be constructed in three-dimensions without self intersection. Another way to model a projective plane is to connect each point on the rim of a hemisphere to a corresponding point on the opposite side with a twist. The third way is to attach a disc to a Möbius strip making it a closed surface without boundaries.

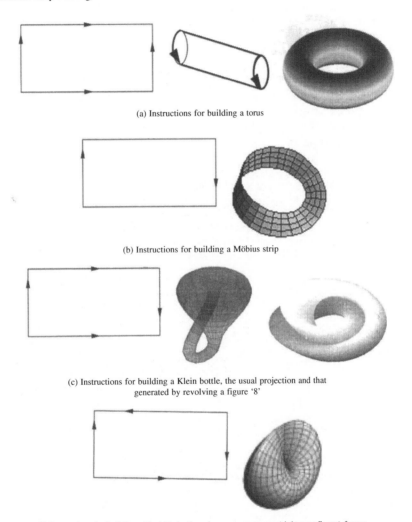

(a) Instructions for building a torus

(b) Instructions for building a Möbius strip

(c) Instructions for building a Klein bottle, the usual projection and that
generated by revolving a figure '8'

(d) Instructions for building a Real Projective plane or a cross-cap (rightmost figure) from a
rectangular paper

Figure 8.9   Some basic surfaces.

With some basic surfaces aforementioned, the main classification theorem for surfaces states that *every closed surface is homeomorphic to a sphere with some handles*[3] *or crosscaps attached*, that is, every single surface is one of the following: (a) a sphere, (b) a connected sum of tori or (c) a connected sum of projective planes or cross caps. The scope of this chapter shall be restricted to closed connected non-intersecting surfaces that can be constructed in three-dimensions. Thus, we would deal with surfaces homeomorphic to a sphere or a connected sum of tori.

8.4   Invariants of Surfaces

To better understand surfaces, we would require some characteristics that capture the essential qualitative properties, and remain invariant under homeomorphic transformations. These are

(a) **Number of boundary components.** This number is represented by an integer c. For instance, for a sphere or a torus, $c = 0$, for a disc or a hemisphere $c = 1$, for an open-ended cylinder $c = 2$ which is the case as well for the surface shown in Figure 8.6.

(b) **Orientability.** Consider a sphere in Figure 8.10(a) with a circle of arbitrary radius drawn about the center Q, a point on the sphere. Let an outward normal **n** be drawn at Q suggesting the orientation of the circle to be anticlockwise, and let C be any arbitrary closed path on the sphere. If Q traverses along C, the orientation of the normal would be preserved if Q returns to its original position. Such surfaces are termed orientable. Like a sphere, a hexahedral surface (a polyhedral surface in general) and torus are both orientable. Consider the sketch of a Möbius surface in Figure 8.10(b). If the circle at point Q commences to traverse towards the left along the closed path shown, the direction of the normal is reversed (dotted line) when Q reaches its original position. Such surfaces are termed non-orientable. Orientability is a Boolan value ε such that $\varepsilon = 1$ for all orientable surfaces while 0 for the non-orientable ones.

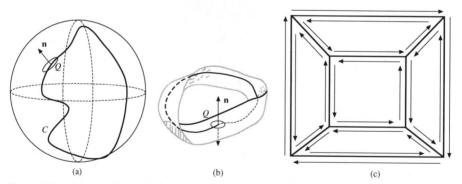

(a)   (b)   (c)

Figure 8.10   (a) Orientable sphere, (b) non-orientable Möbius surface and (c) Möbius rule for orientability of closed polyhedral surfaces

An orientable surface has two distinguishable sides which can be labeled as the inside and outside. Being closed is not a sufficient condition for a surface to be orientable. An example is the Klein bottle

---

[3]A handle is analogous to that in a coffee mug. Note that a torus (doughnut) and a coffee mug are homeomorphic to each other

(Figure 8.9c) that is closed but not orientable. For closed polyhedral surfaces, Möbius suggested a way to determine whether they are orientable. The edges enclosing a face may be traversed clockwise such that the normal to the face points into the solid and the face is to the right of the direction. For a closed surface, each edge will receive two arrows, one for each face that it bounds. The surface is orientable, if and only if, the directions of the two arrows are opposite to each other. An example is shown in Figure 8.10 (c) for a hexahedral topology. Perimeter edges bounding a face may also be traversed anticlockwise so long as this sense of traverse is maintained for all faces.

(c) **Genus.** It is an integer g that counts the number of handles (or voids) for closed orientable surfaces ($\varepsilon = 1$) or crosscaps for closed non-orientable surfaces ($\varepsilon = 0$). For instance, g for a sphere is zero and for a torus is 1. For surfaces with boundary components, one sets the genus to be equal to that corresponding to a closed-up surface. A disc and a sphere have the same genus as zero, and the genus of a Möbius band is the same as that of the projective plane, which is 1.

(d) **Euler characteristic.** In addition to the above, another invariant for polyhedrons based on Euler's law is called the Euler characteristic χ given as

$$\chi = v - e + f \tag{8.1}$$

where v is the number of vertices, e the number of edges and f the number of faces. The rule holds for any polydedron that is homeomorphic to a sphere. Thus, for hexahedrons in Figure 8.4 (a-d), $\chi = 8 - 12 + 6 = 2$. For a tetrahedron without holes, $\chi = 4 - 6 + 4 = 2$. For surfaces, the Euler characteristic can be expressed in terms of the above three invariants

$$\chi = 2 - 2g - c \qquad \text{if } \varepsilon = 1$$

and $$\chi = 2 - g - c \qquad \text{if } \varepsilon = 0 \tag{8.2}$$

Since a hexahedral is homeomorphic to a sphere, χ for a sphere with genus 0 is expected to be 2 which is confirmed by Eq. (8.2). Alternatively, we may represent a sphere in discrete form as a soccer ball, for example Figure 8.11 that has 60 vertices, 90 edges and 32 faces. The Euler characteristic is $60 - 90 + 32 = 2$.

Figure 8.11   A soccer ball

(e) **Connectivity number of a surface.** This number is equal to the smallest number of closed cuts, or cuts connnecting points on different boundaries or on previous cuts that can be made to separate a surface into two or more parts. For closed surfaces, the connectivity number is $3 - \chi$ while for a surface with boundaries, it is $2 - \chi$. A surface with connectivity number 1, 2 or 3 is termed *simply, doubly* or *triply* connected respectively. A sphere is simply connected as it needs a single closed cut to be separated into two parts while a torus is triply connected. The first closed cut will render an open cylinder, the second cut joining the two boundaries of this cylinder will result in a plane while the third cut across the plane boundary will separate it into two parts. For the surface in Figure 8.6, $c = 2$ and $g = 0$. Thus, $\chi = 0$ implying that the surface is doubly connected. One can make two cuts, each joining the outer and inner boundaries to separate the surface into two parts.

8.5   Surfaces as Manifolds

Manifolds are local shapes describing the local topology of geometric entities. For a curve, its local topology is a line for which reason curves are termed *one-manifold*. A surface, however curved and complicated so long as it does not intersect with itself, can be thought of as composed of small, two-dimensional Euclidean patches glued together. For this reason, a surface is called *two-manifold*. Mathematically, a surface is two-manifold if and only if at point P on the surface, there exists an open ball B of a sufficiently small radius r with center P such that the intersection of the ball with the surface is homeomorphic to an open disc. For instance, a sphere and a torus (Figure 8.12a) are both two-manifolds throughout since their intersection anywhere with an open ball yields a closed curve homeomorphic to an open disc. However, a closed surface of two cubes sharing an edge shown in Figure 8.12 (b) is not a two-manifold at the site shown as the intersection with an open ball yields two intersecting discs that cannot be morphed into a single disc without gluing.

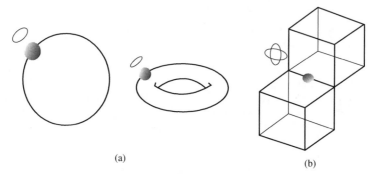

(a)   (b)

Figure 8.12   (a) Two-manifolds and (b) not a two-manifold

8.6   Representation of Solids: Half spaces

Boundary determinism in section 8.1 (c) is a strong property which suggests that a solid V can be identified by a closed and orientable surface $b(V)$ which may either be analytical (a cube or a sphere for example) or may be composed of different generic patches (Coon's, Bézier, B-spline and others) discussed in Chapter 7. We can locally control the shape of such patches to design the desired solid model. A marked advantage achieved for representation schemes is that they are only required to store the boundary surface information and not the points enclosed within $b(V)$. In this regard, half spaces contribute elegantly in the representation scheme for bounded solids, in that by combining half spaces using set operations (union, intersection and difference discussed in section 8.9), various solids can be constructed.

Half-spaces are unbounded geometric entities such that they divide the representation space E^3 into two infinite portions, one filled with material while the other empty. A half-space H can be defined as

$$H = \{P \mid P \in E^3 \text{ and } f(P) < 0\}$$

where P is a point in E^3 and $f(P) = 0$ is the equation of the surface. Most widely used half-spaces amongst analytical are planar, cylindrical, spherical, conical and toroidal defined below

Planar half-space: $H = \{(x, y, z)\mid z < 0\}$

Cylindrical half-space with radius R: $\boldsymbol{H} = \{(x, y, z)|\ x^2 + y^2 < R^2\}$

Spherical half-space with radius R: $\boldsymbol{H} = \{(x, y, z)\ |\ x^2 + y^2 + z^2 < R^2\}$

Conical half-space with cone angle α: $\boldsymbol{H} = \left\{(x, y, z)|\ x^2 + y^2 < \left[\left(\tan\dfrac{\alpha}{2}\right)z\right]^2\right\}$

Toroidal half-space with radii R_1 and R_2

$$\boldsymbol{H} = \{(x, y, z)\ |\ (x^2 + y^2 + z^2 - R_2^2 - R_1^2)^2 < 4R_2^2(R_1^2 - z^2)\}$$

Intricate solids can be represented using half spaces by treating them as lower level primitives and combining them using set operations like those in Constructive Solid Geometry (Section 8.9). For example, a block B with a cylindrical void C shown in Figure 8.13 is represented using seven half spaces. Six of those are planar half spaces with their material sides pointing into the solid. The block is the union of six intersecting half planes. The 7th half space is cylindrical with its material side pointing towards the axis. The complement of the cylindrical half space has the material direction pointing away from the axis. If this half plane is intersected with the block and then a union is taken, Figure 8.13 results, that is, in general, any solid may be considered as the union of intersections of the half planes or their complements.

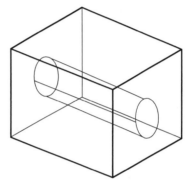

Figure 8.13 A block with a cylindrical void requiring seven half spaces for its representation

Any representation scheme for computer modeling of solids should: (a) be versatile and capable of modeling a generic solid, (b) generate valid solids having characteristics described in Section 8.1, (c) be complete such that every valid representation (solid) produced is unambiguous, (d) generate unique solids in that no two different representations should generate the same object, (e) have closure implying that permitted transformation operations on valid solids would always yield valid solids and (f) be compact and efficient in matters of data storage and retrieval. This chapter discusses three solid modeling techniques, namely, wireframe modeling, boundary representation method and constructive solid geometry. The schemes, by themselves, may not have all the features described above and thus most commercially existing solid modelers employ them in combination as required by the application.

8.7 Wireframe Modeling

This method is perhaps one of the oldest to represent solids. The representation is essentially through a set of key vertices connected by key edges. Consequently, two tables are generated for data storage, one storing the topology (connectivity) and other the geometry. The edges may be straight or curved. In former, the coordinates of the end points are stored. For curved edges, the control points, slopes and knot vector may be stored depending on the Ferguson, Bézier or B-spline segments modeled. For example, the data tables for a tetrahedron are given in Figure 8.14 with the edges numbered within parenthesis.

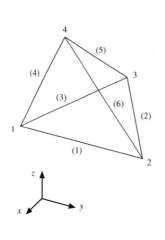

Edge table

Edge Number	Vertex 1	Vertex 2
1	1	2
2	2	3
3	1	3
4	1	4
5	3	4
6	2	4

Vertex table

Vertex Number	x	y	z
1	0	0	0
2	1	4	0
3	−3	2	0
4	−1	2	4

Figure 8.14   Data tables for a tetrahedron wireframe

While data structures used in wireframe models are simple, wireframes are nonunique and ambiguous, which is because the models do not include the facet information. Consider a wireframe in Figure 8.15 left which can either represent a solid on top right or bottom right.

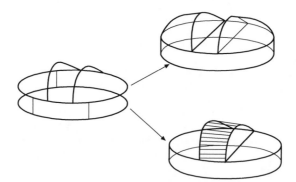

Figure 8.15   Wireframe (left) that may represent two solids on the right

Another example is a block void within a block (Figure 8.16). While we understand that each quadrilateral (or a square) represents a (flat) face, the opening of the void is not quite discernable and the wireframe can represent any of the three possibilities shown below. Note that the three objects would be identical only for a regular cubic void within a cube.

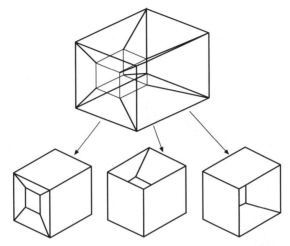

Figure 8.16   A wireframe representing a block void within a block and its solid model interpretations (below)

Due to ambiguous representation, wireframes are limited in use in solid modeling though they are popular in applications like *preview* or *animations* since one does not need to render a complex model or an animated sequence which could be very time consuming.

8.8   Boundary Representation Scheme

B-rep for short, this solid modeling scheme can be regarded as an extension of wireframe modeling to include the face information. The faces, individually, can either be analytical surfaces or design patches discussed in Chapter 7. B-rep directly employs the Jordan's curve theorem on boundary determinism stating that a closed, connected, orientable and nonself-intersecting surface determines the interior of a solid. As in wireframe models, both topological and geometric information is stored in B-rep as well wherein relationships among vertices, edges, faces and orientations form a part of the topological data while design equations of edges and faces are stored as geometric input. Face orientations may be recorded such that a normal to the face points into the solid. This can be ensured by the clockwise ordering of vertices (right-handed rule) associated with the face. Once done for all faces, we can then inspect the normal vectors to distinguish the interior of the solid from its exterior. Thus, for a tetrahedron in Figure 8.14, the vertices of the front face may be ordered as 2, 1 and 4. For the face on the right, the order should be 3, 2 and 4. Likewise, for the back and bottom faces, the order should be 1, 3, 4 and 1, 2, 3, respectively.

8.8.1   Winged-Edge Data Structure

A data structure in wide use for a B-rep model is the Baumgart's winged-edge data structure for polyhedrons, which is also applicable to homeomorphic solids that can be achieved by stretching the straight edges to curved ones to have curved faces. An advantage is that the winged-edge structure employs only edges to document the connectivity. First, the data structure is described for polyhedrons with no voids. Consider a tetrahedron (Figure 8.17a) which shows the edges numbered within parenthesis and polygonal faces denoted alphabetically. Consider edge (3) in Figure 8.17 (b) formed by vertices 2 and 4. The edge has two associated faces **A** and **B**, and if the direction of the edge is from vertex 2 to vertex 4, **A** is on the left and **B** on the right of the edge. When viewed from outside the solid, the faces are traversed clockwise so that the normal to the faces using the right-handed rule points into the solid. As shown in Figure 8.17(b), edge (3) is traversed twice in two different directions, once while traversing **A** and the other while traversing **B**. On face **A**, edge (5) precedes edge (3) while edge (1) succeeds it. Similarly on **B**, (2) is the preceding edge while (6) is the succeeding edge. Note that the schematic in Figure 8.17(b) is suggestive of the winged-edge with edges (1), (2), (5) and (6) being the wings of (3) for which reason, the data structure is named so. We can tabulate the above information for edge (3) as shown in the edge table. Likewise, similar information can be tabulated for all the edges. We would note that the order of the two vertices comprising the edge would determine the direction of the latter, which in turn would help decide the faces on the left and right of the edge. In case the order of the vertices is changed, the entries in the edge table would get altered accordingly.

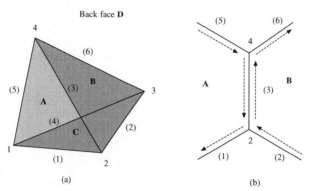

Figure 8.17   Winged-edge data structure for a tetrahedron

Edge table

Edge Name	Vertices		Faces		Clockwise traverse on left face		Clockwise traverse on right face	
	Start	End	Left	Right	Preceding edge	Suceeding edge	Preceding edge	Suceeding edge
(3)	2	4	A	B	(5)	(1)	(2)	(6)
(1)	1	2	A	C	(3)	(5)	(4)	(2)
(2)	2	3	B	C	(6)	(3)	(1)	(4)
(4)	1	3	D	C	(5)	(6)	(2)	(1)
(5)	1	4	D	A	(6)	(4)	(1)	(3)
(6)	3	4	B	D	(3)	(2)	(4)	(5)

In addition to the edge table, the data structure also requires a vertex table and a face table. Both tables document the edges associated with the vertices and faces, respectively. Since more than one edge may be associated with a vertex (or face), the following tables are not unique.

Vertex table		Face table	
Vertex	Edge	Face	Edge
1	(1)	**A**	(1)
2	(2)	**B**	(2)
3	(3)	**C**	(4)
4	(6)	**D**	(5)

With these tables, we can know what vertices, edges and faces are adjacent to a face, edge or a vertex, which are 9 adjacency relations. Note that once a polyhedron is given with fixed number of edges, faces and vertices, the sizes of the three tables containing topological information get fixed. All now remains is to store the geometric data which corresponds to the coordinates of the data points, slopes and/or knot vectors of the boundary curves and faces (surface patches) forming a solid.

In case a solid, e.g. in Figure 8.16, has voids that penetrate the solid partially or completely, there are two ways in which one could apply the Baumgart's data structure. For a face with inner loops as in Figure 8.18(a), we can retain the clockwise order for the outer loops while the inner loops would be ordered counterclockwise. The alternative is to add an auxiliary edge between an inner loop and the outer loop as shown in Figure 8.18(b) with dashed lines. The auxiliary edge will have the same face to its left and right, one way to identify them in the data structure, and the number of loops will get reduced to 1 for a face with inner loops. The topological tables can be constructed in a manner similar to those explained for the tetrahedron above.

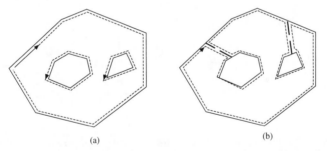

(a)   (b)

Figure 8.18   Two schemes to treat the inner loops for the Baumgart's data structure

Baumgart's winged-edge data structure is efficient in performing certain operations, a reason why it is more popular compared to other schemes. For instance, it helps in identifying all loops on each face quickly and also allows traverse along each edge on a face. Thus, many algorithms in computational geometry (Chapter 9) benefit from this data structure. Baumgart's scheme also helps in quickly determining the orientation (inward pointing normal) of each face. This is important when a computer is rendering the image of an object on the screen. Once the object is transformed to a particular orientation, its visible and invisible faces can be determined quickly. Also, for a given face, finding its neighboring faces is easier.

8.8.2   Euler-Poincaré Formula
The formula relates the number of vertices, edges and faces of a polyhedral solid and has been generalized to include the potholes and through voids penetrating the solid. For V as number of

vertices, E the number of edges, F the number of faces, G the number of holes (or genus) penetrating the solid, S the number of shells and L as the total number of outer and inner loops, the Euler-Poincaré formula is given as

$$V - E + F - (L - F) - 2(S - G) = 0 \qquad (8.3)$$

Here, a shell is an internal void of a solid bounded by a closed connected surface that can have its own genus value. The solid itself is counted as a shell. Euler-Poincaré formula is employed to test the topological validity of a solid, that is, if the right hand side of Eq. (8.3) is non-zero, the solid is an invalid solid. However, the vice-versa is not true, that is, a zero value of the formula does not necessarily mean that the solid is valid.

Example 8.1. Verify the Euler-Poincaré formula for the solids shown in Figure 8.19.

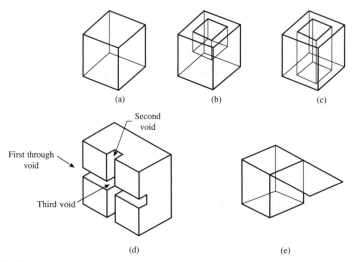

Figure 8.19   (a) A cube, (b) cube with a partial void, (c) cube with penetrating void, (d) half section of the cube with three orthogonally through voids and (e) cube with a rectangular hanging face.

    A cube has 8 vertices, 12 edges, 6 faces and therefore 6 loops with a shell value 1. Euler-Poincaré formula results in $8 - 12 + 6 - (6 - 6) - 2(1 - 0) = 0$. A cube with a partial void in Figure 8.19 (b) has 16 vertices, 24 edges, 11 faces (6 outer and 5 inner) and 12 loops (one inner loop on top surface) for which the formula gives $16 - 24 + 11 - (12 - 11) - 2(1 - 0) = 0$. For a through void in Figure 8.19(c), the solid has 16 vertices, 24 edges, 10 faces, 12 loops (2 inner loops on the top and bottom surfaces, respectively) and one hole for which $16 - 24 + 10 - (12 - 10) - 2(1 - 1) = 0$. For a solid in Figure 8.19 (d) which is a cube with three voids orthogonal to each other, there are 40 vertices, 72 edges, 30 faces, 36 loops, 1 shell and, say, x voids. The Euler-Poincaré result for this solid is $40 - 72 + 30 - (36 - 30) - 2(1 - x) = 0$ implying that $x = 5$, that is, the number of voids is five which seems counter intuitive and can be explained. The first void is a through hole as shown in the figure, and in an orthogonal direction, there are two voids (as opposed to one) as shown. Likewise, in the third direction, there are two voids making a total of 5. The solid in Figure 8.19(e) is a cube with a protruding flat face with 10 vertices, 15 edges, 7 faces and no holes with a shell value 1. Even though the Euler-Poincaré formula gives $10 - 15 + 7 - (7 - 7) - 2(1 - 0) = 0$, it is not a valid solid because of the protruding face that has zero thickness and thus no interior within itself.

8.8.3 Euler-Poincaré Operators

Given a polyhedron model, we may want to edit it by adding or deleting edges, vertices, faces and genus to create a new polyhedron using Euler operators. The operators are designed such that the Euler-Poincaré formula in Eq. (8.3) is always satisfied for the intermediate results. Two groups of Euler operators are put to use, the *Make* and *Kill* groups for adding and deleting, respectively. Euler operators are written as *Mxyz* or *Kxyz* for the Make and Kill groups, respectively, where *x*, *y* and *z* represent a vertex, edge, face, loop, shell or genus. For instance, *MEV* implies making (or adding) an edge and a vertex while *KEV* means killing or deleting an edge and a vertex. Euler operators form a complete set of modeling primitives in that any polyhedron satisfying Euler-Poincaré relation can be constructed using a finite sequence of operators. Euler operators, thus, are significant from the viewpoint of constructing B-rep solid models. The make group table shows some operators of the Make group used to add elements in the existing polyhedral topology and one for the Make-Kill group used to add and delete some elements at the same time.

Operator	Implication	V	E	F	L	S	G	Change in Euler-Poincaré formula
MEV	Make an edge and a vertex	+1	+1					0
MFE	Make a face and an edge		+ 1	+1	+1			0
MSFV	Make a shell, a face and a vertex	+1		+1	+1	+1		0
MSG	Make a shell and a genus					+1	+1	0
MEKL	Make an edge, Kill and loop		+1		−1			0

Note that the operations above are designed such that they do not cause any change in the Euler-Poincaré relation as shown in the rightmost column. *MEV* implies adding an edge and a vertex. A face and an edge are added via *MFE*. When adding a face, a loop also gets added which causes no change in the expression $(L - F)$ of Eq. (8.3). *MSG* makes a shell with a hole and *MEKL* makes an edge and kills a loop. *MEKL* operation is commonly employed when connecting the outer loop with the inner one through an auxiliary edge as suggested in Figure 8.18(b).

The Kill group of Euler operators performs the deletion operations, and exchanging *M* with *K* in the Make group table yields the operators of the Kill group given below. With these operations, we can reduce, for instance, a cube to its non-existence. Otherwise, we may partially delete the entities of an existing polyhedron and then use the Make group operators to reconstruct a new one.

Operator	Implication	V	E	F	L	S	G	Change in Euler-Poincaré formula
KEV	Kill an edge and a vertex	−1	−1					0
KFE	Kill a face and an edge		− 1	−1	−1			0
KSFV	Kill a shell, a face and a vertex	−1		−1	−1	−1		0
KSG	Kill a shell and a genus					−1	−1	0
KEKL	Kill an edge, Make and loop		−1		+1			0

Example 8.2. Construct a cube using the Euler operators.

Following are the operations for constructing a cube afresh. Vertices and edges added are shown in thick and faces added are shaded. Note however that, in general, the intermediate results may not be topologically valid polyhedra though we expect the final result to be one.

Operator	V	E	F	L	S	G	Result
MSFV	+1		+1	+1	+1		(Bottom face made)
MEV	+1	+1					
MEV	+1	+1					
MEV	+1	+1					
MEV	+1	+1					
MEV	+1	+1					
MFE		+1	+1	+1			(Top face made)
MEV	+1	+1					

	V	E	F	L	S	G
MFE		+1	+1	+1		

(Right face made)

	V	E	F	L	S	G
MFE		+1	+1	+ 1		

(Back face made)

	V	E	F	L	S	G
MFE		+1	+1	+1		

(Left face made)

	V	E	F	L	S	G
MFE		+1	+1	+1		

(Front face made)

	V	E	F	L	S	G
MEV	+1	+1				

8.9 Constructive Solid Geometry (CSG)

This scheme is another way of representing solids which can be generated by combining primitives using Boolean operations. Primitives can include solids like block, cone, cylinder, sphere, triangular prism, torus and many others. Solids participating in CSG need not only be bounded by analytical surfaces but also by generic surface patches developed in previous chapters. Some primitive solids are shown in Figure 8.20. Boolean operators include those used in set operations, for instance, union, intersection and difference.

Solids used in the CSG approach are first instantiated, transformed and then combined to form more complex solids. Instantiation involves making available a copy of the primitive (if existing) from the database. Transformation of a primitive is then required to scale or position (translate and/ or rotate) itself with respect to others (or their Boolean result) as desired in design. The primitive in its resulting size and/or position may then be joined with, cut from or intersected with an existing solid to get the desired features. Consider, for instance, the design of an L-shaped bracket which may be treated as a union of two blocks shown in Figure 8.21(a). The blocks can be instantiated, scaled, transformed relatively and then joined to form the bracket as shown in Figure 8.21(b).

For computer modeling, the block primitives above may be treated as objects named Block 1 and

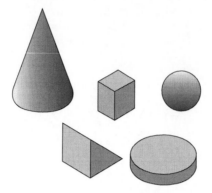

Figure 8.20   Primitive solids used in constructive solid geometry

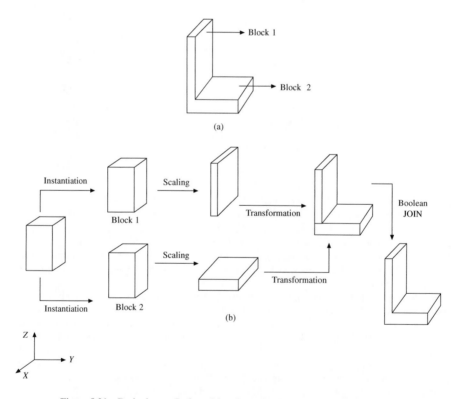

Figure 8.21   Designing an L-shaped bracket using constructive solid geometry

Block 2, respectively which may be identified by the three dimensions (length, width and height) and the location of their respective reference points or local coordinate origins. Initially of standard size, we may scale the three dimensions of the blocks by factors, say, x, y and z using the scale command. Thus, for Block 1, the first CSG operation would be scale (Block 1, x_1, y_1, z_1). This object would then require to be translated in that the reference point of the block would be shifted by, say, (a, b, c) with respect to a global coordinate frame as shown. This may be accomplished using the translate command, translate (scale (Block 1, x_1, y_1, z_1), a_1, b_1, c_1). Similar operations for Block 2 would be translate (scale (Block 2, x_2, y_2, z_2), a_2, b_2, c_2). At this stage, the origins (and local axes) of the two blocks would be positioned appropriately and the blocks would be united using the Boolean union or JOIN command that would appear as

JOIN (translate (scale(Block 1, x, y_1, z_1), a_1, b_1, c_1), translate (scale (Block 2, x, y_2, z_2), a_2, b_2, c_2)) or if the union is expressed using the '+' sign then

translate (scale (Block 1, x_1, y_1, z_1), a_1, b_1, c_1) + translate (scale (Block 2, x_2, y_2, z_2), a_2, b_2, c_2)

$$(E1)$$

8.9.1 Boolean Operations

Given two sets (solids) A and B, their union (A $\cup$ B or A + B) consists of all points belonging to A and B. Their intersection (A $\cap$ B) consists of points common to both A and B and the difference A $-$ B consists of points in A but not in B. Similarly, B$-$A would consist of points only in B and not in A. Consider, for instance, the Boolean interactions between a sphere A and a block B (Figure 8.22 a) which is a cube of side length the same as the diameter of the sphere. The sphere is placed over the cube such that the center of the sphere coincides with that of the top face of the cube. Figures 8.22 (b-e) show the union, intersection and difference operations A–B and B–A, respectively.

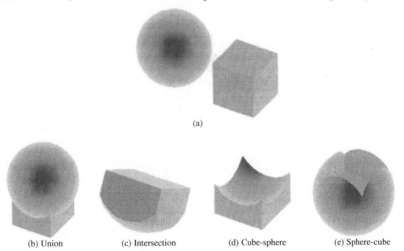

(a)

(b) Union   (c) Intersection   (d) Cube-sphere   (e) Sphere-cube

Figure 8.22   Boolean operations using a cube and a sphere

Note that expression (E1) for the bracket design can be expressed graphically in the form of a history tree or the CSG tree shown in Figure 8.23 (a). In addition, to cut holes in the bracket as shown in Figure 8.23 (b), the way is to instantiate two cylinders, transform (scale and translate) them appropriately and cut (represented using '–' sign) them from the respective blocks before uniting the latter. The graphic representation of the procedure then appears as shown in Figure 8.23(c). For a cylindrical object with radius and length as defining features, and scale factors as r and l, respectively, the CSG expression for a bracket with two holes may be written as

[translate(scale(Block 1, x_1, y_1, z_1), a_1, b_1, c_1) – translate (scale(Cylinder 1, r_1, l_1), a_3, b_3, c_3)]

+ [translate(scale(Block 2, x_2, y_2, z_2), a_2, b_2, c_2) – translate(scale(Cylinder 2, r_2, l_2), a_4, b_4, c_4)]

$$(E2)$$

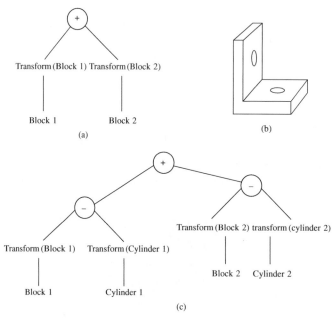

Figure 8.23   The CSG tree representations for a bracket without and with holes

Every solid constructed using the CSG scheme has a corresponding design expression and thus a CSG tree associated with it. We may note, however, that a CSG solid may not necessarily be represented by a unique tree as the operations for constructing the solid may not be unique. For instance, the bracket frame above may result by cutting a block from another block. Alternatively, we may join the two blocks first and then cut holes at respective sites. A CSG tree is concise, unambiguous, closed and easy to create and edit. Its domain depends on the available set of primitive objects, as well as the set of transformational and combinational operators.

8.9.2   Regularized Boolean Operations

The interior and boundary of a solid V have been defined in Section 8.1. Intuitively, the interior $I(V)$ of a solid comprises all points within the solid and not those on its boundary. A point Q is exterior to the solid if there exists an open ball B of radius r centered at Q such that the ball does not intersect with the solid. That is, if any point $p \in B$ ($|p - Q| < r$), then $p \notin V$. A set of all exterior points is termed the exterior of the solid represented as $E(V)$. Points that neither belong to the interior or exterior constitute the boundary $b(V)$ of the solid. The closure of a solid $C(V)$ is then defined as the union of its interior and the boundary, that is, $C(V) = I(V) \cup b(V)$ or $I(V) + b(V)$. In other words, the closure of a solid is the complement $E(V)$ of its exterior and contains all points that do not belong to the exterior of the solid. In a manner, V and $C(V)$ are the same with $C(V)$ as the formal definition of the solid. The above discussion seems necessary to circumvent certain pitfalls of the Boolean operations as given by an example in Figure 8.24. For a block and cylinder shown adjacent to each other, their intersection yields a common disc (a one-manifold) that is not a valid solid and the Boolean operation, as is, violates the closure property in Section 8.1(f).

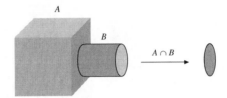

Figure 8.24   Boolean intersection operation with a block and a cylinder

To eliminate the lower dimensional results of set operations, we need to regularize the Boolean operations as follows:

We first compute the result as usual wherein the lower dimensional features (like the disc above) may be generated. Then, the interior of the result is computed that eliminates all lower dimensional components. In this step, we achieve only the interior of the solid which is united with its boundary in the subsequent step by computing the closure. The regularized Boolean operations for solids A and B can be summarized as

Regularized union: $C[I\ (A \cup B)]$

Regularized intersection: $C[I(A \cap B)]$

Regularized difference: $C[I\ (A - B)]$

Based on the above, the regularized intersection between the block and the cylinder shown in Figure 8.24 is an empty set. The two examples that illustrate the modeling procedure using constructive solid geometry are: (i) a hexagonal bolt, different parts of which are shown in Figure 8.25 as components of the history tree and (ii) a more complex one is of a Robosloth, the CSG model of which is shown in Figure 8.26(a) with the realized prototype in Figure 8.26(b).

8.10   Other Modeling Methods

Many engineering components are such that the cross-section is uniform in the depth direction. Also, many are axisymmetric. To model such components, solid modelers employ different sweep methods. A planar wireframe cross-section composed of a simple (nonself-intersecting) closed contour of edges (linear or curved) can be *extruded* along the vector perpendicular to the plane containing the contour. This is called *translational sweep* an example of which is shown in Figure 8.27 (a). A simple closed contour may also be revolved by a known angle about an axis to result in a solid of revolution (Figure 8.27b). This is called *rotational sweep*. In many instances, the wireframe cross-section need not follow a linear path and the sweep path may be represented by a curve. An example of a solid obtained using *nonlinear sweep* is depicted in Figure 8.27 (c). A sweep path is often termed as the *directrix*. In a *hybrid sweep*, we can combine two or more sweep solids using the regularized set operations discussed above.

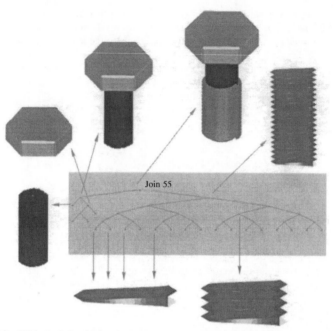

Figure 8.25 **The CAD of a bolt with its history tree (arrows relate between the nodes of the history tree and the corresponding features)**

The cross-section contour is often chosen to lie on a plane. The contour should be closed and nonself-intersecting in that it should bind a single domain with finite area. However, the wireframe may not be connected. That is, it can have an outer ring of edges and a few inner ones to depict the voids in sweep solids, as shown in Figures 8.27 (a) and (b). Further, the directrix should be such that self-intersection does not occur at any instant when a contour is swept along it.

The other modeling approach is an extension of the tensor product method for surface patches to three-dimensional parametric space. The resulting solid is called a *hyperpatch* since it is bounded by surface patches. For three normalized parameters u, v and w, points on or inside the hyperpatch are expressed using an ordered Cartesian triple. That is

$$P(u, v, w) = [x(u, v, w)\, y(u, v, w)\, z(u, v, w)] = \sum_{i=0}^{m} \sum_{j=0}^{n} \sum_{k=0}^{p} \mathbf{C}_{ijk} u^i v^j w^k \qquad (8.4)$$

where $\mathbf{C}_{ijk}$ are the data points for Bézier or B-spline hyperpatches in three-space. A tricubic Bézier hyperpatch for example is obtained for $m = n = p = 3$. The face surfaces, edge curves, and corner vertices can be obtained by substituting appropriate values of the parameters into the above equation. The face surfaces are given by $P(0, v, w)$, $P(1, v, w)$, $P(u, 0, w)$, $P(u, 1, w)$, $P(u, v, 0)$, and $P(u, v, 1)$. Similarly, the equation of any curve is obtained by fixing two parametric variables. $P(u, 0, 0)$ and $P(u, 1, 0)$ are two of the 12 boundary curves. The eight corner vertices correspond to the values of u, v and w as 0 and 1 only.

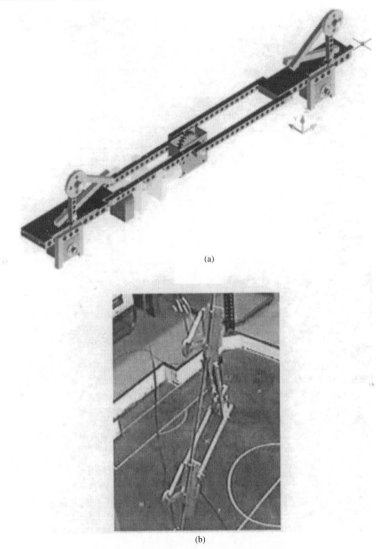

(a)

(b)

Figure 8.26   (a) CSG model of a Robosloth and (b) Robosloth in working

8.11   Manipulating Solids

Manipulations on primitives are performed on a routine basis in solid modeling. Set operations in constructive solid geometry cannot be performed without transformations, that is, re-positioning or scaling of the primitives. Further, determining intersection, union or difference involves computing

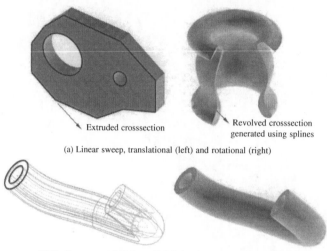

Extruded crosssection

Revolved crosssection
generated using splines

(a) Linear sweep, translational (left) and rotational (right)

(b) Nonlinear sweep (wireframe in left showing crosssection and path)

Figure 8.27   Sweep solids

intersection between the bounding surfaces of two primitives. Related are the issues of determining points and curves on the bounding surface. Segmentation and trimming operations may also be seen as intersection problems between the segmenting or trimming surface and the solid. Segmentation involves splitting a solid into two parts using a plane. Note that post segmentation, each sub-solid must have its own topology and geometry. Trimming entails intersecting the solid with the trimming surfaces followed by the removal of solid portions outside these surfaces. Determining inersection between freeform curves, surfaces or solids is computationally involved and algorithmically complex. The following chapter discusses some ways to find intersections within the realm of computational geometry.

An important manipulating mode is the editing of solid primitives. Most new designs are alterations of existing ones, and editing a solid involves changing its existing topological (rarely) and geometric (mostly) information. Generating solid models for complex engineering parts and assemblies can be arduous, and for a few changes, one may have to regenerate the entire set in absence of the editing capability. It thus seems imperative that solids are represented in their *parametric* form wherein design dimensions and relations between features (constraints) are also stored in the data structure. Consider, for instance, the CAD model of a bolt in Figure 8.25. For different applications, one may require the bolt to have different nominal diameters, thread and bolt lengths, and head sizes. These seem notable features in bolt design and can be treated as its parameters. Among this set, some parameters may depend on others via some design rules, for instance, the head size (diameter of the circumcircle of the hexagon and head thickness) may be governed by the nominal diameter. One may then treat the nominal diameter, and thread and bolt lengths as *independent parameters* which may be altered as required. The user would expect that if the nominal diameter is altered, appropriate changes would be carried through the bolt. That is, the shank and helix diameter (for threads) would change and so would the head size. Note, however, that the relative positions between the shank and head, and shank and threads should not change, i.e., the three features should remain glued appropriately along the common geometric axis. Parametric design of CAD models, being a noteworthy concept, has and is receiving considerable attention of commercial solid modelers.

EXERCISES

1. Verify the Euler characteristic for the following polyhedrons:
   A block with a through block void
   A tetrahedron
   An open cylinder
   A torus
2. Construct the edge and vertex tables for a cube as a wireframe.
3. Construct the winged edge data structure for a cube as a B-rep solid.
4. From a cube, construct a tetrahedron using Euler operators.
5. A Mechanical component made by assembly of three parts is shown in the Figure P8.1 along with dimensions. A CSG Representation is to be made. Define the minimum basic primitives to be used for constructing the component. Give Details of the CSG tree for the given component. Include details of primitives, transformation involved (scaling translation, rotation) and the Boolean operations. Model the components shown in Figure P8.1 using any of the available solid modelers.

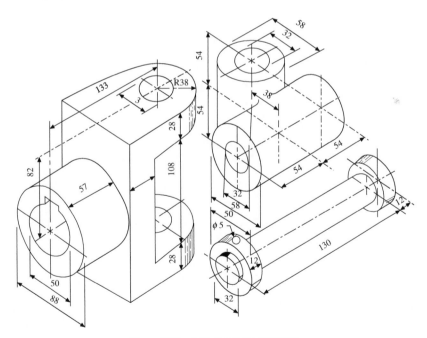

Figure P8.1   A Mechanical component.

6. Choose two machine parts (one component and one assembly) of reasonable complexity from any of the drawing books.
   a. Discuss the topological and geometrical aspects of the components in a coherent manner (point wise).
   b. Discuss steps to create the component by B-rep method.
   c. Discuss steps to create the components by CSG.
   d. Use any of the available solid modelers to create the components.
7. Can half spaces be modeled using freeform patch definitions discussed in Chapter 7. Are such patches unbounded? If not, can the intersection between freeform surface patches still constrain the material to lie within a finite volume?
8. For the solid shown in Figure 8.19(d), draw the CSG tree.
9. Review the literature to learn about how solid modelers display various features.

Chapter 9

Computations for Geometric Design

Finding intersection between curves, surfaces and solids are much used operations in computer aided geometric design and other applications like robotics. Intersection determination is primarily used in computing Boolean relations between two solids in constructive solid geometry. Herein, we are interested in computing the portions common to the two objects (if any). In path planning in robotics, collision detection requires computing the proximity between two objects (robot and obstacles) wherein, it may be required to determine if the robot is colliding (in contact) with the obstacle or not. In case not, then how far is the robot from the obstacle. Virtual assembly simulation is another application domain. For instance, a mechanical assembly has to be checked for service accessibility by a technician. Virtual simulation can verify accessibility by checking the movements of a virtual technician to reach the appropriate parts of the engine without colliding with the other parts. Rendering models (display) in computer graphics requires computation of ray collisions with the object to determine the hidden faces, depth of field and shading. The collection of algorithms to compute various relations like proximity, intersection, decomposition and relational search between geometric entities (points, lines, planes, and solids) lies within the realm of computational geometry. This chapter discusses the implementation of a few such algorithms notwithstanding their complexity or robustness.

A Euclidean space $\mathbf{R}^d$ of dimension d, has a family of natural distance metrics, known as the $\mathbf{L}^p$ norms, which are defined so that the distance between two points $\mathbf{x} = (x_1, x_2, \ldots, x_d)$ and $\mathbf{y} = (y_1, y_2, \ldots, y_d)$ is given as

$$d(\mathbf{x}, \mathbf{y}) = |x_1 - y_1|^p + |x_2 - y_2|^p + \ldots + |x_d - y_d|^p \tag{9.1}$$

The Euclidean distance between two points in a three dimensional space is given by the $\mathbf{L}^2$ norm. To compute the distance between any two geometric entities one algorithmically computes the distances between respectively belonging points.

9.1 Proximity of a Point and a Line

Consider a point $\mathbf{C}$ (x_3, y_3) and a line $\mathbf{AB}$ with end points (x_1, y_1) and (x_2, y_2). The area of triangle $\mathbf{ABC}$ is computed by calculating the determinant Δ as

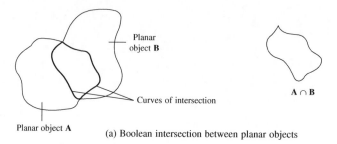

Planar object **B**

Curves of intersection

Planar object **A**

$A \cap B$

(a) Boolean intersection between planar objects

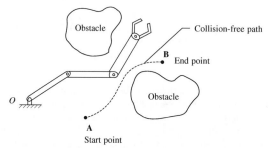

Obstacle

Collision-free path

B
End point

Obstacle

O

A
Start point

(b) Collision free path planning for a manipulator

Figure 9.1   Some examples requiring intersection/proximity analysis.

$$\Delta = \begin{vmatrix} x_1 & y_1 & 1 \\ x_2 & y_2 & 1 \\ x_3 & y_3 & 1 \end{vmatrix}$$

(9.2)

Point **C** can relate to **AB** in three possible ways (Figure 9.2):

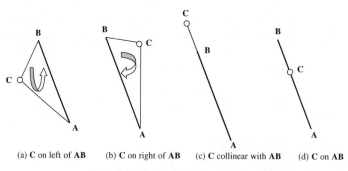

(a) **C** on left of **AB**   (b) **C** on right of **AB**   (c) **C** collinear with **AB**   (d) **C** on **AB**

Figure 9.2   Proximity of a point and a line

(a) Point **C** lies to the left of **AB** (following the convention of moving from **A** to **B**) if Δ is positive.
(b) Point **C** lies to the right of **AB** if Δ is negative.
(c) Point **C** is collinear to **AB** if Δ is zero. Further, **C** lies within **AB** if the x and y coordinates of **C** lie between the x and y coordinates of **A** and **B**.

Example 9.1. Find the proximity of the points $(0, 0)$, $(5, 1)$ and $(1.6, 0)$ with respect to the line whose end points are **A** $(1, 1)$ and **B** $(4, -4)$.

Determinant Δ for point $(0, 0)$ is

$$\Delta = \begin{vmatrix} 1 & 1 & 1 \\ 4 & -4 & 1 \\ 0 & 0 & 1 \end{vmatrix} = -8 \text{ implying that the point is to the left of } \mathbf{AB}.$$

Similarly, Δ for point $(5, 1)$ is 20 and for point $(1.6, 0)$ is 0. Thus, $(5, 1)$ is to the right to **AB** and $(1.6, 0)$ is collinear with **AB**. Further treating $C(1.6, 0)$ as a linear combination of **A** and **B**,

$$1.6 = 1(1 - u_1) + 4u_1 = 1 + 3u_1 \Rightarrow u_1 = 0.2$$

$$0 = 1(1 - u_2) - 4u_2 = 1 - 5u_2 \Rightarrow u_2 = 0.2$$

implying that $u_1 = u_2$ and that $0 < u_1 = u_2 < 1$ for which $(1.6, 0)$ lies within the segment **AB**. For a three-dimensional space, if $\mathbf{AB} \times \mathbf{AC} > 0$, C lies to the left of AB. If $\mathbf{AB} \times \mathbf{AC} < 0$, C lies to the right and if the cross product is 0, C lies on **AB**.

9.2 Intersection Between Lines

Given two lines **AB** and **CD** on a plane, we may find if they intersect and if yes, find the point or line segment (in case of overlap) of intersection. The possibilities are shown in Figure 9.3 and the following algorithm may be used to explore the above.

Check if the segments are intersecting

(a) If the determinants for triangle ABC and ABD have the same sign, then **C** and **D** both lie on the same side of **AB** and hence **AB** and **CD** cannot intersect (Figure 9.3 a). A similar check has to be performed for triangles ACD and BCD. Even though **C** and **D** may lie on either side of **AB**, if **A** and **B** lie on the same side of **CD**, the lines **AB** and **CD** will not intersect as shown in Figure 9.3 (b).
(b) If both determinants for triangles ABC and ABD are zero, then the two lines are collinear, else, the lines intersect at a point.

If the lines intersect, we can solve for the point of intersection using the parametric equations of **AB** and **CD**.

If the lines are collinear, we can find if they overlap. In that case, we can find the segment of intersection by checking each end point of **AB** and **CD** to find whether they lie on the other line, and then finally determine the common segment. The possibilities are shown in Figure 9.3 (d), (e) and (f), respectively.

Example 9.2. Find the intersection of the following lines with line **AB** whose end points are $(2, 0)$ and $(5, 0)$. The end points of the lines are: (a) **C** $(0, -4)$ and **D** $(0, 4)$, (b) **C** $(3, -4)$ and **D** $(3, 4)$ and (c) **C** $(0, 0)$ and **D** $(3, 0)$.

(a) For intersection, ΔABC and ΔABD as well as ΔACD and ΔBCD should be of opposite sign in pairs.

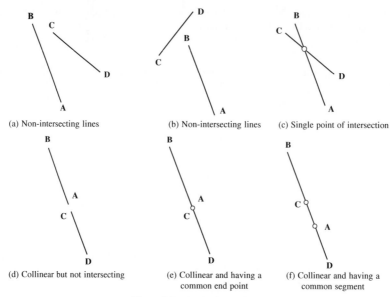

(a) Non-intersecting lines   (b) Non-intersecting lines   (c) Single point of intersection

(d) Collinear but not intersecting   (e) Collinear and having a common end point   (f) Collinear and having a common segment

Figure 9.3   Intersection of lines

$$\Delta ABC = \begin{vmatrix} 2 & 0 & 1 \\ 5 & 0 & 1 \\ 0 & -4 & 1 \end{vmatrix} = -12. \text{ Similarly } \Delta ABD = 12, \Delta ACD = -16 \text{ and } \Delta BCD = -40$$

Since ΔABC and ΔABD are of opposite signs, **C** and **D** lie on opposite sides of **AB**. Since ΔACD and ΔBCD are of same sign, **A** and **B** lie on same side of **CD**. Thus, **CD** and **AB** do not intersect (Figure 9.4 (a)).

(b) The determinants ΔABC, ΔABD, ΔACD and ΔBCD are −12, 12, 8 and −16, respectively. Since ΔABC and ΔABD as well as ΔACD and ΔBCD are of opposite signs in pairs, **AB** and **CD** intersect (Figure 9.4 (b)). The parametric equation of **AB** is $x = 2 + t(5 - 2)$, $y = 0 + t(0 - 0)$, $0 \le t \le 1$ and

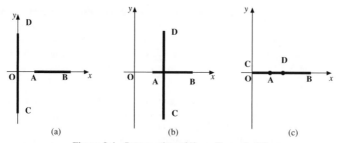

(a)   (b)   (c)

Figure 9.4   Intersection of lines, Example 9.2.

that of **CD** is $x = 3 + s(3 - 3)$, $y = -4 + s(4 + 4)$, $0 \leq s \leq 1$. Solving for t and s gives $t = 1/3$ and $s = 1/2$. Substituting the same in the parametric equation of **AB** or **CD** gives the point of intersection as $(3, 0)$.

(c) The determinants $\triangle ABC$, $\triangle ABD$, $\triangle ACD$ and $\triangle BCD$ are all 0 implying that the lines are collinear. Next, the common line segment is determined (if any). For this, the y coordinates of **A**, **B**, **C** and **D** are examined. They all being equal, further, the x coordinates are checked. **A** lies between **C** and **D**. Also, **D** lies between **A** and **B**. Thus, the common line segment is between **A** and **D**.

9.2.1 Intersection Between Lines in Three-dimensions

Consider two line segments **AB** and **CD** and let **P** and **Q** be the points on **AB** and **CD** such that

$$\mathbf{P} = (1 - t) \, \mathbf{A} + t\mathbf{B}$$

$$\mathbf{Q} = (1 - s) \, \mathbf{C} + s\mathbf{D}$$

for parameters $0 \leq t, s \leq 1$. The distance d between **P** and **Q** may be given by

$$d^2 = (\mathbf{P} - \mathbf{Q}) \cdot (\mathbf{P} - \mathbf{Q})$$

$$= [\mathbf{A} + (\mathbf{B} - \mathbf{A})t - \mathbf{C} - (\mathbf{D} - \mathbf{C})s] \cdot [\mathbf{A} + (\mathbf{B} - \mathbf{A})t - \mathbf{C} - (\mathbf{D} - \mathbf{C})s]$$

The minimum distance between **P** and **Q** can be obtained using $\dfrac{\partial d^2}{\partial t} = \dfrac{\partial d^2}{\partial s} = 0$. Or

$$\frac{\partial d^2}{\partial t} = 2[\mathbf{A} + (\mathbf{B} - \mathbf{A})t - \mathbf{C} - (\mathbf{D} - \mathbf{C})s] \cdot [(\mathbf{B} - \mathbf{A})] = 0$$

$$\frac{\partial d^2}{\partial s} = 2[\mathbf{A} + (\mathbf{B} - \mathbf{A})t - \mathbf{C} - (\mathbf{D} - \mathbf{C})s] \cdot [- (\mathbf{D} - \mathbf{C})] = 0$$

which gives

$$(\mathbf{B} - \mathbf{A}) \cdot (\mathbf{B} - \mathbf{A})t - (\mathbf{D} - \mathbf{C}) \cdot (\mathbf{B} - \mathbf{A})s = \mathbf{C} \cdot (\mathbf{B} - \mathbf{A}) - \mathbf{A} \cdot (\mathbf{B} - \mathbf{A})$$

$$- (\mathbf{B} - \mathbf{A}) \cdot (\mathbf{D} - \mathbf{C})t + (\mathbf{D} - \mathbf{C}) \cdot (\mathbf{D} - \mathbf{C})s = - \mathbf{C} \cdot (\mathbf{D} - \mathbf{C}) + \mathbf{A} \cdot (\mathbf{D} - \mathbf{C})$$

Or in matrix form

$$\begin{bmatrix} (\mathbf{B} - \mathbf{A}) \\ - (\mathbf{D} - \mathbf{C}) \end{bmatrix} \cdot [(\mathbf{B} - \mathbf{A}) \ - (\mathbf{D} - \mathbf{C})] \begin{bmatrix} t \\ s \end{bmatrix} = \begin{bmatrix} (\mathbf{B} - \mathbf{A}) \\ - (\mathbf{D} - \mathbf{C}) \end{bmatrix} \cdot (\mathbf{C} - \mathbf{A})$$

If points **A**, **B**, **C** and **D** are expressed in triples (x_A, y_A, z_A), (x_B, y_B, z_B), (x_C, y_C, z_C) and (x_D, y_D, z_D) then the above system of equations in component form becomes

$$\begin{bmatrix} x_B - x_A & x_C - x_D \\ y_B - y_A & y_C - y_D \\ z_B - z_A & z_C - z_D \end{bmatrix}^T \begin{bmatrix} x_B - x_A & x_C - x_D \\ y_B - y_A & y_C - y_D \\ z_B - z_A & z_C - z_D \end{bmatrix} \begin{bmatrix} t \\ s \end{bmatrix} = \begin{bmatrix} x_B - x_A & x_C - x_D \\ y_B - y_A & y_C - y_D \\ z_B - z_A & z_C - z_D \end{bmatrix}^T \begin{bmatrix} x_C - x_A \\ y_C - y_A \\ z_C - z_A \end{bmatrix}$$

After solving the above set of equations for t and s, if $0 \leq t, s \leq 1$ **P** and **Q** lie within **AB** and **CD** respectively. Further if $d^2 = 0$, **P** = **Q** is the point of intersection satisfying

$$\begin{bmatrix} x_B - x_A & x_C - x_D \\ y_B - y_A & y_C - y_D \\ z_B - z_A & z_C - z_D \end{bmatrix} \begin{bmatrix} t \\ s \end{bmatrix} = \begin{bmatrix} x_C - x_A \\ y_C - y_A \\ z_C - z_A \end{bmatrix}$$

If s or $t \notin [0, 1]$, then $d = \min (| \mathbf{AC} |, | \mathbf{AD} |, | \mathbf{BC} |, | \mathbf{BD} |)$. Note that for a unique solution of $[t \ s]^T$, the respective coefficient matrix should not be singular. That is, **AB** or **CD** should not represent a point or **AB** and **CD** should not be parallel.

9.3 Relation between a Point and a Polygon

A closed polyline or polygon can be a representation of many planar objects. A polygon is an area enclosed by a series of lines connected end-to-end. The convention is to traverse through the boundary lines in the counterclockwise fashion. This ensures that the area immediate to the left of any line (Section 9.1) is interior to the polygon (Figure 9.5a). A polygon is said to be convex if a line joining any two points within the polygon lies completely inside it. For computational check, in a convex polygon, at any vertex, the edges loop counterclockwise. The concave vertex in a non-convex polygon has the edges looping clockwise (Figure 9.5b).

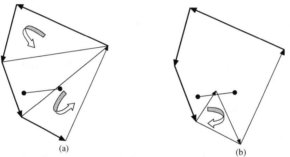

(a)   (b)

Figure 9.5   (a) A convex polygon and (b) a non-convex polygon

9.3.1 Point in Polygon

For any given polygon, one way to find whether a point is inside it is the Jordan's curve theorem (see Chapter 8). An alternate description is that a point is inside a polygon if, for any ray from this point, there is an odd number of crossings or intersections of the ray with the polygon's edges (Figure 9.6). This requires a crossings test as follows:

A ray is shot from the test point along a specified line (+X is commonly used) and the number of crossings is computed. For odd number of intersection points, the point lies within the polygon, else outside. If the test ray passes through any vertex of the polygon, the test point is shifted up or down by a very small distance and the new ray is intersected (Figure 9.7).

The algorithm returns the status of the test point as either 'in' or 'out' of the polygon. Before the crossings test, the point in query is tested whether it lies on the polygonal boundary or not in two stages: (a) the test point is compared with the vertices for coincidence and (b) if it does not coincide with any vertex, then it is tested for its belonging on the polygonal edge. The area of the triangle made by the point and the end points of an edge is computed for the purpose (Section 9.1).

Convex polygons can be intersected faster due to their geometric properties. A point lies inside the

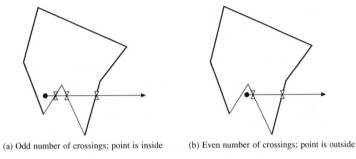

(a) Odd number of crossings; point is inside   (b) Even number of crossings; point is outside

Figure 9.6   The crossings check

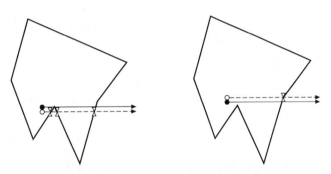

Figure 9.7   Crossings test for a ray passing through a vertex

convex polygon if it lies to the left of all the edges (maintaining counterclockwise traverse). This is illustrated in Figure 9.8 (a). A faster version is to check whether the point lies outside the bounding box before the actual queries described above for both convex and non-convex polygons are made (Figure 9.8 b).

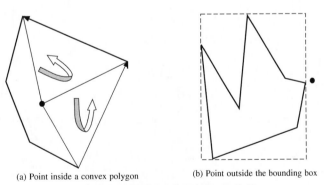

(a) Point inside a convex polygon   (b) Point outside the bounding box

Figure 9.8   Faster point-in-polygon checks

Example 9.3. A quadrilateral is defined by **A** $(1, -1)$, **B** $(0, 5)$, **C** $(-1, -1)$ and **D** $(0, 3)$. Determine if point **E** $(0, 1)$ lies within this polygon or not.

(a) Determine if test point **E** is coinciding with any of the vertices of polygon **ABCD**.

Distance from **E** to **A** is $\sqrt{(1 - 0)^2 + (1 + 1)^2} = \sqrt{5} > 0$, thus implying **E** is not coincident with **A**. Similarly, distance from **E** to **B**, **C** and **D** are > 0. Thus test point **E** is not coincident with any of the vertices of the polygon **ABCD**.

(b) Determine if test point is lying on boundary of polygon **ABCD**.
Find determinant Δ of triangle *ABE*.

$$\Delta ABE = \begin{vmatrix} 1 & -1 & 1 \\ 0 & 5 & 1 \\ 0 & 1 & 1 \end{vmatrix} = 4 \neq 0$$

This implies that **E** is not on **AB**. Similarly, ΔBCE, ΔCDE and ΔDAE are not equal to zero and thus **E** does not lie on the boundary of **ABCD**.

(c) Determine the 'in' or 'out' status by ray shooting algorithm.
Determine a point outside polygon **ABCD** along x-axis to throw a ray to the test point **E**. The x_{max} of the polygon is 1 and thus a ray can be shot from point **F** ($2 (>x_{max})$, 1 (y-coordinate of **E**)).

Determine the number of crossings the ray **EF** makes with the polygon **ABCD** to reach **F** from **E**. Initialize a counter count = 0.

Find if **EF** intersects with the edge **AB**.
Since $\Delta ABE = 4$, $\Delta ABF = -8$ and $\Delta EFA = -4$, $\Delta EFB = 8$ are pair wise opposite in sign, **EF** and **AB** intersect. Increment the counter (count = 1).

Similarly, **EF** does not intersect **BC**, **CD** and intersects **DA** (count = 2). Since number of crossings is even, **E** is outside **ABCD**. Figure 9.9 illustrates the procedure.

9.4 Proximity between a Point and a Plane

Consider a point **D** (x_4, y_4, z_4) and a plane specified three points **A** (x_1, y_1, z_1), **B** (x_2, y_2, z_2) and

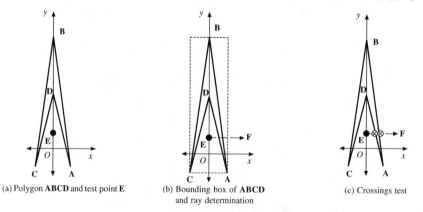

(a) Polygon **ABCD** and test point **E**   (b) Bounding box of **ABCD** and ray determination   (c) Crossings test

Figure 9.9   Example 9.3 on point in polygon

$\mathbf{C}$ (x_3, y_3, z_3) such that the normal to the plane is given by $\mathbf{n} = \mathbf{AB} \times \mathbf{BC}$. The volume of the tetrahedron $ABCD$ is computed by calculating the determinant Δ as

$$\Delta = \begin{vmatrix} x_1 & y_1 & z_1 & 1 \\ x_2 & y_2 & z_2 & 1 \\ x_3 & y_3 & z_3 & 1 \\ x_4 & y_4 & z_4 & 1 \end{vmatrix} \tag{9.3}$$

Point $\mathbf{D}$ can be placed in three possible ways with respect to the plane ABC.

(a) Point $\mathbf{D}$ lies on the same side of the plane as its normal if Δ is negative.
(b) Point $\mathbf{D}$ lies on the opposite side of the normal if Δ is positive.
(c) Point $\mathbf{D}$ lies in the plane if Δ is zero.

The normal $\mathbf{n} = \mathbf{AB} \times \mathbf{BC}$ is given by

$$\mathbf{n} = \begin{vmatrix} \mathbf{r} & \mathbf{j} & \mathbf{k} \\ x_2 - x_1 & y_2 - y_1 & z_2 - z_1 \\ x_3 - x_2 & y_3 - y_2 & z_3 - z_2 \end{vmatrix}$$

If $\mathbf{D}$ is placed on the same side of the normal, then

$$\mathbf{n} \cdot \mathbf{BD} > 0 \text{ or } \Delta_1 = \begin{vmatrix} x_4 - x_2 & y_4 - y_2 & z_4 - z_2 \\ x_2 - x_1 & y_2 - y_1 & z_2 - z_1 \\ x_3 - x_2 & y_3 - y_2 & z_3 - z_2 \end{vmatrix} > 0$$

Performing a few row operations in Δ in Eq. (9.3) results in

$$\Delta = \begin{vmatrix} x_1 & y_1 & z_1 & 1 \\ x_2 - x_1 & y_2 - y_1 & z_2 - z_1 & 0 \\ x_3 - x_2 & y_3 - y_2 & z_3 - z_2 & 0 \\ x_4 - x_2 & y_4 - y_2 & z_4 - z_2 & 0 \end{vmatrix} = \begin{vmatrix} x_1 & y_1 & z_1 & 1 \\ x_4 - x_2 & y_4 - y_2 & z_4 - z_2 & 0 \\ x_2 - x_1 & y_2 - y_1 & z_2 - z_1 & 0 \\ x_3 - x_2 & y_3 - y_2 & z_3 - z_2 & 0 \end{vmatrix} = -\Delta_1$$

Thus, for $\mathbf{D}$ on the side of the normal, Δ is negative and vice-versa.

Example 9.4. Three points $\mathbf{A}$ $(1, 0, 0)$, $\mathbf{B}$ $(0, 1, 0)$ and $\mathbf{C}$ $(0, 0, 1)$ define a triangular lamina (Figure 9.10a). Find how the points: (a) $\mathbf{D}$ $(0, 0, 0)$, (b) $\mathbf{D}$ $(1, 1, 1)$, (c) $\mathbf{D}$ $(1/3, 1/3, 1/3)$ and (d) $\mathbf{D}$ $(1, 1, -1)$ are placed with respect to this lamina.

  (a) $\mathbf{D}$ $(0, 0, 0)$. Find $\Delta ABCD$

$$\Delta = \begin{vmatrix} 1 & 0 & 0 & 1 \\ 0 & 1 & 0 & 1 \\ 0 & 0 & 1 & 1 \\ 0 & 0 & 0 & 1 \end{vmatrix} = 1 \ (> 0)$$

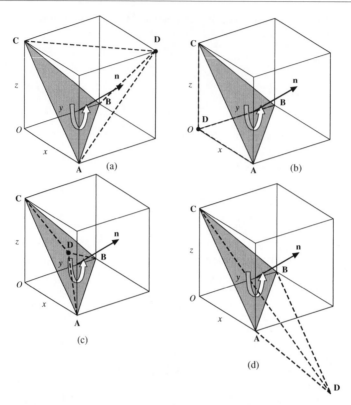

Figure 9.10 **Proximity of a point and a plane (a) point on the same side of the normal, (b) point to the opposite side of the normal, (c) point within the triangular lamina and (d) point coplanar with the given plane but not within the lamina.**

Also, the normal **n** to the lamina is given by $\mathbf{AB} \times \mathbf{BC} = \mathbf{i} + \mathbf{j} + \mathbf{k}$.
This implies that **D** is positioned opposite to that of the normal of lamina **ABC**.

(b) **D** (1, 1, 1), $\Delta ABCD = -2$, thus **D** is positioned on the same side as of the normal of lamina **ABC**.

(c) **D** (1/3, 1/3, 1/3), $\Delta ABCD = 0$. Thus, **D** is coplanar with lamina **ABC**. Further we can check if **D** lies within or outside the lamina by doing a point in polygon test (Section 9.3.1) on one of the projections. If the *xy* plane is chosen, the projected points are **A′** (1, 0), **B′** (0, 1), **C′**(0, 0) and **D′** (1/3, 1/3). Since a triangular lamina is a convex polygon, we can use the special property and confirm a point to be inside if it lies to the left of all three edges of the lamina.

Determine $\Delta A'B'D' = \begin{vmatrix} 1 & 0 & 1 \\ 0 & 1 & 1 \\ 0 & 0 & 1 \end{vmatrix} = 1 \; (> 0)$, implying **D′** is to the left of **A′B′**.

Similarly, $\Delta B'C'D'$ and $\Delta C'A'D'$ are positive implying **D'** is to the left of both **B'C'** and **C'A'**. Thus **D** lies within the lamina **ABC**.

(d) For **D** (1, 1, –1), $\Delta ABCD = 0$. Thus **D** is coplanar with **ABC**. Further, as in the previous case, the determinants $\Delta A'B'D'$ is negative, $\Delta B'C'D'$ and $\Delta C'A'D'$ are positive implying that **D** lies outside **ABC**.

9.4.1   Point within a Polyhedron

Along with the B-rep data structure, polyhedral representation of solids is also common in computer graphics. They have a simple representation and are easy to display. A polyhedral model is a collection of planar faces constituting the boundary. Each face is represented by a sequence of planar vertices in a three-dimensional space. The edge loop formed by these vertices is counterclockwise in direction when viewed from outside the solid. This ensures that the face normal points towards the exterior of the object.

The method described in the previous section works well for convex polyhedrons. For such cases, the query point will lie in the interior if it is in the direction opposite to the normals of all faces. However, to interrogate for a test point to lie within a generic polyhedron, the ray-shooting algorithm discussed earlier can be modified. An example polyhedron BOX is interrogated for the presence of a point **Q** in it (Figure 9.11) for illustration.

(a) Determine x_{max}, an x coordinate outside the bounding box of the polyhedron. Extend a ray parallel to the x-axis from the test point **Q** (x_q, y_q, z_q) to point **X**$_{max}$ (x_{max}, y_q, z_q).

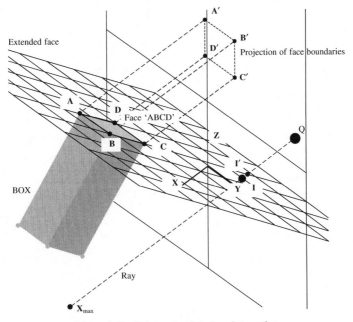

Figure 9.11   Point and polyhedron interaction

(b) Find the point of intersection (if any) for each face (plane) extended indefinitely with this ray. A face plane is represented by the equation $Ax + By + Cz + D = 0$. The coefficients A, B, C and D can be computed using any three bounding vertices. The ray has the parametric equation of the form $x = x_q + t(x_{max} - x_q)$, $y = y_q$ and $z = z_q$. At the point of intersection $t = -[Ax_q + By_q + Cz_q + D]/A/(x_{max} - x_q)$. For existence of the point of intersection, t should lie in the interval [0 1]. Consider a face **f** of the BOX defined by vertices **A**, **B**, **C** and **D**. The ray intersects the extended plane at **I**.

(c) Check if this point of intersection lies within the face being interrogated. This can be solved in two steps. First, an axis plane (x-y, y-z or x-z whose normal is the best approximation of the face's normal) is chosen, and vertices and edges of the face and the intersection point are projected. Second, query if the projection of 'point of intersection' lies within the projection of the 'face'. This is a point in polygon query described in section 9.3.1. For the example illustration, the vertices **A**, **B**, **C** and **D** that bound the face **f** as well as the point of intersection **I** are projected on y-z plane. The projected points are **A′**, **B′**, **C′** and **D′** and **I′** respectively. **I′** lies outside of polygon **A′B′C′D′** implying that the ray does not intersect **f**.

(d) Special cases like the ray passing through a vertex/edge can again be handled by perturbing the ray infinitesimally such that the ray does not pass through any vertex or edge.

(e) Count the number of intersections the ray makes with all the faces to determine the status of the point. For odd number of intersections, the querry point lies within the polyhedron.

9.5 Membership Classification

A majority of operations and queries on three-dimensional geometry involve determining the common portion of interaction between two or more objects in space. Consider two intersecting objects **A** and **B**. To compute the common volume of intersection, initially the boundaries of **A** are intersected with that of **B**. This requires computation of intersection between curves and surfaces. The original boundaries are snipped (trimmed) at points (in case of curves) or curves (in case of surfaces) of intersection (if any). We then evaluate which of these segmented boundaries bound the intersecting volume. This involves what is known as *membership classification*. Discussions presented in sections 9.3.1 and 9.4.1 are examples of point membership classification (PMC) where we are interested in finding the status of a test point with respect to a lamina/volume represented by the first order boundary elements (lines and planes). The point membership classification is basic to the membership classification of curves and surfaces in a generic B-rep model with higher order boundary elements (e.g., B-spline curves and surfaces).

9.6 Subdivision of Space

The intersection algorithms and related membership classification problems are not very easily dealt with in case of B-rep models whose boundary elements are composed of higher order parametric elements like splines. This is because algorithms on intersection of parametric curves and surfaces are iterative in nature and not so robust. Representation of a generic object as a group of cellular sub-domains reduces the computations involved in membership classification. But we pay the cost in terms of accuracy, which depends on cell size. The cellular models are unambiguous, unique and valid representations. These models are also used for mesh generation (Appendix) for the Finite Element Methods discussed in Chapter 11.

A cellular model of a circular disc is shown in Figure 9.12(a). A grid of uniform cell size is superposed on top of the geometry. Like in case of point membership algorithm, the cells are classified as 'in', 'out' or 'on' depending on whether a cell is placed within or outside the geometry, or on the boundary. This is a one-time computation and once the cells have been classified, given a test point, all one needs to determine is the cell that holds the point. The status of the test point is the same as that of the cell that holds it. To improve the accuracy, we need to decrease the grid size. This, however, makes the computations more involved. Quadrees and octrees (in two-dimensions and three-dimensions respectively) are grid structures having cells of different sizes that become smaller as the level of decomposition increases. This grid structure has cells of smaller size near the boundary and large sized cells within the interior. The degree of accuracy at the boundary is proportional to the level of decomposition in the quadtree. A quadtree structure for the circular disc is shown in Figure 9.12 (b). A procedure to generate the quadtree structure and using the same for point membership classification of complex shapes is discussed further.

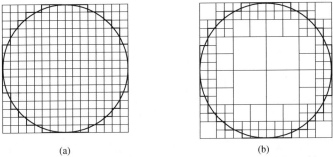

(a)   (b)

Figure 9.12   (a) A circular disc and its grid and (b) its quadtree decomposition

9.6.1   Quadtree Decomposition

Figure 9.13 shows the schematic of a quadtree structure generation. Each node represents a square. The node (parent square) at 'level 0' represents the initial bounding square within which the planar object is enclosed. The parent square is divided into four children squares, and four corresponding

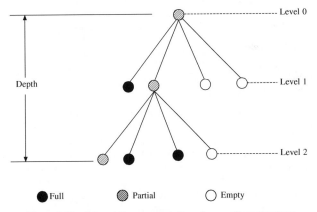

Figure 9.13   Schematic representation of a quadtree structure

nodes at level 1 represent them. The shades represent whether a square is completely inside the object (black), is completely outside (white) or is a leaf node that encompasses the boundary (gray). For black or white cells, we are quite sure that all four children squares would lie within the geometry or outside the geometry, respectively. Thus, only gray nodes are sub-divided further as the level increases to facilitate robustness and minimum data storage. Since the parent cell encompasses the planar object, the cell is neither completely inside nor completely outside the object, and therefore is a leaf node ⬤.

Understanding the data structure storing the quadtree is equally essential. This data structure may be grouped into two parts, the tree part and the queue part, as shown in Figure 9.14. The two parts are uniquely connected with pointers for information in the tree part to be referred in the queue part and vice versa.

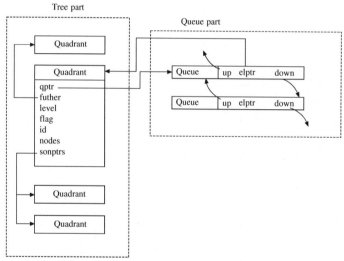

Figure 9.14   Data structure used to store a quadtree structure

The algorithm to generate a quadtree is illustrated using an example in Figure 9.15. Four levels of decomposition for a semi circular disc are shown for the first quadrant at each level. The squares in the first level are enumerated as **0**, **1**, **2** and **3** moving clockwise. The children of square **1** in the second level are numbered **10**, **11**, **12**, and **13** moving clockwise.

(a) Generate the bounding square. Initialize the quadtree data structure by inserting the root square into the queue part for decomposition, inserting the root square into tree part and classify as 'leaf' node and interconnect with the queue. For the example shown, the bounding square for the 2D semi-circular lamina is shown in Figure 9.15 (a). The queue is **root square** ⇔ **NULL**.

(b) Pull out the first 'leaf' square not yet decomposed from the queue part and divide it into four children. Their nodes and position in the tree are calculated and pushed into the queue from the end. Root square is divided into four children **0**, **1**, **2** and **3** (Figure 9.15 (b)). The queue part consists of the following squares: **root square** ⇔ **0** ⇔ **1** ⇔ **2** ⇔ **3** ⇔ **NULL**.

(c) Pull out a square not yet classified from the queue and check its status. If there exists an

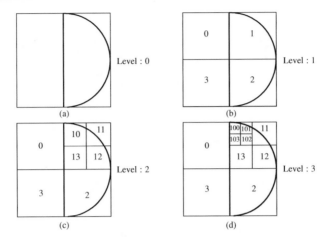

Figure 9.15   Quadtree decomposition of a semi-circular disc

intersection between the square and the boundary (use line intersection recursively for all the edges in the polygon with the edges in square), then it is a leaf node, else it may be either of the following three cases:

(i)   It is leaf node if it still envelopes the object
(ii) It is completely inside the object
(iii) It is completely outside the object

At this stage, the status of a child square is assigned. For the semi-circular lamina, checking the queue part for the first square not yet classified, square **0** is taken for classification. It lies outside the circular lamina and thus classified as *out*. Similarly, **1**, **2** and **3** are classified as *leaf*, *leaf* and *out* respectively. Update the tree part (Figure 9.16a) to complete the quadtree generation till level 1.

(d)   The above two steps are performed recursively till the required depth of the tree is obtained. For example, the first *leaf* square not yet decomposed is **1** and the same is decomposed into squares **10**, **11**, **12** and **13**. Similarly, square **2** is decomposed and the queue becomes **root square** ⇔ **0** ⇔ **1** ⇔ **2** ⇔ **3** ⇔**10** ⇔ **11** ⇔ **12** ⇔ **13** ⇔ **20** ⇔ **21** ⇔ **22** ⇔ **23** ⇔ **NULL**. Classify the newly generated queue entries with respect to the object (in this case the semi-circular lamina) and update the tree. Figure 9.16(b) shows the tree for only the first *leaf* quadrant till level 2. At level 3 (Figure 9.15d), the queue part becomes **root square** ⇔ **0** ⇔ **1** ⇔ **2** ⇔ **3** ⇔**10** ⇔ **11** ⇔ **12** ⇔ **13** ⇔ **20** ⇔ **21** ⇔ **22** ⇔ **23** ⇔ **100** ⇔ **101** ⇔ **102** ⇔ **103** ⇔ **110** ⇔ **111** ⇔ **112** ⇔ **113** ⇔ **120** ⇔ **121** ⇔ **122** ⇔ **123** ⇔ **NULL**. The tree part is shown in Figure 9.16(c) (again only for the first *leaf* quadrant in each level).

It is not necessary to use the square cells for quadtree decomposition. Triangular cells can also be employed for the same and an example is shown in Figure 9.17(a) where the root is the equilateral triangle sub-divided into four children triangles (Figure 9.17b) after every decomposition level.

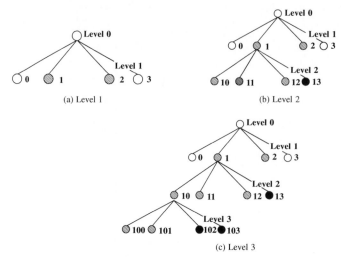

(a) Level 1

(b) Level 2

(c) Level 3

Figure 9.16   The quadtree structure for semi-circular lamina in Figure 9.15

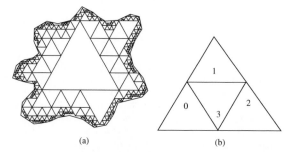

(a)

(b)

Figure 9.17   (a) Quadtree decomposition with equilateral triangles and (b) scheme of decomposition

9.7   Boolean Operations on Polygons

Boolean operations are set theoretical operations performed on basic shapes to evolve more complex definitions. Some basic Boolean operations are union, intersection and negation denoted by $\cup$, $\cap$ and $-$, respectively as discussed in Chapter 8. All sets in the Euclidean space $\mathbf{E}^3$ are not suitable for geometrical representation and set theoretical operations. A subset of $\mathbf{E}^3$ that is bounded, closed, regular and semi analytic are only suitable for geometrical representation. They are called *regularized sets* or *r-sets*. Under the conventional Boolean operations, *r-sets* are not algebraically closed, but they are closed under the regularized set union, intersection and difference denoted by $\cup^*$, $\cap^*$ and $-^*$ as explained in Chapter 8.

An algorithm for determining the regularized Boolean for polygons is given below. Consider two given polygons **A** and **B** (Figure 9.18) constituting of vertices and connecting edges such that the boundaries are traversed in the counterclockwise fashion.

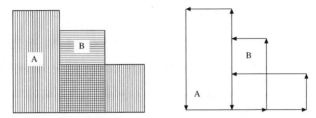

Figure 9.18   Boolean interaction between polygons A and B

(a) Intersect all edges in **A** with that of **B** and vice versa and split them into segments if they intersect. Place all these edge segments keeping the order of vertices consistent with the polygon representation (left side representing the interior) into two lists corresponding to polygons **A** and **B**, respectively. This is illustrated in Figure 9.19 (a).

(b) For all edges in list 1 (that of polygon **A**), classify them as *in*, *on* or *out* of polygon **B**. This may be accomplished by doing a point membership classification of the mid-point of each edge segment in polygon **B**. Do a similar classification of edges in list 2 with polygon **A**. The step is illustrated in Figure 9.19 (b). The line code followed is thick solid for *on*, thin solid for *in* and dashed for *out*.

(c) Collect the edges representing the closed region after a regularized Boolean operation as following:

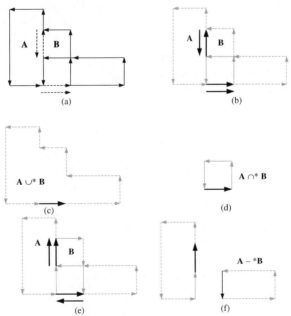

Figure 9.19   Regularized Boolean operations on polygons A and B

(i) $\mathbf{A} \cup^* \mathbf{B}$: collect all edges from list 1 and 2 that are marked *out*. If there are edges marked *on*, select one edge of the pair (since *on* segment will be in duplicate, one in list 1 and the other in list 2) if they have the same direction (Figure 9.19 c).

(ii) $\mathbf{A} \cap^* \mathbf{B}$: collect all edges from list 1 and 2 that are marked *in*. If there are edges marked *on*, select one edge of the pair if they have the same direction (Figure 9.19 d).

(iii) $\mathbf{A} -^* \mathbf{B}$: first, change the direction of all the edges in list 2 (since $\mathbf{B}$ is representing a hole in this operation). This is illustrated in Figure 9.19(e). Collect edges, from list 1 that are marked *out*, and from list 2 that are marked *in*. If there are edges marked *on*, select one edge of the pair if they have the same direction (Figure 9.19 f).

(d)   Chain all the collected edges to form a valid edge loop running counterclockwise to represent a regular set.

The problem of accurate and robust implementation of geometric algorithms is still of considerable research attention. Much difficulty arises from the fact that reasoning about geometry most naturally occurs in the domain of real numbers, which can only be represented approximately on a digital computer. Many times, the correctness of geometric algorithms depends on correctly evaluating the signs of arithmetic expressions, and errors due to rounding or imprecise input can lead to incorrect results or failure to run to completion. Another problem is that of dealing with degeneracies such as the intersection of a line with a polygon only at one vertex, or along an edge. Degeneracies can be a source of non-robustness on one hand, or of serious implementation difficulties on the other. For simplicity, algorithms often assume that primitives are arranged so that there are no degeneracies (i.e. they are placed in a general setting). In practice, however, primitives often are not in general position, causing implementations to fail. Recasting an algorithm to handle degeneracies tends to result in a situation in which much of the code is designed to handle special cases. In devising an algorithm for solving a geometric problem, one thus has to keep in mind both robustness as well and the feasibility of implementation.

9.8   Intersection between Free Form Curves

Let $\mathbf{b}(t)$ and $\mathbf{c}(s)$ be the two free form curves defined by

$$\mathbf{b}(t) = \sum_{i=0}^{m} \phi_i(t)\,\mathbf{b}_i \text{ and } \mathbf{c}(s) = \sum_{j=0}^{n} \psi_j(s)\,\mathbf{c}_j$$

where $\mathbf{b}_i$ and $\mathbf{c}_j$ are the control points and $\phi_i(t)$ and $\psi_j(s)$ are the barycentric weights, for instance the Bernstein polynomials or the Basis splines. The distance or residual $g(s, t)$ between the two curves can be computed as

$$g(s,\, t) = [\mathbf{b}(t) - \mathbf{c}(s)][\mathbf{b}(t) - \mathbf{c}(s)]^T$$

The necessary conditions for the minimum of $g(s, t)$ are $\dfrac{\partial g(s,t)}{\partial t} = \dfrac{\partial g(s,t)}{\partial s} = 0$ incorporating which yields

$$g_1(s,\, t) \equiv [\mathbf{b}(t) - \mathbf{c}(s)]\left[\frac{\partial \mathbf{c}(s)}{\partial s}\right]^T = 0$$

$$g_2(s,\, t) \equiv [\mathbf{b}(t) - \mathbf{c}(s)]\left[\frac{\partial \mathbf{b}(t)}{\partial t}\right]^T = 0$$

The above expression is in the implicit form in both s and t. For some values of s and t, let $g_1(s, t)$ and $g_2(s, t)$ both not be equal to zero. Define $G(s, t) = [g_1(s, t)\ g_2(s, t)]^T$ and consider its first order linear expansion, that is

$$G(s + \Delta s, t + \Delta t) = G(s, t) + \begin{bmatrix} \dfrac{\partial g_1}{\partial s} & \dfrac{\partial g_1}{\partial t} \\[2mm] \dfrac{\partial g_2}{\partial s} & \dfrac{\partial g_2}{\partial t} \end{bmatrix} \begin{bmatrix} \Delta s \\[2mm] \Delta t \end{bmatrix} = \begin{bmatrix} 0 \\[2mm] 0 \end{bmatrix}$$

The intent in the above expression is that for some (iterative) revision $(\Delta s, \Delta t)$ in the values of (s, t), $G(s, t)$ becomes **0**. Rearranging above yields

$$\begin{bmatrix} \dfrac{\partial g_1}{\partial s} & \dfrac{\partial g_1}{\partial t} \\[2mm] \dfrac{\partial g_2}{\partial s} & \dfrac{\partial g_2}{\partial s} \end{bmatrix} \begin{bmatrix} \Delta s \\[2mm] \Delta t \end{bmatrix} = - \begin{bmatrix} g_1(s, t) \\[2mm] g_2(s, t) \end{bmatrix} \quad \text{or} \quad \begin{bmatrix} \Delta s \\[2mm] \Delta t \end{bmatrix} = - \begin{bmatrix} \dfrac{\partial g_1}{\partial s} & \dfrac{\partial g_1}{\partial t} \\[2mm] \dfrac{\partial g_2}{\partial s} & \dfrac{\partial g_2}{\partial s} \end{bmatrix}^{-1} \begin{bmatrix} g_1(s, t) \\[2mm] g_2(s, t) \end{bmatrix}$$

where

$$\frac{\partial g_1(s, t)}{\partial s} = [\mathbf{b}(t) - \mathbf{c}(s)] \left[\frac{\partial^2 \mathbf{c}(s)}{\partial s^2} \right]^T - \left[\frac{\partial \mathbf{c}(s)}{\partial s} \right] \left[\frac{\partial \mathbf{c}(s)}{\partial s} \right]^T$$

$$\frac{\partial g_1(s, t)}{\partial t} = \left[\frac{\partial \mathbf{b}(t)}{\partial t} \right] \left[\frac{\partial \mathbf{c}(s)}{\partial s} \right]^T = - \frac{\partial g_2(s, t)}{\partial s}$$

$$\frac{\partial g_2(s, t)}{\partial t} = [\mathbf{b}(t) - \mathbf{c}(s)] \left[\frac{\partial^2 \mathbf{b}(t)}{\partial t^2} \right]^T + \left[\frac{\partial \mathbf{b}(t)}{\partial t} \right] \left[\frac{\partial \mathbf{b}(t)}{\partial t} \right]^T$$

Thus starting with the initial values of s and t, Δs and Δt can be computed using the above expressions. Parameters can be updated as $s = s + \Delta s$ and $t = t + \Delta t$ and using these new values, $G(s, t) = [g_1(s, st)\ g_2(s, t)]^T$ can be computed. The procedure can be iterated until $G(s, t)$ is desirably close to **0**. Note here that s and t values should not be allowed to assume values outside the interval $[0, 1]$. An intersection point is obtained when $d^2 = g(s, t)$ is adequately close to zero.

EXERCISES

1. Given a line $\mathbf{A} + t\mathbf{d}$ and a plane with base point $\mathbf{B}$ and normal vector $\mathbf{n}$, what is the condition for the line to be perpendicular to the plane? What is the condition for the line to be parallel to the plane?
2. Find the proximity of the points $(0, 0)$, $(1, 5)$ and $(1, 0)$ with respect to the line whose end points are $\mathbf{A}$ $(1, 1)$ and $\mathbf{B}$ $(1, 8)$.
3. Consider the line segments whose end points are $\mathbf{AB}$ $(0, 0)$ $(5, 0)$; $\mathbf{BC}$ $(5, 0)$ $(5, 5)$; $\mathbf{CD}$ $(5, 5)$ $(0, 5)$ and $\mathbf{DA}$ $(0, 5)$ $(0, 0)$. Find the positioning of the point $\mathbf{P}$ $(1, 1)$ with respect to these lines. Comment on the membership (inside/outside/on) of $\mathbf{P}$ in polygon $\mathbf{ABCD}$.
4. A quadrilateral is represented by the vertices $\mathbf{A}$ $(2, -2)$, $\mathbf{B}$ $(0, 15)$, $\mathbf{C}$ $(-2, -2)$ and $\mathbf{D}$ $(0, 4)$. Determine if point $\mathbf{E}$ $(0, -2)$ lies within this polygon. (Hint: The ray passes through the vertex $\mathbf{A}$, thus infinitesimally shift the y coordinate of the ray and then perform the crossings test).

5. Given n points in a plane, devise an algorithm to construct a non self intersecting polygon. (Hint: Choose an extreme point and start connecting the immediate neighbors, keeping track of already connected vertices. There may be many possible solutions.)

6. Given a concave polygon and two points **A** and **B** inside the polygon, write a procedure to find the shortest path between the points. Consider all possibilities of how **A** and **B** are placed inside the polygon.

7. Consider a unit cube placed in the first octant of the coordinate frame. Find separately using the point in polyhedron algorithm the membership of points $(-2, 0)$, $(0.5, 5)$, $(0.6, 0.8)$.

8. Consider a unit square placed in the first quadrant with two edges along the x and y axes. Also, consider an inscribing circle. Generate a quadtree data structure for the inscribed circle using the unit square as the root square. Generate to a depth of three levels. The quadtree thus generated can be used to find the membership of a point with respect to the circle. (Hint: Say for example a point is inside the circle, if it is inside/on any one of the node elements (squares) of the quadtree that are marked "in". Thus, the computation of intersection reduces from ray tracing to searching the quadtree). Specifically, comment using the above quadtree representation on the placement of points, $(0.6, 0.6)$, $(0.98, 0.98)$ and $(2, 5)$. Also give the node number of the quadrant (e.g. Figure 9.15) in case the point is 'inside' the circle. Solve using graph paper.

9. Schematically, using the method presented in section 9.7, find the intersection $(\mathbf{A} \cap^* \mathbf{B})$, union $(\mathbf{A} \cup^* \mathbf{B})$ and negation $(\mathbf{A} - \mathbf{B})$ for the arrangements of polygons A and B shown in Figures P9.1.

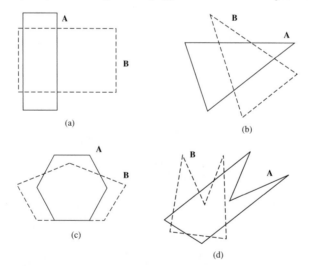

Figure P9.1   Polygons A and B requiring Boolean operations

Chapter 10

Geometric Modeling Using Point Clouds

Chapter 7 discussed the modeling of surfaces with regard to designing solid shapes as a collection of closed, simply connected surface patches. When designing products with free form shapes such as the aircraft wings and fuselage, car body and its doors, seats and windshields, the shape information has to be acquired as a set of data points and then the surface patches have to be designed from the same. Data points can either be user-specified for an entirely new design or can result as a large discrete set called *point cloud* from say laser scanning of an existing product or its prototype. Reverse engineering alludes to constructing surface patches from the point cloud data. This chapter describes some methods on point cloud acquisition and surface/solid modeling using the acquired data.

10.1 Reverse Engineering and its Applications

Reverse engineering is the process of creating engineering design from existing components or their prototypes. The existing part is recreated by acquiring its surface data using a scanning or measurement device. For a new component whose original design data is not available, a CAD model is created using conceptual clay or wood model for further analysis and possible form changes. Coordinate measuring machines (CMMs) have been used to extract surface data but their data capturing operation is very slow for parts having intricate free-form surfaces. In recent years, laser scanning has become a powerful tool in capturing the geometry of complicated models. With present and upcoming range of sensing devices and associated software, surface modeling from point cloud samples of physical objects is a rapidly evolving discipline. A few of the many applications of reverse engineering are listed below.

(a) Generation of custom fits to human surfaces is plausible with reverse engineering to design products like helmets, shoes, arm and knee guards, space suits and others. The inside hull can be obtained by scanning the human surface (head, feet or body) while the outside hull can be designed keeping weight, safety, aerodynamics and many other such factors in mind.

(b) Custom prosthetic design is a medical application of reverse engineering. Custom prosthetics help in better and faster post operative recovery as well as provide better cosmetic appearance to an amputee. In a case where an amputee has a natural limb, the point cloud data can be determined for the skeleton and the outer form from the intact limb using X-ray and laser scanning respectively.

Kinematics can be extracted using the skeletal data while cosmetic design can be accomplished using the outer form. For custom design of both limbs, heuristic design based on the existing data bank may be suggested and incorporated. An orthopedic may prefer stress simulation to foresee the effects of a range of loads, say during gait, on joints and links for which the reverse engineered CAD model shall be useful.

(c) Three dimensional models of internal organs can enhance a surgeon's pre-operative planning, especially in life saving situations involving a single procedure. Point cloud data generated from the biological form is used for its shape synthesis.

(d) An artist/archeologist can also benefit from reverse engineering. Archeologists can reconstruct fossils, archaeological collections from fragmentary material to view and analyze more accurately without damaging the original artifact. They can even reproduce artifacts in absence of the original object by creating 3-D model archives. Artists/sculptors can reproduce their creations in the original form for a larger customer base.

(e) Graphics and multimedia personnel can create enhanced quality computer models of comic/real life creations for animation, movie, virtual reality and show renderings from physical models.

(f) A city planner/geologist can model terrain surfaces for analysis and presentation. The availability of three-dimensional computer models enhances better planning of civil infrastructure based on terrain characteristics with minimal requirement of terrain modification.

The above list provides only a glimpse of the application domain of surface/solid modeling from point samples. The availability of a variety of digitizing equipments and spectrum of software has expanded the application of reverse engineering to almost every area where 3-D modeling of free-form shapes may play a significant role. The procedure to acquire shape information in discrete form to build geometric models depends on the type of the physical object and the purpose for which the model is being created.

10.2   Point Cloud Acquisition

A broad classification and listing of different methods for acquiring point cloud data is given in Figure 10.1. Each method uses some mechanism or phenomenon for interacting with the surface or volume of the object of interest. Non-contact methods use light, sound or magnetic fields while in tactile or contact methods, the surface is probed mechanically. In each case, appropriate analysis of the data acquired has to be performed to locate the positions of points on the surface. For example, in laser range finders, the time of flight is used to determine the distance traveled and subsequently the point location. Each method has pros and cons, which require that the data acquisition system should be carefully chosen for shape capturing.

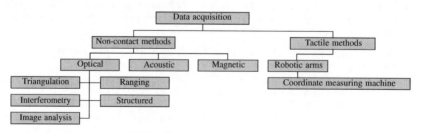

Figure 10.1   Data acquisition methods

Optical scanners are non-contact devices that use light to interact with the object. They are probably the broadest and most popular with relatively fast acquisition rates. Optical methods can be classified based on the principle of operation used:

- *Triangulation* uses location and angles between light sources and photo sensing devices to deduce position. The light source is a high-energy laser and the sensor is a video camera. The method can acquire data at a very fast rate and the accuracy depends on the resolution of the camera.
- *Ranging methods* measure distances by sensing the time of flight of the light beams. Popular methods are based on lasers and pulsed beams.
- *Structured lighting* involves projecting patterns of light upon the surface and capturing an image of the resulting pattern as reflected by the surface. The image is then analyzed to determine the coordinates of a data point on the surface. This method can acquire large amount of data with a single image, but the analysis to determine the coordinates is quite intricate.
- *Image analysis* and *photogrametry* methods are similar to the analysis of image performed in structured lighting. The difference is that instead of structured images, stereo pairs are used to provide the information to determine height and coordinate position.
- *Interferometry* methods measure distances in terms of wavelengths using interference patterns. These methods are very accurate, the accuracy being of the order of the wavelength (hundred of a nanometer).

Acoustic range finders use sound reflected from the surface and magnetic methods use magnetic field measurements using probes. These methods are prone to high noise and thus have not yet been commercially used for engineering measurements. However, they are reasonably accurate for geological survey where the tolerance on measurement accuracy is relatively high. Magnetic Resonance Imaging (MRI) and Computed Tomography (CT) use images obtained by magnetic field/radiation for sensing the internal geometry of the object being scanned. These methods are primarily apt for biomedical applications and the like where data points need to be acquired for the internal geometry not accessible to light.

Tactile methods contact the surface to be digitized using mechanical arms. Sensing devices at arm joints determine the relative coordinate locations. Robotic arms like a coordinate measuring machine (CMM) can be programmed to follow a path along the surface and collect accurate data. These methods are among the most accurate but are slow in data acquisition.

To summarize, every acquisition method interacts with the surface of interest by some phenomenon. The speed at which the phenomenon operates and the speed of response of the sensor determines the speed of data acquisition. The amount of analysis needed to compute the measured data and the accuracy are also determined by the sensor type selected. There are many practical issues when acquiring usable data, the major ones being calibration, accuracy, accessibility, occlusion, fixturing, multiple views, noise, incomplete data and surface finish. Nevertheless, it is possible to obtain adequate point cloud data in reasonably short time period using methods appropriate to an application.

10.3   Surface Modeling from a Point Cloud

Shape characterization from the point cloud is a key step in converting discrete data into a set of piecewise continuous surfaces. The organization of data and neighborhood information are important issues to address at this stage. The procedure for constructing of piecewise surface patches depends not only on the type of data (and hence the source) but also the type of model required. The type of model to be created depends on the intended use. For example, some applications may only require generation of a set of locally planar surfaces (having tangent plane discontinuity at the boundaries). The result would be a meshed or faceted surface (e.g., Figure 10.2b) with local sub-regions being triangular or quadrilateral, primarily former that can be constructed by combining any three neighboring points in the cloud. In case the point cloud results in sets of two-dimensional slices from MRI or CT scans, the object boundary at each plane can be modeled as a closed contour (Figure 10.2c). Lofting or skinning may then be performed using the contours to model the enclosing surface. In a more general case, we may need to determine connected higher order surface patches (Figure 10.2d) with or without enforcing smooth continuity. Following sections discuss the aforementioned three modeling techniques to obtain the bounding surface patches and thus the solid model from the acquired point cloud data.

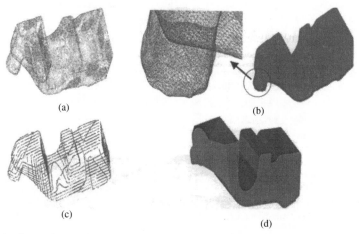

(a)

(b)

(c)

(d)

Figure 10.2   Geometrical models from point cloud data of a mechanical component: (a) point cloud acquired, (b) triangulated model, (c) contour model and (d) surface model

10.4   Meshed or Faceted Models

Mesh/faceted models, for example in Figure 10.2 (b), are simpler to construct yielding local planar facets with barely any intervention of the user. After having obtained a point cloud from any digitizer (contact or noncontact type), mesh models are generated with faces primarily of triangular topology though other polygonal faces are also possible. Triangular faceted models are used in graphics, animation, CAD, CAM as well as prototyping to name a few. A triangular facet data model in STL format is extensively used in CAD data transfer for downstream applications like tooling for manufacturing and rapid prototyping. Delaunay triangulation (Appendix 1) is a predominant method for generating triangular mesh models from point cloud.

A rectangular faceted model is used in Geographical Information System (GIS). Images of geographical terrains are processed using photographs taken from aircrafts or satellites. A series of digitized photographs of a region taken from several angles and at different times of the day are combined and processed to form a digital elevation model (DEM format). This model gives the terrain surface as a mesh of rectangular grids with each grid point associated with the latitude, longitude and altitude information.

10.5 Planar Contour Models

MRI and CT both yield object boundary information as data points placed in parallel planes or slices, which can be arranged as contours to represent a 3-D object. In medical applications, computed tomography is used to scan the interior of the body. It is effective for capturing images of bones and dense organs such as the brain and abdominal region. CT scanners produce images by firing X-rays at the region of interest and measuring the intensity of the rays after they have passed through the body. Industrial CT scanners can be used for image engineering of artifacts to detect cracks or holes. MRI is effective for producing images of soft tissues and is especially useful for detecting tumors in the human body. A set of 2-D CT or MRI scans is treated as a single 3-D image. Usually, the images produced by CT and MRI scanners are noisy, and in medical applications they capture densities for more organs than are needed for the study. These images must be filtered to reduce noise, and further must be thresholded and segmented to isolate regions or organs of interest. CT and MRI scans are available in a variety of vendor-specific formats. For medical images, the American College of Radiology and the National Electronics Manufacturers Association have set a standard called Digital Imaging and Communications in Medicine (DICOM). For displaying and manufacturing these medical models, we need to develop a mesh of facets across contours to represent the bounding surface of the object.

10.5.1 Points to Contour Models

CT or MRI scanning yields a series of cross-sectional intensity images. Each such 2-D image is composed of pixels. A pixel with value 1 (a black pixel) represents void while that with value 0 (a white pixel) represents material. Pixels may also have values between 0 and 1 in the grey range. First, planar contours are constructed from the data contained within a 2-D slice of the 3-D image. In each 2-D slice, there are one or more material blobs. The edges of those material blobs are located and from them an ordered list of points is formed. If the points are connected with straight-line segments, we obtain a polygonal contour representing the cross-section of the object. Two stages are involved in extracting contours from a 2-D image. The first stage, called *component labeling* involves labeling all blobs in the image. The second stage, called *edge following*, requires to follow the edge of each blob and form a list of points describing the contour.

A stepwise illustration of the contour extraction process is as follows:

Initially, the image quality is improved by removing the noise. These images may be cleaned using linear and non-linear filtering techniques applied to several kinds of noise. A Mexican hat low pass band filter function is shown in Figure 10.3(a) and an example illustration of noise filtering is shown in Figure 10.3(b). Once noise is filtered, the grey scale image is converted to an intensity image using threshold. An appropriate threshold value or a range is chosen and all pixels above this threshold or in the range are flagged as 1, otherwise as 0. For I_{th} as the intensity threshold chosen, the intensity

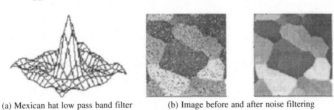

(a) Mexican hat low pass band filter   (b) Image before and after noise filtering

Figure 10.3   An example image processed for noise removal

after threshold operation for a pixel i is given as $I_i = 1$ if $I_i \leq I_{th}$ else $I_i = 0$. Figure 10.4 (a) shows the intensity values before and after the threshold operation and the corresponding images are shown in Figure 10.4(b).

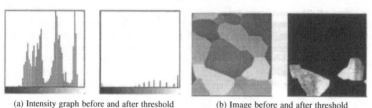

(a) Intensity graph before and after threshold   (b) Image before and after threshold

Figure 10.4   Threshold operations on an image

The subsequent step is of edge detection. The operation takes an intensity image I procured previously as its input, and returns a binary image BW of the same size as I, with 1's where the function finds edges in I and 0's elsewhere. Edge pixels are those where the intensity gradients are above a fixed threshold. A gradient threshold function is shown in Figure 10.5(a) and the edge detection operation is illustrated with an example in Figure 10.5(b). Once the edge points are identified, they are tracked to form a closed loop forming the contour on a plane. A traditional neighborhood

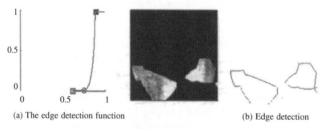

(a) The edge detection function   (b) Edge detection

Figure 10.5   Illustration of edge detection

detection algorithm would work well except for regions where some branching may occur. A way to overcome this is to follow the edge point having the lower angle of curving (Figure 10.6 a). This ensures that the wrong edge points are not selected to form the edge loop. Once the edge pixels are detected, in many applications, a smooth closed B-spline curve is interpolated through them (Figure 10.6b).

After reconstructing all contours, a tiling or skinning surface is created. In some applications, the reconstructed contours can be used as the CAD representation as in case of layered manufacturing. However, most other CAD/CAM operations require a B-rep model. For an object with M contours, $C_i(u_i)$, $1 \leq i \leq M$ with respective parameterization u_i, we wish to fit a surface $S(u, v)$ through the collection of contours. The surface will be closed and periodic in the u direction (direction along the given contours) and open in the v direction (direction through the contours). We may reparameterize the contours using a common global parameter u to fit the tiling surface. An elementary method to determine parameter correspondence is to assume a one-to-one relationship between given contours. We may assume all the contours to be parameterized between 0 and 1 (or any other bounds) and that

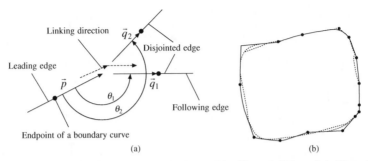

Figure 10.6 **Edge tracking with edge points: (a) edge tracking illustrated at a point of branching and (b) illustration of closed B-spline contour fitting through the tracked edge points**

a parameter value on one contour corresponds to the same parameter value on all the other contours, as illustrated in Figure 10.7, in which a line segment connects corresponding points on two adjacent contours. However, as shown in Figure 10.8(a), this may lead to twisting or shearing of the tiling surface. A better approach, since the contours are closed, may be to use the angular parameter for all contours, using cylindrical coordinates by placing the origin at the geometric center of one of the contours, say C_i. Parameter correspondence between two adjacent contours C_i and C_{i+1} may then be established by taking equal angular spacing. This angular correspondence between points on adjacent contours would avoid twisting of the spline surface fitted through them. The approach is illustrated in Figure 10.8(b). Surface reconstruction from planar contours gets complicated if the object branches wherein more than one contour may be present on any slice. We must then connect either a single contour on one slice with several contours on an adjacent slice (one-to-many tiling or skinning), or many contours on one slice to many contours on an adjacent slice (many-to-many tiling or skinning).

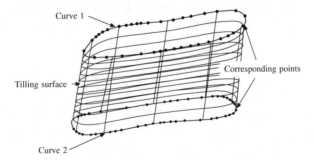

Figure 10.7 **Linear parameter correspondence in tiling surface fitting**

10.6 Surface Models

Surface models (Figure 10.2 d) are B-rep models that represent an object as a set of closed connected higher order surface patches. In majority of CAD/CAM applications, such surface models are required. They offer better continuity across patch boundaries compared to meshed or faceted models. Further, contour models become inapplicable when the point cloud is not arranged in the sliced form. Mathematical forms of surface patches discussed in Chapters 6 and 7 can then be applied here. The degree of

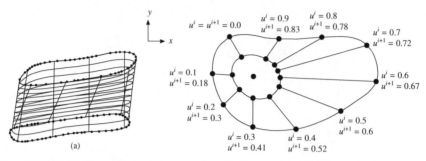

Figure 10.8 **(a) A bad linear parameter correspondence leading to twisting and (b) parameter correspondence determined by angular spacing (viewing into the z direction)**

patches and the continuity across boundaries vary depending on the nature of the object and hence its point cloud. Point cloud of prismatic objects can generally be modeled using quadric surfaces, e.g., planes with C^0 (position) continuity, or in some cases conics with C^1 (slope) continuity. When designing the exterior of automobiles, household appliances, cellular phones and aerospace components, free form patches (e.g., Bézier or B-spline) are usually employed to enhance the design features and functionality.

A broad classification of bounding surface types used is presented in Figure 10.9. Hierarchy of patches (planes, quadrics, sweep surfaces, B-splines) that are in the order of geometric complexity has to be defined at this stage, which a user can recognize and specify to the computer to determine the final model. Free form surface reconstruction from point cloud requires to perform the following:

- *Segmentation*: To divide the original point cloud into subsets of points, one for each natural[1] surface, so that each subset contains just those points sampled from a particular natural surface.
- *Classification*: To determine the type of surface each subset of points may belong to.
- *Fitting*: To find the surface of the chosen type, as the best fit through the points in a given subset.

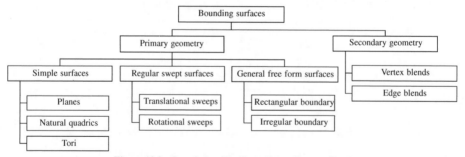

Figure 10.9 **Broad classification of bounding surfaces**

Note that the segmentation and classification of the point cloud above are not sequential but parallel.

---

[1] A surface patch (planar, cylindrical, conical and spherical) that can be easily identified by a user by inspection.

Also, the point cloud for a prismatic object is different from that of a free form object and so the aforementioned treatment would be different for the two point clouds.

10.6.1   Segmentation and Surface Fitting for Prismatic Objects

There are two different methods, *edge-based* and *face-based*, for segmentation and surface fitting of prismatic surfaces. In edge-based methods, we try to determine possible patches by determining their boundaries, and later patches are inferred from the implicit segmentation provided by these boundary curves. Sharp edges are locations where the first difference (derivative) estimated from the point cloud changes rapidly, for example, two intersecting orthogonal planes. For smooth edges on the other hand, we look for sites where the second difference (surface curvature) has discontinuity.

A procedure to estimate curvature from a point set may be as follows. For each point in the set, a local neighborhood is defined based on a limiting distance. Then a local quadratic surface is fit using least square minimization. A quadratic surface in algebraic form is given by

$$f(x, y, z) = a_1x^2 + a_2y^2 + a_3z^2 + a_4xy + a_5xz + a_6yz + a_7x + a_8y + a_9z + a_{10} = 0 \qquad (10.1)$$

The surface curvatures (principle curvatures) and the slope directions can be computed from this locally approximated surface. By inspecting the magnitude (very large) of the principle curvatures, or the change in their sign, we may identify edge points. After all the edge points are determined from the point cloud, they are linked to form the closed boundaries.

In the segmentation stage, to partition the digitized points to regions, all points are tested for belonging to each boundary loops using the scan line algorithm (Chapter 9) as shown in Figure 10.10 with an example point cloud showing one quarter of a cylinder in the first quadrant (Figure 10.10a). Cloud curvature can be estimated and the points are identified where curvature extremes occur as shown in Figure 10.10 (b). The detected edge points and the boundary loops joining them are shown in Figure 10.10 (c).

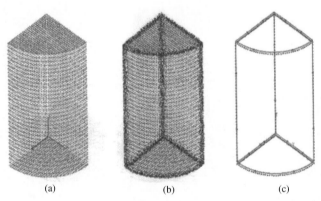

(a)	(b)	(c)

Figure 10.10   **Edge detection using cloud curvature: (a) point cloud of an object, (b) edge points on the cloud are colored; identified (darkened) on curvature estimation from points and (c) the edge loop detected**

The edge-based methods have the following limitations.

(a) Sensor data, particularly from optical scanners, are often unreliable near sharp edges.

(b) Number of data points used for segmentation is small, implying that the information from much of the data is not used to assist in reliable segmentation.
(c) Detection of smooth edges is unreliable since computation of derivatives from the noisy point cloud is error prone.

The face based approach, on the other hand, attempts to infer connected regions of points with similar properties as belonging to the same surface (e.g. groups of neighboring points having the same normal belong to the same plane). This method is more reliable since it works on a larger number of points, using all the available data. The procedure is illustrated in Figure 10.11 for the same object as in Figure 10.10. A seed point is chosen and neighboring points are checked to have the same property as the seed point. If yes, they are added and the region around the seed point grows. Else, the surface definition is appropriately changed only in the initial stage. After a sufficiently large number of data points are checked to belong to a surface definition, the latter is retained. In case of Figure 10.11(b), since the underlying surface is a plane, the definition does not change as the region grows. The iterative evaluation ends after no more points could be found on the plane having an upward normal. The region is shown with dark normal needles in Figure 10.11 (b). The segmented cloud is shown in Figure 10.11(c). Another seed point is chosen from the remaining cloud and the process continues till all points are classified. The problems associated with the face-based approach are:

(a) Choosing good seed points and surface definition for a sub-region in a cloud is often difficult.
(b) Adaptive change in the surface type has to be performed as the region grows.
(c) Bad points, if accidentally added to the region, may change the surface definition.
(d) Deciding whether or not to add to a region can be difficult since these data are susceptible to noise.

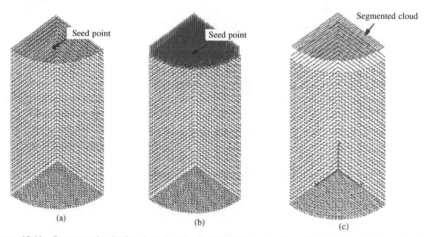

Figure 10.11 **Segmentation by face based methods: (a) point cloud of an object and a selected seed point, (b) region identified as belonging one base surface (the cloud normal are shown in dark needles and (c) segmented region.**

In the face-based method, segmentation and surface fitting are simultaneous and that additional surface fitting techniques may not be required. However, in edge-based methods, surface fitting can be performed using the least square approach. In case of prismatic parts, algebraic surfaces (planar, cylindrical) are fit. Note that the edges determined during segmentation may be error prone and that they may not be considered as the bounding edges of the surface patches, The new bounding edges may be taken as the intersection of these surface patches, illustrated in Figure 10.12. The segmented cloud is shown in Figure 10.12(a). The unbounded algebraic surfaces fitted to these clouds by least squares method are shown in Figure 10.12(b). The edges obtained by intersecting these surfaces are used to trim the surfaces and the resulting bounded surface network is shown in Figure 10.12(c).

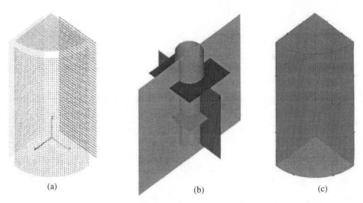

(a)   (b)   (c)

Figure 10.12   (a) Segmented cloud (offset), (b) algebraic surfaces fitted to segmented cloud and (c) solid model after intersection and trimming.

10.6.2   Segmentation and Surface Fitting for Freeform Shapes

Both edge-based and face-based methods cannot be used for representing a complex free form surface, as encountered in sculpted objects, since this will result in many small pieces of say planar or quadratic surfaces, which is not the desired result. However, edge-based methods can be used for segmentation of the point cloud. An alternative approach may be that a user segments the cloud interactively. A general methodology to obtain free form geometry from a point cloud is as follows:

(a) Segment the cloud into regions, each of which are representable by parametric surfaces such as Bézier or B-spline patches.
(b) Parameterize the points.
(c) Determine B-spline patches using least square fit maintaining appropriate continuity between the adjacent patches.

For surface fitting, a large crudely approximated four-sided patch is chosen by the user. Boundaries are chosen such that the points of interest in the cloud lie within the boundary of the surface. The points are then projected on to the surface to find corresponding points on the approximating surface. The distance between the points in the cloud and the corresponding points on the surface is minimized using the least-square method. If the error is too large after least square fitting, iterative parameterization and refit are performed with better approximating surfaces. Figure 10.13 schematically illustrates an iterative step of the approach.

B-splines are used predominantly in free form curve and surface fitting. Two approaches are usually employed for surface fitting of point cloud. The first is to fit directly a B-spline patch to the cloud

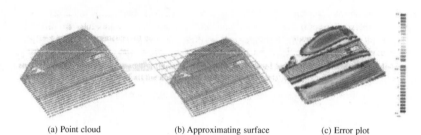

(a) Point cloud   (b) Approximating surface   (c) Error plot

Figure 10.13   An approximating B-spline patch for a segmented point cloud.

discussed above. The second is to fit B-spline curves at the boundaries of the segmented cloud and later define a Coon's patch using the same. Following the concepts developed in Chapters 5 and 7, given M control vertices $\mathbf{B}_i$ ($i = 0, 1,...,M-1$), a B-spline curve of order k is defined as

$$r(u) = \sum_{i=0}^{M-1} B_i N_{k,k+i}(u), \quad u \in [u_{\min}, u_{\max}] \tag{10.2}$$

Let a smooth parametric curve $\mathbf{r}(u)$ defined by the above equation pass through a sequence of data points $\{\mathbf{P}_i, i = 0,..., j\}$. If a data point lies on the B-spline curve, it must satisfy Eq. (10.2). Writing the same for each of j data points yields

$$\mathbf{P}_1(u_1) = N_{k,k}(u_1)\mathbf{B}_0 + N_{k,k+1}(u_1)\mathbf{B}_1 + \ldots + N_{k,k+M-1}(u_1)\mathbf{B}_{M-1}$$

$$\mathbf{P}_2(u_2) = N_{k,k}(u_2)\mathbf{B}_0 + N_{k,k+1}(u_2)\mathbf{B}_1 + \ldots + N_{k,k+M-1}(u_2)\mathbf{B}_{M-1}$$

$$. \ . \ . \ .$$

$$\mathbf{P}_j(u_j) = N_{k,k}(u_j)\ \mathbf{B}_0 + N_{k,k+1}(u_j)\mathbf{B}_1 + \ldots + N_{k,k+M-1}(u_j)\mathbf{B}_{M-1} \tag{10.3}$$

where $2 \le k \le M \le j$. This set of equations is written in matrix form as

$$\mathbf{P} = \mathbf{CB} \tag{10.4}$$

with

$$\mathbf{C} = \begin{bmatrix} N_{k,k}(u_1) & \cdots & \cdots & N_{k,k+M-1}(u_1) \\ \vdots & \vdots & & \vdots \\ \vdots & & \vdots & \vdots \\ N_{k,k}(u_j) & \cdots & \cdots & N_{k,k+M-1}(u_j) \end{bmatrix}$$

where $\mathbf{P}$, $\mathbf{C}$ and $\mathbf{B}$ are the point data, basis and defining polygon matrices, respectively. In case of curve approximation, $\mathbf{C}$ is not a square matrix. The problem is over-specified and can be solved using some mean sense. Noting that a matrix times its transpose is square, the defining polygon vertices for a B-spline curve that smoothes the data is given by

$$\mathbf{B} = [\mathbf{C}^T \mathbf{C}]^{-1} \mathbf{C}^T \mathbf{P} \tag{10.5}$$

Least square fitting technique of B-spline curves described above assumes that $\mathbf{C}$ is known. Given the order k of the B-spline basis, the number of defining polygon vertices M, the parameter values

$u_1, \ldots, u_j$ corresponding to each data point and the knot vectors, the matrix $\mathbf{C}$ can be obtained. A user may want to specify the degree of a B-spline curve as cubic. The number of control vertices M is chosen depending on the complexity of the point cloud shape. The values $u_1, u_2, \ldots, u_j$ may be determined using some parameterization technique. Also, the knot vector may be determined consistent with the above parameterization to minimize the computation for convergence. The three kinds of parameterization mostly used are uniform, centripetal and chord length methods as discussed in Section 5.10. A generalization of all the above parameterization models is

$$u_1 = 0; u_i = u_{i-1} + \frac{\mid \mathbf{P}_i - \mathbf{P}_{i-1} \mid^e}{\sum\limits_{s=1}^{j-1} \mid \mathbf{P}_{s+1} - \mathbf{P}_s \mid^e} \text{ with } 2 \le i \le j, \ 0 \le e \le 1 \tag{10.6}$$

where j is the total number of data points specified on the curve. For $e = 0$, 1, and 0.5, the above equation yields uniform, chord length and centripetal parameterization, respectively.

Least square minimization may commence with the parameter values in Eq. (10.6). Using these, a B-spline curve may be determined using Eqs. (10.2) and (10.5). Thereafter, the error between the fitted B-spline curve and the corresponding points in the cloud is computed. If the error is large, parameter values u_i are optimized by iteratively improving them using a first order Taylor correction for the error expression. An example B-spline curve fitting is illustrated in Figure 10.14(a). Once four such boundary curves for a segmented cloud are determined, a linear or cubic (Hermite) blended Coon's patch (section 7.2.1) can be developed as illustrated in Figure 10.14(b). Note that the cross boundary tangents and twist vectors can be determined in a manner such that the Coon's patch represents the best fit for the segmented point cloud in the least square sense.

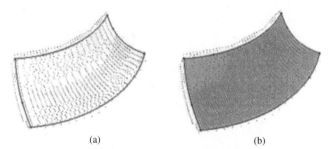

(a)   (b)

Figure 10.14   (a) Boundary B-spline curves with the segmented cloud (b) Coon's patch fit using the boundary curves

After surface fitting is performed for all points in the cloud, the final step is to fine tune the patches to obtain the required continuity across the patch boundaries and to address other engineering constraints like symmetry. The procedure for enforcing tangent plane continuity between adjacent patches is discussed in Section 7.3. Other important geometric properties such as symmetry, parallelism, orthogonality and concentricity, which are essential attributes, have to be enforced to convert the B-rep model to a valid solid model for further downstream applications. This stage requires user interaction, or, artificial intelligence techniques may also help.

10.7 Some Examples of Reverse Engineering

Reverse engineering has been applied widely for rapid product development. An application[1] on reducing the design-to-manufacturing cycle for automobile illustrates how Ford with the assistance of a measuring system, photogrammetry and optical scanning, was able to accomplish this for a new concept car. Some studies[2] demonstrate the creation of physical models of biological forms through reverse engineering for medical treatment planning. The steps in these applications are acquiring CT data, creating a faceted model and physical replication through rapid prototyping. Use of physical models for treatment planninng/visualization instead of solely using computer software generated display images based on X-ray Computed Tomography (CT) or Magnetic Resonance imaging (MRI) data results in better treatment with lowered risk rates and recovery period.

Application of reverse engineering in heritage preservation is another interesting development. The Afghanistan Institute and Museum, Bubendorf (Switzerland) and the *New 7 Wonders* Society and Foundation, Zurich (Switzerland) have launched a campaign to reconstruct the Buddha in Bamiyan to original shape, size and place through photogrammetry[3]. One of the digital archiving projects using laser-scanning technology took place in Kamakura to model the Great Buddha[4]. In another example of heritage preservation application using optical scanning technology, a team of 30 faculty, staff and students from Stanford University and the University of Washington spent the 1998-99 academic year in Italy scanning the sculptures and architecture of Michelangelo[5].

Reverse Engineering of physical objects to extract their three-dimensional features from point clouds for CAD/CAM application is a fast developing technology. The state of the art in reverse engineering and concurrent commercial software systems allow for point cloud processing and single surface modeling with interactive help. The automatic replication of complete B-rep models is possible in simple cases at this time. Key research areas, which still need further work, include improving data capture and coping with noise and missing data, and reliable segmentation and surface fitting to obtain the desired geometric model. This chapter is an effort towards understanding some methods in reverse engineering. The field being an active research area, an interested reader may refer to the ongoing developments in literature.

---

[1]Sherry L. Baranek, "Designing the Great American Supercar," Time-Compression Technologies Magazine, September 2002.

[2]S. Swann, "Integration of MRI and Stereolithography to build medical models. A case study," Rapid Prototype Journal Vol 2. No. 4, 1996 p. 41–46.

[3]Gruen, A., Remondino, F. and Zhang, L., "Reconstruction of the great Buddha of Bamiyan," Afganistan. *ISPRS Commission V Symposium*, Corfu (Greece) 2002.

[4]Daisuke Miyazaki, Takeshi Ooishi, Taku Nishikawa, Ryusuke Sagawa, Ko Nishino, Takashi Tomomatsu, Yutaka Takase and Katsushi Ikeuchi, "The Great Buddha Project: Modelling Cultural Heritage through Observation," in http://www.cadcenter.co.jp/en/webgallery/webgallery_cg24.html.

[5]Marc Levoy, Kari Pulli, Brian Curless, Szymon Rusinkiewicz, David Koller, Lucas Pereira, Matt Ginzton, Sean Anderson, James Davis, Jeremy Ginsberg, Jonathan Shade and Duane Fulk, "The Digital Michelangelo Project: 3D Scanning of Large Statues," *Proc. Siggraph 2000*. pp. 131–144, and http://graphics.stanford.edu/projects/mich.

Chapter 11

Finite Element Method

11.1 Introduction

The design procedure does not cease after accomplishing a solid model. With analysis and optimization, design of a component may further be improved. Real life components are quite intricate in shape for the purpose of stress and displacement analysis using classical theories. An example is the analysis of the wing of an aircraft. Approximations like treating it as a cantilever with distributed loads can yield inaccurate results. We then seek a numerical procedure like the finite element analysis to find the solution of a complicated problem by replacing it with a simpler one. Since the actual problem is simplified in finding the solution, it is possible to determine only an approximate solution rather than the exact one. However, the order of approximation can be improved or refined by employing more computational effort.

In the finite element method (FEM), the solution region is regarded to be composed of many small, interconnected subregions called the *finite elements*. Within each element, a feasible displacement interpolation function is assumed. Strain and stress computations at any point in that element are then performed following which the stiffness properties of the element are derived using elasticity theories. Element stiffnesses are then assembled to represent the stiffness of the entire solution region.

Between solid modeling and the finite element analysis lies an important intermediate step of mesh generation. Mesh generation as a preprocessing step to FEM involves discretization of a solid model into a set of points called *nodes* on which the numerical solution is to be based. Finite elements are then formed by combining the nodes in a predetermined topology (linear, triangular, quadrilateral, tetrahedral or hexahedral). Discretization is an essential step to help the finite element method solve the governing differential equations by approximating the solution within each finite element. The process is purely based on the geometry of the component and usually does not require the knowledge of the differential equations for which the solution is sought. The accuracy of an FEM solution depends on the fineness of discretization in that for a finer mesh, the solution accuracy will be better, that is, for the average finite element size approaching zero, the finite element solution approaches the classical (or analytical) solution, if it exists. We would always desire to seek the 'near to classical' solution. However, the extent of computational effort involved poses a limit on the number of finite elements (and thus their average size) to be employed. A relatively small number of finite elements in a coarse mesh would yield a solution at a much faster rate, though it will be less accurate compared to that obtained using a large number of elements in a fine mesh. In the latter, however, the solution time taken will be more. Thus, there is a trade off involved between the average element size and solution time taken which a designer should keep in mind when performing mesh generation which, by itself, is a very vast and active field of research. Appendix 1 is provided to familiarize the reader with some preliminary methods and algorithms on discretization, mostly in two dimensions. As this chapter deals with discrete (truss, beam and frame) and continuum (triangular and quadrilateral) elements in two dimensions, some methods pertaining to only the abovementioned elements are discussed in the appendix noting that most methods may be extended for use in three dimensions. We may realize at this stage that discrete representation of solids is another approach in solid modeling wherein a solid's volume may be regarded as the sum total of the volumes of constituting tetrahedral or hexahedral elements. To create a discrete representation using mesh generation would, however, require the B-rep information of the solid.

With regard to the finite element analysis, there are many texts available for an in-depth study. This chapter, however, introduces preliminary concepts to the reader by presenting linear elastic analysis using some widely used basic elements. The finite element method as known today was investigated in the papers of Turner, Clough, Martin and Topp, Argyris and Kelsey and many others. The name *finite element* was coined by Clough. The advent of digital computers in the 1960s and 1970s provided a rapid means of performing intricate calculations involved in the analysis that made the method practically viable. With the development of high speed digital computers, the application of the finite element method also progressed at a very impressive rate. Zienkiewicz and Cheung presented a broad interpretation of the method and its applicability to any general field problem. As a result, the finite element equations could also be derived using general methods like the weighted residual (Galerkin) method. This led to a widespread interest among other researchers working with generic nonlinear differential equations.

11.2 Springs and Finite Element Analysis

Preliminary concepts of the finite element analysis are presented here using linear springs. Consider a spring of stiffness k_p shown in Figure 11.1(a). The nodal displacements are allowed along the horizontal direction which makes the spring a two degree-of-freedom system, one at each node. Let the displacements at nodes i and j be u_i and u_j and the external forces acting along the axis be f_i and f_j, respectively. Considering the equilibrium at nodes i and j using Newtonian mechanics, we have

$$f_i = k_p (u_i - u_j)$$

$$f_j = k_p (u_j - u_i) \tag{11.1}$$

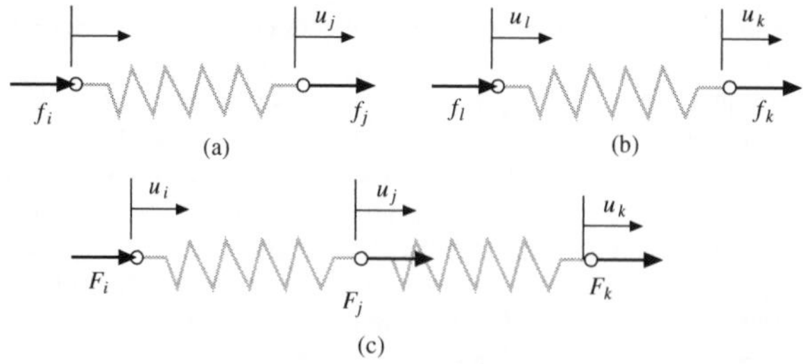

Figure 11.1 Springs: (a) and (b) with stiffness k_p and k_q and (c) assembled in series

Writing Eq. (11.1) in the matrix form, we get

$$\begin{bmatrix} k_p & -k_p \\ -k_p & k_p \end{bmatrix} \begin{Bmatrix} u_i \\ u_j \end{Bmatrix} = \begin{Bmatrix} f_i \\ f_j \end{Bmatrix} \tag{11.2a}$$

or in compact notation

$$\mathbf{k}_p \mathbf{u}_p = \mathbf{f}_p \tag{11.2b}$$

In the finite element nomenclature, matrix $\mathbf{k}_p = \begin{bmatrix} k_p & -k_p \\ -k_p & k_p \end{bmatrix}$ is called the *element stiffness matrix*,

$\mathbf{u}_p = \begin{Bmatrix} u_i \\ u_j \end{Bmatrix}$ the *element displacement vector* and $\mathbf{f}_p = \begin{Bmatrix} f_i \\ f_j \end{Bmatrix}$ the *element force vector*. Note from
Eq. (11.1) that $f_j = -f_i$ which suggests spring equilibrium. It is for this reason that the matrix $\mathbf{k}_p$ is
singular (its determinant is zero) since Eq. (11.1) is, in a way, a single equation. It may further be
noted that $\mathbf{k}_p$ is symmetric and is positive semi-definite as one of the eigen-values is zero and the
other is positive. Consider

$$\mathbf{u}_p^T \mathbf{k}_p \mathbf{u}_p = \begin{Bmatrix} u_i \\ u_j \end{Bmatrix}^T \begin{bmatrix} k_p & -k_p \\ -k_p & k_p \end{bmatrix} \begin{Bmatrix} u_i \\ u_j \end{Bmatrix} = \begin{Bmatrix} u_i \\ u_j \end{Bmatrix}^T \begin{bmatrix} k_p(u_i - u_j) \\ k_p(u_j - u_i) \end{bmatrix} = k_p(u_i - u_j)u_i + k_p(u_j - u_i)u_j$$

$$= k_p(u_i - u_j)(u_i - u_j) \tag{11.2c}$$

which is twice the strain energy stored in the spring. Thus, $\mathbf{u}_p^T \mathbf{k}_p \mathbf{u}_p$ is related to the strain energy
which can never be negative. The singularity, symmetry and positive semi-definiteness are inherent
properties of finite element stiffnesses.

We prefer, however, the elaborate form in Eq. (11.2a) for convenience in matrix assembly. Consider
now another spring of stiffness k_q as shown in Figure 11.1(b). The element equations in matrix form
can be written by inspection from Eq. (11.2a), that is

$$\begin{bmatrix} k_q & -k_q \\ -k_q & k_q \end{bmatrix} \begin{Bmatrix} u_l \\ u_k \end{Bmatrix} = \begin{Bmatrix} f_l \\ f_k \end{Bmatrix} \tag{11.3}$$

If nodes j and l are to coincide such that the two springs are in series with 3 degrees of freedom
as in Figure 11.1(c), then

$$u_j = u_l \tag{11.4}$$

Expressing Eqs. (11.2a) and (11.3) in all three degrees of freedom, we have for springs p and q,
respectively

$$\begin{bmatrix} k_p & -k_p & 0 \\ -k_p & k_p & 0 \\ 0 & 0 & 0 \end{bmatrix} \begin{bmatrix} u_i \\ u_j \\ u_k \end{bmatrix} = \begin{bmatrix} f_i \\ f_j \\ 0 \end{bmatrix} \tag{11.5a}$$

and

$$\begin{bmatrix} 0 & 0 & 0 \\ 0 & k_q & -k_q \\ 0 & -k_q & k_q \end{bmatrix} \begin{bmatrix} u_i \\ u_j \\ u_k \end{bmatrix} = \begin{bmatrix} 0 \\ f_l \\ f_k \end{bmatrix} \tag{11.5b}$$

Adding Eqs. (11.5a) and (11.5b) yields

$$\begin{bmatrix} k_p & -k_p & 0 \\ -k_p & k_p + k_q & -k_q \\ 0 & -k_q & k_q \end{bmatrix} \begin{bmatrix} u_i \\ u_j \\ u_k \end{bmatrix} = \begin{bmatrix} f_i \\ f_j + f_l \\ f_k \end{bmatrix} = \begin{bmatrix} F_i \\ F_j \\ F_k \end{bmatrix} \qquad (11.5c)$$

or in compact form

$$\mathbf{KU = F}$$

where $\mathbf{F} = \begin{bmatrix} F_i \\ F_j \\ F_k \end{bmatrix}$ is the vector of net external forces acting on the nodes with respective subscripts,

$\mathbf{K}$ the global stiffness matrix and $\mathbf{U}$ the global displacement vector. Note that the element stiffness properties are inherited by the global stiffness matrix $\mathbf{K}$, in that the latter is also singular, symmetric and positive semi-definite. Singularity of the stiffness matrix implies that the linear system in Eq. (11.5c) has at least one rigid-body degree of freedom and the system cannot be solved unless some displacements are known or constrained *a priori*. We can further simplify Eq. (11.5c) as

$$\begin{bmatrix} k_p \\ -k_p \\ 0 \end{bmatrix} u_i + \begin{bmatrix} -k_p \\ k_p + k_q \\ -k_q \end{bmatrix} u_j + \begin{bmatrix} 0 \\ -k_q \\ k_q \end{bmatrix} u_k = \begin{bmatrix} F_i \\ F_j \\ F_k \end{bmatrix} \qquad (11.5d)$$

Assuming that node i is fixed so that $u_i = 0$, Eq. (11.5d) becomes

$$\begin{bmatrix} -k_p \\ k_p + k_q \\ -k_q \end{bmatrix} u_j + \begin{bmatrix} 0 \\ -k_q \\ k_q \end{bmatrix} u_k = \begin{bmatrix} F_i \\ F_j \\ F_k \end{bmatrix} \qquad (11.5e)$$

F_i represents the reaction force at node i that depends on displacements u_j and u_k (only u_j in this case). To determine only the displacements, we need to solve

$$\begin{bmatrix} k_p + k_q \\ -k_q \end{bmatrix} u_j + \begin{bmatrix} -k_q \\ k_q \end{bmatrix} u_k = \begin{bmatrix} F_j \\ F_k \end{bmatrix} \quad \text{or} \quad \begin{bmatrix} k_p + k_q & -k_q \\ -k_q & k_q \end{bmatrix} \begin{bmatrix} u_j \\ u_k \end{bmatrix} = \begin{bmatrix} F_j \\ F_k \end{bmatrix}$$

Alternatively, the above set of equations can also be obtained by eliminating the first row (entirely) and first column of the coefficient/stiffness matrix (corresponding to the fixed degree of freedom u_i) in Eq. (11.5c). Further solving gives

$$\begin{bmatrix} u_j \\ u_k \end{bmatrix} = \begin{bmatrix} k_p + k_q & -k_q \\ -k_q & k_q \end{bmatrix}^{-1} \begin{bmatrix} F_j \\ F_k \end{bmatrix} = \frac{1}{\{(k_p + k_q)\, k_q - k_q^2\}} \begin{bmatrix} k_q & k_q \\ k_q & k_p + k_q \end{bmatrix} \begin{bmatrix} F_j \\ F_k \end{bmatrix}$$

or $\qquad u_j = \dfrac{1}{k_p}\, (F_j + F_k)$

$$u_k = \frac{1}{k_p}\left[F_j + \left(\frac{k_p}{k_q} + 1\right)F_k\right]$$ (11.5f)

The reaction force F_i from Eq. (11.5e) is $-k_p u_j = -(F_j + F_k)$ as expected. The displacements can be verified for the case when $F_j = 0$. The effective spring constant using Newtonian mechanics is then

$$k_{\text{eff}} = \frac{k_p k_q}{k_p + k_q}$$ and hence $u_k = F_k/k_{\text{eff}} = F_k\left(\frac{1}{k_p} + \frac{1}{k_q}\right)$ while u_j is $\frac{F_k}{k_p}$.

11.3   Truss Elements

Plane trusses are often used in construction, particularly for roofing of residential and commercial buildings and in short span bridges (Figure 11.2). Trusses, whether two- or three-dimensional, belong to the class of skeletal structures consisting of elongated components called *members* connected at joints. A member or a truss element of elastic modulus E, cross-section area A and length l is shown in Figure 11.3. Like in a spring, two degrees of freedom, namely, u_i and u_j are permitted along the bar axis under the action of external loads f_i and f_j. Let the extension be $dl = (u_j - u_i)$ so that the strain is $\frac{(u_j - u_i)}{l}$ and thus the stress is $E\frac{(u_j - u_i)}{l}$. The internal force $EA\frac{(u_j - u_i)}{l}$ must balance the external forces at the nodes. Thus, at node $i, f_i = -\ EA\frac{(u_j - u_i)}{l}$ while at node $j, f_j = \ EA\frac{(u_j - u_i)}{l}$. Comparing with Eq. (11.1) yields $k_p = AE/l$ and hence the stiffness matrix $\mathbf{k}_t$ for a truss element becomes

$$\mathbf{k}_t = \begin{bmatrix} \dfrac{AE}{l} & -\dfrac{AE}{l} \\ -\dfrac{AE}{l} & \dfrac{AE}{l} \end{bmatrix} = \frac{AE}{l}\begin{bmatrix} 1 & -1 \\ -1 & 1 \end{bmatrix}$$ (11.6)

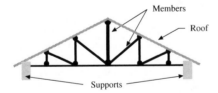

Figure 11.2   Schematic of a roof with truss members

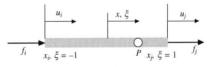

Figure 11.3   A truss element

Although the derivation of the stiffness matrix for a truss element is straightforward using the spring analogy, the same is derived using a more formal finite element procedure. Let the initial positions of nodes i and j be x_i and x_j. A local coordinate measure ξ can be introduced so that at $x = x_i, \xi = -1$ and at $x = x_j, \xi = 1$. The displacement $u(x)$ at any point P in the element can be expressed in terms of the unknown nodal displacements. Since there are only two such displacements, u_i and u_j, it behooves to use a linear interpolating relation, that is

$$u(x) = c_1 + c_2 x \quad \text{or} \quad u(\xi) = d_1 + d_2 \xi$$ (11.7a)

We know that $u(x_i) = u_i$ and $u(x_j) = u_j$ (or $u(\xi = -1) = u_i$ and $u(\xi = 1) = u_j$ in terms of the local measure) substituting which we can solve for the unknown constants c_1 and c_2 (or d_1 and d_2). Solving for these constants gives

$$c_2 = \frac{u_j - u_i}{x_j - x_i} \quad \text{and} \quad c_1 = \frac{u_i x_j - u_j x_i}{x_j - x_i} \quad \left(\text{or} \quad d_1 = \frac{u_i + u_j}{2} \quad \text{and} \quad d_2 = \frac{u_j - u_i}{2} \right)$$

Thus, Eq. (11.7a) becomes

$$u(x) = \left(\frac{u_i x_j - u_j x_i}{x_j - x_i} \right) + x \left(\frac{u_j - u_i}{x_j - x_i} \right) = \left(\frac{x_j - x}{x_j - x_i} \right) u_i + \left(\frac{x - x_i}{x_j - x_i} \right) u_j = N_i(x) u_i + N_j(x) u_j$$

$$(11.7b)$$

which can be expressed in the matrix form as

$$u(x) = [N_i(x) \ N_j(x)] \begin{pmatrix} u_i \\ u_j \end{pmatrix} = \mathbf{N}(x)\mathbf{u} \qquad (11.7c)$$

Alternatively,

$$u(\xi) = \frac{u_i + u_j}{2} + \xi \left(\frac{u_j - u_i}{2} \right) = \left(\frac{1 - \xi}{2} \right) u_i + \left(\frac{1 + \xi}{2} \right) u_j = N_i(\xi) u_i + N_j(\xi) u_j \quad (11.7d)$$

which in matrix form is

$$u(\xi) = [N_i(\xi) \quad N_j(\xi)] \begin{pmatrix} u_i \\ u_j \end{pmatrix} = \mathbf{N}(\xi)\mathbf{u} \qquad (11.7e)$$

Here, $N_i(x) = \dfrac{x_j - x}{x_j - x_i}, N_j(x) = \dfrac{x - x_i}{x_j - x_i}$ or $\left(N_i(\xi) = \dfrac{1 - \xi}{2}, N_j(\xi) = \dfrac{1 + \xi}{2} \right)$ are termed as the
shape or *interpolating* functions. Note that $N_i(x_i) = 1$ while $N_i(x_j) = 0$. Similarly, $N_j(x_i) = 0$ while
$N_i(x_j) = 1$, that is, generically in any finite element the value of the shape function, $N_j(x)$ (or $N_j(\xi)$)
is one at node j and is zero at all the other nodes of that element. The functions are positive and at
any point within the element, they sum to 1. In other words, the finite element shape functions are
barycentric similar to Bernstein polynomials or B-spline basis functions discussed in Chapters 4 and
5, respectively. In fact, we can relate the local coordinates x and ξ by comparing the coefficients in
Eqs. (11.7b) and (11.7d). Comparing $N_j(x)$ with $N_j(\xi)$ gives

$$\frac{x - x_i}{x_j - x_i} = \frac{1 + \xi}{2} \quad \Rightarrow \quad x = \left(\frac{1 - \xi}{2} \right) x_i + \left(\frac{1 + \xi}{2} \right) x_j = \mathbf{N}(\xi)\mathbf{x} \qquad (11.7f)$$

We can either use Eq. (11.7c) or Eq. (11.7e) to compute the axial strain ε_x at location P in the element.
Using the latter we have

$$\varepsilon_x = \frac{\partial u}{\partial x} = \frac{\partial}{\partial x} \mathbf{N}(\xi)\mathbf{u} = \frac{\partial}{\partial \xi} \mathbf{N}(\xi) \frac{d\xi}{dx} \mathbf{u} = \left[-\frac{1}{2} \quad \frac{1}{2} \right] \frac{d\xi}{dx} \mathbf{u} \qquad (11.7g)$$

From Eq. (11.7f), $\dfrac{dx}{d\xi} = \left[-\dfrac{1}{2} \quad \dfrac{1}{2} \right] \begin{pmatrix} x_i \\ x_j \end{pmatrix} = \dfrac{(x_j - x_i)}{2} = \dfrac{l}{2}$

so that

$$\varepsilon_x = \frac{\partial u}{\partial x} = \left[-\frac{1}{2} \quad \frac{1}{2} \right] \frac{2}{l} \mathbf{u} = \left[-\frac{1}{l} \quad \frac{1}{l} \right] \mathbf{u} = \mathbf{Bu} \qquad (11.7\text{h})$$

Here l is the length of the member and $\mathbf{B}$ is termed as the *strain displacement matrix* that relates the strain at a point to the nodal displacements of an element. The stress σ_x in the element is

$$\sigma_x = E\varepsilon_x = E\mathbf{Bu} \qquad (11.7\text{i})$$

At this stage, an alternative weak form of the equilibrium equations in Eq. (11.2b) is introduced. A scalar termed as the work potential is defined as the difference between the strain energy stored in an element and work done be external loads. From Eq. (11.2c), the strain energy in a spring (or truss element) is $\frac{1}{2}\,\mathbf{u}^T\mathbf{ku}$ while the work done by the external loads is $\mathbf{f}^T\mathbf{u}$, where $\mathbf{f} = \begin{pmatrix} f_i \\ f_j \end{pmatrix}$. The work potential WP is then

$$\text{WP} = \frac{1}{2}\,\mathbf{u}^T\mathbf{ku} - \mathbf{f}^T\mathbf{u} \qquad (11.7\text{j})$$

minimizing which with respect to $\mathbf{u}$ yields

$$\frac{\partial\,\text{WP}}{\partial\mathbf{u}} = \mathbf{ku} - \mathbf{f} = \mathbf{0}$$

which is the strong form of the equilibrium condition. The strain energy stored in a truss element is

$$\text{SE} = \frac{1}{2} \int_V \sigma_x^T \varepsilon_x dV = \frac{1}{2} \int_V \mathbf{u}^T \mathbf{B}^T E \mathbf{Bu} dV = \frac{1}{2} \mathbf{u}^T \left(\int_V \mathbf{B}^T E \mathbf{B} dV \right) \mathbf{u} \qquad (11.7\text{k})$$

Comparing with $\frac{1}{2}\,\mathbf{u}^T\mathbf{ku}$, we realize that $\mathbf{k} = \int_V \mathbf{B}^T E\mathbf{B} dV$ or

$$\mathbf{k} = \int_V \begin{pmatrix} \frac{-1}{l} \\ \frac{1}{l} \end{pmatrix} E \left(\frac{-1}{l} \quad \frac{1}{l} \right) dV = E \begin{pmatrix} \frac{1}{l^2} & \frac{-1}{l^2} \\ \frac{-1}{l^2} & \frac{1}{l^2} \end{pmatrix} Al = \frac{AE}{l} \begin{pmatrix} 1 & -1 \\ -1 & 1 \end{pmatrix} \qquad (11.7\text{l})$$

which agrees with Eq. (11.6). Note that $\mathbf{k}$ above is termed as the *local stiffness matrix* as the displacements $\mathbf{u}$ are along the axis of the truss element.

11.3.1 Transformations and Truss Element

Quite often, a truss element may be oriented arbitrarily in the x-y plane and may have different stiffnesses for external loads along the two directions. Consider a truss element oriented at an angle θ (Figure 11.4) where the displacements $\mathbf{u}_e = [u_{ix}, u_{iy}, u_{jx}, u_{jy}]^T$ are to be determined along the x and y axes for external loads $\mathbf{f}_e = [f_{ix}, f_{iy}, f_{jx}, f_{jy}]^T$.

Relating the displacements $\mathbf{u}_e$ with $\mathbf{u}$ along ξ, we have in the matrix form

$$\mathbf{u}_e = \begin{pmatrix} u_{ix} \\ u_{iy} \\ u_{jx} \\ u_{jy} \end{pmatrix} = \begin{pmatrix} \cos\theta & 0 \\ \sin\theta & 0 \\ 0 & \cos\theta \\ 0 & \sin\theta \end{pmatrix} \begin{pmatrix} u_i \\ u_j \end{pmatrix} = \boldsymbol{\lambda}^T \mathbf{u} \qquad (11.7\text{m})$$

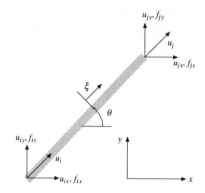

Figure 11.4   A truss element oriented at an angle θ

where $\boldsymbol{\lambda}^T = \begin{pmatrix} \cos\theta & 0 \\ \sin\theta & 0 \\ 0 & \cos\theta \\ 0 & \sin\theta \end{pmatrix}$. We can realize that $\boldsymbol{\lambda}\mathbf{u}_e = \boldsymbol{\lambda}\boldsymbol{\lambda}^T\mathbf{u} = \mathbf{I}_{2\times2}\mathbf{u} = \mathbf{u}$. Substituting $\mathbf{u} = \boldsymbol{\lambda}\mathbf{u}_e$ in

Eq. (11.7l) gives

$$\text{SE} = \frac{1}{2}(\boldsymbol{\lambda}\mathbf{u}_e)^T\left(\int_V \mathbf{B}^T EB dV\right)\boldsymbol{\lambda}\mathbf{u}_e = \frac{1}{2}\mathbf{u}_e^T\left(\boldsymbol{\lambda}^T\left\{\int_V \mathbf{B}^T EB dV\right\}\boldsymbol{\lambda}\right)\mathbf{u}_e = \frac{1}{2}\mathbf{u}_e^T\mathbf{k}_e\mathbf{u}_e \qquad (11.7n)$$

where $\mathbf{k}_e = \boldsymbol{\lambda}^T\left\{\int_V \mathbf{B}^T EB dV\right\}\boldsymbol{\lambda} = \boldsymbol{\lambda}^T\mathbf{k}\boldsymbol{\lambda}$ is the transformed stiffness matrix of the truss element. Using Eqs. (11.7l) and (11.7m), we have

$$\mathbf{k}_e = \begin{pmatrix} \cos\theta & 0 \\ \sin\theta & 0 \\ 0 & \cos\theta \\ 0 & \sin\theta \end{pmatrix} \frac{AE}{l}\begin{pmatrix} 1 & -1 \\ -1 & 1 \end{pmatrix}\begin{pmatrix} \cos\theta & \sin\theta & 0 & 0 \\ 0 & 0 & \cos\theta & \sin\theta \end{pmatrix}$$

$$= \frac{AE}{l}\begin{bmatrix} \cos^2\theta & \cos\theta\cdot\sin\theta & -\cos^2\theta & -\cos\theta\cdot\sin\theta \\ \cos\theta\cdot\sin\theta & \sin^2\theta & -\cos\theta\cdot\sin\theta & -\sin^2\theta \\ -\cos^2\theta & -\cos\theta\cdot\sin\theta & \cos^2\theta & \cos\theta\cdot\sin\theta \\ -\cos\theta\cdot\sin\theta & -\sin^2\theta & \cos\theta\cdot\sin\theta & \sin^2\theta \end{bmatrix} \qquad (11.7o)$$

which satisfies the equilibrium condition $\mathbf{k}_e\mathbf{u}_e = \mathbf{f}_e$, where $\mathbf{f}_e = \boldsymbol{\lambda}^T\mathbf{f}$ analogous to the transformation for displacements. The reader may verify that the rank of $\mathbf{k}_e$ as given in Eq. (11.7o) is one even though it is of 4×4 size. This is because a truss element in two dimensions has three degrees of rigid-body motion, i.e., translation along x and y axes, and rotation in the xy plane.

Example 11.1. An assemblage of truss members is shown in Figure 11.5. It is required to determine the horizontal and vertical displacements at point A which is subjected to the loads of 200 kN and 100 kN as shown. Points B and C are fixed. Length of members AB and BC is 2 m each, and area of cross section and elastic modulus are given in Table 11.1.

First, node numbers are assigned to each node as shown in the figure. The x axis is taken along BA and y along BC. Coordinates of A, B and C (nodes 1, 2 and 3, respectively) are (2, 0), (0, 0) and (0, 2). Table 11.1 provides the connectivity information for each element.

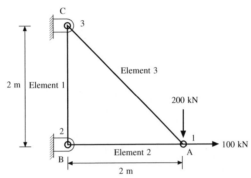

Figure 11.5   Truss assemblage for Example 11.1

Table 11.1 Element data for Example 11.1

Truss member	Element number	Node connectivity	Cross-sectional area (cm^2)	Young's modulus (GPa)
BC	1	2, 3	20	70
AB	2	2, 1	20	200
AC	3	1, 3	50	200

Note that the structure has 6 degrees of freedom, 3 along x and 3 along y axes respectively. Let **U** be the global displacement vector such that its odd entries are assigned x displacements while even are assigned y displacements, that is, for node k, $\mathbf{U}(2k-1) = u_{kx}$ while $\mathbf{U}(2k) = u_{ky}$. Then, **U** will have the form

$$\mathbf{U} = [\mathbf{U}(1) \ \ \mathbf{U}(2) \ \ \mathbf{U}(3) \ \ \mathbf{U}(4) \ \ \mathbf{U}(5) \ \ \mathbf{U}(6)]^T = [u_{1x} \ \ u_{1y} \ \ u_{2x} \ \ u_{2y} \ \ u_{3x} \ \ u_{3y}]^T \qquad (11.8a)$$

and the global stiffness matrix **K** will be of size 6×6. To get **K**, we would assemble the local stiffness matrices as follows:

For element 1, $\theta = 90°$, $l = 2$ m, $A = 20 \times 10^{-4}$ m^2 and $E = 70 \times 10^9$ Nm^{-2}. Using Eq. (11.7o), we have

$$\mathbf{k}_e^1 = 7 \times 10^7 \begin{matrix} & \begin{matrix} 3 & \ 4 & \ 5 & \ 6 \end{matrix} & \\ \begin{bmatrix} 0 & 0 & 0 & 0 \\ 0 & 1 & 0 & -1 \\ 0 & 0 & 0 & 0 \\ 0 & -1 & 0 & 1 \end{bmatrix} & \begin{matrix} 3 \\ 4 \\ 5 \\ 6 \end{matrix} \end{matrix} \qquad (11.8b)$$

For convenience in assembly, the numbers assigned to the degrees of freedom are represented along the rows and columns. For element 2, $\theta = 180°$, $l = 2$ m, $A = 20 \times 10^{-4}$ m^2 and $E = 200 \times 10^9$ Nm^2. Hence

$$\mathbf{k}_e^2 = 20 \times 10^7 \begin{array}{c} \begin{array}{cccc} 1 & 2 & 3 & 4 \end{array} \\ \begin{bmatrix} 1 & 0 & -1 & 0 \\ 0 & 0 & 0 & 0 \\ -1 & 0 & 1 & 0 \\ 0 & 0 & 0 & 0 \end{bmatrix} \begin{array}{c} 1 \\ 2 \\ 3 \\ 4 \end{array} \end{array} \qquad (11.8c)$$

For element 3, $\theta = 135°$, $l = 2\sqrt{2}$ m, $A = 50 \times 10^{-4}$ m^2 and $E = 200 \times 10^9$ Nm^{-2}. Thus

$$\mathbf{k}_e^3 = \frac{10 \times 10^8}{4\sqrt{2}} \begin{array}{c} \begin{array}{cccc} 1 & 2 & 5 & 6 \end{array} \\ \begin{bmatrix} 1 & -1 & -1 & 1 \\ -1 & 1 & 1 & -1 \\ -1 & 1 & 1 & -1 \\ 1 & -1 & -1 & 1 \end{bmatrix} \begin{array}{c} 1 \\ 2 \\ 5 \\ 6 \end{array} \end{array} \qquad (11.8d)$$

Combining the three local stiffness matrices appropriately to form the global stiffness matrix, we have the force-displacement relation as

$$\begin{Bmatrix} F_{1x} \\ F_{1y} \\ F_{2x} \\ F_{2y} \\ F_{3x} \\ F_{3y} \end{Bmatrix} = 10^8 \begin{bmatrix} 3.77 & -1.77 & -2 & 0 & -1.77 & 1.77 \\ -1.77 & 1.77 & 0 & 0 & 1.77 & -1.77 \\ -2 & 0 & 2 & 0 & 0 & 0 \\ 0 & 0 & 0 & 0.7 & 0 & -0.7 \\ -1.77 & 1.77 & 0 & 0 & 1.77 & -1.77 \\ 1.77 & -1.77 & 0 & -0.7 & -1.77 & 2.47 \end{bmatrix} \begin{Bmatrix} u_{1x} \\ u_{1y} \\ u_{2x} \\ u_{2y} \\ u_{3x} \\ u_{3y} \end{Bmatrix} \qquad (11.8e)$$

Noting that $u_{2x} = u_{2y} = u_{3x} = u_{3y} = 0$, we can discount the 3rd, 4th, 5th and 6th rows and columns in the above system to compute only the unknown displacements. Thus

$$\begin{Bmatrix} F_{1x} \\ F_{1y} \end{Bmatrix} = 10^8 \begin{bmatrix} 3.77 & -1.77 \\ -1.77 & 1.77 \end{bmatrix} \begin{Bmatrix} u_{1x} \\ u_{1y} \end{Bmatrix}$$

Substituting for $F_{1x} = 100$ kN and $F_{1y} = -200$ kN and solving for displacements, we get $u_{1x} = -5 \times 10^{-4}$ m and $u_{1y} = -1.63 \times 10^{-3}$ m. The reaction forces at the supports, that is, F_{2x}, F_{2y}, F_{3x} and F_{3y} can all be computed by substituting for u_{1x} and u_{1y} in Eq. (11.8e). The final force vector can be computed as $[100 \; -200 \; 100 \; 0 \; -200 \; 200]^T$ with the last four entries representing the respective reaction forces.

11.4 Beam Elements

A beam element shown in Figure 11.6 has four degrees of freedom, two at each node i and j, respectively, and exhibits bending, that is, transverse deflections (v_i and v_j) and rotations (θ_i and θ_j) under transverse loads (F_i and F_j) and end moments (M_i and M_j). Treating these degrees of freedom as unknowns, a cubic interpolation function for transverse displacements $v(x)$ at any point in the beam may be assumed, that is

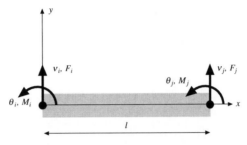

Figure 11.6   A beam element

$$v(x) = a_1 + a_2 x + a_3 x^2 + a_4 x^3 \tag{11.9a}$$

where constants $a_1, \ldots, a_4$ can be determined using the conditions

$$v(x) = v_i \text{ and } \frac{dv(x)}{dx} = \theta_i \text{ at } x = 0,$$

$$v(x) = v_j \text{ and } \frac{dv(x)}{dx} = \theta_j \text{ at } x = l \tag{11.9b}$$

Solving and rearranging yields

$$v(x) = \left(\frac{2x^3}{l^3} - \frac{3x^2}{l^2} + 1 \right) v_i + \left(\frac{x^3}{l^2} - \frac{2x^2}{l} + x \right) \theta_i + \left(\frac{3x^2}{l^2} - \frac{2x^3}{l^3} \right) v_j + \left(\frac{x^3}{l^2} - \frac{x^2}{l} \right) \theta_j$$

or
$$v(x) = N_1(x)v_i + N_2(x)\theta_i + N_3(x)v_j + N_4(x)\theta_j$$

or
$$v(x) = [N_1(x)\ N_2(x)\ N_3(x)\ N_4(x)][v_i\ \theta_i\ v_j\ \theta_j]^T = \mathbf{Nv} \tag{11.9c}$$

with $\mathbf{N}$ as the shape function matrix and $\mathbf{v}$ the displacement vector. From linear beam theory, plane cross-sections remain plane after deformation and hence the axial displacement u due to transverse displacement v can be expressed as (Figure 11.7)

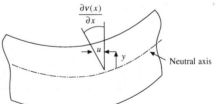

$$u = -y \frac{\partial v}{\partial x} \tag{11.9d}$$

where y is the distance from the neutral axis. The axial strain is given by

$$\varepsilon_x = \frac{\partial u}{\partial x} = -y \frac{\partial^2 v}{\partial x^2} = \mathbf{Bv} \tag{11.9e}$$

Figure 11.7   Axial displacement u in the beam

where $\mathbf{B}$ is the strain displacement matrix. From Eq. (11.9c), we note

$$\frac{\partial^2 v}{\partial x^2} = \left(\frac{12x}{l^3} - \frac{6}{l^2} \right) v_i + \left(\frac{6x}{l^2} - \frac{4}{l} \right) \theta_i + \left(\frac{6}{l^2} - \frac{12x}{l^3} \right) v_j + \left(\frac{6x}{l^2} - \frac{2}{l} \right) \theta_j \tag{11.9f}$$

and thus comparing with Eq. (11.9e) yields

$$\mathbf{B} = -\frac{y}{l^3}\,[12x - 6l \quad l(6x - 4l) \quad -(12x - 6l) \quad l(6x - 2l)]$$

From truss analysis, the local stiffness matrix for a beam element may directly be written as

$$\mathbf{k} = \int_V \mathbf{B}^T E \mathbf{B}\, dV$$

$$= E \int_A \frac{y^2\, dA}{l^6} \int_0^l \begin{bmatrix} 12x - 6l \\ l(6x - 4l) \\ -(12x - 6l) \\ l(6x - 2l) \end{bmatrix} [12x - 6l \quad l(6x - 4l) \quad -(12x - 6l) \quad l(6x - 2l)]\, dx$$

$$= \frac{EI}{l^6} \int_0^l \begin{bmatrix} 36(4x^2 - 4xl + l^2) & 12l(6x^2 - 7xl + 2l^2) & -36(4x^2 - 4xl + l^2) & 12l(6x^2 - 5xl + l^2) \\ 12l(6x^2 - 7xl + 2l^2) & 4l^2(9x^2 - 12xl + 4l^2) & -12l(6x^2 - 7xl + 2l^2) & 4l^2(9x^2 - 9xl + 2l^2) \\ -36(4x^2 - 4xl + l^2) & -12l(6x^2 - 7xl + 2l^2) & 36(4x^2 - 4xl + l^2) & -12l(6x^2 - 5xl + l^2) \\ 12l(6x^2 - 5xl + l^2) & 4l^2(9x^2 - 9xl + 2l^2) & -12l(6x^2 - 5xl + l^2) & 4l^2(9x^2 - 6xl + l^2) \end{bmatrix} dx$$

$$= \frac{EI}{l^3} \begin{bmatrix} 12 & 6l & -12 & 6l \\ 6l & 4l^2 & -6l & 2l^2 \\ -12 & -6l & 12 & -6l \\ 6l & 2l^2 & -6l & 4l^2 \end{bmatrix} \qquad (11.9\text{g})$$

where I is the second moment of area of the cross-section given as $I = \int_A y^2\, dA$. The assembly procedure for beam elements is similar to that in truss elements with the difference that two degrees of freedom are considered in the local coordinate system per node. Note that the stiffness assembly is based on: (a) the interelement continuity of primary variables (deflection and slope) and (b) the interelement equilibrium of secondary variables (shear forces and bending moments) at nodes common to the elements.

Example 11.2. Given is a composite beam with varying cross sections as shown in Figure 11.8 with external loads and displacement boundary conditions. Solve for transverse deflections and slopes and compare with the analytical result.

For the three elements, there are four nodes and two degrees of freedom per node. The global displacement vector $\mathbf{U}$ is such that for node j, $\mathbf{U}(2j - 1) = v_j$ and $\mathbf{U}(2j) = \theta_j$. The global stiffness matrix is of size 8×8 which is determined as follows:

For elements 1, 2 and 3, using Eq. (11.9g), the stiffness matrices are

$$\mathbf{k}_1 = \frac{10^6}{1} \begin{bmatrix} 12 & 6 & -12 & 6 \\ 6 & 4 & -6 & 2 \\ -12 & -6 & 12 & -6 \\ 6 & 2 & -6 & 4 \end{bmatrix} \begin{matrix} 1 \\ 2 \\ 3 \\ 4 \end{matrix}$$

$$\begin{matrix} 1 & 2 & 3 & 4 \end{matrix}$$

$$\mathbf{k}_2 = 10^6 \begin{bmatrix} 6 & 3 & -6 & 3 \\ 3 & 2 & -3 & 1 \\ -6 & -3 & 6 & -3 \\ 3 & 1 & -3 & 2 \end{bmatrix} \begin{matrix} 3 \\ 4 \\ 5 \\ 6 \end{matrix}$$

$$\begin{matrix} 3 & 4 & 5 & 6 \end{matrix}$$

$$\mathbf{k}_3 = 10^6 \begin{bmatrix} 12 & 3 & -12 & 3 \\ 3 & 1 & -3 & 0.5 \\ -12 & -3 & 12 & -3 \\ 3 & 0.5 & -3 & 1 \end{bmatrix} \begin{matrix} 5 \\ 6 \\ 7 \\ 8 \end{matrix}$$

$$\begin{matrix} 5 & 6 & 7 & 8 \end{matrix}$$

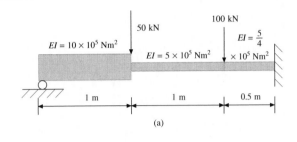

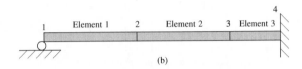

Figure 11.8 (a) A composite beam model and (b) the finite element model

The assembled matrix **K** is

$$
\mathbf{K} = 10^6
\begin{array}{cccccccc}
1 & 2 & 3 & 4 & 5 & 6 & 7 & 8
\end{array}
$$

$$
\mathbf{K} = 10^6
\begin{bmatrix}
12 & 6 & -12 & 6 & 0 & 0 & 0 & 0 \\
6 & 4 & -6 & 2 & 0 & 0 & 0 & 0 \\
-12 & -6 & 12+6 & -6+3 & -6 & 3 & 0 & 0 \\
6 & 2 & -6+3 & 4+2 & -3 & 1 & 0 & 0 \\
0 & 0 & -6 & -3 & 6+12 & -3+3 & -12 & 3 \\
0 & 0 & 3 & 1 & -3+3 & 2+1 & -3 & 0.5 \\
0 & 0 & 0 & 0 & -12 & -3 & 12 & -3 \\
0 & 0 & 0 & 0 & 3 & 0.5 & -3 & 1
\end{bmatrix}
\begin{matrix}
1 \\ 2 \\ 3 \\ 4 \\ 5 \\ 6 \\ 7 \\ 8
\end{matrix}
$$

or

$$
\mathbf{K} = 10^6
\begin{bmatrix}
12 & 6 & -12 & 6 & 0 & 0 & 0 & 0 \\
6 & 4 & -6 & 2 & 0 & 0 & 0 & 0 \\
-12 & -6 & 18 & -3 & -6 & 3 & 0 & 0 \\
6 & 2 & -3 & 6 & -3 & 1 & 0 & 0 \\
0 & 0 & -6 & -3 & 18 & 0 & -12 & 3 \\
0 & 0 & 3 & 1 & 0 & 3 & -3 & 0.5 \\
0 & 0 & 0 & 0 & -12 & -3 & 12 & -3 \\
0 & 0 & 0 & 0 & 3 & 0.5 & -3 & 1
\end{bmatrix}
\begin{matrix}
1 \\ 2 \\ 3 \\ 4 \\ 5 \\ 6 \\ 7 \\ 8
\end{matrix}
$$

Consistent with the definition of **U** for this example, the external force vector is given as

$$\mathbf{F} = 10^4 \begin{bmatrix} F_1 \\ 0 \\ -5 \\ 0 \\ -10 \\ 0 \\ F_7 \\ F_8 \end{bmatrix} \begin{matrix} 1 \\ 2 \\ 3 \\ 4 \\ 5 \\ 6 \\ 7 \\ 8 \end{matrix}$$

with F_1, F_7 and F_8 as the unknown reaction forces and moments at the supports for which **U**(1), **U**(7) and **U**(8) are zero, respectively. To solve only for the unknown displacements, eliminating the 1st, 7th and 8th rows and columns, we have the linear system of size 5×5 as

$$10^6 \begin{matrix} & 1 & 2 & 3 & 4 & 5 & 6 & 7 & 8 \\ & & 4 & -6 & 2 & 0 & 0 \\ & & -6 & 18 & -3 & -6 & 3 \\ & & 2 & -3 & 6 & -3 & 1 \\ & & 0 & -6 & -3 & 18 & 0 \\ & & 0 & 3 & 1 & 0 & 3 \end{matrix} \begin{bmatrix} 1 \\ 2 \\ 3 \\ 4 \\ 5 \\ 6 \\ 7 \\ 8 \end{bmatrix} \begin{bmatrix} 0 \\ \mathbf{U}(2) \\ \mathbf{U}(3) \\ \mathbf{U}(4) \\ \mathbf{U}(5) \\ \mathbf{U}(6) \\ 0 \\ 0 \end{bmatrix} = 10^4 \begin{bmatrix} F_1 \\ 0 \\ -5 \\ 0 \\ -10 \\ 0 \\ F_7 \\ F_8 \end{bmatrix}$$

solving which we get

$$\mathbf{U} = 10^{-2} \times [0 \quad -3.58 \quad -3.00 \quad -1.84 \quad -1.86 \quad 3.62 \quad 0 \quad 0]^T$$

Post-multiplying the above displacement vector with the original 8×8 stiffness matrix, we get the force vector as

$$\mathbf{F} = 10^3 \times [34.87 \quad 0 \quad -50 \quad 0 \quad -100 \quad -0 \quad 115.13 \quad -37.83]^T$$

and thus the reaction forces are $F_1 = 34.87$ kN, $F_7 = 115.13$ kN and $F_8 = -37.83$ kNm.

It is left as an exercise for the reader to solve this example analytically. Note that the beam is statically indeterminate (with 3 reactions and two equations) and that the equilibrium and deflection equations should be solved together.

11.5 Frame Elements

Frame elements are extended beam elements wherein axial displacements are incorporated as well. A frame element has three degrees of freedom per node, namely, the horizontal and vertical displacements and rotation perpendicular to the plane of the element (Figure 11.9). For $\mathbf{u} = [u_i,\ v_i,\ \theta_i,\ u_j,\ v_j,\ \theta_j]^T$ chosen as the displacement vector for the element, the stiffness matrices of the truss and beam elements in Eqs. (11.7l) and (11.9g) can be combined. Expressing the stiffness in all *six* degrees of freedom, for the truss element, we have

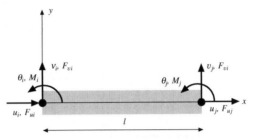

Figure 11.9 A frame element

$$
\mathbf{k}_{\text{truss}} =
\begin{array}{cccccc}
u_i & v_i & \theta_i & u_j & v_j & \theta_j
\end{array}
\left[
\begin{array}{cccccc}
\dfrac{AE}{l} & 0 & 0 & -\dfrac{AE}{l} & 0 & 0 \\[2mm]
0 & 0 & 0 & 0 & 0 & 0 \\[2mm]
0 & 0 & 0 & 0 & 0 & 0 \\[2mm]
-\dfrac{AE}{l} & 0 & 0 & \dfrac{AE}{l} & 0 & 0 \\[2mm]
0 & 0 & 0 & 0 & 0 & 0 \\[2mm]
0 & 0 & 0 & 0 & 0 & 0
\end{array}
\right]
\begin{array}{c}
u_i \\[2mm] v_i \\[2mm] \theta_i \\[2mm] u_j \\[2mm] v_j \\[2mm] \theta_j
\end{array}
$$

and for beam element

$$
\mathbf{k}_{\text{beam}} =
\begin{array}{cccccc}
u_i & v_i & \theta_i & u_j & v_j & \theta_j
\end{array}
\left[
\begin{array}{cccccc}
0 & 0 & 0 & 0 & 0 & 0 \\[3mm]
0 & 12\dfrac{EI}{l^3} & 6\dfrac{EI}{l^2} & 0 & -12\dfrac{EI}{l^3} & 6\dfrac{EI}{l^2} \\[3mm]
0 & 6\dfrac{EI}{l^2} & 4\dfrac{EI}{l} & 0 & -6\dfrac{EI}{l^2} & 2\dfrac{EI}{l} \\[3mm]
0 & 0 & 0 & 0 & 0 & 0 \\[3mm]
0 & -12\dfrac{EI}{l^3} & -6\dfrac{EI}{l^2} & 0 & 12\dfrac{EI}{l^3} & -6\dfrac{EI}{l^2} \\[3mm]
0 & 6\dfrac{EI}{l^2} & 2\dfrac{EI}{l} & 0 & -6\dfrac{EI}{l^2} & 4\dfrac{EI}{l}
\end{array}
\right]
\begin{array}{c}
u_i \\[3mm] v_i \\[3mm] \theta_i \\[3mm] u_j \\[3mm] v_j \\[3mm] \theta_j
\end{array}
$$

Adding the two matrices, we get the frame element stiffness matrix as

$$
\begin{array}{cccccc}
u_i & v_i & \theta_i & u_j & v_j & \theta_j
\end{array}
$$

$$
\mathbf{k} =
\begin{bmatrix}
\dfrac{AE}{l} & 0 & 0 & -\dfrac{AE}{l} & 0 & 0 \\[2mm]
0 & 12\dfrac{EI}{l^3} & 6\dfrac{EI}{l^2} & 0 & -12\dfrac{EI}{l^3} & 6\dfrac{EI}{l^2} \\[2mm]
0 & 6\dfrac{EI}{l^2} & 4\dfrac{EI}{l} & 0 & -6\dfrac{EI}{l^2} & 2\dfrac{EI}{l} \\[2mm]
-\dfrac{AE}{l} & 0 & 0 & \dfrac{AE}{l} & 0 & 0 \\[2mm]
0 & -12\dfrac{EI}{l^3} & -6\dfrac{EI}{l^2} & 0 & 12\dfrac{EI}{l^3} & -6\dfrac{EI}{l^2} \\[2mm]
0 & 6\dfrac{EI}{l^2} & 2\dfrac{EI}{l} & 0 & -6\dfrac{EI}{l^2} & 4\dfrac{EI}{l}
\end{bmatrix}
\begin{matrix}
u_i \\ v_i \\ \theta_i \\ u_j \\ v_j \\ \theta_j
\end{matrix}
\qquad (11.9h)
$$

11.5.1 Frame Elements and Transformations

For a frame element to be oriented at an angle θ in the x-y plane (Figure 11.10), the displacements along the x and y global coordinate axes, namely, u_{ix}, u_{iy} for node i and u_{jx}, u_{jy} for node j can be extracted using the following transformation. Note that the rotations θ_i and θ_j remain invariant.

$$
\mathbf{u}_e =
\begin{Bmatrix}
u_{ix} \\ u_{iy} \\ \theta_i \\ u_{jx} \\ u_{jy} \\ \theta_j
\end{Bmatrix}
=
\begin{pmatrix}
\cos\theta & -\sin\theta & 0 & 0 & 0 & 0 \\
\sin\theta & \cos\theta & 0 & 0 & 0 & 0 \\
0 & 0 & 1 & 0 & 0 & 0 \\
0 & 0 & 0 & \cos\theta & -\sin\theta & 0 \\
0 & 0 & 0 & \sin\theta & \cos\theta & 0 \\
0 & 0 & 0 & 0 & 0 & 1
\end{pmatrix}
\begin{Bmatrix}
u_i \\ v_i \\ \theta_i \\ u_j \\ v_j \\ \theta_j
\end{Bmatrix}
= \boldsymbol{\lambda}^T \mathbf{u}
\qquad (11.9i)
$$

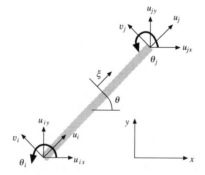

Figure 11.10 A frame element oriented at an angle θ

The stiffness of the frame element for the displacements (and forces) along the x and y directions, and rotations (moments) perpendicular to the plane containing the frame element is given similar to the truss element as

$$\mathbf{k}_e = \boldsymbol{\lambda}^T \mathbf{k} \boldsymbol{\lambda} \qquad (11.9\text{j})$$

with **k** and **λ** defined in Eqs. (11.9h) and (11.9i), respectively.

Example 11.3. Consider a slightly involved example of an over-bridge modeled using frame elements. The horizontal and vertical lengths are of 10 m each except the vertical element at the center of length 20 m. The flexural rigidity EI is taken for all elements as 10^7 Nm^2 and cross section areas A are 10^{-2} m^2. The example is solved using a frame finite element implementation in MATLAB^{TM} for displacements. The assembly procedure for the global stiffness matrix is similar to Example 11.2 with element stiffness matrices computed using Eqs. (11.9h) and (11.9j). The displaced configuration (dashed lines) is given in Figure 11.11(b) with the maximum downward displacement at the center node on bottom edge as 0.0012 m.

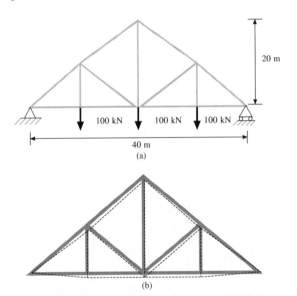

Figure 11.11   (a) An over-bridge modeled with frame elements and (b) displacement profile (scaled)

11.6   Continuum Triangular Elements

Truss, beam and frame elements are often considered discrete elements as they approximate a region only partially. A more comprehensive discretization for a two-dimensional region is performed using triangular or quadrilateral elements. Consider, for instance, a triangular element shown in Figure 11.12 with three nodes i, j and k, each having two degrees of freedom (u_x, v_x) along x and y directions respectively, and the same number of external forces (f_x, f_y) as shown.

To interpolate the displacements (u, v) at point P in the element, we assume that u is dependent only on the nodal displacements in the x direction. Since there are three such unknowns u_i, u_j and u_k, displacement u at P can be interpolated as

$$u = a_0 + a_1 x + a_2 y \qquad (11.10\text{a})$$

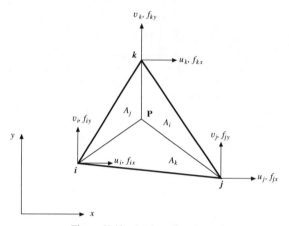

Figure 11.12   A triangular element

Note that the interpolation above is equally biased along the x and y directions. To determine the coefficients a_0, a_1 and a_2, we need to solve

$$u_i = a_0 + a_1 x_i + a_2 y_i$$

$$u_j = a_0 + a_1 x_j + a_2 y_j$$

$$u_k = a_0 + a_1 x_k + a_2 y_k \tag{11.10b}$$

to get

$$a_0 = \frac{u_i(x_j y_k - x_k y_j) + u_j(x_k y_i - y_k x_i) + u_k(x_i y_j - x_j y_i)}{(x_j - x_i)(y_k - y_j) - (x_k - x_j)(y_j - y_i)}$$

$$a_1 = \frac{u_i(y_j - y_k) + u_j(y_k - y_i) + u_k(y_i - y_j)}{(x_j - x_i)(y_k - y_j) - (x_k - x_j)(y_j - y_i)} \tag{11.10c}$$

$$a_2 = \frac{u_i(x_k - x_j) + u_j(x_i - x_k) + u_k(x_j - x_i)}{(x_j - x_i)(y_k - y_j) - (x_k - x_j)(y_j - y_i)}$$

and thus

$$u = \frac{x(y_j - y_k) + x_j(y_k - y) + x_k(y - y_j)}{x_i(y_j - y_k) + x_j(y_k - y_i) + x_k(y_i - y_j)} u_i + \frac{x_i(y - y_k) + x(y_k - y_i) + x_k(y_i - y)}{x_i(y_j - y_k) + x_j(y_k - y_i) + x_k(y_i - y_j)} u_j$$

$$+ \frac{x_i(y_j - y) + x_j(y - y_i) + x(y_i - y_j)}{x_i(y_j - y_k) + x_j(y_k - y_i) + x_k(y_i - y_j)} u_k \tag{11.10d}$$

or

$$u = \frac{A_i}{A} u_i + \frac{A_j}{A} u_j + \frac{A_k}{A} u_k = N_i(x, y) u_i + N_j(x, y) u_j + N_k(x, y) u_k \tag{11.10e}$$

where A_i, A_j and A_k are the triangular areas shown in Figure 11.12 and A is the area of the triangular element ($A = A_i + A_j + A_k$). Note that for P in the interior of the triangle, the shape functions $N_i(x, y)$

$= A_i/A$, $N_j(x, y) = A_j/A$ and $N_k(x, y) = A_k/A$ are all greater than or equal to zero. Further, if P is at node j, $A_j = A$ and so $N_j(x, y) = 1$ while $N_i(x, y) = N_k(x, y) = 0$. Following the same argument to choose the interpolation scheme for u in Eq. (11.10a), we can use a similar expression for y displacements as well, that is

$$v = \frac{A_i}{A} v_i + \frac{A_j}{A} v_j + \frac{A_k}{A} v_k = N_i(x, y)v_i + N_j(x, y)\, v_j + N_k(x, y)v_k \qquad (11.10f)$$

If $\mathbf{u} = [u_i \quad v_i \quad u_j \quad v_j \quad u_k \quad v_k]^T$ is the displacement vector for the element, then

$$\begin{pmatrix} u \\ v \end{pmatrix} = \begin{pmatrix} N_i(x, y) & 0 & N_j(x, y) & 0 & N_k(x, y) & 0 \\ 0 & N_i(x, y) & 0 & N_i(x, y) & 0 & N_i(x, y) \end{pmatrix} \mathbf{u} = \mathbf{Nu} \quad (11.10g)$$

The strain ε_x along the x direction is defined as $\dfrac{\partial u}{\partial x}$ while that in the y direction, ε_y is given by $\dfrac{\partial v}{\partial y}$.

The shear strain γ_{xy} is expressed as $\dfrac{\partial u}{\partial y} + \dfrac{\partial v}{\partial x}$. The strain vector $\boldsymbol{\varepsilon}$ may be written as

$$\boldsymbol{\varepsilon} = \begin{Bmatrix} \varepsilon_x \\ \varepsilon_y \\ \gamma_{xy} \end{Bmatrix} = \begin{pmatrix} \dfrac{\partial}{\partial x} & 0 \\ 0 & \dfrac{\partial}{\partial y} \\ \dfrac{\partial}{\partial y} & \dfrac{\partial}{\partial x} \end{pmatrix} \begin{pmatrix} u \\ v \end{pmatrix} = \begin{pmatrix} \dfrac{\partial}{\partial x} & 0 \\ 0 & \dfrac{\partial}{\partial y} \\ \dfrac{\partial}{\partial y} & \dfrac{\partial}{\partial x} \end{pmatrix} \mathbf{Nu}$$

$$= \begin{pmatrix} \dfrac{\partial}{\partial x} N_i(x, y) & 0 & \dfrac{\partial}{\partial x} N_j(x, y) & 0 & \dfrac{\partial}{\partial x} N_k(x, y) & 0 \\ 0 & \dfrac{\partial}{\partial y} N_i(x, y) & 0 & \dfrac{\partial}{\partial y} N_j(x, y) & 0 & \dfrac{\partial}{\partial y} N_k(x, y) \\ \dfrac{\partial}{\partial y} N_i(x, y) & \dfrac{\partial}{\partial x} N_i(x, y) & \dfrac{\partial}{\partial y} N_j(x, y) & \dfrac{\partial}{\partial x} N_j(x, y) & \dfrac{\partial}{\partial y} N_k(x, y) & \dfrac{\partial}{\partial x} N_k(x, y) \end{pmatrix}$$

$$(11.10h)$$

Now

$$\frac{\partial N_i(x, y)}{\partial x} = \frac{y_j - y_k}{2A}, \; \frac{\partial N_i(x, y)}{\partial y} = \frac{x_k - x_j}{2A}$$

$$\frac{\partial N_j(x, y)}{\partial x} = \frac{y_k - y_i}{2A}, \; \frac{\partial N_j(x, y)}{\partial y} = \frac{x_i - x_k}{2A}$$

$$\frac{\partial N_k(x, y)}{\partial x} = \frac{y_i - y_j}{2A}, \; \frac{\partial N_k(x, y)}{\partial y} = \frac{x_j - x_i}{2A}$$

so that

$$\boldsymbol{\varepsilon} = \frac{1}{2A} \begin{pmatrix} y_j - y_k & 0 & y_k - y_i & 0 & y_i - y_j & 0 \\ 0 & x_k - x_j & 0 & x_i - x_k & 0 & x_j - x_i \\ x_k - x_j & y_j - y_k & x_i - x_k & y_k - y_i & x_j - x_i & y_i - y_j \end{pmatrix} \mathbf{u} = \mathbf{Bu} \quad (11.10i)$$

where **B** is the strain-displacement matrix which is a constant and depends on the position of nodal coordinates. Thus, a triangular element is sometimes referred to as the constant strain triangular or CST element. From linear elasticity, strains are related to stresses in three dimensions as

$$\varepsilon_x = \frac{1}{E}\left[\sigma_x - v(\sigma_y + \sigma_z)\right]$$

$$\varepsilon_y = \frac{1}{E}\left[\sigma_y - v(\sigma_z + \sigma_x)\right] \qquad (11.10j)$$

$$\varepsilon_z = \frac{1}{E}\left[\sigma_z - v(\sigma_x + \sigma_y)\right]$$

$$\gamma_{xy} = \frac{\tau_{xy}}{G}, \gamma_{yz} = \frac{\tau_{yz}}{G}, \gamma_{xz} = \frac{\tau_{xz}}{G}$$

where E is the elastic modulus, v is the Poisson's ratio and G is the shear modulus defined as $G = \dfrac{E}{2(1 + v)}$. Also σ_x, σ_y and σ_z are the normal stresses along the subscript directions and τ_{xy} is the shear stress in the x plane along the y direction. For a plane stress case, where the stresses are non-zero only in a plane, say, the xy plane (that is, $\sigma_z = \tau_{xz} = \tau_{yz} = 0$), we have

$$\varepsilon_x = \frac{1}{E}[\sigma_x - v(\sigma_y)] \qquad \sigma_x = \frac{E}{2}\left[\left(\frac{1}{1-v} + \frac{1}{1+v}\right)\varepsilon_x + \left(\frac{1}{1-v} - \frac{1}{1+v}\right)\varepsilon_y\right]$$

$$\varepsilon_y = \frac{1}{E}[\sigma_y - v(\sigma_x)] \quad \text{or} \quad \sigma_y = \frac{E}{2}\left[\left(\frac{1}{1-v} - \frac{1}{1+v}\right)\varepsilon_x + \left(\frac{1}{1-v} + \frac{1}{1+v}\right)\varepsilon_y\right] \quad (11.10k)$$

$$\gamma_{xy} = \frac{2(1+v)\tau_{xy}}{E} \qquad \tau_{xy} = \frac{E\gamma_{xy}}{2(1+v)}$$

which in matrix form becomes

$$\boldsymbol{\sigma} = \begin{pmatrix} \sigma_x \\ \sigma_y \\ \tau_{xy} \end{pmatrix} = \frac{E}{(1-v^2)}\begin{pmatrix} 1 & v & 0 \\ v & 1 & 0 \\ 0 & 0 & \frac{1-v}{2} \end{pmatrix}\begin{pmatrix} \varepsilon_x \\ \varepsilon_y \\ \gamma_{xy} \end{pmatrix} = \mathbf{D}\boldsymbol{\epsilon} \qquad (11.10l)$$

where **D** is the elasticity matrix for the plane stress case. The strain energy stored in the element then is

$$SE = \frac{1}{2}\int_V \boldsymbol{\sigma}^T \boldsymbol{\epsilon}\, dV = \frac{1}{2}\int_V \boldsymbol{\epsilon}^T \mathbf{D}\boldsymbol{\epsilon}\, dV = \frac{1}{2}\int_V \mathbf{u}^T \mathbf{B}^T \mathbf{D}\mathbf{B}\mathbf{u}\, dV = \frac{1}{2}\mathbf{u}^T\left\{\int_V \mathbf{B}^T \mathbf{D}\mathbf{B}\, dV\right\}\mathbf{u}$$

The element stiffness matrix $\mathbf{k}_e$ is

$$\mathbf{k}_e = \int_V \mathbf{B}^T \mathbf{D}\mathbf{B}\, dV = At\mathbf{B}^T \mathbf{D}\mathbf{B} \qquad (11.10m)$$

where t is the out-of-plane thickness and the constant matrices **B** and **D** are given by Eqs. (11.10i) and (11.10l).

Example 11.4. Consider a rectangular plate cantilevered at one edge as shown in Figure 11.13. The loads of 1 kN and 2 kN act along the vertical and horizontal directions at node 2. Take the elastic modulus as 2.24 GPa, Poisson's ratio as $\frac{1}{4}$ and the out-of-plane thickness as 10 mm.

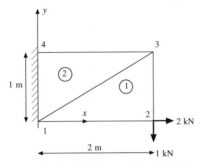

Figure 11.13 Displacement analysis of a rectangular plate

Noting that there are two degrees of freedom per node, the global displacement vector $\mathbf{U}$ can be composed such that for node j, $\mathbf{U}(2j-1)$ represents x displacement while $\mathbf{U}(2j)$ the y displacement. For element (1), the stiffness matrix can be calculated as follows:

The strain displacement matrix $\mathbf{B}_1$ from Eq. (11.10i) is

$$\mathbf{B}_1 = \frac{1}{2}\begin{pmatrix} -1 & 0 & 1 & 0 & 0 & 0 \\ 0 & 0 & 0 & -2 & 0 & 2 \\ 0 & -1 & -2 & 1 & 2 & 0 \end{pmatrix}$$

and the elasticity matrix $\mathbf{D}$ from Eq. (11.10*l*) is

$$\mathbf{D} = \frac{16}{15} \times 224 \times 10^7 \begin{pmatrix} 1 & \frac{1}{4} & 0 \\ \frac{1}{4} & 1 & 0 \\ 0 & 0 & \frac{3}{8} \end{pmatrix}$$

so that $\mathbf{k}_{e1}$ from Eq. (11.10m) is

$$\mathbf{k}_{e1} = A_1 t \mathbf{B}_1^T \mathbf{D} \mathbf{B}_1 = \frac{16}{15} \times 10^7 \begin{array}{cccccc} & 1 & 2 & 3 & 4 & 5 & 6 \\ \begin{pmatrix} 0.56 & 0 & -0.56 & 0.28 & 0 & -0.28 \\ & 0.21 & 0.42 & -0.21 & -0.42 & 0 \\ & & 1.4 & -0.7 & -0.84 & 0.28 \\ & & & 2.45 & 0.42 & -2.24 \\ & & & & 0.84 & 0 \\ sym & & & & & 2.24 \end{pmatrix} & \begin{matrix} 1 \\ 2 \\ 3 \\ 4 \\ 5 \\ 6 \end{matrix} \end{array}$$

For element 2

$$\mathbf{B}_2 = \frac{1}{2} \begin{pmatrix} 0 & 0 & 1 & 0 & -1 & 0 \\ 0 & -2 & 0 & 0 & 0 & 2 \\ -2 & 0 & 0 & 1 & 2 & -2 \end{pmatrix}$$

Hence

$$\mathbf{k}_{e2} = A_2 t \mathbf{B}_2^T \mathbf{D} \mathbf{B}_2 = \frac{16}{15} \times 10^7 \begin{array}{cccccc} 1 & 2 & 5 & 6 & 7 & 8 \\ \begin{pmatrix} 0.84 & 0 & 0 & -0.42 & -0.84 & 0.42 \\ & 2.24 & -0.28 & 0 & 0.28 & -2.24 \\ & & 0.56 & 0 & -0.56 & 0.28 \\ & & & 0.21 & 0.42 & -0.21 \\ & & & & 1.40 & -0.7 \\ sym & & & & & 2.45 \end{pmatrix} \begin{array}{c} 1 \\ 2 \\ 5 \\ 6 \\ 7 \\ 8 \end{array} \end{array}$$

The assembled matrix $\mathbf{K}$ is

$$\mathbf{K} = \frac{16}{15} \times 10^7 \begin{array}{cccccccc} 1 & 2 & 3 & 4 & 5 & 6 & 7 & 8 \\ \begin{pmatrix} 1.40 & 0.00 & -0.56 & 0.28 & 0.00 & -0.70 & -0.84 & 0.42 \\ 0.00 & 2.45 & 0.42 & -0.21 & -0.70 & 0.00 & 0.28 & -2.24 \\ -0.56 & 0.42 & 1.40 & -0.70 & -0.84 & 0.28 & 0.00 & 0.00 \\ 0.28 & -0.21 & -0.70 & 2.45 & 0.42 & -2.24 & 0.00 & 0.00 \\ 0.00 & -0.70 & -0.84 & 0.42 & 1.40 & 0.00 & -0.56 & 0.28 \\ -0.70 & 0.00 & 0.28 & -2.24 & 0.00 & 2.45 & 0.42 & -0.21 \\ -0.84 & 0.28 & 0.00 & 0.00 & -0.56 & 0.42 & 1.40 & -0.70 \\ 0.42 & -2.24 & 0.00 & 0.00 & 0.28 & -0.21 & -0.70 & 2.45 \end{pmatrix} \begin{array}{c} 1 \\ 2 \\ 3 \\ 4 \\ 5 \\ 6 \\ 7 \\ 8 \end{array} \end{array}$$

and the force vector $\mathbf{F}$ is $[0\ 0\ 1000\ -2000\ 0\ 0\ 0\ 0]^T$. After applying the displacement boundary conditions that nodes 1 and 4 (degrees of freedom 1, 2, 7 and 8) are fixed, the relevant entries to compute the unknown displacements $[\mathbf{U}(3)\ \mathbf{U}(4)\ \mathbf{U}(5)\ \mathbf{U}(6)]$ are (in bold).

$$\frac{16}{15} \times 10^7 \begin{pmatrix} 1.40 & 0.00 & -0.56 & 0.28 & 0.00 & -0.70 & -0.84 & 0.42 \\ 0.00 & 2.45 & 0.42 & -0.21 & -0.70 & 0.00 & 0.28 & -2.24 \\ -0.56 & 0.42 & \mathbf{1.40} & \mathbf{-0.70} & \mathbf{-0.84} & \mathbf{0.28} & 0.00 & 0.00 \\ 0.28 & -0.21 & \mathbf{-0.70} & \mathbf{2.45} & \mathbf{0.42} & \mathbf{-2.24} & 0.00 & 0.00 \\ 0.00 & -0.70 & \mathbf{-0.84} & \mathbf{0.42} & \mathbf{1.40} & \mathbf{0.00} & -0.56 & 0.28 \\ -0.70 & 0.00 & \mathbf{0.28} & \mathbf{-2.24} & \mathbf{0.00} & \mathbf{2.45} & 0.42 & -0.21 \\ -0.84 & 0.28 & 0.00 & 0.00 & -0.56 & 0.42 & 1.40 & -0.70 \\ 0.42 & -2.24 & 0.00 & 0.00 & 0.28 & -0.21 & -0.70 & 2.45 \end{pmatrix} \begin{pmatrix} 0 \\ 0 \\ \mathbf{U}(3) \\ \mathbf{U}(4) \\ \mathbf{U}(5) \\ \mathbf{U}(6) \\ 0 \\ 0 \end{pmatrix} = \begin{pmatrix} F_1 \\ F_2 \\ \mathbf{2000} \\ \mathbf{-1000} \\ \mathbf{0} \\ \mathbf{0} \\ F_7 \\ F_8 \end{pmatrix}$$

Solving for displacements gives $\mathbf{U} = \dfrac{15}{16} \times 10^{-3} \times [0 \ 0 \ 0.18 \ -0.24 \ 0.18 \ -0.24 \ \ 0 \ 0]^T$. The displaced plate is shown in Figure 11.14 (dashed lines). The vector containing the applied and support loads can be determined as $\mathbf{KU} = 10^3 \times [0 \ 0 \ 2 \ -1 \ 0 \ 0 \ -2 \ 1]^T$.

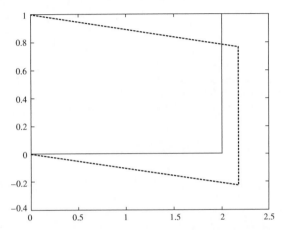

Figure 11.14   Resultant displacements (scaled) for Example 11.4

11.7   Four-Node Elements

Triangular elements are easy to implement with the drawback that the strain throughout the element is a constant. To get better stress field approximations, we may have to use a fine mesh of triangular elements. Otherwise, four-node quadrilateral elements, shown in Figure 11.15, may be employed. The procedure to determine the finite element stiffness matrix is similar to that for a triangular element. The first step is to determine the interpolation or shape functions. Like in case of a triangular element, the displacement u would depend on the nodal displacements along the x direction. Thus, u would depend on $u_1, \ldots, u_4$ which could be modeled using the polynomial approximation

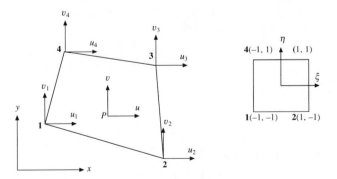

Figure 11.15   A four-node quadrilateral element

$$u = a_0 + a_1 x + a_2 y + a_3 xy \tag{11.11a}$$

Note again that the polynomial basis $[1 \ x \ y \ xy]$ is chosen such that there is no relative bias along the x and y directions. Alternatively, to ease calculations, the four-node element may be mapped onto a square shown on the right in Figure 11.15 with local coordinates (ξ, η). We can first determine the mapping between the global and local coordinate systems. To map the x coordinate of P on the square, we may use

$$x = c_0 + c_1 \xi + c_2 \eta + c_3 \xi \eta \tag{11.11b}$$

employing the same reasoning that we need to map four nodes for which we need four unknowns and the basis of four polynomial functions (as chosen above) should not have any bias towards any local coordinate. Also the x coordinate of P would depend only on the x nodal coordinates, i.e. x_1, x_2, x_3 and x_4. Employing the conditions in Eq. (11.11b) at four nodes, we have

$$x_1 = c_0 - c_1 - c_2 + c_3$$
$$x_2 = c_0 + c_1 - c_2 - c_3$$
$$x_3 = c_0 + c_1 + c_2 + c_3 \tag{11.11c}$$
$$x_4 = c_0 - c_1 + c_2 - c_3$$

Solving Eqs. (11.11c) gives

$$c_0 = \tfrac{1}{4}(x_1 + x_2 + x_3 + x_4) \qquad c_1 = \tfrac{1}{4}(-x_1 + x_2 + x_3 - x_4) \tag{11.11d}$$

$$c_2 = \tfrac{1}{4}(-x_1 - x_2 + x_3 + x_4) \qquad c_3 = \tfrac{1}{4}(x_1 - x_2 + x_3 - x_4)$$

Substituting the coefficients in Eq. (11.11b) and rearranging, we get

$$x = \tfrac{1}{4}(1 - \xi)(1 - \eta)x_1 + \tfrac{1}{4}(1 + \xi)(1 - \eta)x_2 + \tfrac{1}{4}(1 + \xi)(1 + \eta)x_3 + \tfrac{1}{4}(1 - \xi)(1 + \eta)x_4 \tag{11.11e}$$

or

$$x = N_1(\xi, \eta)x_1 + N_2(\xi, \eta)x_2 + N_3(\xi, \eta)x_3 + N_4(\xi, \eta)x_4 \tag{11.11f}$$

where $N_1(\xi, \eta), \ldots, N_4(\xi, \eta)$ are the shape functions. Note that within the square, all shape functions are greater than 0. At node 1, $N_1(-1, -1) = 1$ while all the other functions are zero. Likewise, at node 2 only $N_2(1, -1) = 1$, and so on. To map the y coordinate of P on the square, we can proceed in a manner similar to steps given in Eqs. (11.11b)-(11.11f) to get

$$y = N_1(\xi, \eta)y_1 + N_2(\xi, \eta)y_2 + N_3(\xi, \eta)y_3 + N_4(\xi, \eta)y_4 \tag{11.11g}$$

Eqs. (11.11f) and (11.11g), therefore, generically interpolate any nodal information onto a point interior to the quadrilateral element. We can use identical functions to interpolate the u and v displacements at point P, that is,

$$u = N_1(\xi, \eta)u_1 + N_2(\xi, \eta)u_2 + N_3(\xi, \eta)u_3 + N_4(\xi, \eta)u_4$$
$$v = N_1(\xi, \eta)v_1 + N_2(\xi, \eta)v_2 + N_3(\xi, \eta)v_3 + N_4(\xi, \eta)v_4$$

which can be combined in the matrix form as

$$\begin{pmatrix} u \\ v \end{pmatrix} = \begin{bmatrix} N_1 & 0 & N_2 & 0 & N_3 & 0 & N_4 & 0 \\ 0 & N_1 & 0 & N_2 & 0 & N_3 & 0 & N_4 \end{bmatrix} [u_1 \ v_1 \ u_2 \ v_2 \ u_3 \ v_3 \ u_4 \ v_4]^T = \mathbf{Nu} \tag{11.11h}$$

Since the interpolation functions for the coordinates (x, y) and displacements (u, v) are the same, the element is sometimes called the *isoparametric element*. Because of the term $\xi\eta$ (or xy) in interpolation (Eq. 11.11a or 11.11b), the element is also known as the *bilinear element*. The strain vector can be computed in a similar manner as in Eq. (11.10h), that is

$$\boldsymbol{\epsilon} = \begin{pmatrix} \varepsilon_x \\ \varepsilon_y \\ \gamma_{xy} \end{pmatrix} = \begin{pmatrix} \dfrac{\partial}{\partial x} & 0 \\ 0 & \dfrac{\partial}{\partial y} \\ \dfrac{\partial}{\partial y} & \dfrac{\partial}{\partial x} \end{pmatrix} \begin{pmatrix} u \\ v \end{pmatrix} = \begin{pmatrix} \dfrac{\partial}{\partial x} & 0 \\ 0 & \dfrac{\partial}{\partial y} \\ \dfrac{\partial}{\partial y} & \dfrac{\partial}{\partial x} \end{pmatrix} \mathbf{N}u = \mathbf{AGu} = \mathbf{Bu} \qquad (11.11\mathrm{i})$$

where $\mathbf{B} = \mathbf{AG}$ is the strain-displacement matrix with matrices $\mathbf{A}$ and $\mathbf{G}$ given as

$$\mathbf{A} = \frac{1}{\det(\mathbf{J})} \begin{bmatrix} \mathbf{J}_{22} & -\mathbf{J}_{12} & 0 & 0 \\ 0 & 0 & -\mathbf{J}_{21} & \mathbf{J}_{11} \\ -\mathbf{J}_{21} & \mathbf{J}_{11} & \mathbf{J}_{22} & -\mathbf{J}_{12} \end{bmatrix} \qquad (11.11\mathrm{j})$$

and

$$\mathbf{G} = \frac{1}{4} \begin{bmatrix} -(1-\eta) & 0 & (1-\eta) & 0 & (1+\eta) & 0 & -(1+\eta) & 0 \\ -(1-\xi) & 0 & -(1+\xi) & 0 & (1+\xi) & 0 & (1-\xi) & 0 \\ 0 & -(1-\eta) & 0 & (1-\eta) & 0 & (1+\eta) & 0 & -(1+\eta) \\ 0 & -(1-\xi) & 0 & -(1+\xi) & 0 & (1+\xi) & 0 & (1-\xi) \end{bmatrix} \qquad (11.11\mathrm{k})$$

where $\mathbf{J}$ is the Jacobian matrix given using Eqs. (11.11f) and (11.11g) as

$$\mathbf{J} = \begin{bmatrix} \frac{1}{4}(1-\eta)(x_2 - x_1) + \frac{1}{4}(1+\eta)(x_3 - x_4) & \frac{1}{4}(1-\eta)(y_2 - y_1) + \frac{1}{4}(1+\eta)(y_3 - y_4) \\ \frac{1}{4}(1-\xi)(x_4 - x_1) + \frac{1}{4}(1+\xi)(x_3 - x_2) & \frac{1}{4}(1-\xi)(y_4 - y_1) + \frac{1}{4}(1+\xi)(y_3 - y_2) \end{bmatrix} \qquad (11.11l)$$

For a plane stress, the elasticity matrix $\mathbf{D}$ is defined in Eq. (11.10l) using which we can compute the stress vector $\boldsymbol{\sigma}$. The element stiffness matrix $\mathbf{k}_e$ is given similar to that of the triangular element as

$$\mathbf{k}_e = \int_V t\mathbf{B}^T \mathbf{DB}\, dx dy = t \int_{-1}^{1} \int_{-1}^{1} \det(\mathbf{J})\mathbf{B}^T \mathbf{DB}\, d\xi\, d\eta \qquad (11.11\mathrm{m})$$

where t is the out-of-plane thickness of the element. Note that the matrices $\mathbf{B}$ and $\mathbf{J}$ are functions of the local coordinates ξ and η. The expression within the integral is usually computed using the four-point Gauss integration approach, that is

$$\mathbf{k}_e = t \int_{-1}^{1} \int_{-1}^{1} \det(\mathbf{J})\mathbf{B}^T \mathbf{DB}\, d\xi\, d\eta = t \sum_{i=1}^{4} w_i \det(\mathbf{J}(\xi_i, \eta_i))\mathbf{B}(\xi_i, \eta_i)^T \mathbf{DB}(\xi_i, \eta_i) \qquad (11.11\mathrm{n})$$

where the weights $w_i = 1$, $i = 1, \ldots, 4$ and $\xi_i = \eta_i = \pm \dfrac{1}{\sqrt{3}}$ are the four Gauss points.

Example 11.5. Solve the rectangular plate example above using a quadrilateral element.

The nodal coordinates are $(0, 0)$, $(2, 0)$, $(2, 1)$ and $(0, 1)$. The Jacobian matrix in Eq. (11.11*l*) is

$$\mathbf{J} = \frac{1}{4} \begin{bmatrix} 2(1 - \eta) + 2(1 + \eta) & 0 \\ 0 & (1 - \xi) + (1 + \xi) \end{bmatrix} = \frac{1}{4} \begin{bmatrix} 4 & 0 \\ 0 & 2 \end{bmatrix}$$

It turns out for this example that the Jacobian matrix is a constant which may not be so in general. The matrix **A** in Eq. (11.11j) is

$$\mathbf{A} = \frac{1}{0.5} \begin{bmatrix} 0.5 & 0 & 0 & 0 \\ 0 & 0 & 0 & 1 \\ 0 & 1 & 0.5 & 0 \end{bmatrix} = \begin{bmatrix} 1 & 0 & 0 & 0 \\ 0 & 0 & 0 & 2 \\ 0 & 2 & 1 & 0 \end{bmatrix}$$

From Eqs. (11.11i) and (11.11k), the strain displacement matrix **B** can be determined as

$$\mathbf{B} = \frac{1}{4} \begin{bmatrix} -1 + \eta & 0 & 1 - \eta & 0 & 1 + \eta & 0 & -1 - \eta & 0 \\ 0 & -2 + 2\xi & 0 & -2 - 2\xi & 0 & 2 + 2\xi & 0 & 2 - 2\xi \\ -2 + 2\xi & -1 + \eta & -2 - 2\xi & 1 - \eta & 2 + 2\xi & 1 + \eta & 2 - 2\xi & -1 - \eta \end{bmatrix}$$

and the elasticity matrix from Example 11.4 is

$$\mathbf{D} = \frac{16}{15} \times 224 \times 10^7 \begin{pmatrix} 1 & \frac{1}{4} & 0 \\ \frac{1}{4} & 1 & 0 \\ 0 & 0 & \frac{3}{8} \end{pmatrix}$$

For thickness $t = 10^{-2}$ m, and for Gauss point $\left(\dfrac{1}{\sqrt{3}}, \dfrac{1}{\sqrt{3}} \right)$, the stiffness matrix $\mathbf{k}_e^1$ is given as

$$\mathbf{k}_e^1 = \frac{16}{15} \times 10^6 \begin{bmatrix} 0.3126 & 0.1563 & 0.5750 & 0.1396 & -1.1667 & -0.5833 & 0.2791 & 0.2875 \\ 0.1563 & 0.5471 & 0.2875 & 1.8198 & -0.5833 & -2.0417 & 0.1396 & -0.3252 \\ 0.5750 & 0.2875 & 2.7375 & -0.5833 & -2.1458 & -1.0729 & -1.1667 & 1.3687 \\ 0.1396 & 1.8198 & -0.5833 & 7.0134 & -0.5208 & -6.7915 & 0.9646 & -2.0417 \\ -1.1667 & -0.5833 & -2.1458 & -0.5208 & 4.3541 & 2.1770 & -1.0416 & -1.0729 \\ -0.5833 & -2.0417 & -1.0729 & -6.7915 & 2.1770 & 7.6196 & -0.5208 & 1.2136 \\ 0.2791 & 0.1396 & -1.1667 & 0.9646 & -1.0416 & -0.5208 & 1.9292 & -0.5833 \\ 0.2875 & -0.3252 & 1.3687 & -2.0417 & -1.0729 & 1.2136 & -0.5833 & 1.1533 \end{bmatrix}$$

For Gauss point $\left(-\dfrac{1}{\sqrt{3}}, \dfrac{1}{\sqrt{3}}\right)$, $\mathbf{k}_e^2$ is

$$
\mathbf{k}_e^2 = \frac{16}{15} \times 10^6
\begin{bmatrix}
2.7375 & 0.5833 & 0.5750 & -0.2875 & -1.1667 & -1.3687 & -2.1458 & 1.0729 \\
0.5833 & 7.0134 & -0.1396 & 1.8198 & -0.9646 & -2.0417 & 0.5208 & -6.7915 \\
0.5750 & -0.1396 & 0.3126 & -0.1563 & 0.2791 & -0.2875 & -1.1667 & 0.5833 \\
-0.2875 & 1.8198 & -0.1563 & 0.5471 & -0.1396 & -0.3252 & 0.5833 & -2.0417 \\
-1.1667 & -0.9646 & 0.2791 & -0.1396 & 1.9292 & 0.5833 & -1.0416 & 0.5208 \\
-1.3687 & -2.0417 & -0.2875 & -0.3252 & 0.5833 & 1.1533 & 1.0729 & 1.2136 \\
-2.1458 & 0.5208 & -1.1667 & 0.5833 & -1.0416 & 1.0729 & 4.3541 & -2.1770 \\
1.0729 & -6.7915 & 0.5833 & -2.0417 & 0.5208 & 1.2136 & -21770 & 7.6196
\end{bmatrix}
$$

For $\left(-\dfrac{1}{\sqrt{3}}, -\dfrac{1}{\sqrt{3}}\right)$, $\mathbf{k}_e^3$ is

$$
\mathbf{k}_e^3 = \frac{16}{15} \times 10^6
\begin{bmatrix}
4.3541 & 2.1770 & -1.0416 & -1.0729 & -1.1667 & -0.5833 & -2.1458 & -0.5208 \\
2.1770 & 7.6196 & -0.5208 & 1.2136 & -0.5833 & -2.0417 & -1.0729 & -6.7915 \\
-1.0416 & -0.5208 & 1.9292 & -0.5833 & 0.2791 & 0.1396 & -1.1667 & 0.9646 \\
-1.0729 & 1.2136 & -0.5833 & 1.1533 & 0.2875 & -0.3252 & 1.3687 & -2.0417 \\
-1.1667 & -0.5833 & 0.2791 & 0.2875 & 0.3126 & 0.1563 & 0.5750 & 0.1396 \\
-0.5833 & -2.0417 & 0.1396 & -0.3252 & 0.1563 & 0.5471 & 0.2875 & 1.8198 \\
-2.1458 & -1.0729 & -1.1667 & 1.3687 & 0.5750 & 0.2875 & 2.7375 & -0.5833 \\
-0.5208 & -6.7915 & 0.9646 & -2.0417 & 0.1396 & 1.8198 & -0.5833 & 7.0134
\end{bmatrix}
$$

while for $\left(\dfrac{1}{\sqrt{3}}, -\dfrac{1}{\sqrt{3}}\right)$, $\mathbf{k}_e^4$ is

$$
\mathbf{k}_e^4 = \frac{16}{15} \times 10^6
\begin{bmatrix}
1.9292 & 0.5833 & -1.0416 & 0.5208 & -1.1667 & -0.9646 & 0.2791 & -0.1396 \\
0.5833 & 1.1533 & 1.0729 & 1.2136 & -1.3687 & -2.0417 & -0.2875 & -0.3252 \\
-1.0416 & 1.0729 & 4.3541 & -2.1770 & -2.1458 & 0.5208 & -1.1667 & 0.5833 \\
0.5208 & 1.2136 & -2.1770 & 7.6196 & 1.0729 & -6.7915 & 0.5833 & -2.0417 \\
-1.1667 & -1.3687 & -2.1458 & 1.0729 & 2.7375 & 0.5833 & 0.5750 & -0.2875 \\
-0.9646 & -2.0417 & 0.5208 & -6.7915 & 0.5833 & 7.0134 & -0.1396 & 1.8198 \\
0.2791 & -0.2875 & -1.1667 & 0.5833 & 0.5750 & -0.1396 & 0.3126 & -0.1563 \\
-0.1396 & -0.3252 & 0.5833 & -2.0417 & -0.2875 & 1.8198 & -0.1563 & 0.5471
\end{bmatrix}
$$

Adding the four matrices above yields the element stiffness matrix, that is

$$\mathbf{k}_e = \frac{16}{15} \times 10^7 \begin{bmatrix} (1) & (2) & (3) & (4) & (5) & (6) & (7) & (8) \\ 0.9333 & 0.3500 & -0.0933 & -0.0700 & -0.4667 & -0.3500 & -0.3733 & 0.0700 \\ 0.3500 & 1.6333 & 0.0700 & 0.6067 & -0.3500 & -0.8167 & -0.0700 & -1.4233 \\ -0.0933 & 0.0700 & 0.9333 & -0.3500 & -0.3733 & -0.0700 & -0.4667 & 0.3500 \\ -0.0700 & 0.6067 & -0.3500 & 1.6333 & 0.0700 & -1.4233 & 0.3500 & -0.8167 \\ -0.4667 & -0.3500 & -0.3733 & 0.0700 & 0.9333 & 0.3500 & -0.0933 & -0.0700 \\ -0.3500 & -0.8167 & -0.0700 & -1.4233 & 0.3500 & 1.6333 & 0.0700 & 0.6067 \\ -0.3733 & -0.0700 & -0.4667 & 0.3500 & -0.0933 & 0.0700 & 0.9333 & -0.3500 \\ 0.0700 & -1.4233 & 0.3500 & -0.8167 & -0.0700 & 0.6067 & -0.3500 & 1.6333 \end{bmatrix} \begin{matrix} (1) \\ (2) \\ (3) \\ (4) \\ (5) \\ (6) \\ (7) \\ (8) \end{matrix}$$

That the degrees of freedom (1), (2), (7) and (8) are fixed, removing the corresponding rows and columns gives a 4×4 system as

$$\frac{16}{15} \times 10^7 \begin{bmatrix} 0.9333 & -0.3500 & -0.3733 & -0.0700 \\ -0.3500 & 1.6333 & 0.0700 & -1.4233 \\ -0.3733 & 0.0700 & 0.9333 & 0.3500 \\ -0.0700 & -1.4233 & 0.3500 & 1.6333 \end{bmatrix} \begin{bmatrix} \mathbf{U}(3) \\ \mathbf{U}(4) \\ \mathbf{U}(5) \\ \mathbf{U}(6) \end{bmatrix} = \begin{bmatrix} 2000 \\ -1000 \\ 0 \\ 0 \end{bmatrix}$$

The non-zero displacements are

$$\begin{bmatrix} \mathbf{U}(3) \\ \mathbf{U}(4) \\ \mathbf{U}(5) \\ \mathbf{U}(6) \end{bmatrix} = \frac{15}{16} \times 10^{-3} \begin{bmatrix} 0.1786 \\ -0.2381 \\ 0.1786 \\ -0.2381 \end{bmatrix}$$

which compare well with those in Example 11.4.

The discussion in this chapter is restricted to elementary finite elements to demonstrate the principles in the finite element displacement and stress analysis. The reader may note that numerous advanced books are available on this subject which extend the method to heat transfer and fluid dynamics, and stress and displacement analysis for geometrically and materially nonlinear problems, both in two- and three-dimensions.

EXERCISES

1. Compute the displacements at nodes 2 (treat node 4 the same as node 2) and 5 treating springs as finite elements as in section 11.2. Verify the result using Newtonian equilibrium.

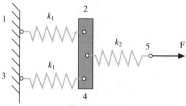

Figure P11.1

2. Determine the horizontal and vertical displacements at node 2 for the truss assemblage shown. Consider the elastic modulus as 65GPa for both trusses.

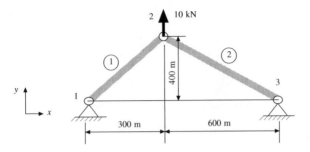

Figure P11.2

3. Determine the deflection at node 2 using beam elements for the problem shown below. Verify the result with the Euler-Bernoulli analysis for small beam deflections. (Hint: One would have to approximate the uniform load distribution with point loads at each node). Will the accuracy improve if the number of beam elements is increased? Explain by solving the same problem using three beam elements. Take the elastic modulus as 10^6 Nm^{-2}. In both cases, take beam elements of equal lengths.

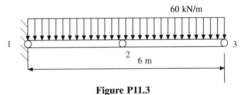

Figure P11.3

4. Assume the structure in Figure P11.2 as an assemblage of frame elements with the joint at node 2 as rigid. Determine the deflections and slope at node 2.
5. Using Eq. (11.10j), derive the elasticity matrix **D** for the plain strain case. Consider non-zero strains in the x-y plane.
6. A triangular lamina is shown in Figure P11.4. Node 4 is the midpoint of nodes 2 and 3. Take the modulus as 2.24 GPa, Poisson's ratio as 0.33 and out-of-plane thickness as 10 mm. Determine the deflections at nodes 3 and 4.

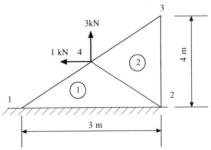

Figure P11.4

7. Gauss point integration is used extensively in computing the integral for the stiffness matrices of four-noded and like finite elements. This problem relates to determining Gauss points and weights. An integral of the form $\int_{-1}^{1} \phi(t)\,dt$ can be computed as $\sum_{i=1}^{n} w_i\phi(t_i)$ where w_i are the non-negative weights and t_i are Gauss points. Let $\phi(t) = a_0 + a_1 t + a_2 t^2$ be *exactly* integrated using an order 2 Gauss rule, i.e., using 2 Gauss points such that the points t_i, $i = 1, 2$ are placed symmetrically in $-1 \leq t \leq 1$. Also, consider the weights w_i, $i = 1, 2$ as symmetric (for a two point rule, $w_1 = w_2$). Determine the weights and Gauss points. Next, determine the weights and Gauss points for an order 3 (or 3 point) rule. (Hint: with weights w_i, $i = 1, ..., 3$ and points t_i, $i = 1, ..., 3$, use symmetry to get $w_2 = w_3$ and $t_2 = -t_3$. Also, t_i, $i = 1, ..., 3$ being symmetrically placed in $-1 \leq t \leq 1$ would suggest that $t_1 = 0$).

8. Solve Problem P11.4 using a single bilinear four-node element with nodes 1, ..., 4. Determine the deflections at the free nodes.

9. Derive the matrices **A** and **G** is Eqs. (11.11i) and (11.11k). Note that we may get different expressions for **A** and **G** though **B** = **AG** will not be altered.

Chapter 12

Optimization

In design, construction and maintenance of any engineering system, engineers have to take many technological (and managerial) decisions at several stages. The goal is to either minimize the effort required or maximize the desired benefit. Both goals are required to be expressed as a function of certain decision variables, optimizing over which would yield better (if not the best) result. Some practical instances of use of optimization are: (a) minimizing material volume (and/or stiffness) when constructing structures like over-bridges, (b) optimizing to determine the material connectivity or topology in such structures, (c) optimizing the shape of an automobile body to minimize aerodynamic drag, (d) optimizing the bumper for crashworthiness, and many more.

Numerical implementation of optimization is usually an iterative procedure wherein at every step the design variables are updated when a better goal value is achieved. An optimization algorithm can either be intuitive, like in the *optimality criteria* method, or can be a result of a rigorous derivation from the zeroth, first or second order approximations of the objective function (or goal) with respect to the design variables, for instance in the mathematical programming schemes. This chapter aims to brief the reader on some existing methods in optimization. Such methods can be classified in numerous ways depending on the number of variables, constraints, their nature (linear or nonlinear), and the nature of solution (generic or problem specific). We brief some generic methods on single-variable and multi-variable optimization.

12.1 Classical Optimization

The necessary condition for optimality for a function $f(x)$ in single variable x is well established in that equating the first derivative $f'(x) = df(x)/dx$ to zero yields the locations of zero slope or the optima. It can be intuitively observed in Figure 12.1 that such locations correspond to sites wherein the function changes its trend of monotonic increase (at maxima) or decrease (at minima). Further, the sign of the second derivative $f''(x)$ conveys, as a sufficiency condition, the nature of the optima at the location of zero slope. We would expect at a maximum that the slope would start decreasing as x increases and vice versa at a minimum, that is, the rate of change of slope $\dfrac{d^2 f(x)}{dx^2} < 0$ at a maximum and > 0 at a minimum.

12.2 Single Variable Optimization

Determining the locations of optima for a function $g(x)$ in single variable amounts to finding the roots

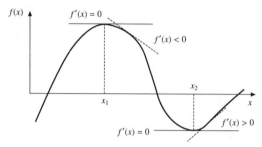

Figure 12.1   Optimality condition for a function of a single variable

of $g'(x) \equiv f(x) = 0$. The roots may be none, one or many and multiple as well. The methods devised to find these roots may depend on their capabilities to find single or all roots (multiple or distinct) at a time. They may also be limited to a class of functions, that is, whether they are applicable to only polynomials, or any generic function. The conventional approach is to plot $f(x)$ and obtain the value(s) of x where it intersects the x axis. A drawback of the plot-and-find method is that the values obtained are not usually very accurate. However, as a quick check, the graphical technique can be employed to determine the initial guesses for many computational procedures. Consider, for instance, a plot of $f(x) = x^3 - 4x + 3$ in Figure 12.2. We may note that at x_l, the value of x to the left of any root x_r, and at x_u, that to the right of x_r, the function value changes sign, that is, $f(x_l) f(x_u) < 0$. Many bracketing methods discussed next are based on this observation.

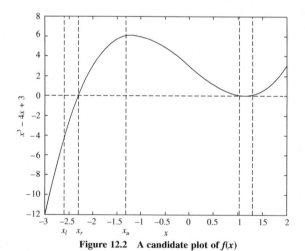

Figure 12.2   A candidate plot of $f(x)$

12.2.1   Bracketing Methods

(a) Method of Bisection

This is implemented by first assigning the lower and upper bound x_l and x_u such that $f(x_l) f(x_u) < 0$. A candidate value of the root is determined as $x_r = \frac{1}{2} (x_l + x_u)$, that is, the root bisects the chosen interval for which reason the method is named so. If sign $(f(x_l))$ = sign $(f(x_r))$, the root lies in the upper sub-interval and so the lower bound is set to x_r. Otherwise, if sign $(f(x_u))$ = sign $(f(x_r))$, the root lies in the lower sub-interval and thus x_u is set to x_r. Note that the bounds get redefined after every step such that the resultant bracket is a subset of the previous one. Thus, certain search region gets eliminated in the process for which reason the bracketing methods are also called the *region elimination* methods. The iterations are terminated when $|x_u - x_l| < \varepsilon$, a prespecified sufficiently small number.

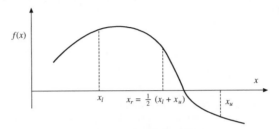

Figure 12.3 Bisection method

The method commences with an initial interval $[x_l, x_u]$ which is shrunk progressively by half into a new interval containing the root. If Δx^0 is the initial interval length and if N steps are required to achieve a desirable interval size, $\varepsilon = x_u - x_l$, then

$$\varepsilon = \Delta x^0/2^N \text{ or } N = \log_2 (\Delta x^0/\varepsilon) \tag{12.1}$$

Thus, if the initial and final interval lengths are known a *priori*, the number of steps required can be computed. Note that at each step, the three function values $f(x_l)$, $f(x_r)$ and $f(x_u)$ may not all be recomputed. If x_l is assigned as x_r in the previous step, $f(x_l)$ may be assigned the value of $f(x_r)$ and $f(x_u)$ may be retained. Otherwise, if $x_u = x_r$, then $f(x_u) = f(x_r)$ and $f(x_l)$ may be reused. All then is required is to compute x_r and $f(x_r)$ in an iteration.

Example 12.1. Find the root of $g'(x) \equiv f(x) = x^3 - 2x^2 - x + 2$. Note that the roots of this equation are -1, 1 and 2. We commence the bisection method with the initial bracket $[-3, 0]$. The results are

x_l	x_u	x_r	$f(x_l)$	$f(x_u)$	$f(x_r)$	$|x_u - x_l|$
−3.0000	0	−1.5000	−40.0000	2.0000	−4.3750	3.0000
−1.5000	0	−0.7500	−4.3750	2.0000	1.2031	1.5000
−1.5000	−0.7500	−1.1250	−4.3750	1.2031	−0.8301	0.7500
−1.1250	−0.7500	−0.9375	−0.8301	1.2031	0.3557	0.3750
−1.1250	−0.9375	−1.0313	−0.8301	0.3557	−0.1924	0.1875
−1.0313	−0.9375	−0.9844	−0.1924	0.3557	0.0925	0.0938
−1.0313	−0.9844	−1.0078	−0.1924	0.0925	−0.0474	0.0469
−1.0078	−0.9844	−0.9961	−0.0472	0.0925	0.0234	0.0234
−1.0078	−0.9961	−1.0020	−0.0472	0.0234	−0.0117	0.0117
−1.0020	−0.9961	−0.9990	−0.0117	0.0234	0.0059	0.0059
−1.0020	−0.9990	−1.0005	−0.0117	0.0059	−0.0029	0.0029
−1.0005	−0.9990	−0.9998	−0.0029	0.0059	0.0015	0.0015

The bisection method can be improved by taking into account the magnitudes of $f(x_l)$ and $f(x_u)$ as well. This helps in making judicious decisions with regard to x_r which is otherwise computed more in a brute force fashion as the average of lower and upper limits of the interval.

(b) Method of Regula falsi or False Positioning or Linear Interpolation

As the name suggests, this method uses linear function interpolation between the lower and upper bounds of the bracket to predict the false position of the root as shown in Figure 12.4. Using similarity of triangles, we have

$$\frac{f(x_l)}{x_r - x_l} = -\frac{f(x_u)}{x_u - x_r}$$

or
$$x_r = x_u - \frac{f(x_u)(x_l - x_u)}{f(x_l) - f(x_u)} \qquad (12.2)$$

The value of x_r so computed is used as in the bisection method described above. The algorithm is continued until the bracket width $x_u - x_l$ is less than the desired value ε or the function value at the root is close to zero. Though the root location is more intuitive in case of the linear interpolation method, it may not guarantee faster convergence when compared with the bisection method. This may be because one of the bracket limits may stay fixed and thus slow down the bracket shrinkage. If one of the limits gets stuck for two or more iterations, it is recommended to reduce the corresponding function value by half, that is, if $f(x_l)$ does not change in two iterations, then $\frac{1}{2}f(x_l)$ is used in place of $f(x_l)$ and the same holds true for $f(x_u)$ as well.

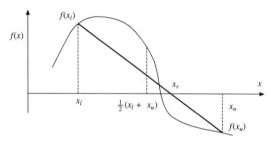

Figure 12.4   Method of Regula falsi

Example 12.2. We solve Example 12.1 with the initial bracket [–0.95, 1.87] using the false positioning (not modified) method. The results are shown in the following table. We ensure that $f(x_l)$ and $f(x_u)$ are of opposite signs when selecting the initial bracket. Also note that the algorithm converges since $f(x_r)$ is close to zero.

x_l	x_u	x_r	$f(x_l)$	$f(x_u)$	$f(x_r)$	$\mid x_u - x_l \mid$
−0.9500	1.8700	0.3749	0.2876	−0.3246	1.3968	2.8200
0.3749	1.8700	1.5881	1.3968	−0.3246	−0.6269	1.4951
0.3749	1.5881	1.2122	1.3968	−0.6269	−0.3698	1.2132
0.3749	1.2122	1.0369	1.3968	−0.3698	−0.0724	0.8374
0.3749	1.0369	1.0043	1.3968	−0.0724	−0.0085	0.6621
0.3749	1.0043	1.0005	1.3968	−0.0085	−0.0009	0.6294

(c) Fibonacci and Golden Section Search

This method determines the maximum (or minimum) of a unimodal function that has only one optimum in the search interval, $x_l \leq x \leq x_u$ by evaluating the function at points placed as per the Fibonacci sequence. Fibonacci numbers are related such that for two consecutive numbers $F(n-2)$ and $F(n-1)$, the third is obtained by the relation

$$F(n) = F(n-1) + F(n-2), \quad n = 2, 3, 4 \ldots$$

with $F(0) = 1$ and $F(1) = 1$. When using this search method, the final interval size ε may be specified in advance. The number of evaluations N are computed such that $F(N) \geq (x_u - x_l) \varepsilon$. Thus, if $x_u - x_l = 1$ and $\varepsilon = 0.001$, then $(x_u - x_l)/\varepsilon = 1000$. The Fibonacci number just greater than 1000 corresponds to $N = 16$ ($F(16) = 1597$), that is, we would need to perform $N = 16$ evaluations to achieve the interval size of 0.001.

With the number of iterations known, the interior points x_1 and x_2 both within the original interval are placed such that $x_1 = x_l + g(N)(x_u - x_l)$ and $x_2 = x_u - g(N)(x_u - x_l)$, where $g(N)$ is the *placement ratio* defined as $F(N-2)/F(N)$. The function $f(x)$ to be maximized is evaluated at the two points which are compared. If $f(x_1) \geq f(x_2)$, the upper limit x_u is set to x_2 implying that the search interval is reduced to $[x_l, x_2]$ and k (the iteration count with 0 initial value) is incremented to $k + 1$. For the subsequent iteration, assignments $x_1 = x_l + (F(N-k-2)/F(N-k)) (x_u - x_l)$ and $x_2 = x_u - (F(N-k-2)/F(N-k)) (x_u - x_l)$ are performed. Otherwise, if $f(x_1) < f(x_2)$, the new interval is reduced to $[x_1, x_u]$ by setting $x_l = x_1$, k is incremented by 1, and $x_1 = x_l + (F(N-k-2)/F(N-k)) (x_u - x_l)$ and $x_2 = x_u - (F(N-k-2)/F(N-k)) (x_u - x_l)$ are assigned. The procedure is continued until $k = N - 1$ and the optimal solution is taken as the midpoint of the final interval.

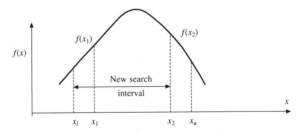

Figure 12.5 Fibonacci and golden section search

The Fibonacci search method minimizes the maximum number of evaluations needed to reduce the interval of uncertainty to within the prescribed length. For instance, an initial unit interval can be reduced to 1/10946 (=0.00009136) with only 20 evaluations. For very large N, $1-g(N)$ approaches the golden mean (0.618) and the method approaches the golden section search.

Example 12.3. Maximize $2 - x^2$ using the Golden section and Fibonacci search method. We can observe by inspection that the function is unimodal and that the optimum lies at $x = 0$. The following tables show how the golden section and Fibonacci method determine this optimum. We choose the interval as $[-1, 6]$ and the number of function evaluations N as 20.

Fibonacci search

$g(N)$	x_l	x_u	$x_1 = x_l +$ $g(N)(x_u - x_l)$	$x_2 = x_u -$ $g(N)(x_u - x_l)$	$f(x_1)$	$f(x_2)$
0.3820	−1.0000	6.0000	1.6738	3.3262	−0.8015	−9.0639
0.3820	−1.0000	3.3262	0.6525	1.6738	1.5743	−0.8015
0.3820	−1.0000	1.6738	0.0213	0.6525	1.9995	1.5743
0.3820	−1.0000	0.6525	−0.3688	0.0213	1.8640	1.9995
0.3820	−0.3688	0.6525	0.0213	0.2624	1.9995	1.9312
0.3820	−0.3688	0.2624	−0.1277	0.0213	1.9837	1.9995
0.3820	−0.1277	0.2624	0.0213	0.1134	1.9995	1.9871
0.3820	−0.1277	0.1134	−0.0356	0.0213	1.9987	1.9995
0.3820	−0.0356	0.1134	0.0213	0.0565	1.9995	1.9968
0.3819	−0.0356	0.0565	−0.0005	0.0213	2.0000	1.9995
0.3820	−0.0356	0.0213	−0.0139	−0.0005	1.9998	2.0000
0.3818	−0.0139	0.0213	−0.0005	0.0079	2.0000	1.9999
0.3824	−0.0139	0.0079	−0.0056	−0.0005	2.0000	2.0000
0.3810	−0.0056	0.0079	−0.0005	0.0027	2.0000	2.0000
0.3846	−0.0056	0.0027	−0.0024	−0.0005	2.0000	2.0000
0.3750	−0.0024	0.0027	−0.0005	0.0008	2.0000	2.0000
0.4000	−0.0024	0.0008	−0.0011	−0.0005	2.0000	2.0000
0.3333	−0.0011	0.0008	−0.0005	0.0002	2.0000	2.0000
0.5000	−0.0005	0.0008	0.0002	0.0002	2.0000	2.0000

Golden section search

$g(N)$	x_l	x_u	$x_1 = x_l +$ $g(N)(x_u - x_l)$	$x_1 = x_u -$ $g(N)(x_u - x_l)$	$f(x_1)$	$f(x_2)$
0.3820	−1.0000	6.0000	1.6740	3.3260	−0.8023	−9.0623
0.3820	−1.0000	3.3260	0.6525	1.6735	1.5742	−0.8005
0.3820	−1.0000	1.6735	0.0213	0.6522	1.9995	1.5746
0.3820	−1.0000	0.6522	−0.3689	0.0211	1.8639	1.9996
0.3820	−0.3689	0.6522	0.0212	0.2622	1.9996	1.9313
0.3820	−0.3689	0.2622	−0.1278	0.0211	1.9837	1.9996
0.3820	−0.1278	0.2622	0.0212	0.1132	1.9996	1.9872
0.3820	−0.1278	0.1132	−0.0357	0.0211	1.9987	1.9996
0.3820	−0.0357	0.1132	0.0211	0.0563	1.9996	1.9968
0.3820	−0.0357	0.0563	−0.0006	0.0211	2.0000	1.9996
0.3820	−0.0357	0.0211	−0.0140	−0.0006	1.9998	2.0000
0.3820	−0.0140	0.0211	−0.0006	0.0077	2.0000	1.9999
0.3820	−0.0140	0.0077	−0.0057	−0.0006	2.0000	2.0000
0.3820	−0.0057	0.0077	−0.0006	0.0026	2.0000	2.0000
0.3820	−0.0057	0.0026	−0.0026	−0.0006	2.0000	2.0000
0.3820	−0.0026	0.0026	−0.0006	0.0006	2.0000	2.0000
0.3820	−0.0026	0.0006	−0.0013	−0.0006	2.0000	2.0000
0.3820	−0.0013	0.0006	−0.0006	−0.0001	2.0000	2.0000
0.3820	−0.0006	0.0006	−0.0001	0.0002	2.0000	2.0000

Other bracketing methods include the *parabolic interpolation* method wherein given three function points, a parabola is interpolated through them and a candidate minimum is taken at a point on the parabola where the slope is zero. An efficient Brent's algorithm that combines parabolic and golden section search is in much use. Bracketing methods are inherently convergent in that the location of the root is gauranteed within the interval specified. However, they are capable of yielding only one root at a time even when there are more than one roots located within the initial interval. Now the issue is how to determine the initial interval. Ad hoc methods that may be employed are: (a) graphical wherein $f(x)$ may be plotted and the bracket $[x_l, x_u]$ may be chosen judiciously about each root by visual inspection and (b) incremental wherein commencing from a value x_l^0, the next value $x_u^0 = x_l^1$ is sought incrementally when the function changes sign. Following this, $x_u^1 = x_l^2$ is recorded upon another function sign change, and so on. The graphical approach is non-automated though it intuitively provides better guesses for the brackets. The incremental method on the other hand is non-intuitive and depends on the initial value x_l^0 and the increment size used.

12.2.2 Open Methods
In contrast to the bracketing methods that employ the bounds to capture a root, open methods employ only the initial guess for the same, or two starting values that do not necessarily bracket the root. Some open methods are discussed as follows:

(a) Single fixed-point iteration or method of successive substitution
We may rearrange the function $f(x) = 0$ as

$$x = h(x)$$

which can be converted to an iterative form as

$$x_{i+1} = h(x_i)$$

Thus, as the name suggests, starting with x_0, $x_1 = h(x_0)$ may be computed. The next guess, x_2 may be evaluated as $h(x_1)$, and so on and the procedure continues up to n steps until x_n and x_{n-1} are desirably close. As trivial as the method seems, it may not converge in every case. Using the Taylor series expansion up to two terms we have

$$h(x_{i+1}) = h(x_i) + h'(x_i)\,(x_{i+1} - x_i)$$

or

$$h(x_{i+1}) - h(x_i) = h'(x_i)\,(x_{i+1} - x_i)$$

$$\Rightarrow \qquad |\, x_{i+2} - x_{i+1}\,| = |\, h'(x_i)\,|\, |\, x_{i+1} - x_i\,| \tag{12.4}$$

Thus, for covergence, it is required that $|\, x_{i+2} - x_{i+1}\,| < |\, x_{i-1} - x_i\,|$ which is possible only if $|\, h'(x_i)\,| < 1$ for all i. Also Eq. (12.4) suggests that the error in each iteration is proportional to that in the previous iteration, one reason why the method of successive substitution is linearly convergent. The condition for convergence, that is, $|\, h'(x)\,| < 1$ may also be verified graphically. In Figure 12.6(a), convergence is guaranteed from both left and right of the root as $|\, h'(x)\,| < 1$ is true in the neighborhood of the root, while in Figure 12.6(b), this is not so.

It may be noted that if $\alpha = h(\alpha)$, then

$$\alpha = (1 - k + k)\,h(\alpha) = (1 - k)\alpha + kh(\alpha)$$

for arbitrary k. Visualizing the above in the iterative form, we have

$$x_{i+1} = (1 - k)x_i + kh(x_i) \tag{12.5}$$

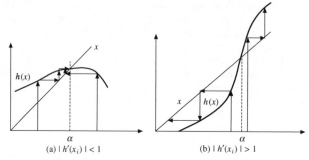

Figure 12.6 Convergence issues in the method of successive substitution

As an alternative relation for successive substitutions, this method can be used for a suitable value of k in case the original relation fails to converge. Here the condition for convergence is $|(1 - k) + kh'(x)| < 1$ in the neighborhood of the root.

Example 12.4. We determine the root of $e^{-x} - x = 0$ using the method of successive substitutions with results shown below. Note that $|h'(x)| = e^{-x} < 1$ for $x > 0$.

x	e^{-x}
2.00	0.14
0.14	0.87
0.87	0.42
0.42	0.66
0.66	0.52
0.52	0.60
0.60	0.55
0.55	0.58
0.58	0.56
0.56	0.57
0.57	0.57

(b) Newton-Raphson method

If $g(x)$ is the function for which the zero is to be determined, from the Taylor series expansion about the guess x_i, we have

$$g(x_{i+1}) = g(x_i) + g'(x_i)(x_{i+1} - x_i) + \tfrac{1}{2} g''(\xi)(x_{i+1} - x_i)^2$$

where $\xi \in [x_i, x_{i+1}]$. An approximate value of x_{i+1} is obtained by considering the Taylor series up to the first derivative and treating it at the root in which case $g(x_{i+1}) = 0$. Thus

$$0 = g(x_i) + g'(x_i)(x_{i+1} - x_i)$$

or
$$x_{i+1} = x_i - \frac{g(x_i)}{g'(x_i)} \qquad (12.6)$$

Graphically, the above Newton-Raphson relation may be interpreted as shown in Figure 12.7.

At point $[x_i, f(x_i)]$, a tangent can be extended to the x axis and the point of intersection of this tangent and the axis becomes the new guess x_{i+1} for the root.

For x_{i+1} as the root x_r, the original relation becomes

$$0 = g(x_i) + g'(x_i)(x_r - x_i) + {}^1\!/_2\, g''(\xi)\, (x_r - x_i)^2$$

Subtracting the Newton-Raphson relation with the one above, we have

$$0 = g'(x_i)(x_r - x_{i+1}) + {}^1\!/_2\, g''(\xi)\, (x_r - x_i)^2$$

Letting $E_i = (x_r - x_i)$, the error in the ith step and $E_{i+1} = (x_r - x_{i+1})$, the error in the $(i + 1)$ step, we have

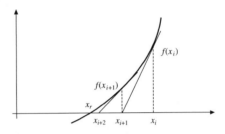

Figure 12.7   Newton-Raphson method

$$E_{i+1} = -[{}^1\!/_2\, g''(\xi)/g'(x_i)]E_i^2$$

which near convergence becomes (since both x_i and ξ approach x_r)

$$E_{i+1} = -[{}^1\!/_2\, g''(x_r)/g'(x_r)]E_i^2 \tag{12.7}$$

that is, near convergence, the error is proportional to the square of the previous error. For this reason, the Newton-Raphson method is quadratically convergent. However, there is no general convergence criterion for this method and the convergence depends mainly on the nature of the function and the location of the initial guess. Moreover, quadratic convergence is an attribute shown by the method only near the location of the root. For other situations where $g'(x_i) \approx 0$ (local maxima or minima in the vicinity of the root or a multiple root), the subsequent guess x_{i+1} may lie far from the root, anywhere on the x axis. For such reasons, the Newton-Raphson method may diverge.

Example 12.5. We estimate a root of $f(x) = x^3 - 2x^2 - x + 2$ with the starting guess of $x = 10$. The results are

x_{old}	x	$f(x)$	Error = $\mid x - x_{old} \mid$
10.00	6.94	792.00	3.06
6.94	4.93	233.23	2.01
4.93	3.62	68.19	1.31
3.62	2.80	19.62	0.82
2.80	2.32	5.44	0.48
2.32	2.08	1.37	0.24
2.08	2.01	0.26	0.07
2.01	2.00	0.02	0.01
2.00	2.00	0.00	0.00

(c) Secant method

A problem in implementing the Newton-Raphson method is the evaluation of the derivative and the related repercussions. In many cases, it may not be easy to compute the derivative analytically for which the derivative may be approximated using the backward difference method as

$$g'(x_i) \approx \frac{g(x_i) - g(x_{i-1})}{x_i - x_{i-1}}$$

substitution of which into the Newton Raphson formula gives

$$x_{i+1} = x_i - \frac{g(x_i)(x_i - x_{i-1})}{g(x_i) - g(x_{i-1})} \tag{12.8}$$

which is the iterative relation for the secant method. Note that two initial estimates are required to initiate the procedure. The above relation is very similar to that in the *regula falsi* approach, however, the difference is that the values x_{i-1} and x_i may not necessarily bracket the root guess x_{i+1}, and thus the secant method may be divergent. Secondly, a derivative of the Newton-Raphson method, the secant method is expected to converge much faster compared to the *regula falsi* approach.

12.3   Multivariable Optimization

In real life situations, there are usually more than one or a set of variables $\{x_1, x_2, x_3, \ldots, x_n\}$ which determine the state in an engineering system. An overview is first given of classical methods in multivariable optimization involving some definitions, mathematical models, theorems and solutions to simple problems by way of Lagrange Multipliers. Subsequent sections deal with linear/nonlinear unconstrained/constrained methods with emphasis on frequently used programming algorithms. However, a detailed treatise on the subject can only be found in a text dedicated to optimization.

12.3.1   Classical Multivariable Optimization

First, the focus is on determining an optimal value of a function dependent on several variables when the latter are not constrained, that is, the variables are not required to adhere to certain conditions. Let f be a function of n variables of the form $f(x_1, x_2, \ldots, x_n) \equiv f(\mathbf{X})$. A point $\mathbf{X}_0$ is an extremum if for all $\mathbf{X}$ in the neighborhood of $\mathbf{X}_0$, $f(\mathbf{X}) \leq f(\mathbf{X}_0)$ (relative maximum at $\mathbf{X}_0$) or $f(\mathbf{X}) \geq f(\mathbf{X}_0)$ (relative minimum at $\mathbf{X}_0$). We can focus on minimization problems noting that maximization of $f(\mathbf{X})$ can be converted into the minimization of $-f(\mathbf{X})$ or $1/f(\mathbf{X})$.

A function $f(\mathbf{X})$ has an extreme point at $\mathbf{X}_0$ if and only if $\dfrac{\partial f(\mathbf{X})}{\partial \mathbf{X}} = 0$ at that point, where $\dfrac{\partial f(\mathbf{X})}{\partial \mathbf{X}}$ is a vector denoting $[(\partial/\partial x_1)f(\mathbf{X}), (\partial/\partial x_2)f(\mathbf{X}), \ldots, (\partial/\partial x_n)f(\mathbf{X})]$. This gives the *first order necessary conditions* for optimality of a multivariate function. Expanding $f(\mathbf{X})$ in the neighborhood of $\mathbf{X}_0$ using the Taylor series till the first term, we have

$$f(\mathbf{X}_0 + \Delta\mathbf{X}) = f(\mathbf{X}_0) + \frac{\partial f(\mathbf{X}_0)}{\partial \mathbf{X}}\Delta\mathbf{X} \tag{12.9}[1]$$

For $f(\mathbf{X}_0 + \Delta\mathbf{X}) \geq f(\mathbf{X}_0)$ for a relative minimum at $\mathbf{X}_0$,

$$\frac{\partial f(\mathbf{X}_0)}{\partial \mathbf{X}}\Delta\mathbf{X} \geq 0 \tag{12.10}$$

for any $\Delta\mathbf{X}$ which will be satisfied if and only if $\dfrac{\partial f(\mathbf{X}_0)}{\partial \mathbf{X}} = 0$. Note that the same can be argued for

---

[1]$\partial/\partial\mathbf{X}f(\mathbf{X}_0)$ implies that the partial derivatives of $f(\mathbf{X})$ w.r.t. $\mathbf{X}$ are evaluated at $\mathbf{X}_0$.

$f(\mathbf{X}_0 + \Delta\mathbf{X}) \leq f(\mathbf{X}_0)$ for a relative maximum at $\mathbf{X}_0$ as well for which reason $\mathbf{X}_0$ is called a *stationary point*. The condition is analogous to the one variable case where the slopes at the extrema are zero. Considering the series expansion up to the second term, we have

$$f(\mathbf{X}_0 + \Delta\mathbf{X}) = f(\mathbf{X}_0) + \frac{\partial f(\mathbf{X}_0)}{\partial \mathbf{X}}\Delta\mathbf{X} + \frac{1}{2}\Delta\mathbf{X}^T \frac{\partial^2 f(\mathbf{X}_0)}{\partial \mathbf{X}^2}\Delta\mathbf{X} \qquad (12.11)$$

At an extremum, after applying the necessary conditions, the above equation becomes

$$f(\mathbf{X}_0 + \Delta\mathbf{X}) - f(\mathbf{X}_0) = \frac{1}{2}\Delta\mathbf{X}^T \frac{\partial^2 f(\mathbf{X}_0)}{\partial \mathbf{X}^2}\Delta\mathbf{X} \qquad (12.12)$$

For $\mathbf{X}_0$ to be a relative minimum, we require from above that $\frac{1}{2}\Delta\mathbf{X}^T \frac{\partial^2 f(\mathbf{X}_0)}{\partial \mathbf{X}^2}\Delta\mathbf{X} > 0$. Note that $\frac{\partial^2 f(\mathbf{X}_0)}{\partial \mathbf{X}^2}$ is a square matrix H_{ij} of dimension $n \times n$ written such that $H_{ij} = \partial^2 f(\mathbf{X}_0)/\partial x_i \partial x_j$, $i, j = 1, \ldots n$, and is known as the *Hessian* of the function $f(\mathbf{X})$. For $\Delta\mathbf{X}^T\mathbf{H}(\mathbf{X}_0)\Delta\mathbf{X} > 0$ for any $\Delta\mathbf{X}$, we require that $\mathbf{H}(\mathbf{X}_0)$ be positive definite by definition[2]. For $\mathbf{X}_0$ to be a relative maximum, using similar arguments, we require that $\Delta\mathbf{X}^T\mathbf{H}(\mathbf{X}_0)\Delta\mathbf{X} < 0$ for any $\Delta\mathbf{X}$ or $\mathbf{H}(\mathbf{X}_0)$ to be negative definite. A way to determine the definiteness of a matrix is by its eignvalues λ such that $\mathbf{A}\mathbf{X} = \lambda\mathbf{X}$ and so $\det(\mathbf{A} - \lambda\mathbf{I}) = 0$. For all positive (and non-zero) eigen values, the matrix is positive definite whereas for all negative eignvalues, the latter is negative definite. A stationary point $\mathbf{X}_0$ is said to be a *saddle point* if the Hessian at $\mathbf{X}_0$ is neither positive nor negative definite.

Example 12.6. Consider the function $f(x, y) = x^2 + y^2$ (Figure 12.8 a). The function is bowl-shaped and has a single global extremum, a minimum. Note that there is a stationary point at $x = 0$, $y = 0$ which can be obtained using $\partial f/\partial x = 2x = 0$ and $\partial f/\partial y = 2y = 0$. Also, note that $\partial^2 f/\partial x^2 = \partial^2 f/\partial y^2 = 2$, whereas $\partial^2 f/\partial x \partial y = 0$ that makes the Hessian $\mathbf{H} = 2\mathbf{I}_{2\times 2}$ which is positive definite. Next, consider the plot of $f(x, y) = (y^3 - 3y)(1 + x^2)^{-1}$ in the region $x \in [-2, 2]$ and $y \in [-2, 2]$ (Figure 12.8b). The partial derivatives can be computed as $\partial f/\partial x = -2x(y^3 - 3y)(1 + x^2)^{-2}$ and $\partial f/\partial y = (3y^2 - 3)(1 + x^2)^{-1}$, and setting both to zero yields $x = 0$ and $y = \pm 1$. Notice the local maximum at $(0, -1)$ and local minimum at $(0, 1)$. Further, $\partial^2 f/\partial x^2 = -2(y^3 - 3y)[(1 + x^2)^{-2} - 4x^2(1 + x^2)^{-3}]$, $\partial^2 f/\partial y^2 = (6y)(1 + x^2)^{-1}$ and $\partial^2 f/\partial x \partial y = -2x(3y^2 - 3)(1 + x^2)^{-2}$. At $(0, -1)$, $\partial^2 f/\partial x^2 = -4$, $\partial^2 f/\partial y^2 = -6$ and $\partial^2 f/\partial x \partial y = 0$ for which the eigenvalues of the resulting Hessian are -4 and -6 (both negative) which confirms the local maximum. At $(0, 1)$, $\partial^2 f/\partial x^2 = 4$, $\partial^2 f/\partial y^2 = 6$ and $\partial^2 f/\partial x \partial y = 0$ for which the eigen values are positive and the point is a relative minimum.

For $f(x, y) = 4 - (x^2 + y^2)$ (Figure 12.8 c), the function is bowl-shaped and has a single global maximum. Note the stationary point at $x = 0$, $y = 0$ for which it is straightforward to show that the Hessian is negative definite. Finally, consider $f(x, y) = x^2 - y^2$ (Figure 12.8 d). The function has a single saddle point (neither a minimum nor a maximum). At the stationary point $(0, 0)$, $\partial^2 f/\partial x^2 = 2$, $\partial^2 f/\partial y^2 = -2$ and $\partial^2 f/\partial x \partial y = 0$ for which the Hessian has one positive and one negative eigenvalue.

12.3.2 Constrained Multivariable Optimization

For engineering design problems, the design goal is almost always associated with some constraints.

---

[2] A matrix $\mathbf{A}$ is positive definite if for all $\mathbf{X} = \{x_1, x_2, \ldots, x_n\}^T$, $\mathbf{X}^T\mathbf{A}\mathbf{X} > 0$ and $\mathbf{X}^T\mathbf{A}\mathbf{X} = 0$ iff $\mathbf{X} = 0$. A negative definite $\mathbf{B}$ satisfies the relation $\mathbf{X}^T\mathbf{B}\mathbf{X} < 0$ for all $\mathbf{X}$ and $\mathbf{X}^T\mathbf{B}\mathbf{X} = 0$ only for $\mathbf{X} = 0$.

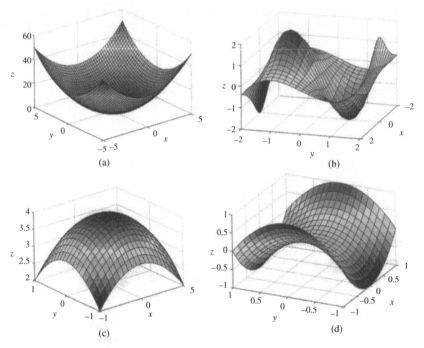

Fig. 12.8   Examples of unconstrained optimization with two variables

Consider, for example, an over-bridge design wherein it may be desired to minimize the overall deflection (strain energy) subject to a stipulated amount of material volume. We may further impose that the stress levels in bridge members do not exceed the yield limit of the material. If the member cross-sections are chosen as design variables, then the latter cannot assume negative values. In other words, some problems may require the variables to be bounded. In this section and the following we discuss multi-variable optimization with equality and inequality constraints.

Consider first, minimizing a function $f(\mathbf{X})$ in n variables $\mathbf{X} \equiv [x_1, x_2, \ldots, x_n]^\mathrm{T}$ with m equality constraints $g_i(\mathbf{X}) = 0$, $i = 1, \ldots, m$, where $m \leq n$. For $m > n$, the problem is overdetermined and there may not exist a solution. A method can be of *direct substitution* wherein by solving the m equality constraints, any set of m variables may be expressed in terms of the remaining $n - m$ variables. The problem then becomes unconstrained in $n - m$ variables and can be solved using the criteria discussed in section 12.3.1. Unfortunately, this method poses difficulties if the constraints are nonlinear in that there is no straighforward way to eliminate the constraints.

The method of Lagrange multipliers works by introducing a variable λ_i for each of the m constraints such that the total number of variables to be determined becomes $n + m$. An augmented Lagrangian $\mathcal{L}$ is constructed such that

$$\mathcal{L}(x_1, x_2, \ldots, x_n, \lambda_1, \lambda_2, \ldots \lambda_m) \equiv \mathcal{L}(\mathbf{X}, \mathbf{\Lambda}) = f(\mathbf{X}) + \sum_{i=1}^{m} \lambda_i g_i(\mathbf{X}) \qquad (12.13)$$

where λ_i are called the Lagrangian multipliers. The Lagrangian $\mathcal{L}(\mathbf{X}, \mathbf{\Lambda})$ is treated as an unconstrained function of $\mathbf{X}$ and $\mathbf{\Lambda}$ to be minimized. Following the derivation in section 12.3.1, the necessary conditions for $\mathcal{L}(\mathbf{X}, \mathbf{\Lambda})$ to have an extremum at $(\mathbf{X}_0, \mathbf{\Lambda}_0)$ are that the first partial derivatives of $\mathcal{L}(\mathbf{X}, \mathbf{\Lambda})$ with respect to the $n + m$ variables are all zero, that is

$$\frac{\partial \mathcal{L}(\mathbf{X}, \mathbf{\Lambda})}{\partial \mathbf{X}} = \frac{\partial f(\mathbf{X}_0)}{\partial \mathbf{X}} + \sum_{i=1}^{m} \lambda_i \frac{\partial g_i(\mathbf{X}_0)}{\partial \mathbf{X}} = 0$$

and
$$\frac{\partial \mathcal{L}(\mathbf{X}, \mathbf{\Lambda})}{\partial \mathbf{\Lambda}} = g_i(\mathbf{X}_0) = 0, i = 1, \dots, m \tag{12.14}$$

The above are $n + m$ equations which can be solved for the same number of variables. The sufficiency condition for $\mathcal{L}(\mathbf{X}, \mathbf{\Lambda})$ to have a relative minimum at $\mathbf{X}_0$ is that the Hessian matrix $\mathbf{H}_{pq} = \partial^2 \mathcal{L}(\mathbf{X}_0, \mathbf{\Lambda}_0)/\partial x_p \partial x_q$, $p, q = 1, \dots, n$ should be positive definite at $\mathbf{X} = \mathbf{X}_0$ for values of $\Delta \mathbf{X}$ for which all the constraints are satisfied.

The above conditions may be derived in a manner similar to that in the unconstrained case. Let $\mathbf{G}(\mathbf{X}) \equiv [g_1(\mathbf{X}), g_2(\mathbf{X}), \dots, g_m(\mathbf{X})]$ so that $\mathcal{L}(\mathbf{X}, \mathbf{\Lambda}) = f(\mathbf{X}) + \mathbf{G}(\mathbf{X})\mathbf{\Lambda}$. Consider the Taylor's expansion of the augmented Lagrangian up to the first derivatives, that is

$$\mathcal{L}(\mathbf{X}_0 + \Delta \mathbf{X}, \mathcal{L}_0 + \Delta \mathbf{\Lambda}) = \mathcal{L}(\mathbf{X}_0, \mathbf{\Lambda}_0) + \left[\frac{\partial}{\partial \mathbf{X}} \mathcal{L}\right]\Delta \mathbf{X} + \left[\frac{\partial}{\partial \mathbf{\Lambda}} \mathcal{L}\right]\Delta \mathbf{\Lambda}$$

or $\mathcal{L}(\mathbf{X}_0 + \Delta \mathbf{X}, \mathbf{\Lambda}_0 + \Delta \mathbf{\Lambda}) - \mathcal{L}(\mathbf{X}_0, \mathbf{\Lambda}_0) = \left[\frac{\partial}{\partial \mathbf{X}} f(\mathbf{X}_0) + \left\{\frac{\partial}{\partial \mathbf{X}}\mathbf{G}(\mathbf{X}_0)\right\}\mathbf{\Lambda}_0\right]\Delta \mathbf{X} + \mathbf{G}(\mathbf{X}_0)\Delta \mathbf{\Lambda}$

$$\tag{12.15}$$

For $\mathcal{L}(\mathbf{X}_0 + \Delta \mathbf{X}, \mathbf{\Lambda}_0 + \Delta \mathbf{\Lambda}) \geq \mathcal{L}(\mathbf{X}_0, \mathbf{\Lambda}_0)$ at a local minimum, we have

$$\left[\frac{\partial}{\partial \mathbf{X}} f(\mathbf{X}_0) + \left\{\frac{\partial}{\partial \mathbf{X}}\mathbf{G}(\mathbf{X}_0)\right\}\mathbf{\Lambda}_0\right]\Delta \mathbf{X} + \mathbf{G}(\mathbf{X}_0)\Delta \mathbf{\Lambda} \geq 0 \tag{12.16}$$

for all small variations $\Delta \mathbf{X}$ and $\Delta \mathbf{\Lambda}$ which is only possible if

$$\frac{\partial}{\partial \mathbf{X}} f(\mathbf{X}_0) + \left\{\frac{\partial}{\partial \mathbf{X}}\mathbf{G}(\mathbf{X}_0)\right\}\mathbf{\Lambda}_0 = \mathbf{0}$$

and
$$\mathbf{G}(\mathbf{X}_0) = \mathbf{0}, i = 1, \dots, m$$

which are the conditions stated in Eq. (12.14). Considering the expansion of the Lagrangian to include the second derivatives and noting that the coefficients of $\Delta \mathbf{X}$ and $\Delta \mathbf{\Lambda}$ are both $\mathbf{0}$ from the necessary condition, we have

$\mathcal{L}(\mathbf{X}_0 + \Delta \mathbf{X}, \mathbf{\Lambda}_0 + \Delta \mathbf{\Lambda}) - \mathcal{L}(\mathbf{X}_0, \mathbf{\Lambda}_0)$

$$= \frac{1}{2} \Delta \mathbf{X}^{\mathrm{T}} \frac{\partial^2}{\partial \mathbf{X}^2} \mathcal{L}(\mathbf{X}_0, \mathbf{\Lambda}_0)\Delta \mathbf{X} + \frac{1}{2} \Delta \mathbf{\Lambda}^T \frac{\partial^2}{\partial \mathbf{\Lambda}^2} \mathcal{L}(\mathbf{X}_0, \mathbf{\Lambda}_0)\Delta \mathbf{\Lambda} + \Delta \mathbf{\Lambda}^T \frac{\partial^2}{\partial \mathbf{X} \partial \mathbf{\Lambda}} \mathcal{L}(\mathbf{X}_0, \mathbf{\Lambda}_0)\Delta \mathbf{X}$$

Since $\mathcal{L}(\mathbf{X}, \mathbf{\Lambda})$ is linear in $\mathbf{\Lambda}$, $\frac{\partial^2}{\partial \mathbf{\Lambda}^2} \mathcal{L}(\mathbf{X}_0, \mathbf{\Lambda}_0) = \mathbf{0}$. Hence

$$\mathcal{L}(\mathbf{X}_0 + \Delta\mathbf{X}, \mathbf{\Lambda}_0 + \Delta\mathbf{\Lambda}) - \mathcal{L}(\mathbf{X}_0, \mathbf{\Lambda}_0) = \frac{1}{2}\Delta\mathbf{X}^{\mathrm{T}}\frac{\partial^2}{\partial\mathbf{X}^2}\mathcal{L}(\mathbf{X}_0, \mathbf{\Lambda}_0)\Delta\mathbf{X} + \Delta\mathbf{\Lambda}^{\mathrm{T}}\left\{\frac{\partial}{\partial\mathbf{X}}\mathbf{G}(\mathbf{X}_0)\right\}\Delta\mathbf{X}$$

Expanding the constraints about the extremum and using the necessary condition for optimality

$$\mathbf{G}(\mathbf{X}_0 + \Delta\mathbf{X}) = \mathbf{G}(\mathbf{X}_0) + \frac{\partial}{\partial\mathbf{X}}\mathbf{G}(\mathbf{X}_0)\Delta\mathbf{X} = \frac{\partial}{\partial\mathbf{X}}\mathbf{G}(\mathbf{X}_0)\Delta\mathbf{X} \qquad (12.18)$$

For $\Delta\mathbf{X}$ satisfying all the constraints, $\mathbf{G}(\mathbf{X}_0 + \Delta\mathbf{X}) = \mathbf{0}$ and so $\frac{\partial}{\partial\mathbf{X}}\mathbf{G}(\mathbf{X}_0)\Delta\mathbf{X} = 0$. Thus

$$\mathcal{L}(\mathbf{X}_0 + \Delta\mathbf{X}, \mathbf{\Lambda}_0 + \Delta\mathbf{\Lambda}) - \mathcal{L}(\mathbf{X}_0, \mathbf{\Lambda}_0) = \frac{1}{2}\Delta\mathbf{X}^{\mathrm{T}}\frac{\partial^2}{\partial\mathbf{X}^2}\mathcal{L}(\mathbf{X}_0, \mathbf{\Lambda}_0)\Delta\mathbf{X} \qquad (12.19)$$

and for the left hand side to be greater than 0 at a local minimum, $\frac{\partial^2}{\partial\mathbf{X}^2}\mathcal{L}(\mathbf{X}_0, \mathbf{\Lambda}_0)$ must be positive definite.

Example 12.7. Calculate the maximum and minimum values of $f = x^2 + y^2$ subject to $g \equiv x^2 + y^2 + 2x - 2y + 1 = 0$ using the Lagrangian multiplier method.

We form the Lagrangian as

$$\mathcal{L} = x^2 + y^2 + \lambda(x^2 + y^2 + 2x - 2y + 1)$$

differentiating which with respect to the variables x and y, respectively, gives

$$2x + \lambda(2x + 2) = 0 \Rightarrow x = -\frac{\lambda}{\lambda + 1}$$

$$2y + \lambda(2y - 2) = 0 \Rightarrow y = \frac{\lambda}{\lambda + 1}$$

Substituting in the constraint and rearranging gives

$$\lambda^2 + 2\lambda - 1 = 0 \Rightarrow \lambda = -1 \pm \sqrt{2}$$

Thus, for $\lambda = -1 + \sqrt{2}$, $x = 1/\sqrt{2} - 1$ and $y = -1/\sqrt{2} + 1$, and for $\lambda = -1 - \sqrt{2}$, $x = -(1/\sqrt{2} + 1)$ and $y = (1/\sqrt{2} + 1)$. The Hessian matrix is

$$\mathbf{H} = \begin{bmatrix} \dfrac{\partial^2\mathcal{L}}{\partial x^2} & \dfrac{\partial^2\mathcal{L}}{\partial x\partial y} \\ \dfrac{\partial^2\mathcal{L}}{\partial x\partial y} & \dfrac{\partial^2\mathcal{L}}{\partial y^2} \end{bmatrix} = \begin{bmatrix} 2\lambda + 2 & 0 \\ 0 & 2\lambda + 2 \end{bmatrix}$$

with eigenvalues as $2\lambda + 2$. For $\lambda = -1 + \sqrt{2}$, the eigenvalues are $2\sqrt{2}$ which are positive and hence the Hessian is positive definite. For $\lambda = -1 - \sqrt{2}$, however, the eignvalues are $-2\sqrt{2}$ and negative, the Hessian for which is negative definite and the point is a relative maximum. Figure 12.9 shows the graphical depiction of the optimal solutions for this example. The function contours are shown in thin lines which are tangent to the constraint curve (thick circle) at points A and B, where A yields the function minimum and B gives the maximum. The corresponding function values at A and B are $3 - 2\sqrt{2}$ and $3 + 2\sqrt{2}$.

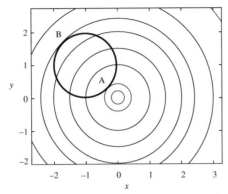

Figure 12.9 **Function (thin lines) and constraint (thick line)
curves for Example 12.7**

Example 12.8. Find the optimal points for $f(x, y) = 2 + xy$ on the circle $(x - 2)^2 + (y - 2)^2 = 4$.
The Lagrangian becomes

$$\mathcal{L} = 2 + xy + \lambda[(x - 2)^2 + (y - 2)^2 - 4]$$

differentiating which yields

$$\partial\mathcal{L}/\partial x = y + 2\lambda(x - 2) = 0$$

$$\partial\mathcal{L}/\partial y = x + 2\lambda(y - 2) = 0$$

solving which we get $y = x$. Using the constraint, we have

$$2(y - 2)^2 = 4$$

or $y = 2 \pm \sqrt{2}$. Thus, the two solution sets are $(2 + \sqrt{2}, 2 + \sqrt{2})$ and $(2 - \sqrt{2}, 2 - \sqrt{2})$. The following summarized solution table is provided along with the function values.

	x	y	λ	$f(x, y)$
(i)	$2 + \sqrt{2}$	$2 + \sqrt{2}$	$-\frac{1}{2} - \frac{1}{\sqrt{2}}$	13.66
(ii)	$2 - \sqrt{2}$	$2 - \sqrt{2}$	$\frac{1}{\sqrt{2}} - \frac{1}{2}$	2.34(

The Hessian is computed as

$$\mathbf{H} = \begin{bmatrix} 2\lambda & 1 \\ 1 & 2\lambda \end{bmatrix}$$

It can be shown that the Hessian for solution (i) is negative definite and for (ii) it is neither positive nor negative definite. Thus, solution (i) (point C in Figure 12.10) provides a local maximum while (ii) (point B) provides a saddle point.

12.3.3 Multivariable Optimization with Inequality Constraints
Consider now, minimizing a function $f(\mathbf{X})$ in n variables with m inequality constraints $g_i(\mathbf{X}) \le 0$,

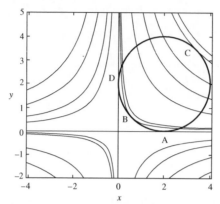

Fig. 12.10 **Function (thin lines) and constraint (thick line)**
curves for Example 12.8

$i = 1, \ldots, m$. These constraints can be converted to equality constraints by adding slack variables y_i such that

$$g_i(\mathbf{X}) + y_i^2 = 0, i = 1, \ldots, m \qquad (12.20)$$

The slack variables ensure that $g_i(\mathbf{X})$, $i = 1, \ldots, m$ are all smaller than or equal to zero. The minimization problem can be solved using the method of Lagrangian multipliers discussed above. For $\mathbf{Y} = [y_1, y_2, \ldots, y_m]$, the augmented Lagrangian can be written as

$$\mathcal{L}(x_1, x_2, \ldots, x_n, \lambda_1, \lambda_2, \ldots, \lambda_m, y_1, y_2, \ldots, y_m) \equiv \mathcal{L}(\mathbf{X}, \mathbf{\Lambda}, \mathbf{Y}) = f(\mathbf{X}) + \sum_{i=1}^{m} \lambda_i [g_i(\mathbf{X}) + y_i^2]$$

$$(12.21)$$

Noting that we have additional m variables $\mathbf{Y}$, employing the necessary conditions in Eq. (12.14) gives

$$\frac{\partial \mathcal{L}(\mathbf{X}, \mathbf{\Lambda}, \mathbf{Y})}{\partial \mathbf{X}} = \frac{\partial f(\mathbf{X}_0)}{\partial \mathbf{X}} + \sum_{i=1}^{m} \lambda_i \frac{\partial g_i(\mathbf{X}_0)}{\partial \mathbf{X}} = 0 \qquad (12.22a)$$

$$\frac{\partial \mathcal{L}(\mathbf{X}, \mathbf{\Lambda}, \mathbf{Y})}{\partial \mathbf{\Lambda}} \equiv g_i(\mathbf{X}_0) + y_i^2 = 0, i = 1, \ldots, m \qquad (12.22b)$$

and $\qquad \dfrac{\partial \mathcal{L}(\mathbf{X}, \mathbf{\Lambda}, \mathbf{Y})}{\partial \mathbf{Y}} \equiv 2\lambda_i y_i = 0, i = 1, \ldots, m \qquad (12.22c)$

which is a system of $n + 2m$ equations in the same number of unknowns $\mathbf{X}$, $\mathbf{\Lambda}$ and $\mathbf{Y}$. From Eq. (12.22c), if $y_i = 0$, then from (12.22b) $g_i(\mathbf{X}_0) = 0$ and the constraint is said to be *active*[3]. In such a case, the corresponding Lagrange multiplier λ_i may or may not be zero. If $y_i \neq 0$, then λ_i has to be zero and $g_i(\mathbf{X}_0)$ is strictly smaller than zero.

---

[3]The inequality constraints satisfied with the equality sign $g_j(\mathbf{X}) = 0$ at the optimum $\mathbf{X}_0$ are called *active* constraints while those satisfied with the strict inequality sign $g_j(\mathbf{X}) < 0$ are called *inactive* constraints.

12.3.4 Karush-Kuhn-Tucker (KKT) Necessary Conditions for Optimality

We may realize from above that the slack variables act as switching mediators between the constraint and the corresponding Lagrange multiplier, that is, for $y_i = 0$ which is when the corresponding constraint is active or $g_i(\mathbf{X}_0) = 0$, the multiplier λ_i need not be zero. On then other hand, if $y_i \neq 0$ ($g_i(\mathbf{X}_0) < 0$, i.e., $\neq 0$), the corresponding multiplier has to be zero. Since it is not important to determine the slack variables, we may eliminate them by restating Eqs. (12.22) as

$$\frac{\partial \mathcal{L}(\mathbf{X}, \mathbf{\Lambda})}{\partial \mathbf{X}} = \frac{\partial f(\mathbf{X}_0)}{\partial \mathbf{X}} + \sum_{i=1}^{m} \lambda_i \frac{\partial g_i(\mathbf{X}_0)}{\partial \mathbf{X}} = 0$$

$$\lambda_i g_i(\mathbf{X}_0) = 0, \ i = 1, \ldots, m \qquad (12.23\text{a})$$

Many texts use the notation

$$\nabla f = \begin{pmatrix} \partial f / \partial x_1 \\ \partial f / \partial x_2 \\ \vdots \\ \partial f / \partial x_n \end{pmatrix} \quad \text{and} \quad \nabla g_j = \begin{pmatrix} \partial g_j / \partial x_1 \\ \partial g_j / \partial x_2 \\ \vdots \\ \partial g_j / \partial x_n \end{pmatrix}$$

where ∇f is termed as the function gradient and ∇g_j is called the constraint gradient with partial derivatives evaluated at $\mathbf{X}_0$ to conveniently represent the above set of equations as

$$\nabla f(\mathbf{X}_0) + \lambda_1 \nabla g_1(\mathbf{X}_0) + \lambda_2 \nabla g_2(\mathbf{X}_0) + \ldots + \lambda_m \nabla g_m(\mathbf{X}_0) = \mathbf{0} \quad (n \text{ equations})$$

$$\lambda_i g_i(\mathbf{X}_0) = 0, \ i = 1, \ldots, m \qquad (12.23\text{b})$$

These are known as the Karush-Kuhn-Tucker necessary conditions for optimality with inequality constraints. The above are $n + m$ equations in $n + m$ variables $\mathbf{X}_0$ and $\mathbf{\Lambda}_0$ that can be computed with the implicit condition $g_i(\mathbf{X}_0) \leq 0, \ i = 1, \ldots, m$. In addition, we may note that for an active constraint $g_i(\mathbf{X}_0) = 0$, the corresponding Lagrange multipliers λ_i will have to be non-negative for $f(\mathbf{X}_0)$ to be a local minimum. To show this, we rearrange Eq. (12.23b) in short notation as

$$-\nabla f = \lambda_1 \nabla g_1 + \lambda_2 \nabla g_2 + \ldots + \lambda_m \nabla g_m \qquad (12.24)$$

Let $\mathbf{S}$ be a feasile search direction such that any step taken along this vector lies within the feasible region, that is, if α is the step size along $\mathbf{S}$, then $\mathbf{X} = \mathbf{X}_0 + \alpha \mathbf{S}$ must satisfy $g_i(\mathbf{X}) \leq 0, \ i = 1, \ldots,, m$. Such a vector has the property that the dot product

$$\mathbf{S}^T \nabla g_i \leq 0 \text{ for all } g_i(\mathbf{X}) = 0 \qquad (12.25)$$

The geometric interpretation is that $\mathbf{S}$ makes an obtuse angle with the normals of active constraints, the minimum angle being 90° for linear or concave constraints. A case is illustrated in Figure 12.11 where two constraints g_1 and g_2 among m constraints are active.

Since the other constraints $g_3, g_4, \ldots, g_m$ are all inactive and strictly smaller than zero, the corresponding multipliers are zero. Eq. (12.24) then becomes

$$-\nabla f = \lambda_1 \nabla g_1 + \lambda_2 \nabla g_2 \qquad (12.26)$$

Premultiplying by $\mathbf{S}^T$ gives

$$-\mathbf{S}^T \nabla f = \lambda_1 \mathbf{S}^T \nabla g_1 + \lambda_2 \mathbf{S}^T \nabla g_2 \qquad (12.27)$$

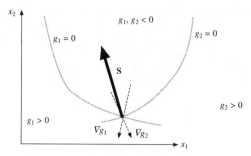

Figure 12.11   Geometric description of a feasible direction vector S

For λ_1 and λ_2 both > 0, $\mathbf{S}^T \nabla f$ may be seen always to be positive. It may be noted that ∇f represents the direction along which the function increases at a maximum rate in an unconstrained case. Thus if $\nabla f(\mathbf{X}_0)$ is chosen as the search direction at $\mathbf{X}_0$ and a scalar α is used as the step size so that the new point is expressed as $\mathbf{X} = \mathbf{X}_0 + \alpha \nabla f(\mathbf{X}_0)$, from Taylor series expansion we have

$$f(\mathbf{X}) = f(\mathbf{X}_0) + \alpha [\nabla f(\mathbf{X}_0)]^T \nabla f(\mathbf{X}_0)$$

Since $[\nabla f(\mathbf{X}_0)]^T \nabla f(\mathbf{X}_0) > 0$, if α is positive, the function value at the new point will be greater. For a constrained case, $\mathbf{S}^T \nabla f$ represents a component of increment of f along the search direction $\mathbf{S}$. If $\mathbf{S}^T \nabla f > 0$, the function value increases as we move along $\mathbf{S}$. Thus, for λ_1 and λ_2 both positive, we may not be able to find any direction in the feasible domain along which the function can be decreased further. Since the point at which Eq. (12.25) is satisfied is assumed to be optimum, λ_1 and λ_2 have to be positive. The reasoning can be extended to cases where more than two constraints are active.

The KKT necessary conditions for a minimum can now be written as

$$\frac{\partial \mathcal{L}(\mathbf{X}, \mathbf{L})}{\partial \mathbf{X}} = \frac{\partial f(\mathbf{X}_0)}{\partial \mathbf{X}} + \sum_{i=1}^{m} \lambda_i \frac{\partial g_i(\mathbf{X}_0)}{\partial \mathbf{X}} = 0$$

$$\lambda_i g_i(\mathbf{X}_0) = 0, \ i = 1, \ldots, m$$

$$\lambda_i \geq 0, \ i = 1, \ldots, m$$

Example 12.9

(a) Minimize: $f(x_1, x_2) = (x_1 - 1)^2 + x_2^2$

$$\text{Subject to: } g_1(x_1, x_2) \equiv (x_2 + 2) - x_1^2 \leq 0$$

We formulate the Lagrangian as

$$\mathcal{L} = (x_1 - 1)^2 + x_2^2 + \lambda[(x_2 + 2) - x_1^2]$$

so that the necessary KKT conditions for an optimum are

$$\frac{\partial \mathcal{L}}{\partial x_1} = 2(x_1 - 1) - 2\lambda x_1 = 0$$

$$\frac{\partial \mathcal{L}}{\partial x_2} = 2x_2 + \lambda = 0$$

and $$\lambda[(x_2 + 2) - x_1^2] = 0$$

Case I: If $\lambda = 0$, then $x_2 = 0$ and $x_1 = 1$ and so $g_1(1, 0)$ is 1 which is greater than 0. Thus, this solution is not feasible.

Case II: For $\lambda \neq 0$ we have from the third condition $(x_2 + 2) - x_1^2 = 0$ or $x_2 = x_1^2 - 2$. Eliminating λ from the other two conditions gives

$$(1 + 2x_2)x_1 - 1 = 0$$

or $$2x_1^3 - 3x_1 - 1 = 0$$

which is a cubic in x_1. The three solutions are 1.36, –1 and –0.36 and the respective values of x_2 are -0.13, –1 and –1.86. Since $\lambda = -2x_2$, the corresponding Lagrangian multipliers are 0.27, 2.00 and 3.73 which are all positive. Thus, $(1.36, -0.13)$, $(-1, -1)$ and $(-0.36, -1.86)$ are the three local minima suggested by the KKT necessary conditions for optimality with function values as 0.15, 5 and 5.35, respectively, as depicted by points A, B and C in Figure 12.12. Note that at these points, the function and constraint contours are tangent as expected since the function and gradient normals are collinear satisfying the relation $-\nabla f = \lambda \nabla g$ for the active constraint. This case is like the minimization of the

objective with an equality (active) constraint. The sufficiency conditions is that $\mathbf{H} = \left[\dfrac{\partial^2 \mathcal{L}}{\partial x_p \, \partial x_q} \right]$, $p = $

1, 2, $q = 1, 2$ should be positive definite. One may compute the Hessian as $[2 - 2\lambda \quad 0; \quad 0 \quad 2]$ and find that the eigen values are $2(1 - \lambda)$ and 2 which are both positive only for $A(1.36, -0.13)$ for $\lambda = 0.27$. Thus, $(1.36, -0.13)$ satisfies both the necessary and sufficient conditions and is a true minimum while the other two solutions do not satisfy the sufficiency condition.

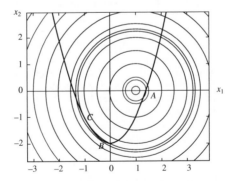

Figure 12.12   **Function (thin lines) and constraint (thick line) contours for Example 12.9**

(b) Consider now an additional constraint $g_2(x_1, x_2) \equiv -x_1 \leq 0$ for which the Lagrangian becomes

$$\mathcal{L} = (x_1 - 1)^2 + x_2^2 + \lambda_1[(x_2 + 2) - x_1^2] - \lambda_2 x_1$$

and the KKT necessary conditions for an optimum are

$$\frac{\partial \mathcal{L}}{\partial x_1} = 2(x_1 - 1) - 2\lambda_1 x_1 - \lambda_2 = 2\left[(1 - \lambda_1)x_1 - 1 - \frac{1}{2}\lambda_2\right] = 0$$

$$\frac{\partial \mathcal{L}}{\partial x_2} = 2x_2 + \lambda_1 = 0$$

$$\lambda_1[(x_2 + 2) - x_1^2] = 0$$

and $\qquad \lambda_2 x_1 = 0$

Case I ($\lambda_1 = \lambda_2 = 0$): Here, $x_2 = 0$ and $x_1 = 1$ for which $g_1(x_1, x_2) = 1$ while $g_2(x_1, x_2) = -1$. Since one constraint is not satisfied, the solution in infeasible.

Case II ($\lambda_1 \neq 0$, $\lambda_2 = 0$): This is equivalent to the solution in Case II of Example 12.9(a). Since $\lambda_1 \neq 0$, $g_1(x_1, x_2)$ is active and satisfied for the three solution points. Evaluating $g_2(x_1, x_2)$ at these points, we get

$$g_2(1.36, -0.13) = -1.36 < 0$$

$$g_2(-1, -1) = 1 > 0$$

and $\qquad g_2(-0.36, -1.86) = 0.36 > 0$

which gives the only feasible minimum at $(1.36, -0.13)$, that is, point A in Figure 12.12. That $g_2(x_1, x_2)$ is inactive at A, the Hessian is still positive definite with eigenvalues $2(1 - 0.27)$ and 2.

Case III ($\lambda_2 \neq 0$, $\lambda_1 = 0$): We have $x_1 = 0$ and $x_2 = 0$ from the second and fourth KKT conditions above. Note that $g_2(x_1, x_2)$ is active while $g_1(0, 0) = 2$ which is greater than 0 and thus this solution is not feasible.

Case IV ($\lambda_1 \neq 0$, $\lambda_2 \neq 0$): We have the two constraints active, that is, $x_1 = 0$ and $(x_2 + 2) - x_1^2 = 0$ which implies that $x_2 = -2$. From the first and second KKT conditions for this problem, computing the multipliers gives $\lambda_1 = 4$ and $\lambda_2 = -2$. Since both multipliers are not positive, $(0, -2)$ is not a minimum for the function.

Combining the Cases I-IV gives $(1.36, -0.13)$ or point A as the constrained minimum which is suggested by Figure 12.12 as points B and C do not comply with $g_2(x_1, x_2) \leq 0$. Also, the sufficiency condition is satisfied for point A.

Section 12.3 discussed the classical methods to solve multi-variable optimization problems without and with equality and inequality constraints. There may be problems wherein we may require handling both equality and inequality constraints for which the KKT conditions get slightly modified. If $f(\mathbf{X})$ is to be minimized subject to m inequality constraints $g_i(\mathbf{X}) \leq 0$, $i = 1, \ldots, m$ and p equality constraints $h_j(\mathbf{X}) = 0$, $j = 1, \ldots, p$, then the KKT necessary conditions can be stated as

$$\nabla f + \sum_{i=1}^{m} \lambda_i \nabla g_i + \sum_{j=1}^{p} \beta_j \nabla h_j = 0$$

$$\lambda_i g_i = 0, \ i = 1, \ldots, m$$

$$g_i \leq 0, \ i = 1, \ldots, m$$

$$h_j = 0, \ j = 1, \ldots, p$$

and $\qquad \lambda_i \geq 0, \ i = 1, \ldots, m \qquad\qquad (12.28a)$

The sufficiency condition may be written as

$$\Delta \mathbf{X}^T[\mathbf{H}] \, \Delta \mathbf{X} > 0 \tag{12.28b}$$

or $\mathbf{H}$ is positive definite at the optimum, where $\mathbf{H}$ is the Hessian of the Lagrangian. Note that the Hessian would comprise only equality and active constraints since the Lagrange multipliers for inactive (and feasible) constraints will be zero.

For solving Eqs. (12.28) when the number of design variables and/or constraints at hand is large, working solutions by hand is quite cumbersome and the numerical implementations are quite involved. Of the many available methods for which the reader is suggested to refer to dedicated texts on optimization, this chapter discusses three generic and often used methods, namely Linear Programming, Sequential Linear Programming (SLP) and Sequential Quadratic Programming (SQP).

12.4 Linear Programming

Linear programming is employed when the function and constraints have linear dependence on the design variables. The constraint equations can either be in the equality or inequality form. In standard form, a linear programming problem can be stated as

Minimize: $\qquad f(x_1, x_2, \ldots, x_n) = c_1 x_1 + c_2 x_2 + \ldots + c_n x_n = \mathbf{c}^T \mathbf{X}$

Subject to:

$$a_{11} x_1 + a_{12} x_2 + \ldots + a_{1n} x_n = b_1$$

$$a_{21} x_1 + a_{22} x_2 + \ldots + a_{2n} x_n = b_2 \equiv \mathbf{aX} = \mathbf{b}$$

$$\ldots$$

$$a_{m1} x_1 + a_{m2} x_2 + \ldots + a_{mn} x_n = b_m$$

with $\qquad x_1, x_2, \ldots, x_n$ all ≥ 0 or $\mathbf{X} \geq \mathbf{0}$ $\qquad\qquad$ (12.29)

Note that for a linear programming problem, the objective should be of the minimization type, the constraints should be of the equality type and all decision (design) variables should be non-negative. The maximization of a function is equivalent to the minimization of its negative and this change should be incorporated accordingly. Constraints of the type

$$a_{p1} x_1 + a_{p2} x_2 + \ldots + a_{pn} x_n \leq b_p$$

can be modified to

$$a_{p1} x_1 + a_{p2} x_2 + \ldots + a_{pn} x_n + x_{n+1} = b_p$$

while those of the type $\qquad a_{q1} x_1 + a_{q2} x_2 + \ldots + a_{qn} x_n \geq b_q$

can be modified to $\qquad a_{q_1} x_1 + a_{q_2} x_2 + \ldots + a_{qn} x_n - x_{n+1} = b_q$

by introducing a non-negative *slack* (for $\leq$ constraints) or *surplus* (for $\geq$ constraints) variable x_{n+1}. For m equality constraints in n variables, we would be interested in an underdetermined set ($m < n$) of linear equations for if $m = n$, the solution, if existing, will be unique and there will not be any scope for optimization. For $m > n$, the solution may not exist at all. Note that for the system $\mathbf{aX} = \mathbf{b}$ with the number of constraints less than the variables, there may exist many solutions. However, they may or may not satisfy $\mathbf{X} \geq \mathbf{0}$. All solutions satisfying $\mathbf{aX} = \mathbf{b}$ and $\mathbf{X} \geq \mathbf{0}$ are called *feasible solutions* and the goal is to determine the optimal solution(s) among the feasible set. The system $\mathbf{aX} = \mathbf{b}$ can be solved using the well known Gauss elimination technique. Using a sequence of row reductions, we may arrive at the canonical form

$$1x_1 + 0x_2 + \ldots + 0x_m + a'_{1,m+1} x_{m+1} + a'_{1,m+2} x_{m+2} + \ldots + a'_{1,n} x_n = b'_1$$

$$0x_1 + 1x_2 + \ldots + 0x_m + a'_{2,m+1} x_{m+1} + a'_{2,m+2} x_{m+2} + \ldots + a'_{2,n} x_n = b'_2$$

$$\ldots$$

$$0x_1 + 0x_2 + \ldots + 1x_m + a'_{m,m+1} x_{m+1} + a'_{m,m+2} x_{m+2} + \ldots + a'_{m,n} x_n = b'_m \qquad (12.30)$$

We can arrive at a solution $[x_1, x_2, \ldots, x_m, x_{m+1}, x_{m+2}, \ldots, x_n]^T = [b'_1, b'_2, \ldots, b'_m, 0, 0, \ldots, 0]^T$ called the *basic solution* by choosing $x_{m+1} = x_{m+2} = \ldots = x_n = 0$, since the solution vector contains no more than m nonzero terms. The pivotal variables $x_1, x_2, \ldots, x_m$ are also called *basic variables* while the rest are called *non-pivotal, non-basic* or *independent* variables. If $b'_1, b'_2, \ldots, b'_m$ are all positive, the condition $\mathbf{X} \geq \mathbf{0}$ is satisfied and the solution then is termed as *basic feasible solution*. It is possible to obtain the other basic solutions from the canonical system in Eq. (12.30) by performing an additional pivotal operation. For this, we choose $a'_{p,q}$ as the pivot term where $q > m$ while p is any row among $1, 2, \ldots, m$. The new canonical system will have x_q as the pivotal variable in place of x_p and thus will yield a new basic solution that may or may not be feasible.

From the foregoing, a basic solution for a system of n variables bound in m constraints ($n > m$) can be obtained by setting any $n - m$ of the n variables to zero and solving for the rest. The resulting solutions may or may not be feasible. Of the feasible ones, checks may then be imposed to determine which basic solution renders a function minimum. The number of basic solutions to be inspected for feasibility and optimality is equal to $\dfrac{n!}{m!(n-m)!}$, that is, the number of ways in which m variables can be selected from the parent set $[x_1, x_2, \ldots, x_n]^T$. For large number of variables and constraints, this value may be large. We thus need a systematic technique to examine the basic feasible solutions such that the subsequent solution renders a lower function value than the previous one, until the function minimum is attained.

12.4.1 Simplex Method

The commencing point in the Simplex method is a set of equations in the canonical form which gives a basic feasible solution. In addition, the objective function is also included in the row reduced form, that is

$$1x_1 + 0x_2 + \ldots + 0x_m + a'_{1,m+1} x_{m+1} + a'_{1,m+2} x_{m+2} + \ldots + a'_{1,n} x_n = b'_1$$

$$0x_1 + 1x_2 + \ldots + 0x_m + a'_{2,m+1} x_{m+1} + a'_{2,m+2} x_{m+2} + \ldots + a'_{2,n} x_n = b'_2$$

$$\ldots$$

$$0x_1 + 0x_2 + \ldots + 1x_m + a'_{m,m+1} x_{m+1} + a'_{m,m+2} x_{m+2} + \ldots + a'_{m,n} x_n = b'_m$$

$$0x_1 + 0x_2 + \ldots + 0x_m - f + c'_{m+1} x_{m+1} + c'_{m+2} x_{m+2} + \ldots + c'_n x_n = -f'_0 \qquad (12.31)$$

Note that $-f$ is treated as a basic variable in the canonical from. If all $b'_i \geq 0$, $i = 1, \ldots, m$, then a basic feasible solution can be deduced such that

$$x_i = b_i', i = 1, \ldots, m$$

$$f = f_0'$$

$$x_i = 0, i = m + 1, \ldots, n$$

The algorithm is intended to move from one basic feasible solution to another using a pivotal operation. Prior to that we ensure that the current solution is non-optimal by inspecting the values of $c_{m+1}', c_{m+2}', \ldots, c_n'$. A basic feasible solution is optimal with a minimum function value f_0' if all $c_{m+1}', c_{m+2}', \ldots c_n'$ are non-negative. From the last row of Eq. (12.31), we have

$$f_0' + c_{m+1}' x_{m+1} + c_{m+2}' x_{m+2} + \ldots + c_n' x_n = f$$

Variables $x_{m+1}, x_{m+2}, \ldots, x_n$ are all zero in the current basic feasible solution and are constrained to be non-negative. In the subsequent basic feasible solution, any one of them, say x_k, $k \in [m + 1, n]$ would enter the set of basic variables by becoming positive. But if $c_k \geq 0$, x_k becoming positive will only increase the function value. Thus, if $c_{m+1}', c_{m+2}', \ldots, c_n'$ are all non-negative, none of the non-basic variables $x_{m+1}, x_{m+2}, \ldots, x_n$ will cause a decrease in the function value by becoming positive or entering the basic variables set, and therefore, the current solution will be the optimal point.

Inspection of $c_{m+1}', c_{m+2}', \ldots, c_n'$ also reveals if there is a multiple optima. Let $c_{m+1}', c_{m+2}', \ldots,$ $c_{k-1}', c_{k+1}', \ldots, c_n'$ be all > 0 except for c_k' which is equal to zero for some non-basic variable x_k. Thus, if x_k enters the basic variable set by becoming positive, no change in the function results in which case there are multiple optima. We can say therefore that a basic feasible solution is uniquely optimal if $c_{m+1}', c_{m+2}', \ldots, c_n'$ are all strictly positive (> 0) for the non-basic variables.

In case the current basic feasible solution is not optimal, it may be improved as follows. If at least one c_k', k belonging to $[m + 1, n]$ is negative, x_k can be made the basic variable to decrease f further. In case more than one c_k's are negative, then x_s is chosen to be the basic variable such that

$$c_s' = \min (c_k' < 0), k = m + 1, \ldots, n$$

If there are more than one c_s's having the same minimum value, then any one among them is arbitrarily chosen. Once x_s is chosen, we make it positive by keeping the rest of the non-basic variables zero and observing the performance of the current basic variables. From Eq. (12.31) we see that

$$x_1 = b_1' - a_{1,s}' x_s, \qquad\qquad b_1' \geq 0$$

$$x_2 = b_2' - a_{2,s}' x_s, \qquad\qquad b_2' \geq 0$$

$$\ldots$$

$$x_m = b_m' - a_{m,s}' x_s, \qquad\qquad b_m' \geq 0$$

$$f = f_0' + c_s' x_s \qquad\qquad c_s' < 0 \qquad\qquad (12.32)$$

That $c_s' < 0$ suggests that x_s should be made as large as possible to reduce the function f. However, in the process of increasing x_s, some existing basic variables may become negative. It can be seen that if all $a_{i,s}' < 0$, $i = 1, \ldots, m$, x_s can be made as large as possible without making any x_i, $i = 1, \ldots, m$ negative in which case the linear programming problem is unbounded. Otherwise, if $a_{i,s}' > 0$, equating x_i to zero in Eqs. (12.32) gives

$$x_s = b_i'/a_{i,s}', a_{i,s}' > 0, i = 1, \ldots, m \qquad (12.33)$$

Since we require the largest possible value for x_s for which all x_i, $i = 1, \ldots, m$ are non-negative, $x_s^* = \min(b_i'/a_{i,s}') = b_r'/a_{rs}'$ (say) for all $a_{i,s}' > 0, i = 1, \ldots, m$. The choice of r in case of a tie is arbitrary. If in case $b_r' = 0$, $x_s = 0$ and cannot be increased in which case the solution is *degenerate*.

For a non-degenerate basic feasible solution, a new basic feasible solution can be constructed with a lower function value as follows. Substituting the value of x_s gives

$$x_s = x_s^*$$

$$x_i = b_i' - a_{i,s}' x_s^*, \qquad i \neq r$$

$$x_r = 0$$

$$x_j = 0, \qquad j = m + 1, m + 2, \ldots, n \text{ and } j \neq s$$

$$f = f_0' + c_s' x_s^* \leq f_0' \qquad (12.34)$$

which is a feasible solution different from the previous one. Since $a_{r,s}' > 0$, a pivot operation involving the rth row will yield a new basic feasible solution that has a function value lower than the previous one. The new solution can be tested for optimality by inspecting the coefficients c_j' and if they are all positive, the procedure should be stopped. Else, a new basic feasible solution should be formed and the method should be repeated. The following example illustrates the working of the Simplex method.

Example 12.10

$$\text{Minimize: } f = -x_1 - 5x_2 - 3x_3$$

$$\text{Subject to: } x_1 + 2x_2 - x_3 \leq 3$$

$$2x_1 + x_2 + x_3 \leq 10, \qquad x_1, x_2, x_3 \geq 0$$

We first convert the inequality constraints to equality constraints by introducing slack variables x_4 and x_5 and reduce the following equations to the canonical form

R_1	x_1	$+ 2x_2$	$-x_3$	$+ x_4$			$= 3$
R_2	$2x_1$	$+ x_2$	$+x_3$		$+x_5$		$= 10$
R_3	$-x_1$	$-5x_2$	$-3x_3$			$-f$	$= 0$

With two rows R_1 and R_2 and five variables, there can atmost be two basic variables. To commence, x_4 and x_5 can be treated as basic variables and $x_1 = x_2 = x_3 = 0$. Then $x_4 = 3$ and $x_5 = 10$. Also, note that the starting value of the function is zero. Following table can be formed:

Basic variables	x_1	x_2	x_3	x_4	x_5	b_i	$b_i/a_{is}, a_{is} > 0$
x_4	1	2	-1	1	0	3	$3/2$ ←
x_5	2	1	1	0	1	10	10
$-f$	-1	-5	-3	0	0	0	

Most negative, x_2 enters the basis

Smallest $b_i/a_{is}, a_{is} > 0$
x_4 leaves the basis

Note that the coefficient of x_2 is most negative in the third row. Next, in the same column, both coefficients a_{12} and a_{22} are positive. However, the one that corresponds to the minimum b_i/a_{is} value is the coefficient in the first row. Thus, the coefficient in the box is made the pivot element allowing x_2 to become the new basic variable replacing x_4. Pivot operations yield

	x_1	x_2	x_3	x_4	x_5	b_i	$b_i/a_{is}, a_{is} > 0$
x_2	1/2	1	−1/2	1/2	0	3/2	
x_5	3/2	0	$\boxed{3/2}$	−1/2	1	17/2	17/3 ←
$-f$	3/2	0	−11/2	5/2	0	15/2	

Most negative, x_3 enters the basis

Smallest b_i/a_{is} for $a_{is} > 0$
x_5 leaves the basis

with the new basic feasible solution as $[0\ 3/2\ 0\ 0\ 17/2]^T$ and the function value as $-15/2$. Note that the coefficient of x_3 in the third row is most negative and in the same column, coefficient x_3 in the second row is positive. Thus, 3/2 as bordered becomes the new pivot coefficient replacing x_3 by x_5 in the basic variable set. Row operations yield

	x_1	x_2	x_3	x_4	x_5	b_i	$b_i/a_{is}, a_{is} > 0$
x_2	1	1	0	1/3	1/3	13/3	
x_3	1	0	1	−1/3	2/3	17/3	
$-f$	7	0	0	2/3	11/3	116/3	

where the new basic feasible solution is $[0\ 13/3\ 17/3\ 0\ 0]^T$ and the function value is $-116/3$. Since now all coefficients pertaining to x_1, x_2 and x_3 in the third row are non-negative, the solution is optimal.

12.5 Sequential Linear Programming (SLP)

Given an optimization problem in standard form, that is

Minimize: $\qquad\qquad f(\mathbf{X})$

Subject to: $\qquad\quad g_j(\mathbf{X}) \le 0, j = 1, \ldots, m$

$\qquad\qquad\qquad h_k(\mathbf{X}) = 0, k = 1, \ldots, p$

the problem is linearized about the solution $\mathbf{X}_i$ using the Taylor series, that is

$$f(\mathbf{X}) = f(\mathbf{X}_i) + \nabla f(\mathbf{X}_i)^T(\mathbf{X} - \mathbf{X}_i)$$

$$g_j(\mathbf{X}) = g_j(\mathbf{X}_i) + \nabla g_j(\mathbf{X}_i)^T(\mathbf{X} - \mathbf{X}_i), \qquad j = 1, \ldots, m$$

$$h_k(\mathbf{X}) = h_k(\mathbf{X}_i) + \nabla h_k(\mathbf{X}_i)^T(\mathbf{X} - \mathbf{X}_i), \qquad k = 1, \ldots, p \qquad (12.35)$$

The above problem is solved using the Simplex method to determine a new solution vector $\mathbf{X}_{i+1}$. The problem is linearized again about $\mathbf{X}_{i+1}$ and the process is continued until the convergence is achieved and a suitable optimal solution $\mathbf{X}^*$ is found. Also note that SLP is a first order method requiring to compute the first derivatives of the function and constraints.

The method commences with an initial guess $\mathbf{X}_0$ which may or may not be feasible. The objective and constraints are linearized as in Eqs. (12.35) and then the linear programming problem is stated as

Minimize: $\quad f(\mathbf{X}_i) + \nabla f(\mathbf{X}_i)^T(\mathbf{X} - \mathbf{X}_i)$

Subject to: $\quad g_j(\mathbf{X}_i) + \nabla g_j(\mathbf{X}_i)^T(\mathbf{X} - \mathbf{X}_i) \leq 0, \qquad j = 1, \ldots, m$

$\qquad\qquad\quad h_k(\mathbf{X}_i) + \nabla h_k(\mathbf{X}_i)^T(\mathbf{X} - \mathbf{X}_i) = 0, \qquad k = 1, \ldots, p$

which is solved using the Simplex method to obtain the new vector $\mathbf{X}_{i+1}$. The original constraints are evaluated at $\mathbf{X}_{i+1}$, that is, $g_j(\mathbf{X}_{i+1})$, $j = 1, \ldots, m$ and $h_k(\mathbf{X}_{i+1})$, $k = 1, \ldots, p$ are determined and if $g_j(\mathbf{X}_{i+1}) \leq \varepsilon$ for all j and $|h_k(\mathbf{X}_{i+1})| \leq \varepsilon$ for all k for some prespecified positive tolerance value ε, the constraints are assumed to be satisfied and the procedure is stopped with $\mathbf{X}^* = \mathbf{X}_{i+1}$.

If some constraints are violated, i.e. $g_j(\mathbf{X}_{i+1}) > \varepsilon$ for some j or $|h_k(\mathbf{X}_{i+1})| > \varepsilon$ for some k, the most violated constraint is determined, for instance

$$g_l(\mathbf{X}_{i+1}) = \max(g_j(\mathbf{X}_{i+1}))$$

which is linearized about $\mathbf{X}_{i+1}$ as

$$g_l(\mathbf{X}_{i+1}) + \nabla g_j(\mathbf{X}_{i+1})^T(\mathbf{X} - \mathbf{X}_{i+1}) \leq 0$$

and included as the $(m + 1)$th inequality constraint in the previous linear programming (LP) problem. The new iteration number is set to $i + 1$ and the new LP problem is solved with $m + 1$ inequality constraints and p equality constraints.

12.6   Sequential Quadratic Programming (SQP)

The sequential quadratic programming is a general-purpose mathematical programming technique that involves solving the quadratic programming sub-problem of the type

Minimize: $\quad \phi = f_k + (\nabla f_k + \mathbf{\Lambda}_k^T \nabla \mathbf{h}_k)\nabla \mathbf{X}_k + \frac{1}{2}\nabla \mathbf{X}_k^T (\nabla^2 f_k + \mathbf{\Lambda}_k^T \nabla^2 \mathbf{h}_k) \Delta \mathbf{X}_k$

$\qquad\qquad\quad \nabla \mathbf{h}_k \Delta \mathbf{X}_k + \mathbf{h}_k = 0 \hfill (12.36)$

Subject to: $\quad \mathbf{X}_l \leq \mathbf{X} \leq \mathbf{X}_u$

where f_k is the objective value in the kth iteration, ∇f_k and $\nabla \mathbf{h}_k$ are the gradients of the function and active constraints, $\nabla^2 f_k$ and $\nabla^2 \mathbf{h}_k$ are the second order derivatives of the function and active constraints and $\mathbf{\Lambda}_k$ are the Lagrange multipliers for active constraints. Here, active constraints refer to both equality constraints and tight inequality constraints inclusive. $\nabla^2 f_k + \mathbf{\Lambda}_k^T \nabla^2 \mathbf{h}_k$ is the Hessian of the Lagrangian, ϕ with respect to the design variables $\mathbf{X}$ and is usually updated using DEP of BFGS methods to ensure its positive definiteness and thus proper local convergence behavior. The reader is referred to texts on optimization for details on such update methods. The algorithm is executed in the following steps:

(a) An initial feasible variable vector and an active constraint set including equality constraints is first chosen. At this time, the Hessian of the function, $\nabla^2 f_k$ is approximated as an identity matrix and Lagrange multipliers are initialized to zero.

(b) A feasible search vector, $\Delta \mathbf{X}_k$ and Lagrange multipliers, $\mathbf{\Lambda}_k$ are computed using the first order KKT necessary conditions.

(c) A step-length, α_k along the direction of $\Delta \mathbf{X}_k$ is determined in the interval $[0, 1]$ that minimizes the objective using line search. If the inequality constraints are violated, α_k is reduced so that the solution lies on the surface of the most violated constraint thus retaining feasibility. The most violated constraint is then added to the set of active constraints.

(d) The variables are updated from the results of the line search.

(e) The signs of the Lagrange multipliers are checked and if they are negative, the constraint corresponding to the most negative multiplier is dropped from the active set.

(f) Termination criteria are evaluated and if they are not satisfied, the algorithm resumes to step (b). Schematic of the implementation is given in Figure 12.13.

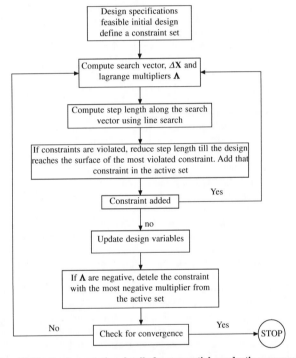

Figure 12.13   Implementation details for sequential quadratic programming

12.7   Stochastic Approaches (Genetic Algorithms and Simulated Annealing)

Quite a few methods exist that are probabilistic in nature and rely only on the function values (and not its gradient or Hessian) to reach an optimal solution, two of them being the genetic algorithms and simulated annealing. A genetic algorithm mimics three operations of nature, namely, reproduction, crossover, and mutation to form a new and better population or the generation of variable vectors **X** from the previous one. The algorithm commences with a constant even size population that is a predetermined number of candidate variable vectors. These vectors are initially generated randomly such that the variable values lie between the lower and upper limits. For random generation of variable values, a normal distribution scheme may be incorporated.

The function value for every vector in the population is evaluated and the fitness of the vector is computed. For maximization problems, fitness of a vector is proportional to its function value. The modulus or magnitude of the function may not be used to account for its negative value. In reproduction, the vectors are selected based on their fitness values for possible inclusion in the new population, that is, the more the vector's fitness, better are the chances that it gets selected for the mating pool, which then is used to create the next generation. This ensures that high-fitness vectors stand better chances of reproducing, while low-fitness ones are more likely to disappear. If the fitness is high, multiple copies of that vector can be included into the mating pool.

Crossover refers to the blending of two vectors in the mating pool to create two new ones for the subsequent generation. This operator selects two vectors at random from the mating pool that may be different or identical. Based on the crossover probability p_c, the operator decides if the mating should occur. If crossover takes place, then the two vectors are split at a random splicing point and the splices are intermixed to create two new vectors (off-springs), which are then placed in the new population. If crossover is not performed, the two vectors are copied to the new population. These vectors are not deleted from the mating pool as they are good candidates and may be used multiple times during crossover. The crossover probability is usually chosen very high to encourage fit vectors to mate. As the crossover operator creates better offsprings, it enables the evolutionary process to move towards the promising regions of the search space.

Reproduction and crossover operators can generate a large number of new vectors. However, depending on the size of the initial population chosen, the entire design space may not be spanned. it may also happen that the algorithm converges to a vector not quite close to an optimum due to an inappropriately chosen initial population. Mutation can then be performed either during selection (reproduction) or crossover, the latter being more usual. For each vector in the mating pool, the operator checks if it should perform mutation with a predetermined probability p_m. In that case, it randomly changes the vector's constituent values to new ones. The mutation probability is normally chosen very low as high mutation rate destroys fit vectors which degenerates the algorithm into a random search. Mutation helps prevent the population from stagnating by introducing fresh vectors into the population while retaining its rich vectors.

Once the new population of vectors is generated, fitness values of the vectors are computed and accordingly, better vectors are sent to the mating pool for crossover and mutation for the geneation of subsequent population. Convergence is achieved when the final population of vectors is adequately uniform and that the objective cannot be further improved significantly.

Simulated annealing is another probabilistic method that emulates slow cooling (annealing) of molten metals thus aiding them reach their lowest possible energy state. The cooling phenomenon is simulated by controlling a temperature like parameter T. As per the Boltzmann distribution relation, a system in thermal equilibrium at a temperature, T has its energy distributed probabilistically as

$$P(E) = \exp(-E/kT)$$

where k is the Boltzmann's constant and E the energy state. The relation suggests that a system at a high temperature has a high probability of being at any energy state. However, at low temperatures, the probability of being at a high-energy state decreases. For function minimization, at any instant t, the energy E is replaced by the function value $f(\mathbf{X}^t)$, where $\mathbf{X}^t$ is a vector of design variables. The probability of the next point being at $\mathbf{X}^{t+1}$ depends on the differences in the function values $\Delta E = E(t + 1) - E(t) = f(\mathbf{X}^{t+1}) - f(\mathbf{X}^t)$ and is computed as

$$P(E(t + 1)) = \min[1, \exp(-\Delta E/kT)] \tag{12.37}$$

If the difference in the function values $\Delta E \le 0$, the probability is one and the point $\mathbf{X}^{t+1}$ is always accepted. When $\Delta E > 0$, $\mathbf{X}^{t+1}$ is worse than $\mathbf{X}^t$ and is consequently rejected by most traditional gradient based searches. In simulated annealing however, there is a finite probability that the point is accepted. If the temperature parameter T is large, the probability is high for $\mathbf{X}^{t+1}$ to be accepted even for largely disparate function values (high ΔE). Thus, at each stage, the system can move either to a state for which the energy is higher than its present state, or to a state of lower energy. Eq. (12.37) allows the system to move consistently towards lower energy states, yet still jump out of local minima due to the probabilistic acceptance of some upward moves.

The initial temperature T, the number of function evaluations n performed at a particular temperature and the cooling schedule are three essentials governing the performance of simulated annealing. If a large initial T is chosen, it takes long to converge. For small initial T, the search is not adequate to thoroughly investigate the problem space before converging to a true (global) optimum. An estimate of the initial temperature parameter can be obtained by computing the average of the function values at randomly generated points in the search space. A large value of n is recommended to achieve the quasi-equilibrium state for which the computation time is more. For most problems, n is usually chosen between 20 and 100 depending on the computation resources and time.

This chapter discussed some of the many optimization methods in use in engineering applications. First, one-variable optimization methods were discussed which are of the bracketing and open type, the former being of zeroth order requiring only function values while the latter being of the first or second order requiring the first or second function derivatives. The classical necessary and sufficient optimality conditions for multi-variable optimization methods with and without constraints were derived followed by a discussion on the Karush-Kuhn-Tucker or KKT conditions. Linear programming which involves linear function and constraints was discussed in some detail with simplex implementation following which sequential linear programming (SLP) was briefed. Numerical implementation of the sequential quadratic programming (SQP) which employs the KKT conditions iteratively was discussed in brief followed by a mention on some stochastic methods like genetic algorithms and simulated Annealing. The scope of this chapter is restricted only to the aforementioned noting that numerous texts and codes are available exclusively on optimization. We close this chapter by citing an example on topology optimization of compliant mechanisms with local stress constraints which involves linear frame finite element analysis discussed in Chapter 11.

Example 12.11. Compliant mechanisms are single-piece devices designed for prescribed motion, force and/or energy transduction through elastic deformation. Consider a design region discretized using linear frame elements shown in Figure 12.14(a) with length 150 mm and width 50 mm for a compliant crimper. The left vertical edge is fixed while the bottom edge is on a roller support. A load of 20 N is applied as input at the top right corner and it is desired to maximize the deformation at point P along the direction shown. To compute the deformation at P, the virtual work principle is used where a unit dummy load is applied at P along the direction of desired deformation and the nodal displacements $\mathbf{V}$ are computed as a response only to the dummy load. If $\mathbf{U}$ is the displacement vector due to the input load only then

$$\Delta_{\text{out}} (P) = \mathbf{V}^{\text{T}} \mathbf{K} \mathbf{U}$$

where $\mathbf{K}$ is the global stiffness matrix. The axial stress in each frame element i can be computed as

$$\sigma_i = E_i \mathbf{B}_i \mathbf{u}_i$$

where E_i is the Young's modulus of the ith frame element, $\mathbf{B}_i$ the strain displacement matrix and $\mathbf{u}_i$ the local nodal displacements due to the input load. The optimization problem can be formulated as

Minimize: $-\mathbf{V}^T\mathbf{K}\mathbf{U}$

Subject to: $\left(\dfrac{x_i}{x_u}\right)^n\left(\dfrac{|\sigma_i|}{\sigma_a}-1\right)\le\varepsilon,\quad i=1,\dots,N$

$$x_l\le x_i\le x_u \tag{12.38}$$

where σ_a is the allowable stress limit (10 N/mm²), x_i the width of the ith frame element treated as a design variable, n is a prespecified exponent (= 3 in this case), N the total number of finite elements and ε (0.01) is a prespecified relaxation parameter having a small positive value. Out-of-plane thickness is taken uniform as 2 mm. The notion in this topology optimization example is that if the widths of frame elements are zero, they are non-existent. However, a very low but positive value $x_l = 0.001$ mm is chosen as the lower limit for the widths to prevent the global stiffness matrix from being singular at any stage in optimization. Thus, a frame element would be considered absent from the topology if its width assumes the lower bound. The widths of the frame elements are also bounded from above such that they cannot exceed a value of $x_u = 4$ mm. The stress constraints are posed so that for $x_i \approx x_l$, the effective upper limit on the stress $|\sigma_i|$ becomes $\sigma_a[\varepsilon(x_u/x_l)^n + 1] = 10[0.01(0.4/0.001)^3 + 1] = 6.4\times10^6$ which is a much larger number compared to 10, that is, stress constraints are effectively not imposed on elements which are non-existent in the topology. However, for $x_i \approx x_u$ the effective upper limit on stress $|\sigma_i|$ is $\sigma_a[\varepsilon + 1] = 1.01\sigma_a$ which is very close to the allowable limit.

Eq. (12.38) is solved using the sequential quadratic programming in MATLAB™ and the optimal solution is given in Figure 12.14(b). Solid lines show the optimal connectivity while the dashed lines depict the deformed configuration.

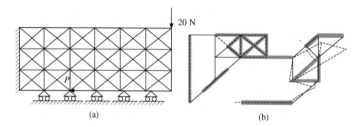

Figure 12.14 **Topology design example of a compliant crimper using SQP**

EXERCISES

1. Using any bisection method discussed in the chapter, determine the roots of

$$y = x^4 - 6x^3 + x^2 + 24x - 20$$

   Note that there are a maximum of five roots of the above polynomial so that the brackets may be chosen accordingly.

2. Using the Newton Raphson and secant methods, try and determine the roots of

$$y = x^5 - x^4 - 5x^3 + x^2 + 8x + 4$$

   What are the possible difficulties one would experience with the two methods? Can these methods be applied in case when a polynomial has multiple roots?

3. Choose a suitable root finding method to determine the roots of $\tan x = x$. How many roots of this equation exist? Can all be determined?

4. Study and suggest methods to determine multiple and/or complex roots of polynomial equations.

5. Find and determine the nature (maximum or minimum) of the optimal point(s) of the function $f(x) = x_1^2 + x_2^3 + x_3^2$ subject to $h(\mathbf{x}) = x_1 + x_2 + x_3 - 1 = 0$ using the Lagrangian multiplier method.

6. Minimize $f(\mathbf{x}) = x_1^2 x_2 + x_2^2 x_3 + x_1 x_3^2$ subject to $h(\mathbf{x}) = x_1 + x_2 + x_3 - 6 = 0$. Verify the sufficiency condition.

7. Minimize: $\qquad\qquad\qquad\qquad x_1^2 + 4x_2^2$

    Subject to: $\qquad\qquad\qquad x_1 - 2 \geq 0$

    $\qquad\qquad\qquad\qquad\qquad x_2 - 5 \geq 0$

    Use KKT necessary conditions and sufficiency criterion to determine the nature of optimal solutions.

8. Minimize: $\qquad\qquad\qquad\qquad x_1^2 + 8x_2^2 + 2x_3^2$

    Subject to: $\qquad\qquad\qquad x_1 + x_2 + x_3 - 15 \leq 0$

    $\qquad\qquad\qquad\qquad\qquad x_1 + x_2 + x_3 - 2 \geq 0$

    Use sufficiency criterion to determine the nature of the optimal point(s).

9. Solve Problem 8 using an additional constraint $x_1 - x_2 + 2x_3 + 2 \leq 0$. Is the optimal solution (if it exists) any different from that in Problem 8?

Appendix

Mesh Generation

A1.1   Mesh Generation with Discrete Elements

Discrete elements like the truss, beam or frame elements are simple in topology (two vertices and an edge) and finite element implementation and perhaps yield faster results compared to their continuum counterparts, that is, the triangular and quadrilateral elements in two dimensions, and tetrahedral and octahedral elements in three dimensions. Two kinds of mesh implementations are employed with discrete finite elements. The first is the *full ground structure* wherein each node is connected to every other node in the region by means of truss, beam or frame elements. Full ground structures are used to capture as much of the region as possible (Figure A1.1 a). However, numerous elements intersect or overlap which may be avoided by using a *partial* or *super ground structure* (Figure A1.1 b) that uses an array of elements arranged in a cell (square or cube). In full ground structures, node placement can either be uniform or random though in partial ground structures, it is usually uniform. The two ground structures can also be generated in three-dimensional solids.

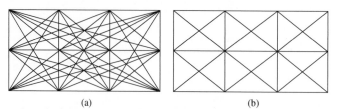

(a)   (b)

Figure A1.1   Full and partial ground structures

A1.2   Mesh Generation with Continuum Elements

Two of the simplest and most used finite elements in two dimensions are the triangular and quadrilateral elements. Triangular meshes can be mapped or freely generated though in the former, the region to be discretized needs to be triangular. The mapping involves blending of a mesh generated in a *parametric* domain into the real domain, as the parametric boundary blends into the real boundary. Thus, a region $Q_1Q_2Q_3$ may be mapped to a parametric equilateral triangle of unit edge $P_1P_2P_3$ as shown in Figure A1.2. If P is a point in the interior of $P_1P_2P_3$, then

$$P = (A_1/A)P_1 + (A_2/A)P_2 + (A_3/A)P_3 \qquad (A1.1)$$

where A_1, A_2 and A_3 are the triangular areas as shown and $A = A_1 + A_2 + A_3$. For a one-to-one map between triangles $P_1P_2P_3$ and $Q_1Q_2Q_3$ if Q is an interior point in the latter, then

$$Q = (A_1/A)Q_1 + (A_2/A)Q_2 + (A_3/A)Q_3 \qquad (A1.2)$$

Thus, if P coincides with P_1, $A_1 = A$, and $A_2 = A_3 = 0$ in which case Q overlaps with Q_1. Likewise, one can argue for P_2 coinciding with Q_2 and P_3 coinciding with Q_3. In other words, the boundary of the parametric triangle blends with that of the real one. The notion is to mesh the region $P_1P_2P_3$ using equilateral (regular) triangles of predetermined size and transport the mesh (node points and element connectivity) back to the original domain. For this reason, the method is termed as the *transport mapping method*.

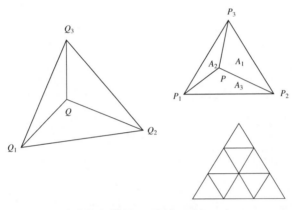

Figure A1.2 Structured triangular meshes

A1.2.1 Unstructured Meshes with Triangular Elements
Any arbitrary complex geometry can be more flexibly filled with unstructured meshes of triangular elements. There exist numerous methods in use for triangulation of generic domains. Most primitive is the *manual mesh generation* wherein the user defines each element by the vertices. The approach is infeasible and time consuming if the number of elements is large. Among the automatic ones are the (a) *advancing front method*, (b) *Delaunay-Voronoi triangulation* and (c) *sweepline method* which are discussed below.

A1.2.2 Triangulation with Advancing Fronts
The advancing front method is used to generate grids having triangular or/and quadrilateral elements. The domain boundaries are initialized as piecewise linear curves with nodes and edges which forms the *front*. As the algorithm progresses, new internal nodes are generated, and triangular and quadrilateral elements are formed at the contour. The front is initialized to the new internal boundary and the algorithm continues until the front is empty, that is, when there exists no internal boundary to be advanced further. A stepwise implementation of the algorithm is provided as follows:

1. The domain boundary is discretised using piecewise linear curves which is initialized as the *front*.
2. The front is updated (edges are deleted and added in the front) as triangulation proceeds.

3. Two consecutive edges are considered from the front for triangulation. For angle α between the two edges, three possibilities may be identified

   (i) $\alpha < \frac{1}{2}\pi$: the edges (bc and cd) form the part of the single triangle created (Figure A1.3a).
   (ii) $\frac{1}{2}\pi \leq \alpha \leq \frac{2}{3}\pi$: an internal point i and two triangles are generated (Figure A1.3b).
   (iii) $\alpha > \frac{2}{3}\pi$: a triangle is created with edge ab and an internal point i

4. The internal nodes are positioned to be *optimal* in that the element with such a node is as regular as possible. For case in Figure A1.3 (b), the internal node i is generated on the angular bisector at a distance dependent on the lengths of edges bc and cd, that is

$$| d_{ic} | = (1/6)(2\,|d_{bc}| + 2|\,d_{cd}| + |\,d_{ab}| + |\,d_{de}|) \tag{A1.3}$$

for angles β and γ between $(1/5)\pi$ and $2\pi - (1/5)\pi$ with $(1/5)\pi$ an empirically chosen value. Otherwise, construction in Step 1 may be incorporated. For case in Figure A1.3 (c), a triangle as equilateral as possible may be formed.

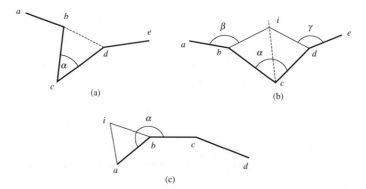

Figure A1.3   Advancing front method with various internal angles between consecutive edges of the front

5. It must be ascertained for an internal point that

   (i) it must be a part of the domain, that is, it must be inside the primary contour of the domain and outside the contour of holes that may be present
   (ii) there should not be any existing node within a certain proximity of the internal point. If there is, the node becomes the internal point
   (iii) the internal point should not be contained within the existing triangular element(s).

6. The node and element connectivity lists are updated
7. A new front is formed by

   • deleting the edges from the present front belonging to a triangle created from the present front
   • adding those edges of triangle(s) created which are not common to two elements, and to the two fronts.

8. The procedure is continued unless the front is empty. Figure A1.4 shows a few steps of the advancing front method.

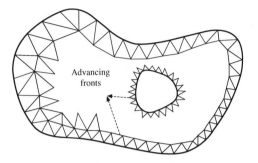

Figure A1.4   Schematic of the advancing front method

The advancing front method can be extended to three dimensions in that the front would comprise a set of triangles at the surface of the solid to start with and interior points would be generated to create almost regular tetrahedral elements.

A1.2.3   Delaunay Triangulation

The method is one of the most widely used as it yields efficient triangulation with relatively easy implementation and provides better results for most applications. Delaunay triangulation is the geometric *dual* of Voronoi tessellation also known as Theissen or Dirichlet tessellation in that one can be derived from the other. For *N* points in a plane, Voronoi tessellation divides the domain into a set of polygonal regions, the boundaries of which are perpendicular bisectors of the lines joining the nodes (Figure A1.5 a). Each polygonal region contains only one of the *N* points.

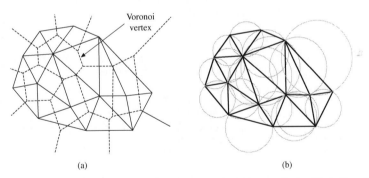

(a)   (b)

**Figure A1.5   (a) Delaunary triangulation (solid) and Dirichlet tessellation (dashed) and
(b) Delaunay triangles with circumcircles**

As per the Delaunay criterion, in a valid triangulation, the circumcircle of each triangle does not contain any node of the mesh. By construction, each Voronoi vertex is the circumcenter of a Delaunay triangle. Delaunay triangulation being the geometric dual of Voronoi tessellation, many methods have been developed to arrive at the former using the latter, though many methods for direct triangulation are also in use. A simple and widely used Watson's algorithm for Delaunay triangulation is briefed here. The algorithm starts by forming a super-triangle that encompasses all the given (boundary) points of the domain. Initially, the super-triangle is flagged as incomplete and the algorithm proceeds by incrementally inserting new points[1] in the existing triangulation. A search is made for all triangles whose circumcircles contain the new point. Such triangles are deleted to give an *insertion polygon* and a new set of triangles is formed locally with the inserted point. This process is continued until point insertion is accomplished and thereafter all triangles having the vertices of the super-triangle are deleted. Figures A1.6 depict the implementation of the Watson's algorithm.

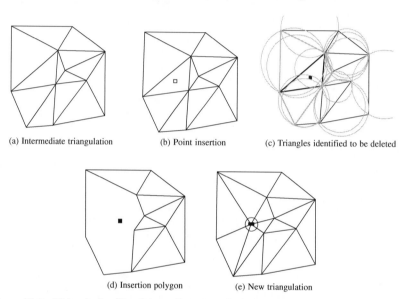

    (a) Intermediate triangulation   (b) Point insertion   (c) Triangles identified to be deleted

    (d) Insertion polygon   (e) New triangulation

Figure A1.6   **Watson's algorithm: intermediate steps of point insertion, local deletion of triangles to form the insertion polygon and new local triangulation**

In three dimensions, a super tetrahedron may initially be formed to encompass all boundary nodes. At an intermediate step, a point may be generated in space and checked whether it lies within the circumspheres of the existing tetrahedra. Such tetrahedra may be deleted to result in an insertion polyhedron within which new tetrahedral elements may be formed.

A1.2.4 Quadtree Approach

For a two dimensional domain approximated using polygonal discretization, a *quadtree* grid is constructed as shown in Figure A1.7 (a) in the following manner.

1. A square cell of minimum size is formed to contain all contour points
2. The quadtree approach consists of cell *splitting* involving each cell (parent) to be divided into four (children) cells (Section 9.6.1).

---

[1]Point insertion itself may be iterative to ensure that the resultant triangles formed are almost regular.

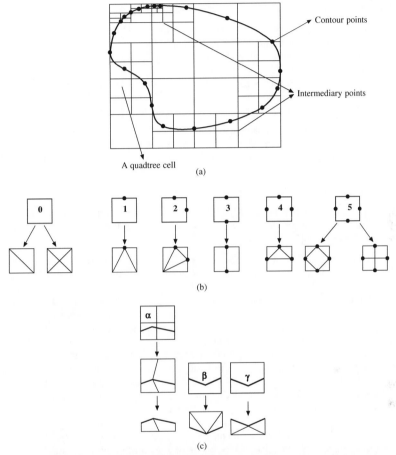

Figure A1.7   **Triangulation using the quadtree approach (a) a quadtree grid including the domain such that at most one contour point lies on or within a cell (b) patterns 0, . . . ,5 for cells with the intermediary points and their discretization schemes (c) internal cells intersecting with contour (dark lines) and discretization based on how the contour point is located and which region within the cell is inside the domain**

3. In a recursive process, cell splitting is performed such that each cell contains at most one contour point. Thus, for a fine mesh around a contour, the latter should be approximated using more nodes to have many children cells encompassing the contour. This, in a way, serves to control the element size around a contour.

4. Once splitting is over, each cell can be analyzed as follows:

   (a) External cell: Any cell not within the domain and also not containing any segment of the contour is not of interest and is eliminated

(b) Internal cell not intersecting with the contour can be of six kinds:

 (i) those without any intermediary point (Quadtree node where cells from two different levels of decomposition meet) on its sides (pattern 0)
 (ii) those with one intermediary point on one side (pattern 1)
 (iii) those with an intermediary point each on two consecutive sides (pattern2)
 (iv) those with an intermediary point each on non consecutive sides (pattern 3)
 (v) those with an intermediary point each on three sides (pattern 4)
 (vi) those with an intermediary point on each of their sides (pattern 5)

Such cells can be discretized as shown in Figure A1.7 (b). An internal cell not including any contour point produces a quadrilateral that can be split into triangles if its sides do not include any intermediary point (pattern 0), or is split into triangles or possibly quadrilaterals for patterns 1, . . . 5.

(c) Cells intersecting with the contour can be as follows:

 (i) contour point included in a cell is close to a vertex of the cell (pattern α in Figure A1.7c).
 (ii) contour point is close to the mid point (or *clearly* internal) of the cell (patterns β, γ, etc. in Figure A1.7c).

The intersection of the contour and the sides of the cell are created. A partitioning of the cell is defined where only the part internal to the domain is retained. The final contour of the mesh is created at this stage.

Element formation in the final mesh is done as a consequence of the enumeration of different patterns possible. Once the final mesh is obtained, the *regularization* of the internal points is then performed. Internal points are the vertices of the cells excluding those on the contour. Regularization or *mesh smoothing* may be performed such that for an internal point, its neighboring points are determined and their *barycenter* or the geometric center is computed, and that internal point is repositioned to this geometric center. Further, to avoid *flat elements*, diagonal swapping between two neighboring triangles can be applied iteratively. For three-dimensional mesh generation, an *octree* type cell decomposition may be incorporated. Cell patterns like those in Figure A1.7 may be categorized and identified, and tetrahedral elements may be generated.

A1.2.5 Meshes with Quadrilateral Elements

Noting that two neighboring triangular elements may be combined to form a quadrilateral element, the result of the triangulation algorithms may be used to generate grids exclusively comprising of the four-noded elements. A goal of triangular-to-quadrilateral mesh conversion is to maximize the number of adjacent triangular pairs and minimize the number of triangular elements in the process. The adjacent triangles may be selected based on how best (close to a square) a quadrilateral element may be formed, and then *fused* at their common diagonal. With such algorithms, not all triangles may be able to participate in quadrilateral formation and as a result, some isolated triangles may appear in the mesh. As the goal is to have a mesh exclusively of quadrilateral elements, a *swapping* scheme may be employed to swap the edges of quadrilaterals lying between two isolated triangles until the triangles become adjacent. Another way may be to subdivide or swap the edges of isolated triangles until they locally get converted to all-quadrilateral elements like in Figure A1.8.

Of the non-conversion or direct approaches for quadrilateral mesh generation is a semi-automatic approach called the *multi-block* method which is based on mapped meshing. The domain to be

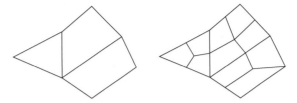

Figure A1.8   Conversion of a quad-dominant mesh to all-quadrilateral mesh

discretized is divided into blocks which are then meshed separately using parametric mapping (Figure A1.9). Though mesh of individual blocks may be regarded as structured, the overall mesh is unstructured because of the likewise decomposition of the block. Human intervention is involved in manually decomposing the geometry into blocks though some algorithms attempt to automate the geometry decomposition using medial-axes and medial-surface techniques with some heuristics. However, automatic decomposition of a complex domain into mappable regions seems non-trivial. Significant disadvantage of the mapped meshing algorithms is the limited flexibility of the mesh size control. To ensure mesh conformity at the common block boundaries, the same division must be used in neighboring blocks.

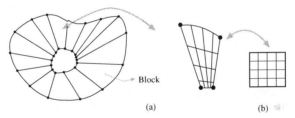

(a)   (b)

Figure A1.9   Semiautomatic quadrilateral mesh generation: (a) decomposition into blocks and (b) mesh generation in a block using mapping

A1.3   Mesh Evaluation

We note that the applicability and accuracy of the finite element analysis is dependent to a large extent on the validity and quality of the elements generated in a mesh. Of the various criteria in use for mesh quality, some for planar/surface meshes are

1. The variation in the element area should not be large. That is, the ratio of the area of the largest/smallest element to all the immediate neighbours should not be drastically low/high.
2. Elements, especially the continuum type, should be as regular in shape as possible. It is desired for triangular elements to be possibly close to equilateral triangles, and for quadrilateral elements to be of square shape. A measure called the *aspect ratio* for elements should be as close to 1 as possible. For triangular elements, the aspect ratio is defined as the ratio of the circumradius of the triangle to twice its inradius. Note that the aspect ratio of an equilateral triangle is 1.
3. The ratio of the largest to the smallest edge/angle (may also be regarded as the aspect ratio) of the element should be close to 1.
4. No two elements in a mesh should intersect. (Full ground structure with discrete elements is an exception).

Suggested Projects

Project 1

Find the configuration of a set of rigid bodies that satisfy a set of geometric constraints. Take an example of a four-bar mechanism and simulate graphically the system based on the vector loop method. Also plot the locus, velocity and acceleration of the point on the coupler.

Project 2

An object B of drop shape is tied to a rope fixed at O (0, 0, 10) and revolved around the Z-axis such that the apex A traces a circle of radius 5 on Z = 0 plane counterclockwise as shown in Figure P1. A point P initially located at (5.9688, 0, – 1.9377) also traces counterclockwise a circular path of radius 5.9688 about Z-axis on the plane Z = –1.9377. At time $t = 0$ the point P and the apex A of the object B lie on the XZ plane. Refer Figure P1 for all dimensions related to the configuration. Assume that the object B revolves with a constant angular velocity $\bar{\omega}_b$ and point P is revolving with an initial angular velocity $\bar{\omega}_p$ and with a constant angular acceleration α_0. The point P which is inside the object B at time $t = 0$ will exit the object B's volume after some time instance t_s and then reenter the drop after time instance t_e.

A 3D object's surface can be approximated by a set of triangular patches. The first order representation of this kind is a very useful tool for computer display as well as computer aided manufacturing. An industry standard for such a representation is STL. An STL file contains a series of triangular patch data consisting of normal information and coordinates for the three vertices. A typical ascii STL file appears as:

solid Sample

        facet normal –0.36970 0.27086 –0.88880
        outer loop
          vertex 122.91010 27.04771 182.30157
          vertex 101.78409 25.09600 190.49422
          vertex 101.80428 38.64565 194.61511
        endloop
        endfacet
        facet normal –0.30950 0.36777 –0.87690
        outer loop
          vertex 122.91010 27.04771 182.30157
          vertex 101.80428 38.64565 194.61511

vertex 119.99031 40.82108 189.10866
endloop
endfacet
:
:
:
:

endsolid Sample

Create an STL file of the object B of drop shape (use any of the solid modeling packages to create a drop (unit radius half sphere and a cone of height 3 units mounted on the circular face, export the geometry in STL file format) with apex A at origin and the major axis aligned to Z-axis. Code an algorithm in MATLAB to do the following. Import the STL file and display the same (help: use *trisurf* function and set 100% transparency for the faces). Apply the necessary initial transformation to the given object so that the drop assumes the orientation as depicted in Figure P1 for time $t = 0$. Apply then the necessary transformation for each time step so as to simulate the motion of object B as well as point P and display the motion as an animated sequence from time $t = 0$ to $t = t_e$. During the animated sequence show the point P with different colors depending on the status whether the point is inside/on or outside the object B. Do not analytically compute t_s and t_e but simulate the motion and find position and orientation of object B and point P for each unit time step and perform a PMC (point membership classification) query for point P in object B.

A small note on how to evaluate whether a point is inside/on an object or not. This is also referred to as PMC (point membership classification) and is one of the most important computations in geometry. The logic described here holds well for this problem only. The object under consideration is convex and also the relative motion between the object and the point is simple. Let P_{xy}, P_{yz} and P_{zx} be the orthographic projections of the point P on to the principal planes. B_{xy}, B_{yz} and B_{zx} be the projections of the object B on to principal planes. The projections B_{xy}, B_{yz} and B_{zx} will be polygons with triangular mesh for a triangulated object as is the present case. Find the bounding polygon for each of the projections as BP_{xy}, BP_{yz} and BP_{zx}. The bounding polygon is the convex hull of the projected vertices in this case since the object drop is convex (help: use *convhull function*). Now perform a 2D PMC for point P_{xy} in BP_{xy}, P_{yz} in BP_{yz} and P_{zx} in BP_{zx}. (help: use *inpolygon* function). The point P is outside the object B if for any one of the three 2D PMC the answer is "out". Else the point is inside/on the object.

Interested readers may refer to text on computational geometry for the *ray-tracing* algorithm, which is a generic PMC but is algorithmically as well as computationally more involved.

The program is to be designed so that the user imports the data, selects the constant angular velocity for the object B, initial angular velocity as well as constant angular acceleration for point P. Use Graphical User Interface (GUI). Report also the time values t_s and t_e in a textbox in the GUI. The program should be self-explanatory (use adequate comments so as to follow the code). Perform two simulation runs, Case 1: $\overline{\omega}_b$, $\overline{\omega}_p$ and $\alpha_0 = 0$, Case 2: $\overline{\omega}_b$, $\overline{\omega}_p$ with positive α_0.

Project 3

Develop a program to generate automatic cutout for tailoring a simple men's shirt. The program shall input the basic feature dimensions such as chest diameter, arm length, wrist diameter, shoulder width, collar diameter and produce the required cut plans as drawings constituting of lines and Bézier curves. The first step is to develop a feature graph constituting the various feature elements (in this

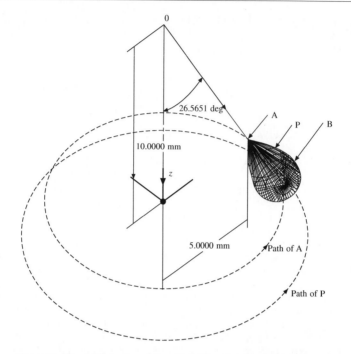

Figure P1. The configuration of the object "Drop" and point P at time $t = 0$

case the cloth cuttings required for stitching the shirt) and then developing the relation between the geometric parameters required to draw these cutting in terms of the user specified feature dimensions.

Project 4

Develop an algorithm to design generic B-spline curves and surfaces. The Input parameters shall be the order, number of control points and their co-ordinates. Write functions to calculate the knot vectors, basis functions and finally the function, for calculating points on a B-spline to display. Also for a parameter value u on the curve, calculate the position vector, tangent and curvature.

Project 5

Develop algorithms for designing surfaces of revolution and sweep surfaces. The Input parameters shall be the order, number of control points, their coordinates for the B-spline curve and a vector defining the axis of revolution or the direction of sweep as the case may be. Use the code developed in Project 4 to design a B-spline curve to be used as curve to be revolved or swept. For simplicity, assume the curve to be lying on the XY plane. In case of surface of revolution, the axis of revolution is the Y-axis and in case of a sweep surface, the direction of sweep is the Z-axis. Also for parameter values 'u' and 'v' on the surface, calculate the position vector, tangents and curvatures.

Project 6

Design a wine glass and a speed breaker bump surface as shown in Figures P2 and P3 using the code developed in Project 5 and show 3D display.

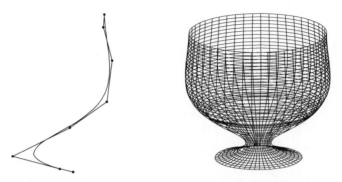

Figure P2   **A B-spline and the wine glass generated by revolving the same about Y-axis. The curve is a cubic uniform B-spline with 8 control points. Assume suitable locations for them.**

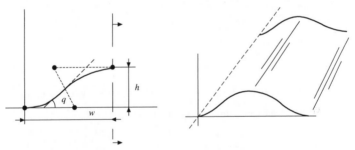

Figure P3   **Cross section profile and the corresponding sweep surface. Input parameters for the cross section profile are $(w, h$ and $q)$. First find the position vectors r_0, r_1, r_2 and r_3 and then proceed.**

Project 7

Create a software to design two cubic Bézier curves and later specify the required continuity to get the final blended shape. The user should be able to design the curves by providing the location of control points (control polyline) preferably by mouse and not by entering the values through key. Once the curves are created, the user specifies the required continuity for blending [C^0 (position continuity), C^1 (tangent continuity) or C^2 (curvature continuity)]. Let the curves be A and B. Keeping the curve A as the reference, modify curve B such that curve B blends with curve A with the required continuity. Now measure the positional difference on curve B before and after blending and plot the difference as a function of the parameter u. An illustrative example on Bézier curves is shown in Figure P4 to give an idea of what may be expected as an out come.

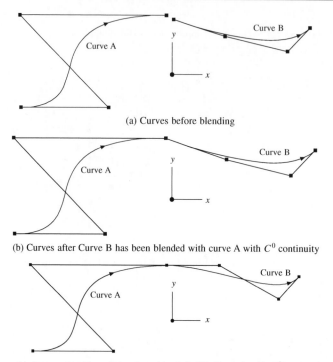

(a) Curves before blending

(b) Curves after Curve B has been blended with curve A with C^0 continuity

(c) Curves after Curve B has been blended with Curve A with C^1 continuity

Figure P4   Illustrative example on Bézier curves.

Project 8

Develop a surface modeler to design an automobile hood. The feature representing the automobile hood can be approximated to a patch layout as illustrated in Figure P5. The feature constitutes of three primary surfaces, three quadratic fillet surfaces and one triangular Bézier patch. The whole set when mirrored about the symmetry plane produces the complete hood. The primary surfaces are to be modeled as bicubic hermite/Bézier/B-spline surface patches. In your software include all options for specifying the type of surface and then the required parameters. The secondary surfaces should be computed based on the primary surfaces. Note that they can be modeled as quadratic fillet surfaces and they need not be of constant radius as is occurring between primary surface one and two. Model the corner formed by fillet surfaces by a triangular quadratic Bézier patch. On the mirror plane at least first order continuity is required on the hood surfaces.

The software should be interactive so that the user can dynamically modify the surfaces till satisfactory results are obtained.

Project 9

Develop a software to model a 2D polygonal object in terms of its polytree. Use quadtree decomposition. The software may require to input the vertex list defining the object boundary and the number of

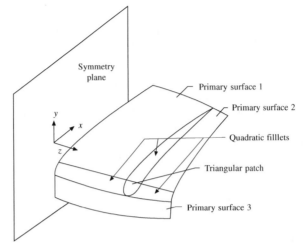

Symmetry plane

Primary surface 1

Primary surface 2

Quadratic filllets

Triangular patch

Primary surface 3

Figure P5   A scheme of surface patches for the automobile hood

levels of decomposition. The output shall be the quadtree data structure to the required levels. One would need a robust point membership classification algorithm for 2D as a pre-requisite.

Project 10

Develop a modeling package incorporating various manufacturing operations required for sheet metal applications. The software should model various operations like stamping, bending, welding and sweeping. Develop a proper data structure, graphical user interface and rendering.

Project 11

Blank nesting is an important job in the tool and die industry. Any effort design stage to minimize scrap be selecting interlocking figures or by using careful layouts may result in substantial material savings. Develop a two-dimensional nesting algorithm for nesting 2D polygonal objects. Refer to the following literature:

Nee A.Y.C., Computer aided layout of metal stamping blanks, Proceedings of Institution of Mechanical Engineers, 1984, Vol. 1, 98B, No. 10.
Prasad, Y.K.D.V. and Somasundaram S., CASNS: A heuristic algorithm for the nesting of irregular-shaped sheet-metal blanks, Computer-Aided Engineering Journal, 1991.

Project 12

This project requires the study of how an expert taylor, with minimum number of measurements on the human torso, creates the shirts, trousers and other apparels. The cut on the cloth-piece are free-form curves optimally made so that various pieces can be accomodated on the standard width of the cloth price.

Create a software to help the taylor in preparing a layout for various types of apparels.

Project 13

Many engineering structures are made up of simple elements like beams, trusses and plates. Create a finite element software to help in the static analysis (stress, deflection) of simple structures.

Project 14

Vibration analysis of beams, shafts and uniform plates and simple spring/dashpot type machine parts are often desirable. Finite element analysis can help the designer in carrying out a simulation. Create software modules for some simple elements with material properties, boundary conditions, force and frequency magnitudes etc. as user defined inputs.

Bibliography

1. Baer, A., Eastman, C. and Henrion, M. *Geometric Modeling: a Survey*", Computer Aided Design, 1979, **11**, 5, pp. 253–272.
2. Barthold, Lichtenbelt, Randy, Crane, Shaz Naqvi. Introduction to Volume Rendering, Prentice-Hall, New Jersey, 1998.
3. bø K., Lillehange F.M. CAD Systems Frame Work, North-Holland Publishing Company, Amsterdam, 1983.
4. Brunet, P., Hoffman, C. and Roller, D. New concepts and approaches in various topics of computer-aided design. CAD tools and algorithms for product design, Springer-Verlag, Berlin Heidelberg, 2000.
5. Charles, E. Knight. A Finite Element Method Primer for Mechanical Design, PWS Publishing Company, Boston, 1994.
6. Charles, M. Foundyller, CAD/CAM, CAE, The Contemporary Technology, Daratech Associates, USA, 1984.
7. Charles, S. Knox, Engineering Documentation for CAD-CAM Applications Marcel Dekker Inc., New York, 1984.
8. Choi, B.K., Surface Modeling for CAD/CAM, Elsevier, Amsterdam, Netherlands, 1991.
9. Cohen, E., Riesenfeld, R.F. and Elber, G., Geometric Modeling with Splines A.K. Peters Ltd., MA, 2001.
10. Cook, R.D., Malkus, D.S. and Plesha, M.E. Concepts and Applications of Finite Element Analysis, John Wiley & Sons (ASIA) Pte Ltd, 1989.
11. Cormen, T.H., Leiserson, C.E. and Rivest, R.L. Introduction to Algorithms, MIT Press, Cambridge, MA, 1990.
12. Cox, M.G. Practical Spline Approximation. In: Turner, P.R. (Ed.) Topics in Numerical Analysis, Springer, USA, 1981, pp. 79–112.
13. Coxeter, Harold S. Macdonald, Regular Polytopes, Macmillan, 1963.
14. Daniel, L. Ryan, Computer Aided Graphics and Design, Marcel Dekker Inc., USA, 1979.
15. David, H. von Seggern, CRC Handbook of Mathematical Curves and Surfaces, CRC press Inc., Florida, 1990.
16. David, Salomon, Computer Graphics & Geometric Modeling, Springer-Verlag, New York, 1999.
17. de Boor C. 1972, On Calculating with B-splines, Journal of Approximation Theory, Vol 6, pp 50–62.
18. de Boor C. A Practical Guide to Splines, Springer, 1978.
19. Deb, K. Optimization for Engineering Design-Algorithms and Examples, Prentice-Hall of India Private Limited, New Delhi, 1996.
20. Dierck'x P., Curve and Surface Fitting with Splines, Oxford University Press, Oxford, 1993.
21. Ellis, T.M.R. and Sewenkov, O.L., Advances in CAD/CAM, North-Holland Publishing Company, Amsterdam, 1983.
22. Encarnacao, J., Schuster, R. and Voge, E. Product Data Interfaces in CAD/CAM Applications, Springer-Verlag, Berlin, 1986.

23. Eric, Teicholz, CAD/CAM Hand Book, McGraw-Hill Book Company, USA, 1985.
24. Eric, W. Weisstein. Homeomorphism. From *MathwWorld*-A wolfram Web Resource. http://mathworld.wolfram.com/Homeomorphism.html.
25. Eric, W. Weisstein. Topology. From *MathWorld*-A wolfram Web Resource. http://mathworld.wolfram.com/Topology.html.
26. Farin, G. Curves and Surfaces for Computer Aided Geometric Design: A Practical Guide, Academic press, San Diego, 1990.
27. Farin, G., Hansford, D., Peters, A.K. and Natick, M.A. (Eds.) The geometry toolbox for graphics and modeling, ISBN 1-56881-074-1.
28. Faux, I.D. and Pratt, M.J., Computational Geometry for Design and Manufacture, Ellis Harwood Limited, West Sussex, England, 1979.
29. Foley, J.D., Van Dam, A., Feiner, S.K. and Hughes, J.F., Computer Graphics: Principles and Practice, Addison Wesley Publishing, Massachusets, 1996.
30. Franco, P. Preparata and Michael Ian Shamos, Computational Geometry, an introduction, Springer-Verlag, New York, 1985.
31. Gallier, J. Curves and Surfaces in Geometric Modeling: Theory and Algorithms, Morgan Kaufmann Publishers, CA, 2000.
32. George, P.L. Automatic Mesh Generation, John Wiley, Chichester, 1991.
33. Gero, J.S. Expert Systems in Computer Aided Design, Elsevier Science Publisher, North-Holland, 1987.
34. Glen, Mullineux. CAD: Computational Concepts and Methods, Kogan Page Ltd., London, 1986.
35. Goetz, A. Introduction to Differential Geometry, Addison Wesley Publishing Co., New York, 1970.
36. Goldberg, D.E. Genetic Algorithms in Search, Optimization and Machine Learning, Addison-Wesley, 1989.
37. Gouri Dhatt, Gilbert and Touzot. The Finite Element Method Displayed, John Wiley & Sons, New York, 1984.
38. Gray, A. Modern Differential Geometry of Curves and Surfaces with Mathematica,® CRC Press, Washington, DC, 1998.
39. Gregory, J.A. The Mathematics of Surfaces, Clarandon Press, Oxford, 1986.
40. Hetem, V. Communication: Computer Aided Engineering in the Next Millennium, Computer Aided Design, 2000, Vol. 32, pp. 389–394.
41. Hill Jr., F.S. Computer Graphics using OpenGL, Pearson Education, Singapore, 2003.
42. Hoggar, S.G. Mathematics for Com puter Graphics, Cambridge University Press, 1992.
43. http://www-gap.dcs.st-and.ac.uk/~history/HistTopics/Topology_in_mathematics.html: A history of topology.
44. Jeong, J. Kim K., Park, H., Cho, H., Jung, M., "B-Spline Surface Approximation to Cross-Sections Using Distance Maps", Int. J. Adv. Manuf. Technol, 1999, 15: pp 876–885.
45. John Stark. What every engineer should know about practical CAD/CAM Applications, Marcel Dekker Inc., New York, 1986.
46. Juneja, B.L. Pujara K.K. and Sagar, R. CAD, CAM, Robotics and Factories of the Future, Tata McGraw-Hill Publishing Company Ltd., New Delhi, 1989.
47. Kapur J.N. Mathematical Modeling, Wiley Eastern Limited, New Delhi, 1998.
48. Kelley, J.L. General Topology, Van Nostrand Reinhold Company, 1955.
49. Kochan, D. Integration of CAD/CAM, North-Holland publishing company, Amsterdam, 1984.
50. Krause, F.L. and Jansen, H. Advanced Geometric Modeling for Engineering Applications, Elsevier Science Publishers, North-Holland, 1990.
51. Kunii, T.L. Shinagama Y. and Gotoda H. Intelligent Design and Manufacturing, New York, Wiley, 1992.
52. Kunwoo Lee. Principles of CAD/CAM/CAE Systems, Addison-Wesley, Reading, MA, 1999.
53. Laszlo, M.J. Computational Geometry and Computer Graphics in C++, Prentice Hall, 1996.
54. Loney, S.L. The Elements of Coordinate Geometry, Macmillan and Company Ltd., London, 1953.
55. Lord, E.A. and Wilson C.B. The Mathematical Description of Shape and Form, Ellis Horwood Ltd., England, 1984.
56. M.A. Armstrong, Basic Topology, Springer, 1980.

57. Mansfield, M.J. Introduction to Topology, D. Van Nostrand Company, 1963.

58. Márta, Szilvási-Nagy. *"Almost Curvature Continuous Fitting of B-Spline Surfaces"*, J. for Geometry and Graphics, Vol. 2, 1998, No. 1, pp. 33–43.

59. Medland, A.J. and Piers Burnett. CAD CAM in Practice – a manager's guide to understanding and using CAD CAM, Kogan Page Ltd., London, 1986.

60. Michael, J. French. Conceptual Design for Engineers (second edition), The Design Council, London, 1985.

61. Mortenson, M.E. Geometric Modeling, John Wiley and Sons, New York, 1985.

62. Mortenson, M.E. Mathematics for Computer Graphics Applications (second edition), Industrial Press Inc., New York, 1999.

63. Parviz, E. Nikravesh. Computer Aided Analysis of Mechanical Systems, Prentice-hall International Edition, USA, 1988.

64. Paul, de Faget and de Casteljau. Mathematics and CAD – Shape Mathematics and CAD, Hermes Publishing, France, 1985.

65. Penna, M.A. and Patterson, R.R. Projective Geometry and its Applications to Computer Graphics, Prentice Hall, NJ, 1986.

66. Piegl, L. On NURBS: a survey, IEEE 1991, pp. 55–71.

67. Prakash, N. Differential Geometry, An Integratad Approach, Tata McGraw-Hill Co. Ltd., New Delhi, 1981.

68. Rao, S.S. Optimization: Theory and Applications, Wiley Eastern Limited, New Delhi, 1984.

69. Rao, P.N. CAD/CAM Principles and Applications, Tata McGraw-Hill Co. Ltd., New Delhi, 2002.

70. Reddy, J.N. Applied Functional Analysis and Variational Methods in Engineering, McGraw-Hill Book Company, Singapore, 1986.

71. Robert, F. Steidel and Jr. Jerald. M. Henderson, The Graphic Languages of Engineering, John Wiley & Sons Inc, New york, 1963.

72. Rogers, D.F. and Adams, J. A. Mathematical Elements of Computer graphics, 2nd edition, McGraw-Hill Publishing, N.Y., 1990.

73. Rogers, D.F. Procedural Elements for Computer Graphics, McGraw-Hill Book Company, USA, 1985.

74. Rooney, J. and Steadman, P. Principles of Computer Aided Design, Prentice Hall, NJ, 1987.

75. Schumaker, L.L. Reconstructing 3D Objects from Cross-sections, Computation of Curves and Surfaces, W. Dahmen et al. (Eds), Amsterdam: Kluwer Academic Publishers, pp. 275–309, 1990.

76. Shene, C.K., CS3621 Introduction to Computing with Geometry—Notes http://www.cs.mtu.edu/~shene/COURSES/cs3621/NOTES/notes.html

77. Singer, I. M. and Thorpe, J. A. Lecture Notes On elementary Topology and Geometry, Springer-Verlag, 1967.

78. Struik, D.J., Lectures on Classical Differential Geometry, Addison Wesley Publishing Co., MA, 1961.

79. Ten Hagen P.J. W., Tomiyama T. Intelligent CAD Systems – theoretical and methodological aspects, Springer-verlag Berlin, 1987.

80. Thomas, E. French, Charles, J. Vierck and Robert, J. Foster, Graphic Science and Design, International Student Edition, McGraw-Hill Book Company, New York, 1984.

81. Tuckey, C.O. and Armistead, W. Coordinate Geometry Longmans, Green and Co. Ltd., London, 1953.

82. Vanderplaats, G.N. Numerical Optimization Techniques for Engineering Design with Applications, McGraw-Hill Book Company, USA, 1984.

83. Wade, T.L. and Taylor, H.E. contemporary Analytic Geometry, McGraw-Hill, New York, 1969.

84. Walter, J. Meyer. Concepts of Mathematical Modeling, McGraw-Hill Book Company, Singapore, 1985.

85. William, M. Newman and Robert, F. Sproull, Principles of Interactive Computer Graphics, McGraw-Hill Book Company, Japan, 1979.

86. Willmore, T.J. and Gerald Farin, Geometric curve approximation, Computer Aided Geometric Design, Volume 14, 1997, pp. 499–513.

87. Wozny, M.J. Turner, J.U. and Preiss, K. Geometric Modeling for Product Engineering, elsevier Science Publishers, North Holland, 1990.

88. Yamaguchi, F. Curves and Surfaces in Computer Aided Geometric Design, Springer-Verlag, Berlin, 1988.

89. Yvon Gardan, Mathematics and CAD-Numerical Methods for CAD, Hermes Publishing, France, 1985.

90. Zeid, I. CAD/CAM Theory and Practice, McGraw-Hill Publishing, N.Y., 1991.

Index